Principles of Behav

To our mentors, our colleagues, and our students

Principles of Behavioral Genetics

Robert R. H. Anholt
Trudy F. C. Mackay

AMSTERDAM • BOSTON • HEIDELBERG • LONDON • NEW YORK • OXFORD
PARIS • SAN DIEGO • SAN FRANCISCO • SINGAPORE • SYDNEY • TOKYO
Academic Press is an imprint of Elsevier

ELSEVIER

Academic Press is an imprint of Elsevier
32 Jamestown Road, London NW1 7BY, UK
30 Corporate Drive, Suite 400, Burlington, MA 01803, USA
525 B Street, Suite 1900, San Diego, CA 92101-4495, USA

First edition 2010

British Library Cataloguing-in-Publication Data
A catalogue record for this book is available from the British Library

Library of Congress Cataloging-in-Publication Data
A catalog record for this book is available from the Library of Congress

ISBN: 978-0-12-372575-2

For information on all Academic Press publications
visit our website at www.elsevierdirect.com

Typeset by Macmillan Publishing Solutions
(www.macmillansolutions.com)

Contents

Behaviors are the ultimate expression of the nervous system. They manifest themselves as spontaneous activities or as appropriate responses to events that take place in our environment, actions that occur after integrating sensory input, physical needs, and social constraints, within the context of our individual personalities and based on previous experience. Truly, our behaviors reflect our humanity. In a much broader context it is clear that, in the animal kingdom, behaviors are essential for survival and procreation. Thus, behaviors provide the stage on which natural selection can act, and represent a vehicle for evolution. Behaviors depend on the developmental history of the individual, its genetic composition, the connectivity of its nervous system, its physiological state, its physical and social environment, and, at the molecular level, ultimately the carefully-orchestrated interplay of a host of biochemical reactions. Indeed, one might fairly say that behavior provides a window through which we can view much of biology.

During most of the twentieth century, studies on behavior were almost entirely descriptive. Understanding the complexity of behavior from molecular detail to organismal integration was considered a daunting, virtually intractable challenge. This changed with rapid advances in neuroscience during the last quarter of the century and the subsequent genomic revolution, which provided scientists with unprecedented opportunities to study the "genes–brain–behavior" axis, which is the underlying genetic principle that enables the nervous system to express behaviors. Neuroscience has led the way in this endeavor, and a variety of excellent textbooks on behavioral neuroscience have become available. Comprehensive textbooks on the genetics of behavior, however, are far and few between, although the pioneering text of 1974 by McClearn and DeFries deserves mention, as does the 2000 edition of *Behavioral Genetics* by Plomin et al., which focuses especially on human cognition.

Here, we have attempted to define the field of behavioral genetics as a comprehensive discipline that encompasses both studies on model organisms and people, with emphasis on unifying principles whenever possible. Whereas this book treats the study of behavior deliberately from a genetics viewpoint, we recognized that neurobiological and ecological aspects could not be ignored, but had to be integrated with the text. Furthermore, behavioral genetics cannot be discussed without touching on evolutionary genetics and gene flow in populations.

Behaviors are complex traits that result from the coordinated action of multiple segregating genes that are sensitive to environmental conditions. Thus, behavioral genetics falls into the realm of quantitative genetics, and a basic understanding of quantitative genetic principles is essential if one is to understand how the genome enables the expression of behavior. Genes that contribute to behavior fall into two classes, those that contribute to the *manifestation* of behavior, and a subset of these genes that give rise to *variation* in behavior. The former can be studied through mutagenesis; the latter requires classical quantitative genetic approaches.

In the first half of the book, we present an introductory chapter that describes the history of the fledgling field of behavioral genetics, followed by chapters that describe the essentials of the function and organization of the nervous system, and chapters that are designed to provide the reader with a basic understanding of quantitative genetic principles. The second part of the book focuses on a range of commonly-studied behaviors, including social behaviors, chemoreception, learning and memory, locomotion, circadian activity and sleep, and addiction. Each chapter has an overview that describes the material to be covered, and a summary that reiterates the major principles. Text boxes with ancillary information are provided throughout, along with study questions that will help the student master the material, and a list of recommended reading of classical and contemporary papers. A glossary of terminology is included for the reader's convenience.

Draft chapters of this book have been used to teach a course on Principles of Behavioral Genetics at North Carolina State University in the fall of 2008. We received valuable feedback, both from students and postdoctoral trainees, which was greatly appreciated. We would like to thank especially Elaine Smith and Susan Harbison for valuable and perceptive comments on Chapter 14 and Chapter 11, respectively. Our colleagues at North Carolina State University John Godwin, Robert Grossfeld, Christina Grozinger, and Jane Lubischer also provided valuable input on several of the chapters. Special thanks are due to Mariana Wolfner from Cornell University for critical reading and detailed editing of many of the chapters, and to Hans Hofmann from the University of Texas at Austin for critical reading of Chapter 16. Numerous colleagues have provided us with reprints or preprints of manuscripts to help in the preparation of this book. They are John Carlson (Yale University), Joshua Dubnau (Cold Spring Harbor Laboratories), Howard Edenberg (Indiana University School of Medicine), Hopi Hoekstra (Harvard

University), Ed Kravitz (Harvard University), Michael Meaney (McGill University), Charalambos Kyriacou (University of Leicester), Peter Mombaerts (Max Planck Institute at Tübingen), Randi Nelson (Ohio State University at Columbus), Catharine Rankin (University of British Columbia), Gene Robinson (University of Illinois at Urbana-Champaign), Dean Smith (University of Texas Southwestern Medical Center at Dallas), Leslie Vosshall (The Rockefeller University), Michael Wade (Indiana University), and Jerry Wilkinson (University of Maryland at College Park). We are grateful for the support of the behavioral genetics community, and hope that this book will be a valuable resource for the next generation of scientists in this field.

Robert R. H. Anholt
Trudy F. C. Mackay

Companion Website:

If you want to see 4 colour versions of any of the images in this book, please go to www.elsevierdirect.com/companions/9780123725752

For instructors, you will also find assessment questions with recommended answers for your convenience.

Chapter 1

Introduction and Historical Perspective

"Nothing in biology makes sense except in the light of evolution."

Theodosius Dobzhansky

OVERVIEW

Behavioral genetics aims to understand the genetic mechanisms that enable the nervous system to direct appropriate interactions between organisms and their social and physical environments. Early scientific explorations of animal behavior defined the fields of experimental psychology and classical ethology. Behavioral genetics has emerged as an interdisciplinary science at the interface of experimental psychology, classical ethology, genetics, and neuroscience. This chapter provides a brief overview of the emergence of experimental psychology and ethology, followed by a historical perspective of how concepts of natural selection and principles of heredity were combined by the founders of the modern evolutionary synthesis into the sciences of population and quantitative genetics. Subsequently, population genetic, quantitative genetic and molecular genetic principles could be applied to experimental psychology, behavioral ecology, and behavioral neuroscience, to give rise to the modern field of behavioral genetics. We will highlight some of the major historical milestones and controversies. We indicate how the past history of the field has laid the foundation for examining the genetic architectures of behaviors in the genomic era. This historical perspective provides an important reference frame for understanding past, current, and future trends and issues in behavioral genetics.

THE RISE OF THE MODERN FIELD OF BEHAVIORAL GENETICS

Behaviors are mediated by the nervous system in response to environmental conditions. From a genetics perspective, behaviors are complex traits determined by networks of multiple segregating genes that are influenced by the environment. Both genetic factors and neural circuits can be modified by the developmental history of the organism, its physiology – from cellular to systems levels – and by the social and physical environment. Finally, behaviors are shaped through evolutionary forces of natural selection that optimize survival and reproduction (Figure 1.1). Truly, the study of behavior provides us with a window through which we can view much of biology.

Understanding behaviors requires a multidisciplinary perspective, with regulation of gene expression at its core. The emerging field of behavioral genetics is still taking shape and its boundaries are still being defined. Behavioral genetics has evolved through the merger of experimental psychology and classical **ethology** with evolutionary biology and genetics, and also incorporates aspects of neuroscience (Figure 1.2). To gain a perspective on the current definition of this field, it is helpful to survey some of the historical milestones of experimental psychology and the study of animal behavior, together with the development of the concept of evolution through natural selection and its coalescence with Mendel's principles of heredity, which gave rise to the fields of quantitative and population genetics. These are the critical cornerstones of today's behavioral genetics. In the following sections we will provide broad overviews of the development of each of these disciplines to show, at the end of this chapter, how they can be brought together to study the link between genes, brain, and behavior.

Experimental Psychology and Animal Behavior

Since time immemorial philosophers and naturalists have been intrigued by animal behaviors. Twenty-three centuries ago Xenophon, a disciple of Socrates, wrote a treatise on *The Art of Horsemanship* which covers basic aspects of horse husbandry and training, while paying great attention to their behaviors. Studies on animal behavior, however, remained descriptive and **anthropocentric** (interpreted in terms of human experience) for much of recorded history. Human emotions and cultural values were projected onto animals, some of which were considered intrinsically noble (lions, horses), loyal (dogs), or repugnant (snakes), and discussions on the relative intelligence of species generated fruitless debates.

R. Anholt and T. Mackay: Principles of Behavioral Genetics
ISBN: 978-0-12-372575-2

1

in which an animal is trained to press a lever to obtain a food reward. Skinner argued that whereas positive reinforcement was an effective way for modifying behaviors, **negative reinforcement**, i.e. punishment, was ineffective for long-term behavioral modification, as, in his view, the subject would not modify the behavior that caused the punishment, but rather seek ways to avoid the punishing consequence. For example, some may argue that the risk of imprisonment does not deter criminal behavior, but rather encourages criminals to devise schemes that avoid the consequence of imprisonment. Skinner advocated the use of operant conditioning in raising children, and argued that positive reinforcement would be a method for improving society. His idealistic and unconventional views of **social engineering** have, however, been controversial.

Like Pavlov, Thorndike, and other experimental psychologists of his time, Skinner did not consider genetic influences on behavior, but espoused the view that behaviors could be entirely controlled by the environment. Skinner's studies provided fuel for the **nature versus nurture** debate that still smolders today, even though the notion that both genetic and environmental factors contribute to the manifestation of behaviors has gained wide acceptance.

Ethology: The Early Years

While Skinner and his contemporaries studied animal behaviors in the laboratory, two Austrian scientists, Konrad Lorenz and Karl von Frisch, and a Dutch biologist, Nikolaas Tinbergen, began to apply careful experimental approaches to animals in the natural environment. Together they laid the foundations for ethology, the study of animal behavior and the modern field of behavioral ecology. The collective contributions of Lorenz, von Frisch, and Tinbergen, for which they shared the 1973 Nobel Prize, was their pioneering experimental approaches to uncover fundamental principles that would apply not only to a single species, but would find widespread relevance.

Born in Vienna, where he worked as professor at the University of Vienna from 1928 to 1935, Konrad Lorenz formulated the idea of **fixed action patterns** of instinctive behaviors. According to Lorenz, such stereotyped behaviors are set in motion by an **innate releasing mechanism**, which elicits a fixed sequence of behavioral events. Courtship and mating rituals, and nest building of birds, are examples of such behaviors. Lorenz also popularized the notion of **imprinting**, originally described by the nineteenth century English scientist Douglas Spalding. (Note that this concept of psychological imprinting, in which an animal learns the characteristics of its parents, should not be confused with the term "imprinting" used by molecular biologists to indicate inactivation of genes on one parental chromosome through DNA modification.) Lorenz observed that when he hatched greylag geese in his laboratory, the goslings would imprint on him instead of on a natural parent, and they would follow him around (Figure 1.5). The major contributions Konrad Lorenz made to experimental psychology are clouded by the history of his political activities. He joined the Nazi party in 1938 and the Wehrmacht (German armed forces) in 1941. He spent four years in a Russian prison camp from 1944 to 1948. In later years he apologized for his Nazi past and spent his remaining scientific career at the Max Planck Institute for Behavioral Physiology at Seewiesen in Bavaria.

Karl von Frisch, working in Munich, dedicated much of his life to the study of bees. He designed clever experiments in which he placed food sources on colored pieces of paper which he surrounded by papers with different matched shades of gray to demonstrate that bees had color vision. In addition to documenting the sensitivity of honeybees to color, ultraviolet, and polarized light, von Frisch was intrigued by the common observation that after a bee had found a distant food source, many more bees would soon gather around the food. It seemed as if the original forager had somehow communicated its location to other members of the hive, and recruited additional foragers. Karl von Frisch set out to discover how this communication might occur, and discovered that when a bee located a food source in the vicinity of the hive it would perform a flight pattern which he described as a "round dance." When the food source was remote from the hive, bees would perform a more intricate flight pattern that consisted of elaborate figure-of-eight movements, known as the "waggle dance" (Figure 1.6). By experimentally manipulating the location of the food source with respect to the hive and carefully observing the behavior of the bees, von Frisch showed that the waggle dance communicates accurate information about both the direction and the distance of the food source. Furthermore, as the position of the sun shifts during the day the angle of the waggle dance would

FIGURE 1.5   Konrad Lorenz and his imprinted geese.

shift accordingly, but the endogenous biological clock of the bees entrained to the light–dark cycle would compensate appropriately, so that the information on the location of the food source could be communicated accurately at different times during the day. This is essential, as bees use polarized light vectors for navigation. The elegant experiments by von Frisch explained a complex social communication system that elicits distinct behavioral patterns.

Nikolaas ("Niko") Tinbergen became motivated to study animal behavior when Konrad Lorenz visited the Dutch University of Leiden, where Tinbergen had a minor faculty position (Figure 1.7). Lorenz would have a lasting influence throughout Tinbergen's career, most of which was spent at Oxford University. Tinbergen became best known for his early studies on reflex behaviors in sea gulls on the Dutch island of Texel, and the breeding behavior of three-spined stickleback fishes. For example, he observed that sea gulls will only recognize an egg when it is in their nest. When the egg is moved only slightly from the nest, the bird will not relate to it. However, birds will accept "fake eggs," even of the wrong shape, such as a square, as long as it has certain color characteristics. Tinbergen became well-known for formulating a set of what he considered critical questions that should be addressed to the study of any behavior, and that relate respectively to function, causation, development, and evolution of the behavior: what is the impact of the behavior on the animal's survival and reproduction; what stimuli elicit the behavior, and how can these behavioral responses be modified by learning; how do behaviors change with age and to what extent are early critical periods essential for development of the behavior; and how do similar behaviors compare between related species and how have they arisen in the course of evolution? These questions poignantly defined the field of ethology, and remain relevant today. Tinbergen's appreciation for comparative studies of behaviors and their relationship to evolution, and his interest in adaptation set him apart from other ethologists of his day and link him, at least conceptually, to the earlier exploits of Charles Darwin, who a century earlier had formulated the theory of natural selection.

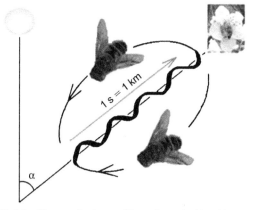

FIGURE 1.6   The waggle dance of honeybees elucidated by Karl von Frisch. The angle from the sun indicates the direction toward the food source. The duration of the waggle part of the dance encodes the distance. Approximately one second of dance corresponds to one kilometer of distance. The bee adjusts the angle of its dance to the position of the sun, depending on the time of day.

Charles Darwin and the Theory of Natural Selection

The first notion of the evolution of species was reported in 1809 by Jean-Baptiste Pierre Antoine de Monet, Chevalier de Lamarck, more commonly known as Jean-Baptiste Lamarck (Figure 1.8). In his *Philosophie Zoologique* Lamarck proposed that beneficial changes acquired during an organism's lifetime could be passed on to its progeny and that over successive generations this process would alter the organism's characteristics. This theory of inheritance of acquired traits invoked a teleological process of evolution (i.e. a process that saw design or purpose in nature). For example, according to Lamarck, giraffes would stretch their necks to be better able to reach increasingly higher foliage and this change would be transmitted to their progeny. Although the Lamarckian theory of evolution has been discredited decisively, it should be noted that at his time no theoretical framework existed to explain the diversity of species and their different adaptations to their environments, and that the notion of gradual changes of species over time was a novel concept.

Charles Darwin was born in 1809, the same year Lamarck published his theory of inheritance of acquired traits. Darwin initially went to the University of Edinburgh to study medicine, but could not stand the sight of surgery and instead became a naturalist under the tutelage

FIGURE 1.7   Nikolaas Tinbergen.

Box 1.1   A political war against science

The rise to power of Trofim Denisovich Lysenko (1898–1976) during the Stalin years of the Soviet Union provides an example of how resistance to scientific progress can have disastrous consequences (Figure A). Born to a Ukrainian peasant family, and working initially at an agricultural station in Azerbaijan, Lysenko claimed that he could increase grain yields by cold-treating seeds and that the benefits of this cold treatment could be inherited by future generations of the grain. This idea is a typical example of the theory of Jean Baptiste Lamarck on "inheritance of acquired characteristics," which had already been discredited by scientists in Western Europe.

FIGURE A   Trofim Denisovich Lysenko.

Under a repressive communist regime that felt pressure to increase agricultural production to feed the large Soviet population, Lysenko's pragmatic approach, which went against already-established genetic principles, was favorably viewed by Stalin, and he was appointed director of the Institute of Genetics of the USSR Academy of Sciences. While bragging about mostly imaginary agricultural successes, Lysenko began a campaign to discredit the science of genetics, and persecuted prominent Soviet biologists, including the previous director of the Institute of Genetics, the well-respected botanist and geneticist Nikolai Vavilov, who was fired, arrested, and eventually died from malnutrition in prison during the German siege of Leningrad in 1943. Lysenko's relentless rule of terror led to the repression, expulsion, imprisonment, and death of hundreds of scientists, and to the demise of genetics in the USSR. His ill-conceived pseudoscientific ideas and farming techniques ultimately had disastrous consequences for both Soviet science and agriculture. Lysenko's influence persisted during the reign of Khrushchev, and it was not until after Khrushchev's fall from power in 1964 that Lysenko was finally ousted and modern science was allowed to reclaim its place in the Soviet Union. The period of Lysenkoism in the USSR is a vivid example of the profound importance of scientific infrastructure to a society, and highlights the dangers of sacrificing sound scientific technology and theory to political ideology.

FIGURE 1.8   Jean-Baptiste Lamarck.

of Robert Edmund Grant, who supported the theories of Lamarck. In 1827, when it became clear that the young Darwin had no interest in the practice of medicine, his father sent him to Christ's College at the University of Cambridge to prepare him for a career as a clergyman. Members of the clergy could look forward to a comfortable lifestyle, and were often avid naturalists as they considered it their duty to "explore the wonders of God's creation." At Cambridge, Darwin fell under the influence of the naturalist John Stevens Henslow, who arranged for him to travel as a gentleman's companion and naturalist with captain Robert FitzRoy on the *HMS Beagle* for a two-year expedition to explore the South American coast. The voyage turned into a five-year expedition during which Charles Darwin collected a vast amount of specimens. Among others, he noted that distinct species of mockingbirds, tortoises, and finches inhabited different islands in the Galapagos Islands. The fauna of the Galapagos Islands would later become a cornerstone of evidence for his theory of evolution, especially the 14 species of finches that displayed distinctly different beak morphologies, functionally adapted to their different food sources – insects, grubs, leaves, seeds or fruit – and which have since become known as "Darwin's finches" (Figure 1.9). After his return to England, Darwin quickly reached a position of

FIGURE 1.9   Drawings of finches from the Galapagos Islands as they appear in Charles Darwin's book *The Voyage of the Beagle* (1839). 1. *Geospiza magnirostris*; 2. *Geospiza fortis*; 3. *Geospiza parvula*; 4. *Certhidea olivacea*.

FIGURE 1.10   Alfred Russell Wallace in 1862.

FIGURE 1.11   The young Darwin (left), already a member of the scientific elite, and a classic image of the old Darwin (right), as he appeared in 1880.

prominence in the scientific and social elite. In subsequent years he would develop his theory that species evolve as a consequence of natural selection that favors survival and procreation of the best adapted individuals. (This process of evolution by natural selection is aptly described by the phrase, **survival of the fittest**, first used by Herbert Spencer in 1864 in reference to Darwin's theory.) Darwin was well aware of the controversies his theory entailed, as it dispensed with the need for a divine creator and made no fundamental distinction between man and other animals. He avoided the term **evolution**, although his book *On the Origin of Species by Means of Natural Selection* ends with the words "endless forms most beautiful and most wonderful have been, and are being, evolved." His theory on the origin of species was not published until 1859 spurred on by his friend, the prominent geologist Charles Lyell.

While Darwin developed his ideas of natural selection, another naturalist, Alfred Russell Wallace, traveled extensively in the East Indies (Figure 1.10). Wallace had previously met and corresponded with Darwin. In 1858, Wallace solicited Darwin's opinion on an essay, entitled *On the Tendency of Varieties to Depart Indefinitely From the Original Type*. Darwin was astounded. Wallace had arrived independently at the same theory of the origin of species. In a letter to Charles Lyell, Darwin wrote: "He could not have made a better short abstract! Even his terms now stand as heads of my chapters!" Fearing that Wallace would forestall publication of his own discoveries, Darwin quickly completed his book, and both Darwin's and Wallace's findings were presented together by Lyell and Sir Joseph Dalton Hooker at a meeting of the Linnaean Society of London on 1 July, 1858. Darwin, being a well-respected member of the scientific establishment, received priority credit (Figure 1.11). Nonetheless, both men remained good friends.

Both Darwin and Wallace were influenced by Thomas Malthus' *An Essay on the Principle of Population*, which was published in 1798 and painted a grim picture for the future of mankind. Malthus predicted that population growth would outrun food supply, creating a scenario of intraspecies competition in line with the ideas of Wallace and Darwin. Darwin's theories were controversial from the moment *The Origin of Species* was published (Figure 1.12). Adam Sedgwick, a Cambridge geologist and former tutor of Darwin, wrote to him: "If I did not think you a good tempered and truth loving man I should not tell you that … I have read your book with more pain than pleasure. Parts of it I admired greatly; parts I laughed at till

FIGURE 1.12 A satirical cartoon of Charles Darwin as a monkey, which appeared in *Hornet* magazine, reflecting the way he was often ridiculed by his adversaries.

my sides were almost sore; other parts I read with absolute sorrow; because I think them utterly false and grievously mischievous."

Among the fierce supporters of Darwin's theory was Thomas Huxley, who would acquire the nickname "Darwin's bulldog" for his eloquent and aggressive defenses of Darwin's theory of natural selection. During subsequent years natural selection became firmly established as a mechanism for the evolution of species. It has become a central tenet in biology, and represents one of few inviolate laws in the life sciences, not unlike the basic law of gravity in physics (Figure 1.13).

Mendel and the Discovery of Heredity

While Darwin and Wallace were developing their theories of natural selection, an obscure Austrian monk, Gregor Mendel, performed experiments that would lay the foundation for the science of genetics in what is today the city of Brno in the Czech Republic. Gregor Mendel (Figure 1.14) was born in Heinzendorf in 1822. Early in his life he developed a great interest in the natural sciences. He was ordained a priest in 1847, and entered the Augustinian monastery of St Thomas in Brno. The monastery of St Thomas was not only a spiritual center, but also an intellectual center with a large library, which encouraged the exploration of art, philosophy, and the natural sciences.

Recognizing his talents, the abbot of the monastery sent Mendel to the University of Vienna to study physics, chemistry, zoology, and botany. In 1856, after his return to the monastery in Brno, Mendel began his famous experiments

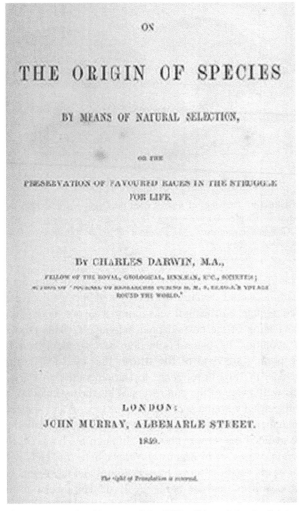

FIGURE 1.13 The title page of the 1859 edition of *On the Origin of Species*.

FIGURE 1.14 Gregor Mendel.

FIGURE 1.15 Monks of the monastery of St Thomas in 1862. Gregor Mendel is the second standing from the right.

on garden peas (*Pisum sativum*), which he cultivated in the small monastery garden (Figure 1.15). Peas were excellent subjects for his experiments on plant hybridization as they are easy to cultivate and the anatomy of their flowers prevents cross-pollination. Another key to Mendel's success was his decision to examine only distinctive physical traits that could be categorically classified. He carefully controlled his experiments, and noted that some of the characters did not permit "sharp and certain separations" as they showed "differences of the 'more or less' nature." (These would later become known as **quantitative traits**.) He therefore selected seven characteristics which "stand out clearly and definitely in the plants." These were round or wrinkled seeds, yellow or green seeds, inflated or wrinkled seed pods, green or yellow seed pods, purple or white flowers, flowers along the stem or at the tip, and tall or dwarf plants.

Mendel observed that these traits are passed from parents to their offspring according to set ratios. He reasoned that individuals possess two sets of factors that underlie each of these traits, one from each parent. He found that a particular characteristic was sometimes expressed (dominant) and sometimes concealed (recessive), but that recessive traits would reappear in the second generation in predictable ratios, and dominant characters appeared in that second generation three times as frequently as recessive characters. He also observed that the different traits generally segregated independently. For example, a pea may develop seeds that are round and yellow, wrinkled and yellow, round and green, or wrinkled and green.

In 1865, Mendel presented his results to the Brno Society for the Study of Natural Science, and in 1866 he published his *Versuche über Pflanzenhybriden* (*Research on Plant Crosses*). He distributed about 40 copies of this paper to individuals he thought would be interested in his observations, but his work was largely ignored. The only person who understood the significance of his findings was Carl Nägeli, Professor of Botany at the University of

FIGURE 1.16 The statue of Mendel in the courtyard of the monastery of St Thomas in Brno.

Munich, who encouraged him to perform similar experiments on his pet plant, hawkweed. Unfortunately, Mendel was not aware that hawkweed also reproduces asexually, and to his disappointment could not reproduce his observations with the peas in hawkweed.

In 1867, Mendel became abbot of the monastery of St Thomas. His work on heredity was forgotten until it was rediscovered in 1900 simultaneously by three European botanists, Carl Correns, Erich von Tschermak, and Hugo de Vries, who independently searched the literature, stumbled on Mendel's *Versuche*, and realized that Mendel had published their observations 34 years earlier. Thereafter, Mendel became known as the "Father of Genetics," (Figures 1.16 and 1.17) even though the term *genetics* (from the Greek "to give birth") was not used until 1905 for the first time by William Bateson in a personal letter to Adam Sedgwick. The terms **gene**, **genotype** and **phenotype** were first coined by Wilhelm Johansen in 1909.

FIGURE 1.17   The plaque above Mendel's grave with an inscription that reads "Scientist and biologist, in charge of the Augustinian monastery in Old Brno. He discovered the laws of heredity in plants and animals. His knowledge provides a permanent scientific basis for recent progress in genetics."

Mendel's success stemmed from his careful note-taking, his aptitude for mathematics and statistical analysis, and his ability to think creatively. He was one of the first to apply statistical data analysis to biological observations.

Pioneers of Biometrics

A half-cousin of Charles Darwin, Sir Francis Galton, was an unusually talented scientist who, at the age of six, studied Latin and Greek and read Shakespeare for pleasure (Figure 1.18). Like Charles Darwin, Galton enjoyed traveling and made extensive journeys through the Middle East and Africa, which he chronicled for the Royal Geographic Society. Galton was strongly influenced by Darwin's book *On the Origin of Species* and, being virtually obsessed with counting and measuring, was intrigued by the observed variation in human characteristics. He was intrigued by the question of whether human abilities were hereditary or environmental, and was the first to coin the phrase "nature versus nurture." He devised questionnaires which he circulated among the 190 Fellows of the Royal Society requesting information about family characteristics to assess whether scientific capability was innate. Recognizing the limitations of this experiment, he used a similar questionnaire approach in twin studies, being the first to recognize twins as an invaluable resource for studies on genetic variation of human characteristics. He also proposed adoption studies, to tease out hereditary effects from environmental effects. Despite his innovative studies, Galton was not able to conclusively resolve the nature versus nurture issue, which would kindle a polarized debate in the scientific community for another century.

FIGURE 1.18   Sir Francis Galton.

Based on his results, Galton became convinced that the human population could be improved by encouraging marriages between "appropriate" individuals. He advocated that incentives should be provided for early marriages between members of higher social rank. The notion that government interference in directing human reproduction to improve society was acceptable became known as **eugenics**, a term Galton invented. As the notion of eugenics is based on subjective judgements of what constitutes a better individual, and since eugenics principles have been

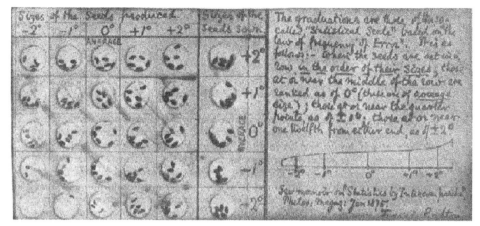

FIGURE 1.19   Francis Galton's description of his experiments with pea seeds, dated 1875.

used to justify genocide, for example by the Nazis during the Holocaust, the concept of eugenics fell into disrepute during the second half of the twentieth century, and is now widely regarded as unethical and scientifically flawed.

In 1877, Galton sent seven batches of sweet pea seeds to a group of friends, who planted them and sent offspring seeds back to Galton (Figure 1.19). When Galton measured the mean diameters of the parents and offspring seeds he noticed that they correlated, and that when these measures were plotted against each other the points were distributed around a straight line. He coined the term **co-relation**, which could be measured by the **coefficient of regression** represented by the slope of the line. Galton's graph of seed sizes represented the first regression line (Figure 1.20).

Galton discovered the statistical principle of **regression toward the mean**, where above-average parental phenotypes tend to regress to average phenotypic values in the offspring, as genes recombine and favorable combinations of alleles are lost. Galton also constructed histograms, and described and analyzed for the first time the **normal distribution**. Another of Galton's notable contributions was his study on fingerprints. Although he was not the first to realize the individual uniqueness of fingerprints and its potential applications to forensic science, Galton estimated the probability of two persons having the same fingerprint, and by placing fingerprinting on a solid scientific basis opened the way for its use in criminal courts.

One might wonder why Galton, despite his meticulous measurements and his mathematical genius, did not make the same breakthrough observations of the principles of heredity as Gregor Mendel. The reason is that whereas Mendel had deliberately chosen categorical traits, Galton examined continuous traits that were complex and arose from multiple interacting genes. His statistical analyses laid the foundation for the field that would become known as **quantitative genetics**.

The analysis of measurements of biological parameters, which became known as **biometrics**, was enthusiastically

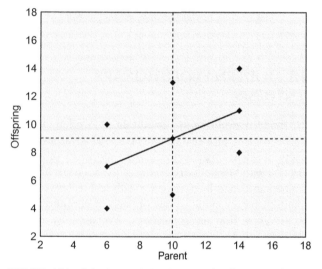

FIGURE 1.20   Galton's correlation between the diameters of sweet pea seeds from parents and offspring. The slope of the regression line is exactly 0.50. Many of the points are closer to the offspring pea size mean of 9 on the y-axis than to the parental pea size mean of 10 on the x-axis, which gave rise to Galton's concept of regression toward the mean.

embraced by Karl Pearson, Galton's protégé and intellectual heir, who after Galton's death in 1911 became the first professor of eugenics at the University of London (Figure 1.21). Pearson was a person of internal contradictions, in that he was a Marxist socialist, yet at the same time a strong advocate for eugenics, to the extent of espousing profoundly racist theories. He loathed the working class and advocated "war" against "inferior races." He remains, however, well-known for his important work which refined Galton's concepts of linear regression and correlation, and his classification of probability distributions. The **Pearson product–moment correlation coefficient**, a frequently-used statistic for estimating correlations, is named after him. The biometrics movement around the turn of the twentieth century provided a statistical basis for the modern evolutionary synthesis, which would soon follow.

FIGURE 1.21   Karl Pearson (left) with the 87-year-old Francis Galton.

FIGURE 1.22   Thomas Hunt Morgan.

THE MODERN EVOLUTIONARY SYNTHESIS

The **modern evolutionary synthesis**, also referred to as neo-Darwinism, provided major conceptual advances in the history of science by combining Darwin's theory of natural selection with Mendel's discoveries of the basic mechanisms of inheritance, and by developing statistical techniques that would enable quantitative descriptions of **complex traits** and of **gene flow** in populations. Evolution by natural selection was predicted to proceed through the occasional occurrence of mutations that would confer advantages to the individuals harboring them. Soon after the rediscovery of Gregor Mendel's work in 1900, a Kentucky-born scientist working at Columbia University, Thomas Hunt Morgan (Figure 1.22), attempted to identify such mutations while studying the fruit fly, *Drosophila melanogaster*. In 1910 Morgan discovered a mutant fly which had white eyes instead of the usual red eyes. He referred to this mutant as "*white*," thereby starting the tradition of naming *Drosophila* genes after their mutant phenotypes. When Morgan crossed white-eyed males with red-eyed females, he noticed that the offspring were all red-eyed, indicating that the mutation was recessive. When he subsequently crossbred the offspring, the white-eyed phenotype reappeared, but only in males, indicating that the trait was **sex-linked** (Figure 1.23). Morgan went on to identify many more fly mutants, and his work established *Drosophila* as an important genetic model organism. Morgan postulated that genes were carried on chromosomes, on which they would be linearly arrayed. Furthermore, he deduced that the amount of crossing-over between adjacent genes differs, and that crossover frequency could be used as a measure of the distance separating genes on a chromosome. His student, Alfred Sturtevant, created

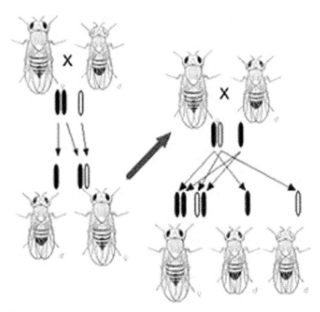

FIGURE 1.23   Sex-linked inheritance of the *white* allele; the smaller flies represent males. The open symbol designates the X-chromosome with the *white* allele. Males are XY (the Y-chromosome is not shown).

what is often considered the first genetic **linkage map**. The English geneticist J.B.S. Haldane later suggested that the unit of genetic recombination should be named the **morgan** in his honor. Morgan's concepts influenced the work of George Beadle and Edward Tatum, who showed in the early 1940s that individual mutations induced by X-rays in the bread mold, *Neurospora crassa*, caused changes in specific metabolic enzymes. This work laid the foundation for their **one gene, one enzyme hypothesis**.

FIGURE 1.24 Barbara McClintock.

Sewall Wright in 1954

J.B.S. (John Burdon Sanderson) Haldane

R.A. Fisher in 1956

FIGURE 1.25 Three prominent contributors to the modern evolutionary synthesis.

In the 1920s, Barbara McClintock, working on maize, developed a technique to visualize maize chromosomes, and studied the process of genetic recombination through crossing-over during meiosis (Figure 1.24). Later, between 1948 and 1950, McClintock discovered mobile elements in maize that would give rise to unstable **mosaic phenotypes**. The discovery of transposition mediated by such mobile elements – also known as jumping genes or **transposons** – would have a significant impact on later studies using other organisms, where transposition would become the principal tool for introducing mutations (see Chapter 9). McClintock understood the importance of transposons on modulation of gene action and in evolution long before transposition was generally recognized as an important genetic mechanism.

The major concept that emerged from the modern evolutionary synthesis was that genes provide the units for natural selection, and that changes in allele frequencies represent the mechanism for evolution. This was clearly formulated by Theodosius Dobzhansky, a Russian immigrant scientist who worked with Morgan at Columbia University and moved with him to the California Institute of Technology in 1928. In his 1937 book *Genetics and the Origin of Species*, Dobzhansky defined evolution as "a change in the frequency of an allele within a gene pool." This concept became a central tenet of **population genetics**, a science that arose as a result of the modern evolutionary synthesis.

Although a number of prominent scientists contributed to the modern evolutionary synthesis, three central protagonists stand out: Ronald A. Fisher; J. B. S. Haldane; and Sewall Wright (Figure 1.25). Fisher devised methods to partition variance observed in experimental data sets in different sources, a technique that became known as **analysis of variance** (ANOVA), which remains a central analytical tool in experimental biology. Fisher also invented the **maximum likelihood estimation method**, a technique extensively used today in gene-mapping studies. One of Fisher's major insights was the notion that the probability that a mutation increases the fitness of an organism decreases with the severity of the mutation. He also pointed out that larger populations that carry more variation have a greater chance of survival than small populations with a restricted gene pool. Fisher developed many of the concepts that form the foundation for the field of population genetics. Like his predecessors, Fisher was a fervent believer in eugenics, although he did not share the extreme prejudices of Karl Pearson.

J.B.S. Haldane, an English geneticist descended from Scottish aristocrats, published a book in 1932, *The Causes of Evolution*, which explained natural selection in evolution as a consequence of Mendelian genetics in mathematical terms. He remains perhaps best known for **Haldane's rule**, which states that when one sex is absent, rare or sterile, in the F1 offspring of two different animal races, that sex is the heterozygous (**heterogametic**) sex. Haldane's rule has implications for speciation, because if a sex-linked gene necessary for fertility or viability in two subspecies is absent from the homozygous chromosome and not transmitted to offspring with the heterozygous sex, fertility and viability of the F1 hybrids will be reduced. As speciation progresses, ultimately, the two subspecies will no longer be able to interbreed and become different species. Whereas Darwin and the early evolutionary biologists' distinguished species based on morphological criteria, the twentieth-century biologist Ernst Mayr defined different species by their inability to interbreed. Darwin's finches, from the Galapagos Islands, can, in fact, interbreed in captivity.

While Fisher and Haldane were developing their theories in Britain, major contributions to the developing field of population genetics were being made at the University of Chicago by Sewall Wright. Wright studied the effect of inbreeding on fitness, and realized that statistical variation of alleles of a gene that change their frequencies in a population during successive generations of inbreeding would lead to fixation to homozygosity of one of the alleles, a phenomenon known as **genetic drift**. The fixed alleles would not necessarily be beneficial but could, in fact, lead to a reduction in fitness, known as **inbreeding depression**. Genetic drift is more pronounced the smaller the population size. For example, this has become a major concern in the management of endangered species. Wright defined the **inbreeding coefficient**, and his work laid important theoretical foundations for investigators like Jay Lush and Alan Robertson, who were instrumental in applying genetic principles to animal and plant breeding. Wright also introduced the concept of **adaptive landscapes** into evolutionary biology. These are three-dimensional graphical representations in which peaks represent genotypes with high reproductive success (fitness) separated by valleys. Wright's concept implies that, for a genotype to evolve from its present fitness value to a phenotype with higher fitness, it must accumulate detrimental mutations that will eventually allow it to cross through an adaptive valley of lower fitness to another nearby peak.

A contentious issue that remained throughout the modern evolutionary synthesis period was the question of whether evolution proceeds through gradual changes, as envisioned by Charles Darwin, or through mutations of large effect, like those observed by Thomas Hunt Morgan, and whether complex traits are determined by a few genes of large effect, or by an infinite number of genes, that each contribute only a small effect (the **infinitesimal model**). The latter issue was addressed convincingly by Alan Robertson at the University of Edinburgh, who proposed that much of the variation in complex traits can be accounted for by a limited number of genes with large effects, and a larger number of genes with minor effects,

striking a now widely-accepted compromise between the two alternative theories.

THE RISE OF MOLECULAR GENETICS

One of the greatest discoveries in modern science was the elucidation of the structure of DNA. DNA (deoxyribonucleic acid) was discovered in 1868 by a Swiss biologist, Friedrich Miescher, who isolated it from nuclei of pus cells he found in discarded surgical bandages. As strange as it may seem today, in the early twentieth century there was debate as to whether DNA or proteins would represent the genetic material. This issue was conclusively resolved in 1943, when Oswald Avery, Colin MacLeod, and Maclyn McCarty at the Rockefeller University showed that they could transform an inactive strain of *Streptococcus pneumoniae* into a virulent strain by incorporating DNA isolated from virulent bacteria into the inactive strain. Furthermore, in 1952, Alfred Hershey and Martha Chase showed that when bacteria are infected with bacteriophage T2, it is the viral DNA and not its protein that is injected into the bacteria and directs viral replication.

The structure of DNA was discovered in 1953 by an American geneticist, James Watson, and an English physicist, Francis Crick, working together at the University of Cambridge (Figure 1.26). In formulating a model for DNA, Watson and Crick first arrived at the wrong conclusion, a triple helix with the bases on the outside. Meanwhile, Linus Pauling in the United States had come up with a similar model. Watson and Crick realized that it would only be a

FIGURE 1.26 A photograph taken by Antony Barrington Brown in 1953 of Watson (left) and Crick (right) with their DNA model.

matter of months for Pauling to realize that the triple helix model was chemically unfavorable, and that they had only a short time to solve the structure of DNA. Pauling had intended to visit London, but was denied a visa by the US State Department as his antinuclear sympathies were viewed unfavorably during the McCarthy period. If Pauling's visit to London had taken place, and if he would have had access to the same critical information as Watson and Crick, he might have derived at the same structure of DNA earlier. This critical information consisted, in part, of an X-ray photograph taken by Rosalind Franklin, who was performing comprehensive X-ray diffraction studies on DNA. Her now-famous photograph 51 (Figure 1.27) was shown to James Watson by Franklin's supervisor Maurice Wilkins without Franklin's knowledge, an act which has been considered by many as unethical. Photograph 51 provided the last bit of information needed by Watson and Crick to derive their model for the structure of DNA. Previously, Erwin Chargaff had reported that the molar concentration of adenine always equaled that of thymidine, and that the amount of guanine always equaled that of cytosine. Using cardboard models of the four bases, Watson had noticed that the sizes of the adenine–thymidine and guanine–cytosine base pairs were similar in size. These facts, together with the X-ray diffraction pattern in Franklin's photograph, allowed Watson and Crick to describe a model in which adenine–thymidine and guanine–cytosine base pairs were located on the inside of a double helix, with the phosphate backbone exposed on the outside. Moreover, the helical strands were antiparallel and, when taken apart, could each serve as a blueprint for replication of the other. Furthermore, based simply on arithmetic considerations, Crick hypothesized that three nucleotides would specify an amino acid and that, consequently, the genetic code would be degenerate as there would be 64 possible three-nucleotide codons to encode only 20 amino acids. The genetic code was subsequently unraveled by Marshall Nirenberg and his coworkers between 1961 and 1965. The notion that proteins are encoded by an mRNA

that is synthesized from a DNA template became known as the **central dogma** in molecular biology. The central dogma has largely held up, although it was modified in 1971 by Howard Temin who showed that RNA viruses can direct the synthesis of DNA from an RNA template, an exception to the central dogma.

In a videotaped message to the participants of the Nineteenth International Congress of Genetics in Melbourne (Australia) in 2003, which marked the fiftieth anniversary of Watson and Crick's discovery of the structure of DNA, Francis Crick described their discovery and asserted that he and Watson were fully aware of the impact their work would have, but that they never could have imagined that one day whole genomes could be sequenced, since they believed at that time that sequencing DNA would be extremely difficult. The first method for DNA sequencing was developed in 1975 by Frederick Sanger. With the advent of improved methodologies and automatic sequencers, high-throughput DNA sequencing for whole-genome analyses became a reality near the end of the twentieth century.

Recombinant DNA technology was pioneered by Paul Berg and others in California in the early 1970s. Propagation of engineered DNA constructs in bacteria allowed such constructs to be multiplied (cloned) in large amounts. Another major breakthrough in DNA technology came in 1983, when a young Californian scientist, Kary B. Mullis, introduced the **polymerase chain reaction** (PCR), which uses a thermostable DNA polymerase that can withstand multiple heat cycles necessary to separate and re-anneal DNA strands, and is active at a temperature of 72°C (Figure 1.28). PCR allows the amplification of precisely-identified DNA sequences, even if present in minute amounts, to sizable quantities. It is now a standard tool in the molecular biologist's toolkit. DNA cloning reached another milestone in 1997, when a Scottish investigator in Edinburgh, Ian Wilmut, announced that he had cloned a sheep from an adult cell from the udder of a six-year-old

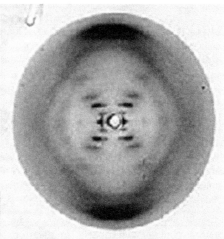

FIGURE 1.27 Rosalind Franklin (left) and her famous photograph 51 showing the X-ray diffraction pattern of DNA that enabled Watson and Crick to deduce its structure (right).

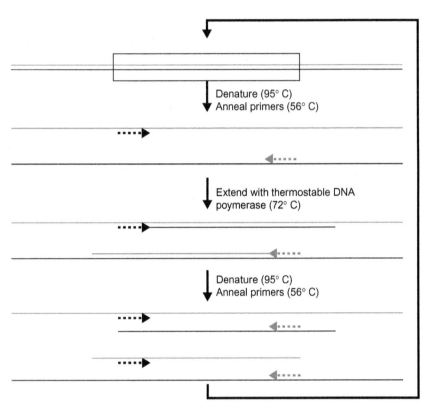

Denature (95° C)
Anneal primers (56° C)

Extend with thermostable DNA
poymerase (72° C)

Denature (95° C)
Anneal primers (56° C)

FIGURE 1.28 The polymerase chain reaction. The black rectangle indicates the target DNA sequence. The DNA strands are separated by heating at 95°C, and oligonucleotide primers, indicated by the dotted arrows, are allowed to anneal to the two antiparallel DNA strands at each side of the target sequence at an optimal temperature, in this example 56°C. The temperature is then raised to 72°C. At this temperature, a thermostable DNA polymerase – isolated from bacteria that grow in hot geothermal springs – is optimally active and can synthesize complementary DNA from each primer. Since this enzyme survives the denaturation temperature of 95°C, the process can be repeated. Repeating this cycle 20 or more times in a thermal cycler will lead to exponential amplification of the target sequence.

ewe. The cloned sheep was named Dolly (after the US country singer Dolly Parton), and her creation was soon followed by that of cloned mice, cows, and pigs. This led to a controversy that is still hotly debated regarding the possibility of human cloning, which many consider unethical as well as premature, in part because potential adverse health effects remain uncertain.

The impressive growth of molecular genetics provided a much-needed molecular framework for the study of quantitative traits, which during the modern evolutionary synthesis were described mostly statistically. As will become clear in this book, behavioral genetics requires both molecular and quantitative genetic approaches.

A BRIEF HISTORY OF NEUROSCIENCE

While the modern evolutionary synthesis was in full swing, the field of neuroscience was being born. In the 1920s and 1930s, classical experiments by Otto Loewi and Sir Henry Hallett Dale led to the identification of the first **neurotransmitters**, acetylcholine and adrenaline, and Dale coined the terms **cholinergic** and **adrenergic** to describe their often antagonistic physiological effects (Figure 1.29). During the following decades the repertoire of neurotransmitters expanded dramatically to include a host of other bioamines, amino acids and their derivatives, and peptides.

Using home-built oscilloscopes and amplifiers, early electrophysiologists studied how neurons generate electrical

FIGURE 1.29 Early pioneers in neuroscience.

impulses to encode and convey information. The first **action potential** was recorded by Kenneth C. Cole and H. J. Curtis in 1939 (Figure 1.30). Working at Cambridge in the 1930s and 1940s, Sir Andrew Fielding Huxley and Alan Lloyd Hodgkin developed the **voltage-clamp method** (explained in Chapter 2) that enabled them to explain the ionic basis of the action potential as the sequential opening and closing of "gates" that are selectively permeable to sodium and potassium. At the same time, a controversy arose as to whether **synaptic transmission**, that is, communication between one cell and another in the nervous

FIGURE 1.30   The first action potential showing a transient increase in membrane conductance, recorded by K.C. Cole.

FIGURE 1.31   Sir Bernard Katz.

FIGURE 1.32   Linda Buck receiving the 2004 Nobel Prize from His Majesty King Carl Gustav XVI of Sweden, for her discovery of odorant receptors and the organization of the olfactory system.

system, was electrical, as proposed by John Eccles, or chemical, as advocated by Sir Bernard Katz (Figure 1.31). The controversy was resolved in the 1950s when Katz demonstrated that synaptic transmission at the neuromuscular junction was due to the release of discrete packages (**quanta**) of acetylcholine from the **presynaptic cell**, which would elicit changes in the membrane potential of the muscle fiber (the **postsynaptic cell**) after diffusing across a **synaptic cleft**. Soon after, the advent of electron microscopy consolidated the existence of a physical separation at the synapse.

The subsequent development of increasingly more sophisticated neuroanatomical tracer and imaging techniques, and extracellular, intracellular, and multicellular recording methods have led to an explosion of information about neural circuitry in the brain, and its relationship to physiological and behavioral functions. Among many noteworthy achievements are the studies by Eric Kandel and his colleagues at Columbia University who developed a seemingly unlikely invertebrate model for learning, the sea slug *Aplysia californica* (see also Chapter 14). This simple organism exhibits a defensive gill withdrawal reflex when a jet of water is sprayed against its siphon. This response can be paired with an electric shock to the tail, so that

subsequent electric shocks will elicit stronger gill withdrawal responses, a characteristic example of reinforcement through classical conditioning, not unlike the salivation responses elicited by Pavlov in his dogs. The neural circuitry that underlies this response, as well as evolutionarily conserved mechanisms that control alterations in gene expression during learning, could be identified and characterized. Landmark discoveries in sensory neurobiology included the organization of the visual cortex and elucidation of how visual information is processed by deconstructing aspects of the image along parallel neural pathways, for which David Hubel and Tornsten Wiesel received the 1981 Nobel Prize, and the discovery of a large multigene family of odorant receptors and the functional organization of the olfactory system, for which Linda Buck and Richard Axel received the 2004 Nobel Prize (Figure 1.32).

In addition to great advances in understanding the neural mechanisms that mediate sensory information processing, **sensory-motor integration** and motor control, complex cognitive brain functions once thought intractable, have become amenable to scientific scrutiny. Such processes include learning and consolidation and retrieval of memory, fear, attention, and even such seemingly esoteric perceptions as empathy and appreciation of music.

Whereas advances in neurobiology paralleled developments in genetics, and occurred contemporaneously with the exploits of the early ethologists, there was relatively little conceptual cross-talk between these disciplines, other than the application of molecular genetic techniques. Neurogenetics – although recognized as a subdiscipline – was not a major focus of behavioral neuroscience.

This changed with the availability of whole-genome sequences and genomics technologies near the end of the twentieth century, which has spurred a profound and widespread interest in exploring the link between the genome and the nervous system. The concept of "genes, brain, and behavior" – that is, how genes enable the nervous system to respond to environmental cues and elicit an appropriate behavior – has become the central focus of behavioral biology. Although this book focuses on genetic aspects of behavior, appreciating the neural circuits that express behaviors is indispensable. Fundamental neurobiological principles are described in greater detail in Chapter 3, and auxiliary neurobiological information is provided in text boxes in subsequent chapters.

THE EMERGENCE OF BEHAVIORAL GENETICS

The classical ethologists and early experimental psychologists demonstrated that behaviors were not of such intractable complexity that they could not be studied using carefully-designed experimental approaches. Behavioral biology has developed since then along two parallel tracks that have only recently begun to interconnect. On the one hand, advances in neurobiology, including the ability to record electrophysiologically from multiple neurons simultaneously in real-time in live behaving animals, have rapidly identified neural circuits that contribute to the manifestation of a wide range of behaviors. On the other hand, geneticists have begun to investigate how spatial and temporal patterns of expression of ensembles of genes enable the activity of such neural circuits to allow the organism to sense and respond to changes in its environment. Until the advent of technologies that enabled whole-genome analyses, genetic studies lagged behind the neurobiological approaches. The present **genomic era** can look forward to the coalescence of both approaches into **neurogenomics** applications that will monitor activity of the nervous system, while at the same time assessing the transcriptional, translational, and post-translational dynamics of the genome that supports this activity.

Artificial selection has long been the standard procedure by which animal breeders improve livestock by selecting individuals of extreme phenotypes as parents for each successive generation (e.g. milk yield in dairy cows). The earliest artificial selection experiment in behavioral genetics was reported by R.C. Tryon in 1942, by selecting rats based on their ability to negotiate a maze. Poor performers were labeled "dull" and good performers were designated "bright." Tryon observed a substantial difference in maze running ability in two selected lines after only seven generations of selecting "bright" and "dull" lines by breeding the best and worst maze-running rats with others of similar abilities. However, subsequent studies showed that the performance of dull rats could be improved dramatically by rearing them in an enriched environment, with a variety of objects to explore and ample social interactions. These studies provided one of the first well-documented examples of **genotype-by-environment interaction** on a behavioral trait, since manifestation of the genetic difference between the dull and bright strains is dependent on the environmental conditions in which the rats are reared.

Another noteworthy early and pioneering experiment in behavioral genetics was performed in the 1960s by J. Hirsch, on **geotaxis** in *Drosophila*. Flies are positively geotactic, that is, they move upwards against the force of gravity. Hirsch constructed a vertical maze, in which he could introduce flies on one side in the center of the maze and watch them move toward a light source at the other side of the maze. The maze consisted of sequential branch points, each of which required flies to make a decision to move upward or downward. They would emerge at nine different vertical locations at the other side of the maze. Hirsch selected flies that emerged near the top of the maze, and flies that emerged near the bottom of the maze, and subjected them to iterative cycles of artificial selection, breeding "high" and "low" flies from each successive generation as parents for the next generation. These experiments resulted in the isolation of distinct "high" and "low" geotactic lines that have been propagated for decades, and still exist today.

Other pioneering behavioral genetic experiments were reported in 1971 by Seymour Benzer and his students at the California Institute of Technology on **circadian activity** of *Drosophila* (see also Chapter 11). Flies, like other animals, show periods of activity and inactivity that are governed by an endogenous clock **entrained** to the daily light–dark cycle. Using a photoelectric monitoring device, Benzer monitored activity of flies during the day and night. Normal wild-type flies showed a 24-hour circadian rhythm. Benzer identified mutants that had altered circadian rhythms. He found three such mutants: one was arrhythmic, displaying random periods of rest and activity; one had a significantly shorter circadian period of only 19 hours; and the third mutant had an unusually long circadian rhythm of 28 hours (Figure 1.33). All of these mutants were out of synch with the daily light–dark cycle. In addition to their locomotor behavior, their rhythms of eclosion (emergence from the pupal case) were similarly disrupted. Flies eclose usually in the early morning hours, the time of day when they are most active. Benzer found that all three mutations were located in a single gene. They mapped it to the *X*-chromosome, and named it "*period (per)*." This gene was subsequently found to be a transcriptional regulator, and only one of many components that make up the biological clock. The experiments by Benzer were critical to developing the field of **chronobiology**, i.e. the study of biological clocks. They were also important, because circadian variation in gene expression impacts the experimental

Box 1.2   Model organisms

Throughout the history of the biological sciences a number of animals have found widespread use as laboratory models, because they are easy to rear in large numbers in the laboratory and sophisticated genetic techniques could be developed that allow scientists to address questions more easily in these laboratory species under controlled genetic and environmental conditions than in wild animals in their natural habitats. Fruit flies (*Drosophila*), rats, mice, and the nematode *Caenorhabditis elegans* are the most frequently used **model organisms** in behavioral genetics. However, questions are often raised as to what extent findings in laboratory animals that have been propagated in captivity under artificial conditions for many generations reflect reality in nature. Whereas this is a legitimate concern, many questions can only be addressed under controlled conditions, and the advantages offered by the resources and techniques available for studies on these organisms are substantial and outweigh their possible disadvantages. With the ever-increasing number of sequenced genomes, the number of "model organisms" is likely to expand. Honeybees (*Apis mellifera*) are now recognized as a model for studies on the genetic architecture of social behavior (sociogenomics), whereas songbirds are becoming an increasingly valued model for studies of genetic control of neural plasticity during song learning (a model for acquisition of language) and pair-bond formation. Sequenced genomes and extensive databases have become available for both species. In the decades ahead there will likely be an upsurge in comparative genomics approaches, which will be accompanied by extended comparisons between behaviors in laboratory strains and natural populations. Thus, the definition of what exactly is a "model system" is likely to become more blurred, as technologies become available to study the genetic underpinnings of behaviors in more species at the whole-genome level. Such developments will be especially welcome for understanding the evolution of analogous behaviors in related species.

design of behavioral experiments. The manifestations of behaviors may be affected by the time of day at which measurements are taken. The most extreme example is, perhaps, courtship and mating in *Drosophila*, which occurs only during specific times of the circadian day, with a major activity peak in the morning and a smaller peak of activity in the late afternoon.

The studies of Benzer and Hirsch are often viewed as contrasting approaches to the study of behavior, with Benzer as the proponent of the "one-gene-at-a-time" approach, and Hirsch the protagonist of the "multiple genes ensemble" approach. In reality, however, both approaches are complementary, rather than mutually exclusive. The "Hirschian" strategy, however, has recently gained especially strong support as whole-genome analyses are becoming increasingly more sophisticated. Genes identified in such an approach can then be studied intensively individually in a "Benzerian" approach.

Behavioral studies in mice have benefited greatly from a major breakthrough in the late 1980s when Mario Capecchi, Sir Martin Evans, and Oliver Smithies developed methods for deleting genes in mice through **homologous recombination**, a discovery that earned them the 2007 Nobel Prize. Such "knock-out" mice proved to be powerful tools for clarifying the functions of specific genes on physiology and behavior. **Transgenic** mouse technologies were further expanded to allow introduction of foreign genes at defined locations in the genome ("knock-in" mice), and by constructing strains in with the expression of transgenes could be targeted to certain cells or tissues, and could even be controlled temporally. These methods will be discussed in greater detail in Chapter 9.

Although transgenic mice have generated many new insights, this "one-gene-at-a-time" approach led to the notion that each gene contributes a specific function, a conceptual framework that began to erode when whole-genomic approaches and transcriptional profiling studies revealed that there is extensive functional **pleiotropy**

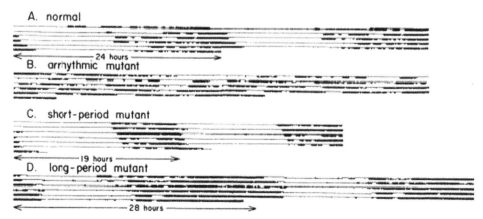

FIGURE 1.33   Locomotor activity recorded for normal flies and arrhythmic period mutants and short- and long-period mutants. (From Konopka and Benzer, (1971). *Proc. Natl. Acad. Sci. USA*.)

(that is, a gene can contribute to the manifestation of multiple phenotypes) with wide-ranging genetic interactions. It is becoming increasingly clear that, with few exceptions, behaviors are emergent properties of complex genetic networks, rather than the result of linear genetic pathways that are composed of genes with single dedicated functions.

The availability of whole-genome sequences of an increasing number of species, including *Homo sapiens*, chimpanzee, rat, mouse, zebrafish, zebra finch, *Drosophila*, honeybee, and the nematode *C. elegans*, to name but a few, together with advanced neurobiological techniques, will in the future enable powerful comparative genomic approaches that will allow us not only to define the genetic architecture of behaviors and understand the relation of genetic networks with neural function, but also to gain insights into the evolutionary processes that have occurred to shape the behavioral phenotypes which we observe today.

SUMMARY

Experimental approaches with predictive value for the study of animal behaviors were first performed by the pioneers of experimental psychology, such as Pavlov, Thorndike, and Skinner, and the early behavioral ecologists, Lorenz, von Frisch, and Tinbergen. A thorough understanding of the genetic basis of behaviors, like most of the foundations of modern biology, rests on Darwin's concept of natural selection, which envisions that the best-adapted individuals in a population will have a more favorable chance to survive and contribute to the next generation. Following the rediscovery of Mendel's rules of inheritance at the turn of the twentieth century, it was recognized that genes provide the physical targets for natural selection during evolution. The modern evolutionary synthesis reformulated Darwin's concept, by viewing evolution as a change of allele frequencies within a gene pool. Whereas the incidence of phenotypes that are determined by a single segregating gene can be readily predicted according to Mendel's rules, complex (or "quantitative") traits that arise from multiple segregating genes that are influenced by the environment require more complex statistical analyses. Such analyses, including analysis of variance and likelihood distributions, were pioneered by Roland Fisher building on earlier work by Francis Galton and his protégé Karl Pearson, who had laid the statistical groundwork for regression and correlation analyses. Fisher, Haldane, Wright, and others studied gene flow in populations and laid the foundation for population genetics. The second half of the twentieth century saw the rise of molecular genetics, with Watson and Crick's discovery of the structure of DNA, elucidation of the genetic code, formulation of the central dogma, and the flow of genetic information from DNA to RNA to protein. These discoveries coincided with major advances in the neurosciences as Huxley, Hodgkin, and Katz explored the mechanisms for neural signal propagation and synaptic transmission. The early experimental psychologists and behavioral ecologists paid little attention to contemporary advances in genetics and neuroscience, but were able to provide independent insights into fundamental aspects of behavior, including instinctive behaviors, such as imprinting and social communication. Tinbergen framed the study of behavior as a series of questions that could be answered through scientific studies, which asked how behaviors are elicited and modified through learning and interactions with the environment, and how they develop and evolve. Some of the first experiments in behavioral genetics were done by Benzer and Hirsch on circadian behavior and geotaxis, respectively, using the fruit fly *Drosophila melanogaster*, developed earlier as a favorable genetic model organism by Thomas Hunt Morgan. Advances in genetics (especially the availability of whole-genome sequences and the development of methods for gene targeting), and the neurosciences around the turn of the twenty-first century are facilitating the integration of the disciplines of behavioral genetics, neuroscience, ecology, and ethology into the modern science of behavioral biology. One of the central goals of behavioral genetics is to explore how regulation of gene expression enables the nervous system to drive and modify appropriate behaviors, and how changes in allele frequencies cause behaviors to evolve.

STUDY QUESTIONS

1. How does classical conditioning described by Pavlov differ from operant conditioning described by Skinner?
2. What contributions did Lorenz and von Frisch make to understanding social behavior?
3. What contribution did Tinbergen make to define the modern field of ethology?
4. Describe the conceptual difference between the evolutionary theories of Lamarck and Darwin.
5. Explain why Mendel was able to derive his fundamental rules for inheritance, whereas Galton fell short of making the same discoveries.
6. What is eugenics? Why did it appear a reasonable concept in the early days of the history of genetics, and why has it fallen into disrepute?
7. What is the fundamental contribution of the modern evolutionary synthesis as formulated by Dobzhansky?
8. What is Haldane's rule?
9. What was the early evidence that DNA represents the genetic material?
10. What critical pieces of information enabled Watson and Crick to develop their model for the structure of DNA?
11. What is the central dogma?
12. What were some of the major contributions made to behavioral genetics by Thomas Hunt Morgan?
13. Describe the contributions of S. Benzer and J. Hirsch to behavioral genetics.

RECOMMENDED READING

Darwin, C. (1859). *On the Origin of Species.* John Murray, London, UK.

Dobzhansky, T. (1937). *Genetics and the Origin of Species.* Columbia University Press, New York, NY.

Fisher, R. A. (1930). *The Genetical Theory of Natural Selection.* Clarendon Press, Oxford, UK.

Gillham, N. W. (2001). *A Life of Sir Francis Galton: From African Exploration to the Birth of Eugenics.* Oxford University Press, New York, NY.

Haldane, J. B. S. (1932). *The Causes of Evolution.* Longman, Green and Co., Princeton University Press, Princeton, NJ.

Hirsch, J., and Erlenmeyer-Kimling, L. (1962). Studies in experimental behavior genetics: IV. Chromosome analyses for geotaxis. *J. Comp. Physiol. Psychol.*, **55**, 732–739.

Konopka, R. J., and Benzer, S. (1971). Clock mutants of *Drosophila melanogaster. Proc. Natl. Acad. Sci. USA*, **68**, 2112–2116.

Lorenz, K. (1965). *Evolution and Modification of Behavior.* University of Chicago Press, Chicago, IL.

Mayr, E., and Provine, W. B. (eds) (1980). *The Evolutionary Synthesis: Perspectives on the Unification of Biology.* Harvard University Press, Cambridge, MA.

Skinner, B. F. (1947). "Superstition" in the pigeon. *J. Exp. Psychol.*, **38**, 168–172.

Sturtevant, A. H. (1965). *A History of Genetics.* Harper & Row, New York, NY.

Tinbergen, N. (1951). *The Study of Instinct.* Clarendon Press, Oxford, UK.

Von Frisch, K. (1973). Decoding the language of the bee 1992. In *Nobel Lectures Physiology or Medicine 1971–1980*, J. Lindsten, (ed.), pp. 76–87. World Scientific Publishing Co, Singapore.

Watson, J. D. (1968). *The Double Helix: A Personal Account of the Discovery of the Structure of DNA.* New American Library, New York, NY.

Watson, J. D., and Crick, F. H. C. (1953). A structure for deoxyribose nucleic acid. *Nature*, **171**, 737–738.

Weiner, J. (1999). *Time, Love, Memory: A Great Biologist and His Quest for the Origins of Behavior.* Vintage Books, New York, NY.

Wright, S. (1931). Evolution in Mendelian populations. *Genetics*, **16**, 97–159.

Chapter 2

Mechanisms of Neural Communication

OVERVIEW

Behaviors are manifestations of the nervous system's functions, and represent the expression of the integrated activity of neural networks in response to the animal's perception of its external and internal environments. Although this book focuses on genetic mechanisms that predispose and enable the nervous system to manifest distinct behaviors, some knowledge of the functional organization of the nervous system is indispensable for a full appreciation of the fundamental mechanisms that direct these behaviors. In this chapter we will describe how electrical signals are elicited and propagated by neurons, and how such signals are transmitted and integrated at synapses in the nervous system.

TRANSMISSION OF INFORMATION IN THE NERVOUS SYSTEM

The nervous system coordinates an animal's physiology and behavior. Neurons consist of a **cell body** (also known as a **soma** or **perikaryon**), which contains the nucleus, mitochondria, endoplasmic reticulum, and Golgi apparatus; organelles that house the machinery for gene transcription, protein synthesis, and intermediary metabolism. Processes that emanate from the cell body are the single **axon**, which conducts signals away from the cell body, called **efferent signals**, and **dendrites**, which conduct signals toward the cell body, called **afferent signals**. The axons can extend over short or long distances, e.g. the sciatic nerve arises from the sacral region of the spinal cord to innervate the lower limbs down to the toes. **Glia** are supporting cells that are closely associated with neurons.

Information in the nervous system is transmitted by a combination of chemical and electrical signals. How are electrical currents generated in a biological system? Electrical currents arise from movements of small numbers of ions across and along cell membranes. Since ions are charged atoms, the movement of ions will carry a current. The distribution of ions across the membrane of any cell is asymmetric, and establishes the electric field potential that drives an ion current through a conducting medium.

The extracellular concentration of sodium ions (Na^+) is maintained at about 145 mM, whereas its intracellular concentration is about 12 mM. Conversely, the extracellular concentration of potassium ions (K^+) is only about 5 mM, whereas its intracellular concentration is about 150 mM. Most cell membranes are virtually impermeable to ions, allowing ions to leak only slowly between the cellular and extracellular fluids. This reciprocal asymmetry of Na^+ and K^+ concentrations is established and maintained by the ATP-dependent Na^+/K^+ pump (Figure 2.1). The leaking current would ultimately lead to the collapse of the Na^+ and K^+ gradients were it not for the Na^+/K^+-ATPase. This pump is **electrogenic**, in that it transports three sodium ions out of the cell in exchange for the entry of only two potassium ions. Inhibition of the Na^+/K^+-ATPase experimentally by the cardiac glycoside inhibitor ouabain causes the ion gradients to collapse, and prevents neuronal function.

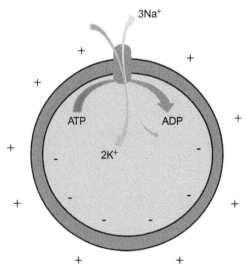

FIGURE 2.1 The function of the Na^+/K^+ pump. The Na^+/K^+-ATPase is a tetrameric enzyme that consists of two α subunits that perform the transport of ions and two glycosylated β subunits. Transport of three intracellular Na^+ ions out of the cell, and simultaneous transport of two K^+ ions into the cell, is accompanied by hydrolysis of ATP. During this process the enzyme becomes transiently phosphorylated, and the energy released by its dephosphorylation drives the exchange of monovalent ions. Electrogenic activity of the pump contributes to building up a net negative charge within the cell compared to the outside.

R. Anholt and T. Mackay: Principles of Behavioral Genetics
ISBN: 978-0-12-372575-2

THE RESTING MEMBRANE POTENTIAL

When a cell is penetrated by an electrode containing a highly-conductive salt solution and connected to a voltage meter, a voltage, or difference in electrical potential, is measured across the membrane. This voltage is known as the **membrane potential**. It results primarily from the differential permeability properties of the cell membrane, but also to a small extent from the electrogenic activity of the Na^+/K^+ pump, and the presence of nondiffusible negatively-charged molecules inside the cell, such as nucleic acids, proteins, and negatively-charged metabolites. Thus, each cell behaves like a tiny battery: the cell's inside is negative with regard to the outside (Figure 2.2). Let us first focus on the movement of potassium across the cell membrane. The cell membrane of neurons and glia is about twenty times more permeable to potassium than to sodium, due to the smaller hydration shell of the potassium ion, and there is thirty times less K^+ outside the cell than inside. Thus, a diffusion force exists for potassium to leak out of the cell. At the same time, the charge distribution across the membrane exerts an electrochemical force that counteracts the outward diffusion of the positively-charged potassium ions. When the electrical potential exactly balances the concentration gradient driving the potassium leak, there will be no net movement of potassium, and equilibrium is established (Figure 2.3, opposite). (Note that this does not mean that there is no movement at all, but simply that the influx of potassium would equal its efflux.) Note that, at equilibrium, the membrane potential would be negative inside, not zero. Zero would be the value of the membrane potential if the ion gradients collapsed, as would be the case if the Na^+/K^+ pump were rendered inactive.

The **equilibrium potential** is defined as the theoretical voltage that would be produced across the cell membrane to counteract the tendency for ions to diffuse across it. For a single ion, e.g. K^+, the equilibrium potential can be calculated using the **Nernst equation**, which in a simplified form is shown as:

$$E_X = 61/z \, \log\left(\frac{[X]_o}{[X]_i}\right)$$

in which z is the valence of permeable ion X, and subscripts o and i refer to the concentrations of X outside and inside a cell at 37°C.

Since the intracellular and extracellular concentrations of potassium are known, we can calculate that the membrane potential at equilibrium would be about $-90\,mV$, if potassium were the only ion able to move across the membrane (the minus sign indicates that the inside of the cell would be negative with respect to the outside). This is called the equilibrium potential for potassium, abbreviated to E_K. Similarly, if sodium were the only ion to diffuse across the cell membrane, it would generate an equilibrium potential for sodium, i.e. E_{Na}, of about $+60\,mV$ under physiological conditions. The Nernst equation is important, in that it allows calculation of the membrane potential expected for diffusion of a single ion when the concentrations on both sides of the membrane are known.

Since potassium is not the only ion to which the cell membrane is permeable, the empirically measured membrane potential is weighted among the equilibrium potentials of all ions that can move across the membrane. This is known as the **resting membrane potential**. In mammalian neurons and glial cells it is approximately $-70\,mV$. This value is closer to E_K than to E_{Na}, because the membrane is more permeant to potassium than to sodium. Note that only a very small fraction of the cellular sodium needs to move to establish a relatively large difference in electrical potential. The resting membrane potential can be estimated using the **Goldman-Hodgkin-Katz equation**, which is an extension of the Nernst equation to include all permeable monovalent ions:

$$E = RT/zF \, \ln\left(\frac{P_{Na}[Na]_o + P_K[K]_o + P_{Cl}[Cl]_i}{P_{Na}[Na]_i + P_K[K]_i + P_{Cl}[Cl]_o}\right)$$

where E is the membrane potential at equilibrium, R is the gas constant ($8.314\,V\,C\,K^{-1}mol^{-1}$ in electrical units), T the temperature in °K, z is the valence of the permeantion, F is Faraday's constant ($9.648 \times 10^4\,C\,mol^{-1}$), and P is the permeability of the membrane to the ion indicated in the subscript.

If the membrane permeability were to be changed by opening channels through which one or more ions can flow, the membrane potential would change. Cells that contain **voltage-gated ion channels** are known as **excitable cells**. Voltage-gated ion channels are proteins that

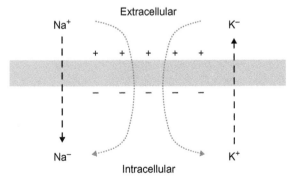

FIGURE 2.2 Diffusion forces (black arrows) and electrochemical forces (gray arrows) that act on the distribution of positively-charged ions across the cell membrane. Both the action of the Na^+/K^+ pump and the permeability barrier of the lipid bilayer of the membrane allow the resulting charge separation to be maintained.

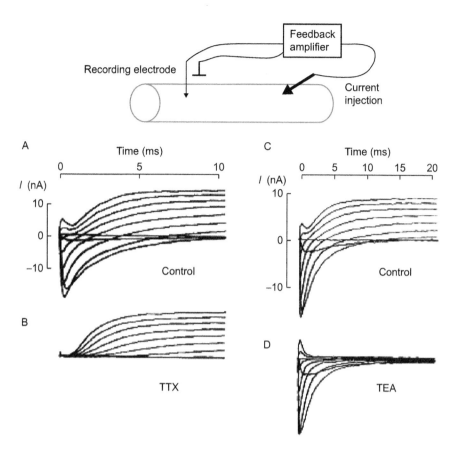

FIGURE 2.3 Measuring current under voltage clamp in an axon. Panels A–D show current recordings following 15 mV voltage steps from −60 mV to +60 mV. Downward deflections indicate net inward currents, and upward deflections indicate net outward currents. Note that voltage steps under controlled conditions show a rapid net inward current, followed by a net outward current. In the presence of the sodium channel blocker tetrodotoxin (TTX; panel B), the inward current is abolished, whereas the net outward current still occurs. The potassium channel blocker TEA (panel D) leaves the inward current intact, but abolishes the net outward current. (Adapted from B. Hille (1984). *Ionic channels of excitable membranes*, Sinauer Associates, Inc., Sunderland, MA.)

change their conformation in response to changes in the voltage across the membrane, thereby opening or closing transmembrane pores through which specific ions can diffuse toward their equilibrium. Excitable cells include nerve cells, muscle cells, cardiac cells, and specialized sensory cells. The ability to change the permeability of the cell membrane selectively to cations under the influence of the membrane voltage forms the basis for the generation and propagation of nerve impulses.

THE MECHANISM OF THE ACTION POTENTIAL

If voltage-gated sodium channels were to open, and positively charged Na^+ ions were to diffuse down their concentration gradient into a neuron, the resting membrane potential would change the resting value and become less negative. This decrease in the charge difference across the cell membrane is known as **depolarization**. An increase in the charge difference across the membrane, on the other hand, is termed **hyperpolarization**. The flow of ions through aqueous channels in the membrane is expected to obey Ohm's law, which states that the amount of current (I) that flows is linearly related to the voltage (V) and

the resistance (R): I = V/R. Ohm's law can also be written as V = I/g, where g is the "conductance" and g = I/R. At a certain potential in an excitable cell, the **threshold potential** (about −55 mV in neurons), the membrane potential changes in an explosive fashion, climbing toward the equilibrium potential for Na^+ (approximately +60 mV), after which it rapidly repolarizes back toward the original resting membrane potential. This event is called an **action potential** or **spike** and lasts for about 3 ms.

The first action potential was recorded in 1939 by Cole and Curtis from the giant axon of the squid *Loligo*, an axon that innervates the animal's muscular mantle to mediate rapid escape responses. Observing the first action potential was a remarkable event, not unlike landing the first man on the moon, as there was no explanation for this phenomenon based on any laws of physics. The mechanism that underlies the action potential remained elusive as the event occurred within 3 ms, until Hodgkin and Huxley (Figure 2.4) in the 1940s used the voltage clamp technique to identify the currents that give rise to this phenomenon. Under voltage clamp, a feedback amplifier compensates continuously for changes in the membrane voltage by delivering an opposing electrical current, thereby forcing the membrane potential to remain constant (Figure 2.3). By applying electrical stimuli and using selective toxins, such as

FIGURE 2.4   Sir Alan Lloyd Hodgkin in 1963 (left) and Sir Andrew Fielding Huxley at Trinity College, Cambridge, UK, July 2005 (right).

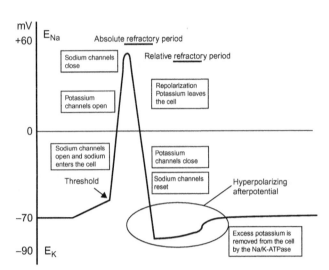

FIGURE 2.5   Diagrammatic representation of an action potential.

tetrodotoxin (TTX), a blocker of voltage-gated Na^+ channels, and tetraethylammonium (TEA), a blocker of some types of voltage-gated K^+ channels, under voltage clamp conditions, Hodgkin and Huxley showed that the rising phase of the action potential is due to a rapid voltage-dependent opening and subsequent closing of Na^+ channels, allowing a burst of sodium to flow into the cell. This is accompanied by a slower voltage-dependent opening of K^+ channels, which mediates efflux of potassium and thereby, causes repolarization of the membrane toward the resting membrane potential as the sodium channels close (Figures 2.3 and 2.5).

The action potential is an example of an electrical charge produced by passive transport, which means that the flow of ions down their concentration gradients through the pores of voltage-gated channels does not require the investment of metabolic energy. However, energy must be invested to establish and maintain the ion gradients that drive the action potentials through the activity of the Na^+/K^+-ATPase. In this sense, pumps and channels are complementary, and both are essential for enabling neural activity. It should be noted that the fluxes of ions needed to carry an action potential are extremely small and cause very little disturbance of the total asymmetric distribution of ions across the membrane. For example, an action potential in the giant axon of the squid would change the ion concentration gradients by about 0.001%. They can, therefore, be easily corrected by the Na^+/K^+ pump, when action potentials occur at low frequency.

The action potential represents an example of a **feed-forward process**, a relatively uncommon mechanism in biological systems. The change in voltage leads to opening of voltage-gated sodium channels, and the flow of sodium

through these channels leads to further depolarization, which promotes opening of more sodium channels. This feed-forward mechanism ensures that action potentials are **all-or-none events**, and that they are all of the same size. At the same time, they are self-limiting, as sodium channels close automatically. Stimulus intensity is encoded by the frequency, not the amplitude, of action potentials. The maximum frequency at which action potentials can be generated is limited by the **refractory period**; while Na^+ channels are already open, a second action potential cannot be generated. This is the **absolute refractory period**. During the hyperpolarizing after-potential an action potential can be evoked, but only at greater stimulus intensities, as more depolarization is needed to reach the threshold potential. This is the **relative refractory period** (Figure 2.5).

At the peak of the action potential, the inside of the cell temporarily will be charged positively relative to the outside. This creates a situation in which a difference in potential now exists between the activated region and an adjacent region, with a conducting salt solution (the intracellular fluid) between them. This causes current to flow between these regions. This spread of current is great enough to cause membrane depolarization in the adjacent region of the cell, thereby generating another action potential of the same size as the original action potential. The sequential regeneration of action potentials in this manner from one region to another is how nerve impulses are propagated. The all-or-none quality of the action potential ensures that action potentials can be propagated over long distances without loss of amplitude. Due to the refractory period, action potentials cannot be conducted backward through a region that is still active. However, when action potentials are evoked artificially by stimulating an axon at any location away from the cell body, action potentials will travel in both directions away from the site of stimulation. Under physiological conditions, action potentials are usually evoked at the junction of the axon with the cell body, a region known generally as the **spike initiation zone** (or **axon hillock** in spinal motor neurons), and will travel unidirectionally through the axon away from the soma (Figure 2.6).

Whereas stimulus intensity is encoded in the frequency of action potentials, the nature of the stimulus is perceived as a result of the "wiring" of the nervous system. Thus, an action potential in the optic nerve is no different from an action potential in the auditory nerve. Yet, the brain interprets the signal coming from the optic nerve as light, and that from the auditory nerve as sound.

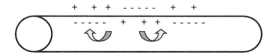

FIGURE 2.6 Spreading of current and depolarization of adjacent regions of the axon around the action potential site.

Myelination and Saltatory Propagation of Action Potentials

If an excitable cell is stimulated weakly, too weakly to activate voltage-gated channels, current will spread only a short distance along the axon. How it does so can be described by the **cable properties** of the cell membrane, i.e. how the cell membrane represents an electrical cable. An excitable membrane can be compared to an electrical circuit in which a resistor and capacitor are linked in parallel (Figure 2.7). The membrane acts like a capacitor in

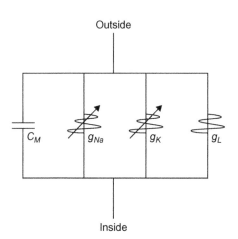

FIGURE 2.7 A simplified diagram of the electrical equivalence circuit of an axon, showing the membrane capacitance (C_M), and voltage-gated sodium and potassium channels as variable resistors (g indicates the conductance for the ion in the subscript). The leak conductance (g_L) is also indicated.

its ability to maintain a charge separation. It also possesses resistance in its ion channels. When the resistance is high, i.e. voltage-gated channels are closed, current only flows from the capacitor, and the distance over which current can be propagated declines exponentially. Capacitative currents flow faster, but over smaller distances, than currents that are propagated via action potentials.

The rapid propagation of action potentials is facilitated by glial cells. Glial cells include Schwann cells that form **myelin** sheaths around axons in the peripheral nervous system and oligodendrocytes that have a similar function in the brain. Myelin, which is wrapped in an onion-like fashion around the axon, insulates the cell membrane, preventing generation of action potentials. The white appearance of nerves (the **white matter** in the brain) is due to myelination (Figure 2.8, overleaf). Myelination of axons occurs in a patchy manner. In myelinated nerve fibers, voltage-gated ion channels are concentrated at bare areas of axonal membrane between repeating segments of myelin. These intermittent unmyelinated regions are called **nodes of Ranvier**. In myelinated axons, action potentials are normally generated only at the nodes of Ranvier.

The short distances between the nodes, and the high density of voltage-gated sodium channels, enable the capacitative cable current from the spike initiation zone or an adjacent node to depolarize the membrane to threshold at the next node of Ranvier, before the current dissipates (Figure 2.9, overleaf). Thus action potentials in myelinated nerves travel in a jumping fashion from node to node, bypassing myelinated segments, a process known as **saltatory conduction**. This greatly enhances the speed with which nerve impulses can be propagated by regenerating action potentials only at intervals of a few millimeters rather than at every site along the axon.

Synaptic Transmission and Postsynaptic Potentials

We can now ask how neurons signal to each other or to the target tissues they innervate. This occurs at synapses, regions where axon terminals closely approach their postsynaptic targets, and release neurotransmitters that will activate or inhibit the postsynaptic cell. Classical studies on synaptic transmission were performed in the early 1950s by Fatt and Katz on the frog **neuromuscular junction** (also known as the **endplate**). When they recorded from the muscle fiber, even without stimulating the nerve, they observed small changes in the resting membrane potential that appeared as little blips on the oscilloscope. These blips appeared at random, and their sizes were unitary multiples of the same amplitude. Fatt and Katz hypothesized correctly that these elementary voltage changes, which they named **miniature endplate potentials (mepps)** were due to the occasional release of discrete packages (**quanta**) of neurotransmitter (Figure 2.10).

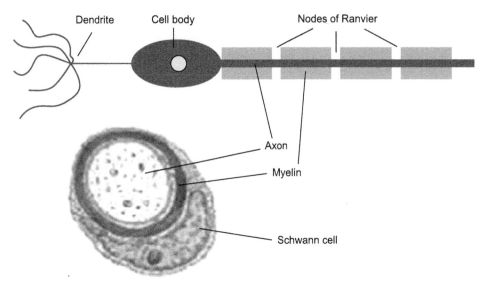

FIGURE **2.8** Diagram of a myelinated neuron (top) and electron micrograph of a Schwann cell with a sheath of myelin wrapped around an axon (bottom).

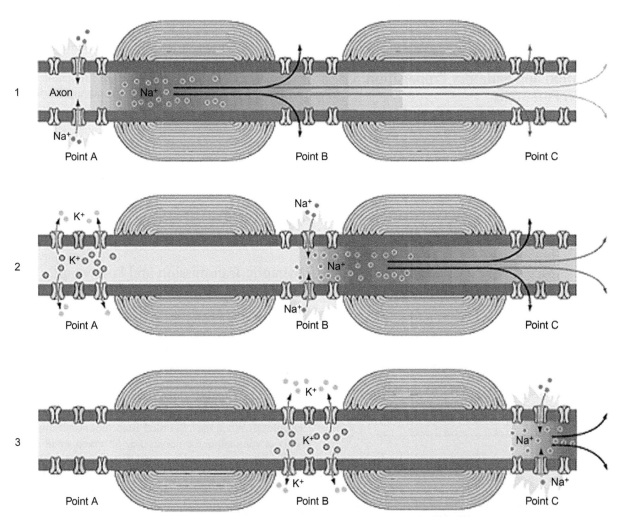

FIGURE 2.9   Schematic representation of saltatory propagation of action potentials between the nodes of Ranvier of myelinated axons. Na$^+$ ions are black, K$^+$ ions are gray.

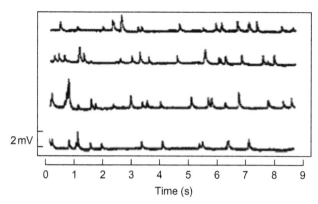

2 mV

FIGURE 2.10 Miniature endplate potentials recorded from the frog neuromuscular junction.

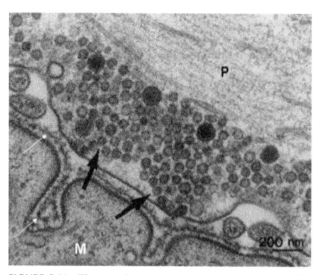

FIGURE 2.11 Electron micrograph of a neuromuscular synapse. The presynaptic terminal (P) contains numerous electron dense vesicles (black arrows), which release neurotransmitter from release sites across invaginations of the postsynaptic muscle (M). An electron dense ribbon of extracellular matrix is apparent in the synaptic cleft (white arrows) that contains acetylcholine esterase.

When they stimulated the nerve, the muscle fiber membrane would depolarize incrementally, depending on the extent of nerve stimulation. This depolarizing potential is known as the **postsynaptic potential**. When it depolarizes the postsynaptic cell to threshold it elicits an action potential. Katz and his colleagues inferred that the postsynaptic potential represents the summation of miniature endplate potentials, due to the simultaneous release of multiple quanta of neurotransmitter as a consequence of stimulation of the presynaptic neuron. Their work elucidated the fundamental mechanism of chemical transmission at synapses prior to the development of electron microscopy, which would soon enable actual visualization of synaptic vesicles in the presynaptic terminal, and corroborate their hypothesis (Figure 2.11).

When the action potential invades the nerve terminal, it activates the opening of voltage-sensitive calcium channels. The influx of calcium triggers the rapid fusion of synaptic vesicles with the presynaptic cell membrane, thereby releasing neurotransmitter into the synaptic cleft, a $150\,\mu m$ space that separates the presynaptic from the postsynaptic cell. At the neuromuscular junction, the neurotransmitter is acetylcholine, which is synthesized by choline acetyltransferase and packaged into synaptic vesicles. The content of a single vesicle is equivalent to a quantum that gives rise to a miniature endplate potential on vesicle release. Acetylcholine diffuses across the synaptic cleft, and binds to receptors at the postsynaptic membrane. This causes a conformational transition in the receptor molecule, thereby opening an ion channel through which cations can diffuse to carry the current that gives rise to the postsynaptic potential. Acetylcholine is rapidly degraded by acetylcholine esterase, a fast-acting enzyme located in the synaptic cleft and at the postsynaptic membrane, which ensures that synaptic transmission is transient (Figure 2.12). Acetylcholine receptors at the neuromuscular synapse are known as **nicotinic acetylcholine receptors**, and are structurally and pharmacologically distinct from **muscarinic acetylcholine**

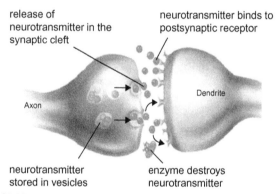

FIGURE 2.12 Schematic representation of synaptic transmission between the axon of a presynaptic cell and the dendrite of a postsynaptic cell.

receptors, which mediate the effects of acetylcholine in the autonomic nervous system (described in Chapter 3). The nomenclature is based on their differential sensitivities to nicotine and the fungal compound muscarine. Furthermore, nicotinic cholinergic receptors can be blocked by curare, a poison used by Amazonian Indians on the tips of their arrows to paralyze their quarry, whereas muscarinic receptors are inhibited by atropine and scopolamine.

The membrane depolarization in the postsynaptic cell is known as an **excitatory postsynaptic potential** (EPSP). In neurons, the EPSP will spread across the dendrite and soma of the cell via cable properties and, when depolarization reaches threshold at the axon hillock, action potentials will be generated. In muscle cells, action potentials trigger contraction of the muscle. The number of ions that flow into the postsynaptic cell depends on the number of receptors that are activated, which is proportional to the amount of neurotransmitter released from the presynaptic terminal. Thus, the EPSP differs from the action potential in that it is a graded potential rather than an all-or-none event. If the frequency of action potentials in the presynaptic cell increases, more neurotransmitter will be released before previously released neurotransmitter is cleared from the synaptic cleft, thereby increasing the concentration of the neurotransmitter. As a consequence, a greater EPSP will be elicited. Furthermore, if a postsynaptic cell makes synaptic contacts with multiple presynaptic neurons, as is common in the central nervous system, simultaneous or near-simultaneous release of neurotransmitter from different presynaptic cells will lead to summation of postsynaptic potentials, and the response of the postsynaptic cell will be the weighted average of all inputs received within a defined time window. Thus, postsynaptic potentials can be summed both in space and over time, thereby allowing synaptic integration (Figure 2.13). Synaptic integration is not only achieved through summation of EPSPs, but can also include **inhibitory postsynaptic potentials** (IPSPs) through the opening of anion channels following the release of an inhibitory neurotransmitter, such as γ-aminobutyric acid (GABA), and subsequent hyperpolarization of the postsynaptic cell.

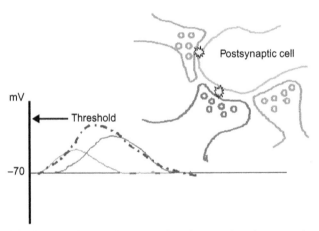

FIGURE 2.13 Synaptic integration through summation of postsynaptic potentials elicited in a postsynaptic cell through the simultaneous release of neurotransmitter from two presynaptic terminals. The corresponding EPSPs are indicated schematically by the solid lines. The dotted line represents their sum, which approaches the threshold potential for activation of the postsynaptic neuron.

ION CHANNELS, G-PROTEIN-COUPLED RECEPTORS, AND SIGNAL TRANSDUCTION

Ligand-Gated Ion Channels

The first neurotransmitter receptor to be purified was the nicotinic acetylcholine receptor. The Taiwanese snake, *Bungarus multicinctus*, and the Indian cobra produce potent snake venoms with which they immobilize their preys.

These toxins bind with high affinity to nicotinic acetylcholine receptors, and were used to purify these receptors from membrane extracts of the electric organs of electric eels and rays, which contain an abundance of nicotinic acetylcholine receptors. The nicotinic acetylcholine receptor consists of five subunits, two α subunits and a β, γ, and δ subunit, which are closely related in terms of their amino acid sequences. Cooperative binding of two molecules of acetylcholine by the α subunits causes a conformational change in the pentameric complex that results in the opening of a 6.5 Å water-filled pore through which cations can diffuse (Figures 2.14 and 2.15). The similar sequences of the subunits suggest that they arose during evolution from an ancestral gene as a consequence of gene duplication events. Comparisons of the sequences of the nicotinic acetylcholine receptor with other **ligand-gated channel proteins**, including the receptors for the neurotransmitters GABA and glycine, and the 5-HT type 3 serotonin receptor, showed that they are evolutionarily related and together form a gene family of ligand-gated ion channels, even though the acetylcholine receptor is permeable to cations, whereas the GABA and glycine receptors are permeable to anions. These receptors are known as the "Cys loop family" of ligand-gated ion channels (Figure 2.16).

There are 19 different, but related, genes that encode different variants of α, β, and γ subunits of the GABA receptor. The products of these genes form diverse pentameric complexes that all mediate GABA-activated opening of a chloride channel, but differ in their pharmacological and kinetic properties. This heterogeneous array of functional GABA complexes allows exquisite tuning of the sensitivity to this

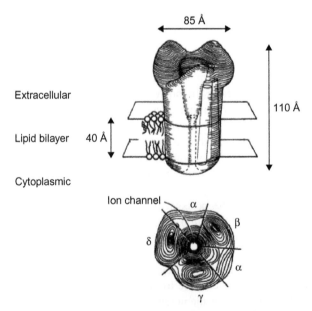

FIGURE 2.14 Model of the three-dimensional structure of the nicotinic acetylcholine receptor containing a funnel-shaped ion channel based on X-ray diffraction studies. The lower densitometric image illustrates the pentameric subunit topography, as viewed from the synaptic side with arbitrarily drawn borders separating the subunits. (Modified from Kistler et al. (1982). *Biophys. J.*, **37**, 371–383.)

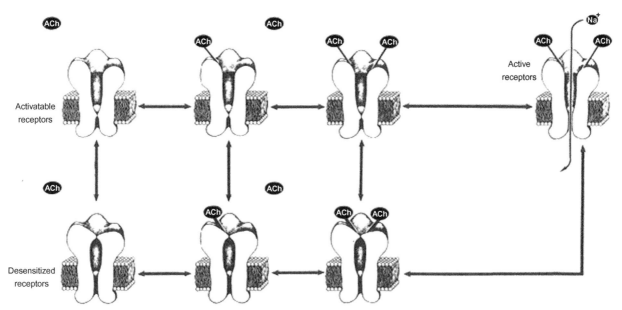

FIGURE 2.15   State transitions during activation of nicotinic acetylcholine receptors. Cooperative binding of two acetylcholine molecules to the α-subunits of the receptor causes a conformational change that results in the opening of a cation-specific channel by removing a hydrophobic plug composed of conserved leucine side chains contributed by the M2 transmembrane segments of each subunit. Prolonged presence of the agonist results in closing of the channel as the receptor transitions into a desensitized state. Removal of acetylcholine will allow relaxation of this desensitized conformation back into an activatable state.

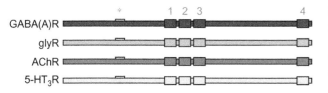

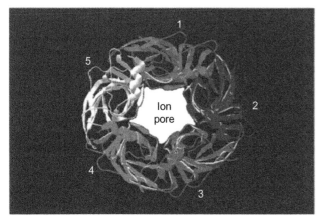

FIGURE 2.16   Structures of the Cys loop family of ion channels. The GABA receptor (GABA(A)R), glycine receptor (glyR), nicotinic acetylcholine receptor (AChR) and 5-HT$_3$ receptor (5-HT$_3$R) share a similar structure. Each subunit contains an extracellular N-terminal domain with the Cys loop signature motif (asterisk) followed by four transmembrane helical domains, indicated 1–4. The subunits form a pentameric complex with the second transmembrane domain of each subunit contributing to the formation of a central ion pore, as seen in this picture "looking down on it" from outside the cell. (Modified from Connolly, C.N. and Wafford, K.A. (2004). *Biochem. Soc. Trans.*, **32**, 529–534).

inhibitory neurotransmitter at individual synapses, depending on the locally-expressed receptor complex.

Other families of ligand-gated ion channels are the glutamate receptors (described in Chapter 3), **transient receptor potential** (TRP) **channels**, and **purinergic receptor** (P2X) **channels**. P2X channels are opened by ATP.

Voltage-Gated Ion Channels

Voltage-gated ion channels are members of yet a different superfamily. The voltage-gated sodium channel and the voltage-gated calcium channel consist of four tandem repeats, each of which contains six transmembrane domains (Figure 2.17). The fourth transmembrane domain contains a series of positively-charged arginine residues, spaced three positions apart, which act as voltage sensors and on depolarization drive the conformational transition that leads to the opening of the channel pore. This pore is formed by "hairpin loops," rich in hydroxylated amino acid residues, which penetrate into the membrane between the fifth and sixth transmembrane domain and contain a binding site for tetrodotoxin (TTX). Arginines in the first three tandem repeats are associated with channel opening, whereas the voltage sensor of the fourth repeat domain has been implicated in channel inactivation.

Potassium channels consist of only a single repeat domain that is similar to the tandem repeats found in voltage-gated sodium and calcium channels. However, potassium channels are composed of tetramers, thus forming

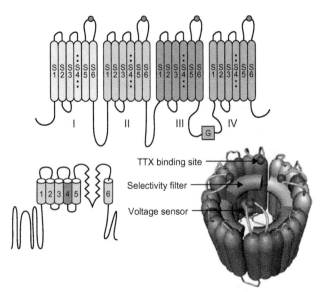

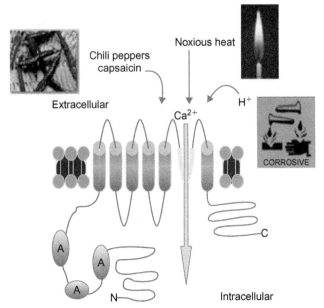

FIGURE 2.17 Structure of the voltage-gated sodium channel. The four homologous repeat domains are indicated with roman numerals. The fourth transmembrane helix of each repeat contributes to the voltage sensor. The loop between S5 and S6 in each domain forms a hairpin that contributes to the lining of the channel (illustrated in the lower left diagram). The circles indicate a binding site for TTX. The channel contains an inactivation gate (G) between the third and fourth repeat domain, which mediates channel inactivation subsequent to voltage-dependent opening. The four domains form a tetrameric structure. Potassium channels are structurally similar, except that the tetrameric complex is formed from four independent subunits, each of which is homologous to the sodium channel domains.

FIGURE 2.18 The vanilloid VR1 receptor. Note the structural similarity of this TRP channel with voltage-gated channels (Figure 2.19). "A" designates ankyrin domains (ankyrins are proteins that link integral membrane proteins to the underlying cytoskeleton). The channel is opened by capsaicin, heat, and acid, and is permeant to calcium.

a quaternary structure similar to that of the other voltage-gated channels. Voltage-dependent inactivation is here mediated by occlusion of the channel pore through movement of the N-terminal region of the channel polypeptide. This has been referred to as the "ball-and-chain" model for inactivation, in which the "ball" is represented by the N-terminal peptide region that swings into the channel.

A wide variety of subtypes of voltage-gated channels exists, including calcium-activated potassium channels, that differ in their kinetic properties, including their inactivation kinetics (i.e. the probability of the open state of the channel declines with time after voltage-gated opening; this is a characteristic of both sodium and potassium channels). These diverse regulatory properties ensure delicate control over the threshold sensitivity and firing frequency of action potentials in different neurons.

Transient Receptor Potential Channels

Transient receptor potential (TRP) channels form a particularly interesting family of ion channels. The first TRP channel was discovered as a light-activated calcium channel that mediates **phototransduction** in *Drosophila*, and is named after the mutant "transient receptor potential" phenotype, in which absence of a functional *trp* gene causes flies to be rapidly blinded under continuous bright light. Following the discovery of the *Drosophila* TRP channel, more than 20

mammalian homologs have been identified. They are categorized into two families, TRP and TRPM. Their activation mechanisms are diverse, and include activation by diacylglycerol and calcium, mechanical force, or intrinsic pH- and temperature-dependent mechanisms. TRP channels are the gateway to sensory perception for many sensory modalities. They have been implicated in the transduction of mechanosensory stimulation in the nematode *Caenorhabditis elegans*, in **auditory transduction** in the inner ear, in sensing osmotic pressure, in the transduction of bitter taste, the perception of mammalian pheromones, and transduction of **nociceptive** stimuli, such as pain, acid pH, and excessive hot and cold temperature, as well as the chemically-induced sensations of hot pepper by the compound capsaicin, and the cooling sensation of menthol. TRP channels that mediate nociception belong to the category of **vanilloid receptors**. The first receptor of this class to be identified was the VR1 receptor, which is activated by capsaicin, by low pH, and by heat. VR1 is expressed in **polymodal** nociceptive sensory neurons located in dorsal root ganglia of the spinal cord. Activation of the VR1 receptor opens a cation channel through which calcium ions enter and depolarize the cell (Figure 2.18).

G-Protein-Coupled Receptors

Purification of the nicotinic acetylcholine receptor in the 1970s raised expectations that this receptor might be the ultimate model for all neurotransmitter receptors. Robert Lefkowitz and his colleagues, however, made the surprising discovery that receptors for the hormone epinephrine (adrenaline) and the neurotransmitter norepinephrine (noradrenaline) are entirely different. Purification of the β_2-adrenergic receptor showed that, in contrast to the nicotinic acetylcholine receptor, this receptor does not form a multimeric complex, but consists of a single polypeptide chain. Analysis of its amino acid sequence predicted that it would contain seven membrane-spanning domains with the N-terminus located extracellularly, three alternating intracellular and extracellular loops, and the C-terminus on the inside of the cell (Figure 2.19, p. 35). This hypothesis has been proven correct. The transmembrane domains are aligned in such a way that they form a binding pocket for the neurotransmitter. Binding of the ligand triggers a conformational shift in the receptor that elicits a response from the target cell. Subsequent to the classic work by Lefkowitz and his colleagues, it became increasingly clear that many receptors have the same structural organization as the β_2-adrenergic receptor, including a host of neurotransmitter and neuropeptide receptors, a large family of odorant receptors, as well as the photoreceptor rhodopsin. Like the

---

Box 2.1 Phototransduction and olfactory transduction: Variations on a theme

Photoreceptor cells in the retina consist of an inner segment, which contains the nucleus and mitochondria, and an outer segment that is a modified cilium, which is filled with a stack of internal disc membranes. The discs contain the photoreceptor, rhodopsin. Rhodopsin consists of two components, opsin and the pigment retinal, which is derived from vitamin A. Opsin is a member of the superfamily of G-protein-coupled receptors, and retinal occupies a binding pocket formed by the transmembrane helical domains. Absorption of a photon of light causes the isomerization of 11-*cis*-retinal to all-*trans*-retinal. This results in movement of about 5 Å of the hydrocarbon chain of retinal around the double bond, and results in dissociation of the pigment from the opsin protein. This elicits a conformational change in the protein, which sets in motion the process of vision. Thus, absorption of a photon of light has been translated into molecular motion. Activated opsin, like any other GPCR, activates a G-protein, known here as "transducin." The activated α-subunit of transducin (with GTP bound) causes activation of a cyclic GMP phosphodiesterase that breaks down cyclic GMP into 5'-GMP. In the dark, cyclic GMP binds to ion channels in the rod's plasma membrane. This binding is required for these channels to remain open. Thus, in the dark a current continuously flows through the cell (the "dark current"). A lowering of the intracellular concentration of cyclic GMP, as a result of activation of rhodopsin, results in closing of a fraction of these cyclic GMP-activated ion channels. This causes a relative hyperpolarization of the cell.

In the dark, inhibitory neurotransmitter is continuously released onto bipolar cells, neurons that connect photoreceptor cells to ganglion cells. The axons of ganglion cells form the optic nerve. A disruption of the release of this neurotransmitter from the rod now enables the postsynaptic bipolar cell to depolarize, and in turn to transmit a neural signal to ganglion cells. Phototransduction achieves a great degree of amplification. A single activated rhodopsin molecule can activate about 500 molecules of transducin, causing the closure of about 200 ion channels (about 2% of channels that are open in the dark) and a hyperpolarization of about 1 mV.

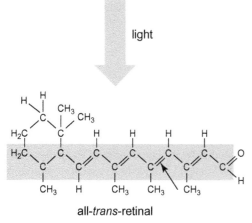

11-*cis*-retinal

light

all-*trans*-retinal

FIGURE A Photoisomerization of retinal.

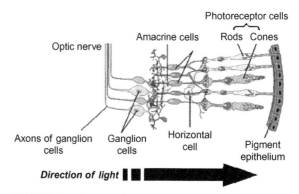

Direction of light

FIGURE B Diagrammatic representation of the retina. Light passes through the retinal cell layers onto photoreceptor cells and the signal is transmitted via bipolar neurons to the ganglion cells. Axons of the ganglion cells form the optic nerve. Amacrine cells and horizontal cells form horizontal modulatory connections.

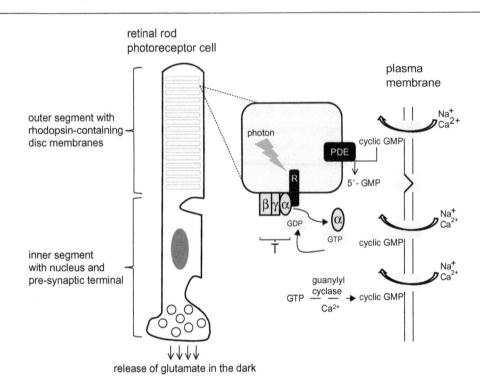

FIGURE C   The phototransduction pathway in photoreceptor cells. Light-induced isomerization of rhodopsin (R) causes activation of transducin (T) and subsequent activation of a phosphodiesterase (PDE) that converts cyclic GMP into 5'GMP, which results in lowering of the intracellular cyclic GMP concentration and closing of a fraction of cyclic GMP-activated channels. This results in hyperpolarization of the cell, and a reduction in the release of neurotransmitter at the synapse with the bipolar neuron. Decrease in intracellular calcium, as a result of channel closing, results in increased activity of the guanylyl cyclase, restoring the level of cyclic GMP.

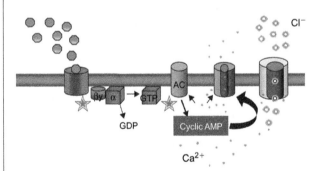

FIGURE D   Olfactory transduction. Odorants (hexagons) bind to a receptor (OR) causing activation of a G-protein, which activates adenylyl cyclase (AC). Calcium, which enters through cyclic AMP-activated channels, opens a chloride channel, and the efflux of chloride amplifies the generator current.

Olfactory signal transduction and phototransduction show remarkable similarities. In olfactory neurons, binding of an odorant to a receptor results in activation of an adenylyl cyclase via a stimulatory G-protein, G_{olf}. Cyclic AMP then opens an ion channel in the ciliary plasma membrane, similar to the cyclic nucleotide-activated channel of retinal photoreceptor cells. Note, however, that here the stimulus results in opening rather than closing of the channel. Opening of ion channels by cyclic AMP results in an influx of cations. Calcium, which enters through the cyclic AMP-gated channel, opens a chloride channel and efflux of chloride through this calcium-activated chloride channel amplifies the depolarizing generator current to elicit action potentials from the axon hillock when it reaches threshold.

ligand-gated and voltage-gated ion channels, these receptors form an evolutionarily-related superfamily of proteins, known as seven transmembrane domain receptors or **G-protein-coupled receptors (GPCRs)**. The latter – and most popular designation – reflects the mechanism by which they elicit their physiological responses, explained in the next paragraph.

In the early 1970s, Martin Rodbell discovered that an impurity in his ATP preparation could enhance the effects of the hormone glucagon in a liver slice preparation. This impurity was identified as guanosine triphosphate (GTP). Subsequent studies by the laboratories of Rodbell and Alfred Gilman showed that GTP binds to a regulatory protein complex, the **G-protein**, which links the activation of a seven transmembrane domain receptor to activation of an effector system in the cell. The G-protein consists of three subunits, designated α, β, and γ, which occur in the resting conformation as a heterotrimeric complex in which GDP is bound to

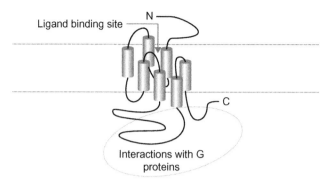

FIGURE 2.19   Diagram of a GPCR with a ligand-binding pocket formed in the transmembrane domains, as is the case for adrenergic receptors. Other members of the GPCR family, e.g. receptors for neuropeptides, may have binding sites formed by the extracellular N-terminal domain.

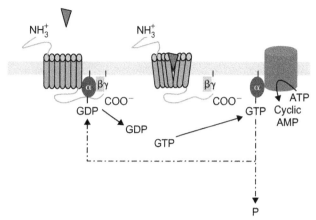

FIGURE 2.20   G-protein-mediated signal transduction. A neurotransmitter or hormone (indicated by the triangle) binds to a receptor and causes dissociation of α- and βγ-subunits concomitant with GDP/GTP exchange. The activated α-subunit can now activate an effector enzyme, illustrated here by activation of adenylyl cyclase. Note that lipid modifications anchor α- and γ-subunits to the membrane. GTP-activating proteins that promote GDP/GTP exchange and proteins that accelerate GTP hydrolysis are not indicated.

the α-subunit. Upon binding of a ligand to its receptor, the G-protein complex dissociates into the α-subunit and a βγ complex. After the α-subunit dissociates, it exchanges the bound GDP for GTP. When GTP is bound to the α-subunit, this complex can interact and activate an effector enzyme in the cell, which in the case of activation of the β_2-adrenergic receptor is an adenylyl cyclase that converts ATP into cyclic adenosine 3′5-monophosphate (cyclic AMP), a **second messenger**, which in turn can activate enzymes that set in motion a process of **phosphorylation** reactions. The α-subunit slowly hydrolyzes its GTP, converting it to GDP. It then reassociates with the βγ-subunit as an inactive complex. This "internal timer" guarantees that activation is transient (Figure 2.20). Furthermore, auxiliary proteins have been identified that can accelerate the exchange of GDP and GTP, and enhance activation or inactivation of G-proteins. A diverse array of G-proteins has been identified that all have the same heterotrimeric structure and share a similar activation mechanism, but couple the activation of different GPCRs to different intracellular signaling pathways. In the case of the β_2-adrenergic receptor, binding of epinephrine ultimately stimulates adenylyl cyclase, via a G-protein α-subunit, designated $G\alpha_s$, where "s" indicates stimulation.

A different G-protein mediates inhibition of adenylyl cyclase (designated G_i, where "i" indicates inhibition). Yet another G-protein has been associated with the opening of potassium channels, designated G_o ("o" designates "other"), which is expressed only in the heart and the nervous system. A G-protein, designated G_q, is associated specifically with activation of phospholipase C, which liberates inositol-1,4,5-triphosphate (IP_3) from phosphatidyl-inositol bisphosphate. Generation of IP_3 results in the release of calcium from the cisternae of the endoplasmic reticulum into the cytosol. Calcium binds to a regulatory protein, calmodulin, and the calcium–calmodulin complex has widespread effects on cellular metabolism, e.g. through the activation of calcium–calmodulin dependent protein kinase. Another product of the generation of IP_3 is diacylglycerol which activates yet another protein kinase enzyme, **protein kinase C**.

Diacylglycerol can undergo further degradation, liberating an unsaturated fatty acid from the 2′ position, which is often arachidonic acid. Arachidonic acid itself is a precursor for many signaling molecules, such as prostaglandins and leukotrienes. Other G-proteins have also been implicated in activation of phospholipase C and release of IP_3. Thus, activation of GPCRs has the potential to unleash an avalanche of intracellular signals (Figures 2.21 and 2.22).

Functions of G proteins		
G protein family	Subunit implicated in activity	Function
G_s	α	Activates adenylyl cyclase, Ca^{2+} channels
G_{off}	α	Activates adenylyl cyclase in offactory sensory neurons
G_i	α	Inhibits adenylyl cyclase
	βγ	Activates K^+ channel
G_o	βγ	Activates K^+ channel, Inactivates Ca^{2+} channels
	α and βγ	Activates phospholipase C-β
G_t (transducin)	α	Activates cyclic GMP Phosphodiesterase in Vertebrate photoreceptors
G_q, G_{11}	α	Activates phospholipase C-β

FIGURE 2.21   G-proteins and their functions. The figure illustrates the diversity of G-protein functions, but the list is not exhaustive; for example, the G_i family contains several members. In addition, G-proteins not indicated here play diverse roles in intracellular trafficking, exocytosis, and cell motility.

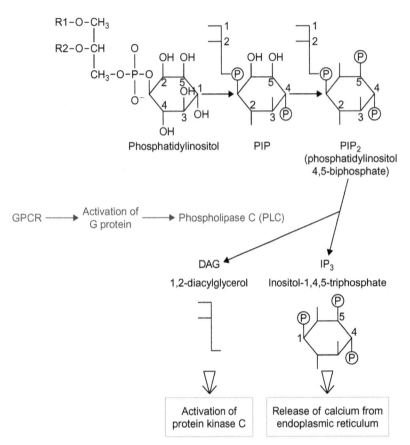

FIGURE 2.22 Generation of IP$_3$ and diacylglycerol (DAG) from phosphatidylinositol-4,5-biphosphate (PIP$_2$) through activation of phospholipase C (PLC). DAG can be further hydrolyzed to liberate arachidonic acid, a precursor for prostaglandins and leukotrienes, components of the endocrine system, known as "paracrine effectors," which often exert local effects on the peripheral tissues that produce them.

Signal amplification is an integral characteristic of signal transduction mechanisms. A single activated receptor can activate multiple G-proteins, and each activated G-protein in turn can activate multiple effector enzymes. In addition to signal termination at the level of the G-protein, receptors can be rendered inactive through phosphorylation by a variety of protein kinases. One group of receptor kinases, the **G-protein receptor kinases** (GRKs), mediates receptor desensitization by phosphorylating only liganded receptors. The phosphorylated receptors then bind an intracellular adaptor protein, β-**arrestin** in the case of the β$_2$-adrenergic receptor, which directs internalization. Receptors can be recycled to the membrane following removal of the phosphate groups by **phosphatases** (Figure 2.23).

In addition to the superfamily of GPCRs and the much smaller, but still substantial, family of G-proteins, effector enzymes, such as adenylyl cyclase and phospholipase C also exist as different isoforms with different kinetic and regulatory properties. Moreover, different GPCRs can couple to different G-protein-regulated pathways in the same cell, and these transduction pathways can influence one another. There is also evidence that, under some conditions, certain GPCRs may switch from coupling to one G-protein to interacting with a different G-protein. The picture that emerges is a complex signaling network that enables cells to respond with exquisite precision to external

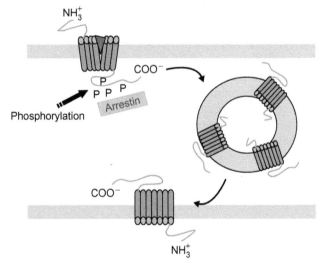

FIGURE 2.23 Desensitization and internalization of G-protein-coupled receptors in the continued presence of ligand (triangle) following phosphorylation and binding of arrestin. After dephosphorylation, receptors sequestered in vesicles can be recycled to the membrane.

stimuli. This is especially important in the nervous system. This notion of interactive networks is a recurrent theme that applies not only to cellular metabolism, but also to neural connectivity and genetic ensembles that contribute to the manifestation of complex traits.

SUMMARY

Information is transmitted in the nervous system as electrical signals in the form of action potentials, which result from the sequential opening of voltage-gated sodium and potassium channels. The driving force for movement of cations through these channels arises from the asymmetric distribution of potassium and sodium across the cell membrane, which is maintained through the action of the Na^+/K^+-ATPase. Propagation of action potentials occurs as a result of depolarization of adjacent membrane regions during the rising phase of the action potential. Myelination enhances the speed by which action potentials can travel, since action potentials can be elicited only at the nodes of Ranvier, while capacitative current can spread rapidly between the nodes. When the action potential invades the synaptic terminal, voltage-gated calcium channels open and the entry of calcium triggers the fusion of neurotransmitter-containing vesicles with the presynaptic membrane. Once released, the neurotransmitter diffuses across the synaptic cleft and binds to receptors at the postsynaptic membrane. Neurotransmitter removal can be achieved, either by enzymatic cleavage – as in the case of acetylcholine – or through reuptake into the presynaptic cell followed by enzymatic inactivation – as in the case of bioamines. Whereas action potentials are all-or-none phenomena that allow signals to be transmitted over long distances without loss of amplitude, postsynaptic potentials are graded events that reflect the concentration of neurotransmitter in the synaptic cleft and allow temporal and spatial summation, which forms the basis for synaptic integration.

Neurotransmitters act either by binding to receptors that belong to the family of ligand-gated ion channels, or by binding to receptors that are members of the superfamily of G-protein-coupled receptors. Activation of the latter results in dissociation of a heterotrimeric G-protein, which is followed by activation of an enzyme which generates an intracellular second messenger, such as cyclic AMP or IP_3, that subsequently directs the response of the postsynaptic cell.

STUDY QUESTIONS

1. What are the values of the equilibrium potentials for sodium and potassium in neurons under physiological conditions?
2. What will be the equilibrium potential for calcium when the extracellular concentration of calcium is 1.25 mM and the intracellular calcium concentration is 100 nM, under conditions in which calcium is the only permeant ion?
3. Explain the ionic mechanisms that give rise to an action potential.
4. Explain why action potentials cannot spontaneously reverse direction as they travel along the axon.
5. Explain why the action potential is an all-or-none event, and why the postsynaptic potential is a graded potential. Why is it essential that the action potential is an all-or-none event and the postsynaptic potential a graded event?
6. An electrophysiologist records currents from an axon at different voltage steps under voltage clamp conditions. What will he observe: (1) if he removes extracellular sodium; (2) in the presence of tetrodotoxin (TTX); and, (3) in the presence of tetraethylammonium (TEA)?
7. The issue of whether synaptic transmission was electric or chemical was hotly debated in the 1950s. Sir Bernard Katz was a protagonist of the chemical synaptic transmission theory, and provided strong evidence for chemical synaptic transmission through his observations of miniature endplate potentials (mepps). What properties of mepps provided support for the notion that release of neurotransmitter is quantal?
8. Explain how alterations in the concentrations of cyclic nucleotides mediate phototransduction and olfactory transduction.
9. Describe common aspects of the structure and signal transduction mechanisms of G-protein-coupled receptors.

RECOMMENDED READING

Bear, M. F., Connors, B., and Paradiso, M. (2006). *Neuroscience. Exploring the Brain*, 3rd edition. Lippincott, Williams and Wilkins, Baltimore, MD.

Fatt, P., and Katz, B. (1952). Spontaneous subthreshold activity at motor nerve endings. *J. Physiol.*, **117**, 109–128.

Hille, B. (2001). *Ion Channels of Excitable Membranes*, 3rd edition. Sinauer Associates, Inc., Sunderland, MA.

Jiang, Y., Lee, A., Chen, J., Ruta, V., Cadene, M., Chait, B. T., and MacKinnon, R. (2003). X-ray structure of a voltage-dependent K^+ channel. *Nature*, **423**, 33–41.

Lefkowitz, R. J. (2004). Historical review: a brief history and personal retrospective of seven- transmembrane receptors. *Trends Pharmacol. Sci.*, **25**, 413–422.

Levitan, I. B. (2006). Signaling protein complexes associated with neuronal ion channels. *Nat. Neurosci.*, **9**, 305–310.

Purves, D., Fitzpatrick, D., and Augustine, G. (eds) (2004). *Neuroscience*, 3rd edition. Sinauer Associates, Inc, Sunderland, MA.

Ramsey, I. S., Delling, M., and Clapham, D. E. (2006). An introduction to trp channels. *Annu. Rev. Physiol.*, **68**, 619–647.

Chapter 3

Functional Organization of the Nervous System

OVERVIEW

Information from the environment is perceived by specialized sensory organs, and conveyed to the central nervous system via the peripheral nervous system. This information is integrated in the brain, where it is placed in the context of the animal's previous experience and physiological needs. Appropriate responses are directed and coordinated via motor pathways that descend in the spinal cord. In addition to responses to environmental changes, cognitive and emotional processes in the brain can initiate behaviors independent of previous environmental stimulation. Furthermore, voluntary activity is sustained by involuntary effectors of the autonomic nervous system, which maintain continuous neural control over fundamental physiological processes, and can generate a rapid "fight or flight" response under threatening or stressful environmental conditions. In this chapter we will provide a brief overview of the major features of the anatomical organization of both mammalian and insect brains, and of the major neurotransmitters that enable communication among brain regions. For a more in-depth description of the nervous system's structure and function the reader is referred to textbooks on neurobiology, some of which are listed at the end of the chapter. Understanding the complex neuroanatomical organization of the mammalian central nervous system is made easier when we consider how the brain and spinal cord develop. In the following sections, we will briefly describe how embryonic structures that emerge as dilatations in the rostral neural tube during early development correspond to regions in the adult mammalian brain. Additional auxiliary information about neural pathways and mechanisms is presented as text boxes in later chapters of this book.

THE ORGANIZATION OF THE MAMMALIAN NERVOUS SYSTEM

The vertebrate nervous system develops from embryonic ectoderm. Closure of the **neural groove** in the early embryo results in the formation of a tube above the notochord, and above it a plate of ectodermal tissue, the **neural crest**. The neural crest contributes, among others, to the formation of the peripheral nervous system and the adrenal medulla. The neural tube will develop into the spinal cord. Its hollow center will become the central canal of the spinal cord, filled with cerebrospinal fluid. In the rostral region of the neural tube three dilatations appear, which form the primitive forebrain (**prosencephalon**), midbrain (**mesencephalon**), and hindbrain (**rhombencephalon**) (Figure 3.1). The prosencephalon further divides into the **telencephalon** and the

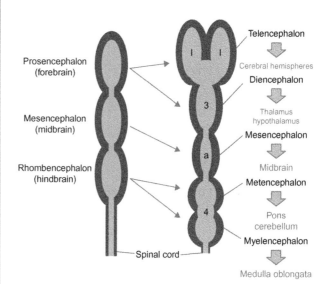

FIGURE 3.1 Schematic representation of early development of the central nervous system. The fluid filled cavities in the embryonic brain (left) will develop into the lateral (l), third (3), and fourth (4) ventricles of the adult brain that are continuous with the central canal of the spinal cord. "a" designates the aqueduct.

R. Anholt and T. Mackay: Principles of Behavioral Genetics
ISBN: 978-0-12-372575-2

diencephalon. The telencephalon gives rise to the two cerebral hemispheres (the **cerebrum**), which in humans become greatly enlarged, covering the rest of the brain and representing approximately 80% of the total mass of the adult human brain. The diencephalon forms the **thalamus** and **hypothalamus**. The mesencephalon develops into the midbrain. The rhombencephalon also divides into two parts: the **metencephalon** and the **myelencephalon**. The former gives rise to the **pons** and **cerebellum**, and the latter to the **medulla oblongata**, which is continuous with the spinal cord.

The embryonic neural canal persists as cavities in the adult central nervous system. These cavities are filled with cerebrospinal fluid, and are known as **ventricles**. The two lateral ventricles are bordered by the cerebral hemispheres that arise from the telencephalon and the third ventricle is found in the diencephalon. The fourth ventricle is continuous with the central canal of the spinal cord, and spans the medulla oblongata and the metencephalon. It is continuous with the third ventricle via a narrow passage in the midbrain, the **aqueduct**. The two cerebral hemispheres are connected by a fiber tract, the **corpus callosum**. Fibers connecting the two hemispheres are referred to as **commissural fibers**.

To accommodate the great expansion of the cerebrum within the cranial cavity, the human brain developed convolutions, giving it a wrinkled appearance. Grooves of the wrinkles are known as **sulci**, and the surface elevations are known as **gyri**. Several deep sulci (fissures) divide each cerebral hemisphere into five lobes: the **frontal lobe**; the **parietal lobe**; the **temporal lobe**; the **occipital lobe**; and the **insula** (Figures 3.2 and 3.3). The outer layer of the cerebrum consists of a 2–4 mm thick **cortex**, which contains the dendrites and cell bodies of neurons (this is the **gray matter**), and the interior consists of myelinated fiber tracts (the

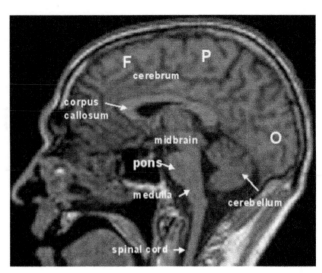

FIGURE 3.3 An MRI (magnetic resonance image) view of the human brain, showing many of the major brain structures. F, P, and O designate the frontal, parietal, and occipital lobes of the cerebrum.

white matter). Groups of neuronal cell bodies are referred to as **nuclei** in the brain and **ganglia** in the periphery.

Deep inside the white matter of the cerebrum are assemblies of neurons important for the control of voluntary movement. These are the **basal ganglia** (Figure 3.4). The largest of these nuclei is the **corpus striatum**, which consists of the **caudate nucleus** and the **lentiform nucleus**. The latter is subdivided into the **putamen** and **globus pallidus**. Certain functions are associated with particular cerebral regions. The frontal lobe is thought to be responsible for higher cognitive functions, personality, and intellectual processes. It also contains the primary motor control area localized to the **precentral gyrus** (rostral to the central sulcus). The parietal lobe contains the **primary somatosensory area** in the **postcentral gyrus** (caudal to the central gyrus). Both the somatosensory and motor areas are organized as topographical maps of the body in proportion to the extent of innervation (e.g. the lips are represented by a disproportionately large area of somatosensory cortex compared to the entire body trunk) (Figure 3.5).

The occipital lobe contains the primary visual cortex, important for processing visual information (Figure 3.6, p. 43). According to a cortical map developed by the early twentieth-century neurosurgeon Brodmann, this area of the occipital lobe also known as Brodmann area 17, or the **striate cortex**, because of its striped appearance.

The temporal lobe is important for auditory perception. The medial temporal lobe includes important subcortical structures, the **hippocampus**, which is intimately involved with the formation of memory, and the **amygdala**, which is associated with aggression and fear. The insula, often referred to as **association cortex** is involved in association and integration of information, and also plays a role in memory.

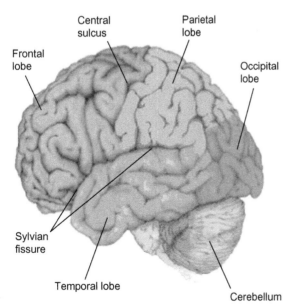

FIGURE 3.2 Morphology of the human brain. The insula contacts the frontal, parietal, and temporal lobes, and is not visible from the surface.

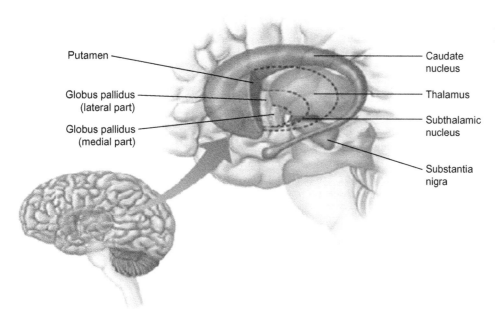

FIGURE 3.4 Diagram of the location and anatomical organization of the basal ganglia.

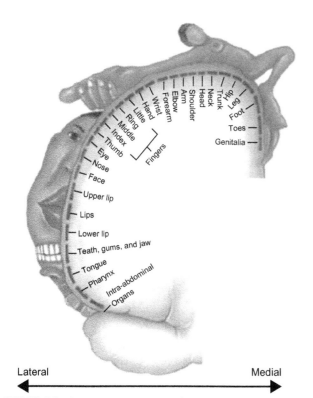

FIGURE 3.5 Somatosensory representations in somatosensory cortex. The postcentral gyrus contains a systematic neural representation of the body surface that is proportional to the extent of innervation rather than physical proportions.

The oldest region of the brain evolutionarily is the **rhinencephalon** (the "smelling" brain), more commonly referred to as the **limbic system**. The term limbic system was originally coined by the French neurologist, Pierre Paul Broca, but its neural circuitry was first elucidated in 1937 by James Papez. The limbic system is a neural circuit that forms a loop ("limbus") around the brain stem, connecting the **cingulate gyrus**, the amygdala, the hippocampus and the hypothalamus (Figure 3.7, p. 44). The limbic system is involved with emotions, motivation, and emotional associations with memory. The limbic system processes olfactory information and regulates instinctive behaviors, such as aggression, fear, feeding, and sex drive (libido). The paucity of cortical connections with the limbic system may explain why we have so little conscious control over our emotions. The hypothalamus has centers that control feeding, circadian activity (the **suprachiasmatic nucleus**), thermoregulation, osmoregulation, and the secretion of hormones from the pituitary through the production of releasing hormones in neurosecretory cells of the paraventricular nucleus.

COMMUNICATION BETWEEN THE BRAIN AND THE PERIPHERY

Information relayed toward the central nervous system is said to be **ascending** (Figure 3.8, p. 44), whereas information that travels away from the brain is **descending** (Figure 3.9, p. 45). Ascending sensory information includes the continuous monitoring of the state of our muscles, our position in space, and the relative positions of our body parts, with respect to one another and with respect to the external environment. This type of sensory input is known as **proprioception**. When a baseball player catches a ball, he performs an impressive neural feat: his eyes track the object while his brain calculates within fractions of a second the trajectory and activates motor pathways to position his arm and hand in the position where the ball is expected to arrive. This accomplishment is achieved by integrating sensory information that is sent from the periphery to the central nervous system and based on this input coordinating descending neural messages that direct movements of appropriate muscles.

Voluntary motor movements are initiated in the motor area of the precentral gyrus, and modified and coordinated through the basal nuclei and the cerebellum. The cerebellum receives proprioceptive information from sensory organs in the muscles together with information from other sensory modalities, including visual and auditory information. The cerebellum integrates this information and communicates with the basal ganglia to orchestrate the appropriate muscle movements. There is no direct connection between the cerebellum and the cortex. All output from the cerebellum is mediated by the inhibitory neurotransmitter, GABA. Hence, suppression of inappropriate neural circuits is critical for enabling muscle coordination.

Cortical neurons form descending projections that cross over in the medulla to the contralateral side of the body. The regions where the fibers **decussate** have a triangular striped appearance, and are known as the **pyramids**. These descending projections are known as pyramidal tracts. The basal nuclei and nerve tracts that originate in the brainstem form extrapyramidal tracts. These tracts can cross over to the contralateral side of the body further down the spinal cord after connections have been formed with interneurons.

Box 3.1   Phineas Gage and the role of the frontal cortex in personality

On September 13, 1848, outside the small town of Cavendish, Vermont, a dramatic accident happened that led to surprising insights into the function of the frontal cortex. Phineas P. Gage, a construction foreman of the Rutland and Burlington Railroad Company, was working on the construction of a railroad track and was using a metal tamping rod to position gunpowder to blast rock to clear terrain. The gunpowder exploded, sending the 3-foot-long, 1.25-inch-thick tamping rod through his skull. The rod entered his skull with great force below his left cheek bone and exited from the top of his skull after passing through his left eye socket and the anterior frontal cortex, landing 30 meters behind him. Remarkably, Gage regained consciousness within minutes, and survived a 45-minute cart ride back to his boardinghouse, where he was treated by Dr J.M. Harlow. Harlow plugged the holes in his skull, and Gage returned to work soon thereafter. Although he appeared physically fully recovered Gage, who had previously been a prudent, hard-working, responsible and popular man, had undergone a profound change in personality. This is described by a famous passage in Harlow's diary:

"Gage was fitful, irreverent, indulging at times in the grossest profanity (which was not previously his custom), manifesting but little deference for his fellows, impatient of restraint or advice when it conflicts with his desires, at times pertinaciously obstinate, yet capricious and vacillating, devising many plans of future operations, which are no sooner arranged than they are abandoned in turn for others appearing more feasible. A child in his intellectual capacity and manifestations, he has the animal passions of a strong man. Previous to his injury, although untrained in the schools, he possessed a well-balanced mind, and was looked upon by those who knew him as a shrewd, smart businessman, very energetic and persistent in executing all his plans of operation. In this regard his mind was radically changed, so decidedly that his friends and acquaintances said he was 'no longer Gage'."

Gage was fired from his job by the railroad company, spent about a year as a freak attraction in a sideshow at P.T. Barnum's New York museum, and later worked as a stagecoach driver in Chile. He died in San Francisco in 1860. Seven years after his death, his skull was exhumed and kept in the Warren Anatomical Medical Museum at Harvard in Boston. In 1994, Damasio and his colleagues performed a computerized reconstruction of Phineas Gage's brain, based on the dimensions of his skull, and determined the trajectory of the rod through the brain (Figure A). Gage's injury, and his resulting Jekyll to Hyde personality transformation, showed that the frontal lobe is important for shaping personality and mediating social skills.

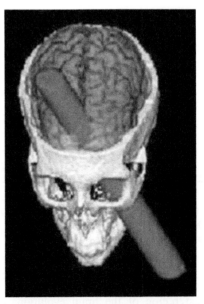

FIGURE A   A model of the skull of Phineas Gage, showing the trajectory of the tamping rod that was blown through his brain during his famous accident.

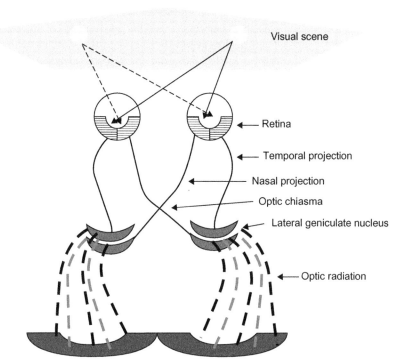

Visual scene

Retina

Temporal projection

Nasal projection

Optic chiasma

Lateral geniculate nucleus

Optic radiation

Primary visual (striate) cortex

FIGURE 3.6 Diagram of the visual projection. Images from the visual scene are seen by the temporal half of the **contralateral** retina and the nasal half of the **ipsilateral** retina. Axons from retinal ganglion cells form the optic nerve, which projects to the **lateral geniculate nucleus (LGN)** of the thalamus. In this projection axons from the nasal half of each retina cross over in the **optic chiasma**, so that the left LGN sees the right visual scene and the right LGN sees the left visual scene. Inputs from the eyes remain separate within distinct layers of the LGN, and this binocular separation is retained in the optic radiation which projects to the primary visual cortex in the occipital lobe. Processing of visual information is accomplished by deconstruction of orientation, color, and motion characteristics of objects in the visual field, and these aspects are analyzed in parallel before being reconsolidated in the perception of the complete visual image.

Ascending and descending pathways between the brain and spinal cord can involve several synapses. For example, pain and temperature perception involve a long ascending interneuron between the nerves from the primary receptor and the sensory neuron that arrives in the postcentral gyrus of the cortex. Thus, three nerves are involved in transmitting this type of sensory information. Ascending sensory information also crosses over to reach the contralateral hemisphere. Thus, the right side of the brain controls the left side of the body and *vice versa*.

The ascending and descending fiber tracts in the spinal cord are organized as six columns, known as **funiculi**. Cross-sections through the spinal cord reveal an H shape, with gray matter (cell bodies) around the **central canal** and white matter containing the myelinated nerve tracts in the periphery. Sensory fibers enter the spinal cord at the **dorsal horn**, and the cell bodies of sensory neurons are located in **dorsal root ganglia** outside the spinal cord. Motor neurons leave the spinal cord from the **ventral horn** and their cell bodies are located within the gray matter of the spinal cord. The spinal cord also contains numerous interneurons that mediate communication between sensory neurons and motor neurons.

Nerves that innervate muscles, such as the sciatic nerve that innervates the thigh muscle, contain both sensory and motor fibers, which separate when they enter the dorsal and ventral horn of the spinal cord, respectively. There are 31 pairs of nerves that originate from the spinal cord, 8 pairs from the **cervical** region, 12 pairs from the **thoracic**

Box 3.2   Recognizing fear and the amygdala

Paul Bucy and Heinrich Klüver removed large parts of both medial temporal lobes from *Rhesus* monkeys, and observed that the monkeys were no longer able to recognize familiar objects, even though their vision was intact. They also explored objects excessively with their mouths, and showed hyperactive and hypersexual behaviors. In addition, monkeys that had been caught in the wild became unusually tame and easy to handle. Whereas normally monkeys would be afraid of snakes, monkeys without their medial temporal lobes would handle snakes without any fear. Subsequently, it was shown that this **Klüver–Bucy syndrome** could be induced by bilateral removal of only the amygdala.

Impairment in the ability to sense or recognize fear was observed also in an anonymous patient, referred to as S.M. She suffered from a rare autosomal recessive disease (Urbach–Wiethe disease), that resulted in bilateral calcification and atrophy of the amygdala and a small portion of nearby entorhinal cortex. S.M. had normal basic perceptions, no memory impairment, and intact language and reasoning skills. However, her processing of emotional information was impaired. S.M. did not show normal conditioned fear responses, and was indiscriminately trusting and friendly. When presented with pictures of facial expressions, she could correctly identify happy, sad, surprised, disgusted, and angry expressions, but could not identify fearful expressions. At least part of this impairment could be explained by her inability to make normal use of information from the eye regions of the pictured image, which appear to be most informative in conveying the emotion of fear, and could be remedied by forcing her to fix on the eyes. Thus a connection appears to exist between the amygdala, directing gaze, and the perception of fear.

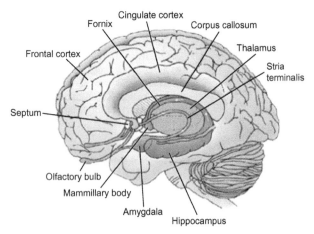

FIGURE 3.7 The limbic system.

region, 5 pairs from the **lumbar** region, 5 pairs from the **sacral** region and 1 pair from the region of the tail bone (the **coccygeal** region).

In addition to the **spinal nerves** there are 12 pairs of **cranial nerves**. They are designated by Roman numerals, and have Latin names that indicate their function or connectivity. The **olfactory nerve** (I), the **optic nerve** (II), and the **vestibulocochlear nerve** (VIII) carry sensory information from the olfactory, visual, and vestibular–acoustic organs, respectively. The **oculomotor nerve** (III), **trochlear**

nerve (IV), and **abducens nerve** (VI) innervate the 6 muscles that coordinate tracking movements of the eyeballs. The **facial nerve** (VII) innervates the muscles of the face, the lacrimal glands, and the salivary glands. The **chorda tympani** branch of the facial nerve innervates the anterior two-thirds of the tongue, dedicated to the perception of sweet and salty tastes. It projects to the **nucleus of the solitary tract** in the medulla oblongata. The **trigeminal nerve** (V) mediates nociceptive sensory perception from the cornea, the nasal and oral cavities, the tongue and the jaws. It is the dentist's nerve, which senses toothaches and also enables the perception of spicy and pungent food qualities, such as mustard, horseradish, and chili peppers. Like small fibers in the periphery that transmit information about pain, trigeminal nerve fibers use the neuropeptide, **substance P**, as neurotransmitter. (The name of this 11-amino acid neuropeptide derives from its original isolation as a white *p*owder, rather than its function in mediating perception of *p*ain.) They project to the large trigeminal nucleus in the medulla oblongata. The **glossopharyngeal nerve** (IX) innervates the pharynx and taste papillae in the posterior third of the tongue, which is dedicated to the perception of bitter taste. The **hypoglossal nerve** (XII) innervates the muscles of the tongue. The **accessory nerve** (XI) innervates muscles of the head, neck, and shoulders. Finally, the **vagus nerve** (X) performs a variety of important visceral functions through **parasympathetic** nerve

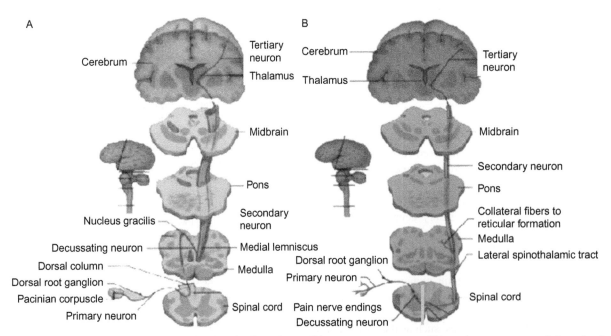

FIGURE 3.8 Ascending somatosensory (A) and nociceptive (B) pathways. Primary sensory neurons, which receive somatosensory information from the skin, such as the **Pacinian corpuscles**, synapse with **interneurons** in **nucleus gracilis** and **nucleus cuneatus** in the medulla, which cross over (**decussate**) to the contralateral side and project via a neural pathway known as the **medial lemniscus** to the thalamus, from where tertiary neurons convey the information to somatosensory cortex. Information from the lower limbs reaches the nucleus gracilis, input from the trunk and upper limbs reaches the nucleus cuneatus, and cranial somatosensory input is routed through the **trigeminal nucleus** in the brainstem. Nociceptive information is also conveyed to somatosensory cortex via three ascending neurons, but **decussation** occurs here earlier in the spinal cord.

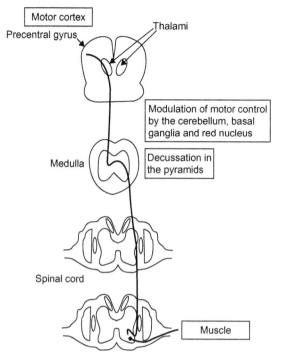

FIGURE 3.9 Diagrammatic representation of the descending (pyramidal) pathway for voluntary motor control.

fibers, e.g. slowing of the heartbeat, swallowing, and intestinal movements. It also receives sensory information from the posterior region of the tongue. Together, the cranial and spinal nerves comprise the peripheral nervous system.

ORGANIZATION OF THE NERVOUS SYSTEM IN INSECTS

Since many genetic studies of behavior have focused on *Drosophila melanogaster* and other arthropods as model systems, we will describe the neuroanatomy of the fly brain as an example of the functional organization of the insect central nervous system. Despite species-specific variations, this organization is representative of all arthropods. The *Drosophila* brain consists of about 250 000 cells (compared to 100 billion neurons in the human brain). Major structures that process sensory inputs include the **optic lobes**, the **antennal lobes**, and the **subesophageal ganglion**. These inputs are relayed to central brain structures for higher processing and directing appropriate motor outputs. The most prominent structures in the central brain are the **central complex** and the **mushroom bodies** (Figure 3.10).

Box 3.3   Sensory–motor reflexes

Simple sensory–motor reflexes are mediated via the spinal cord. During the **knee-jerk reflex**, stretching the **patellar ligament** by striking it with a rubber mallet stretches the **muscle spindle**, a sensory organ that monitors the state of contraction of the muscle, and as a compensatory response the muscle contracts. This reflex is mediated by only two neurons, a sensory neuron from the muscle spindle that synapses with the motor neuron which signals the muscle to contract. Most reflexes, however, involve interneurons in the spinal cord. For example, when touching a hot stove with your hand, the spinal cord reflex mediates instant withdrawal of the hand by causing

one arm muscle to contract while simultaneously preventing the antagonistic muscle from contracting. This inhibition is mediated via interneurons that release glycine as neurotransmitter. At the same time a "pain message" is sent to the brain via ascending tracts, along with proprioceptive information that allows the brain to monitor the position of the hand. A further level of complexity is evident when we consider the everyday activity of walking. Contraction of a leg muscle during walking requires not only inhibition of the contralateral antagonistic muscle of the same leg, but also opposite coordination of muscle contraction and inhibition of the other leg.

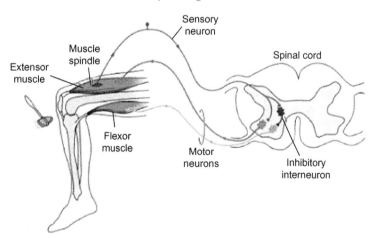

FIGURE A   The knee-jerk reflex is a simple reflex which involves a single synapse between the sensory neuron and the motor neuron that innervates the **flexor** (hamstring) muscle, and an inhibitory interneuron that inhibits the antagonistic **extensor** (quadriceps) muscle.

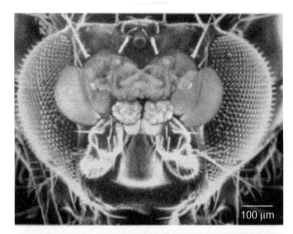

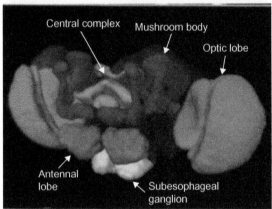

FIGURE 3.10 Diagrammatic representations of the major subdivisions of the *Drosophila* brain.

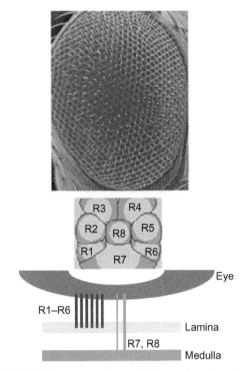

FIGURE 3.11 The compound eye of *Drosophila* and projections of photoreceptor cells to the lamina and medulla of the optic lobe. R1–R6 are most sensitive to light in the blue–UV range, R7 to UV light, and R8 to light in the blue–green spectrum. The R1–R6 projection to the lamina mediates visual motion perception, whereas the R7 and R8 projection to the medulla is important for color vision.

Behaviors in flies rely strongly on visual and chemosensory cues. The *Drosophila* compound eye is composed of about 750 visual units (**ommatidia**), each of which contains eight photoreceptor (retinal) cells. The optic lobes are located adjacent to the eyes, and consist of the **lamina**, **medulla**, **lobula**, and **lobula plate**. Six of the photoreceptor cells (R1–R6) extend axons to the lamina and the other two (R7, R8) to the medulla. These projections remain topographically organized so that the lamina consists of about 750 columns, called neuro-ommatidia, in which the original ommatidial neighborhood relations are preserved. The information derived from individual ommatidia is processed into wide-field visual information in the lobula and lobula plate. The lobula is associated with color processing, whereas the lobula plate contains a small number of giant neurons that integrate information from different ommatidia to enable detection of wide-field visual motion. Projection neurons transmit this preprocessed visual information to the central brain. There are also neural projections that connect the optic lobes of the two hemispheres (Figure 3.11).

Olfactory information is gathered by olfactory sensory neurons in the **antennae** and **maxillary palps**. These neurons project to the antennal lobes. In the antennal lobes, neurons that express the same odorant receptor converge into distinct spherical structures that contain numerous synapses with local interneurons and output neurons and are called **glomeruli**. The number of glomeruli varies greatly among insects, from 43 in *Drosophila melanogaster* to 120 in the honeybee, *Apis mellifera* (Figure 3.12). From the antennal lobes projection neurons travel to the mushroom bodies in the central brain.

Gustatory sensations are mediated by sensory neurons in the **proboscis** and legs. The subesophageal ganglion receives chemosensory and mechanosensory information from the proboscis, and mediates the **proboscis extension reflex**, i.e. the extension of the proboscis in response to a sugar stimulus.

Chemosensory information is integrated in the mushroom bodies. The mushroom bodies are bilaterally symmetric structures, which are responsible for higher-order integration of sensory information, associative learning, and memory formation. Each mushroom body contains two cup-shaped structures called **calyces**, and a stalk which is divided into an α and a β lobe. Each calyx contains the cell bodies of a distinct class of neurons called **Kenyon cells**. The axons of the Kenyon cells descend through the center of the calyx and form three branches, which connect the adjacent calyx and

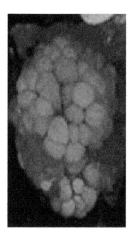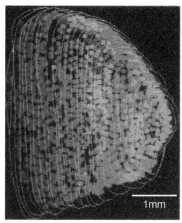

FIGURE 3.12 A confocal image of glomeruli in the antennal lobe of the honeybee, *Apis mellifera*, is shown on the left. The right panel shows a computer-generated three-dimensional reconstruction of optical sections from confocal images of the left olfactory bulb of a 12-week-old mouse. Note the remarkable similarity between the glomerular organization of the insect antennal lobe and the mammalian olfactory bulb. (From Pomeroy et al. (1990). *J. Neurosci.*, **10**, 1952–1966).

project into the α and β lobes, respectively (Figure 3.13). Sensory information is processed in the calyces and the α lobe, whereas the β lobe sends output neurons, including motor neurons, to the **lateral horn**, which is part of a brain region known as the **lateral protocerebrum**.

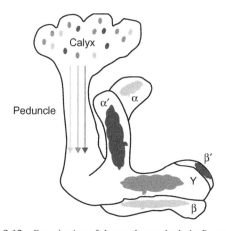

FIGURE 3.13 Organization of the mushroom body in *Drosophila*. The α lobe is oriented vertically, the β lobe medially, and the γ lobe is parallel to the β lobe. The cell bodies of Kenyon cells (circles) are located in the calyx and send axons via the peduncle to the α, α′, β, β′, and γ lobes. Axons that project to the α and α′ lobes bifurcate, sending collaterals to the β land β′ lobes, respectively.

Sensory–motor integration, which enables the fly to mount a behavioral response to an environmental stimulus, depends on the central complex. The central complex lies in the middle of the brain, and contains the main fiber tract that links the two hemispheres. It consists of four

subdivisions: the **protocerebral bridge**; the **fan-shaped body**; the **ellipsoid body**; and the **noduli**. The protocerebral bridge connects the optic system and the antennal lobes to the fan-shaped body. This structure is a center for sensory–motor integration. Disruption of the protocerebral bridge in fly mutants impairs walking behavior, and renders them unable to fly. Single cell recordings from central complex cells and responses of these neurons to polarized light, used by insects as navigational cues, further indicate a role for the central complex in mediating movement based on environmental information.

The fan-shaped body is connected to the protocerebral bridge and the ellipsoid body, as well as to lateral parts of the brain. It is a highly-ordered structure with a repetitive and modular architecture that forms a series of layers, subdivided into rows of 16 columns. It has been implicated in the higher-order control of behavioral activity. The precise functions of the ellipsoid body and the noduli, a pair of ganglia located on the underside of the brain, are still poorly understood. Motor output from the central complex descends to the **ventral nerve cord**.

Having a central nervous system with about four million times fewer neurons than the human brain, the wide repertoire of complex behaviors that can be mediated by the elegant simplicity of the insect brain is truly remarkable. It is surpassed only by the nervous system of the nematode *Caenorhabditis elegans*, which contains exactly 302 individually-identifiable neurons with known interconnections and developmental histories that mediate several behaviors, including chemosensation, mechanosensation, and locomotion, and fundamental physiological functions, such as defecation and mating.

NEUROTRANSMITTERS

Acetylcholine was the first neurotransmitter identified. It was originally discovered by Otto Loewi in 1921 as a substance released from the vagus nerve that would slow the heart rate. The discovery of acetylcholine was followed by the identification of **epinephrine** and **norepinephrine** (known also as **adrenaline** and **noradrenaline**). In the first half of the twentieth century it must have been hard to imagine the vast diversity of neurotransmitters and **neuromodulators** that would be discovered throughout the twentieth century. These neurotransmitters can be categorized in three groups: (1) neurotransmitters that are derived from amino acids, such as **catecholamines** and other **bioamines**, **γ-aminobutyric acid (GABA)**, and amino acids themselves, such as glutamate and glycine; (2) small peptides and polypeptides, including **endorphins** and **enkephalins**, **substance P**, and **bradykinin** (peptides associated with pain perception), and many polypeptides classically described as hormones, such as **glucagon** and **cholecystokinin**; and (3) a variety of small molecules that do not fall

Box 3.4   Receptive fields

The **receptive field** of a neuron is that segment of the environment in which application of a stimulus elicits activity in the cell, in other words that segment of the external world which is "seen" by that particular cell. The concept of receptive field can be readily demonstrated by touching the skin with a pair of calipers. When the ends of the calipers are brought closer together, there comes a moment at which the perception of the two indentations in the skin merges and is perceived as only a single pinprick. At this point, both ends of the stimulus activate dendrites of the same neuron, whereas when they were further apart they activated different neurons. Thus, receptive fields can be compared to pixels on a television screen. The smaller the size of the receptive field is, the better the resolution. Somatosensory neurons at the fingertips have small receptive fields, allowing very sensitive touch discrimination, whereas receptive fields of somatosensory neurons in the trunk and legs are much larger.

When one records from ganglion cells in the retina while moving a spot of light through the visual scene, one observes that in some cells the spontaneous firing rate of action potentials is inhibited when the light hits the periphery of the receptive field and stimulated when the light moves across the center of the receptive field. The opposite phenomenon is also observed: activation of the cell when the light hits the periphery of the receptive field and inhibition in the center. These "center-surround" phenomena are classified as "ON center" and "OFF center" cells, respectively. They arise through the activity of horizontal interneuronal connections in the retina, and contribute to the perception of contours.

Auditory neurons also have receptive fields. They respond to certain frequencies in the sound spectrum. These receptive fields are represented as **tuning curves** that define the sound spectrum that elicits activity in the cell.

In the olfactory system, receptive fields can be characterized as molecular receptive fields of odorant receptors, i.e. the range of chemical compounds that activates a particular receptor, and consequently the cell in which that receptor is expressed.

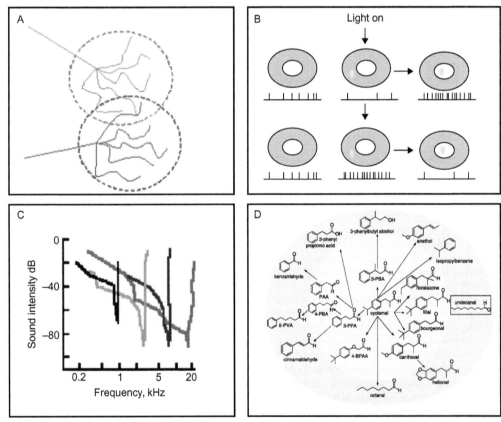

FIGURE A   Somatosensory (A), visual (B), auditory (C), and olfactory (D) receptive fields. In A the circles show the areas on the skin in which a stimulus is sensed by the corresponding neurons. Panel B shows, in the top row, a retinal ganglion cell whose spontaneous activity is inhibited when light is applied to the periphery and activated when the beam of light moves into the center of the receptive field (an "ON center" cell). The bottom row shows an "OFF center" cell, which is activated when the light spot hits the peripheral receptive field, and inhibited when its center is illuminated. Diffuse illumination in both the center and the surround will cancel the inhibitory and excitatory effects and the resulting activity will appear similar to the cell's spontaneous firing rate. Panel C shows four frequency tuning curves from fibers of the cochlear nerve. The graphs show the minimum sound levels necessary to evoke action potentials across all sound frequencies to which the fibers respond. Panel D shows the molecular receptive range of a human testicular odorant receptor hOR17-4, expressed in HEK293 cells grown in cell culture. Compounds that activate the receptor are shown in the inner circle, whereas derivatives that fail to activate this receptor are shown in outer peripheral circle. (Adapted from Spehr et al. (2003). *Science*, **299**, 2054–2058).

within the previous two categories, including acetylcholine, adenosine, ATP, **nitric oxide**, and carbon monoxide.

Bioamines

Catecholamines are bioamines that are derived from **tyrosine**. The rate-limiting step in the biosynthesis of catecholamines is hydroxylation of the tyrosine ring, which generates the characteristic doubly-hydroxylated catechol ring (Figure 3.14). This reaction is catalyzed by tyrosine hydroxylase, and results in the formation of **dihydroxyphenylalanine** (DOPA). Decarboxylation of DOPA results in the formation of **dopamine**. Dopamine is a neurotransmitter in the neural pathway in the brain that projects from the **substantia nigra** to the corpus striatum. This neural projection is important in the initiation of movement, and dopamine deficiencies in this pathway result in **Parkinson's disease**, characterized by rigidity, muscle tremors, and difficulty in initiating movements and speech (see also Chapter 11). Symptoms of Parkinson's disease can be controlled with DOPA, the precursor of dopamine. Dopamine itself cannot cross the **blood–brain barrier**; a protective barrier formed by **astrocytes**, a type of glia that prevents many substances from the circulation gaining direct access to the brain. However, DOPA, the immediate precursor of dopamine, can diffuse across the blood–brain barrier and subsequently be converted into active neurotransmitter.

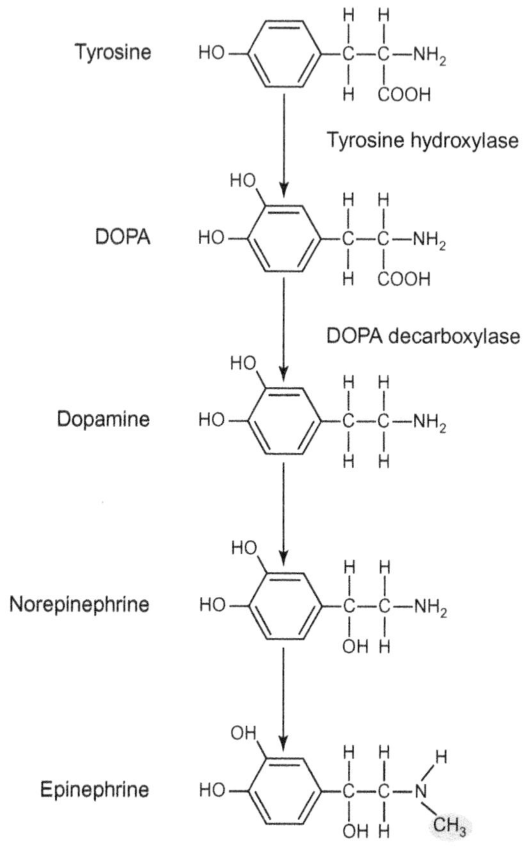

Dopamine is also implicated in schizophrenia and manic depression. These psychiatric disorders involve different dopaminergic projections, the **mesocortical pathway**, which projects from the **ventral tegmentum** in the midbrain to the frontal cortex, and the **mesolimbic pathway**, which connects the ventral tegmentum to the **nucleus accumbens** (Figure 3.15). The nucleus accumbens is associated with pleasurable feelings, reward, and desire. The mesolimbic dopaminergic pathway has been implicated in delusions and hallucinations experienced during schizophrenia, as well as addiction.

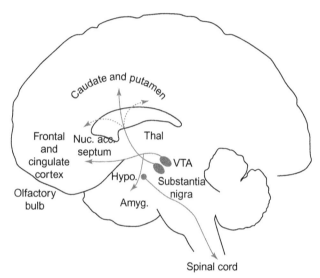

FIGURE 3.15 Dopaminergic pathways in the brain. (Nuc acc, nucleus accumbens; VTA, ventral tegmental area; Thal, thalamus; Hypo, hypothalamus; Amyg, amygdala.)

Dopamine serves as precursor for norepinephrine and epinephrine. Norepinephrine is a neurotransmitter, whereas epinephrine is strictly a hormone, produced by chromaffin cells in the adrenal medulla. Epinephrine is the principal hormone mediating the physiological effects that are known as the **fight or flight response**. Norepinephrine is the neurotransmitter used by the **sympathetic nervous system** and often, but not always, counteracts the effects of acetylcholine released by neurons of the **parasympathetic nervous system**. The terms **adrenergic** and **cholinergic** refer to these actions, respectively.

Catecholamines are removed from the **synaptic cleft** via reuptake in the presynaptic cell (Figure 3.16). After being reabsorbed, they are metabolized by catechol-O-methyltransferase or monoamine oxidase. Inhibitors of monoamine oxidase, such as deprenyl, are used to treat chronic depression.

In the brain there is a widespread adrenergic projection from a nucleus in the brainstem, known as the

FIGURE 3.14 The catecholamine biosynthetic pathway. Note the methyl-group (shaded) that distinguishes norepinephrine from epinephrine.

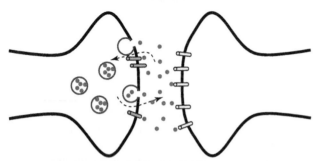

FIGURE 3.16 Diagrammatic representation of a dopaminergic, adrenergic, or serotonergic synapse. Neurotransmitter is released from the presynaptic cell and binds to receptors (open cylindrical symbols) on the postsynaptic membrane. A presynaptic transporter (closed cylindrical symbols) binds the released neurotransmitter for reuptake followed by inactivation or recycling in the presynaptic terminal. Cocaine blocks reuptake at catecholaminergic synapses, thereby potentiating the effects of norepinephrine (creating the "buzz") and dopamine (creating the "pleasure" high).

locus coeruleus, radiating throughout the cerebral cortex. It is thought that this projection is important for general arousal. This may underlie the stimulating effects of amphetamine, which enhances release of norepinephrine, and cocaine, which blocks reuptake of norepinephrine by the presynaptic cell. The actions of both drugs increase the effective synaptic concentration of the neurotransmitter. **Octopamine** is the functional homolog of norepinephrine in insects. Octopamine is derived by hydroxylation of tyramine, which itself results from decarboxylation of tyrosine (Figure 3.17).

Serotonin (5-hydroxytryptamine, 5-HT) is a derivative of the amino acid tryptophan (Figure 3.18). It is also a bioamine, but not a catecholamine, as it does not have a catechol ring. Serotonin has many profound behavioral effects in the brain, and has been linked to arousal, sleep, and aggressive behavior. The hallucinogenic effects of the mood-altering drug lysergic acid diethylamide (LSD) also are mediated via serotonin receptors. The antidepressant drug fluoxetine hydrochloride (Prozac) blocks serotonin reuptake in presynaptic cells.

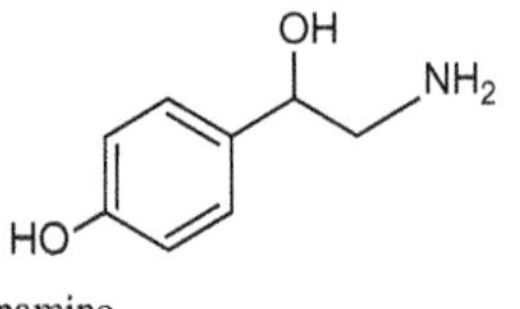

FIGURE 3.17 Octopamine.

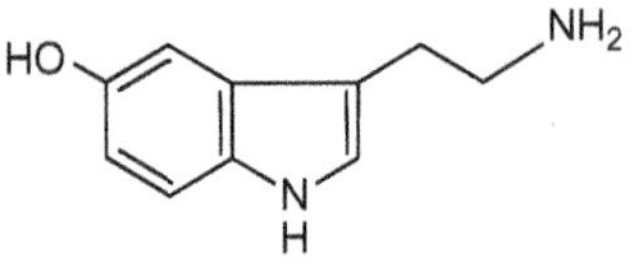

FIGURE 3.18 Serotonin (5-hydroxytryptamine; 5-HT).

Glutamate and Long-term Potentiation

The amino acid glutamate is the most abundant neurotransmitter in the mammalian brain. There are several types of glutamate receptors in the brain. During normal neurotransmission, **kainic acid-** or **AMPA-type glutamate receptors** (kainic acid and AMPA (α-amino-3-hydroxyl-5-methyl-4-isoxazole-propionate) are homologs of glutamate) open channels through which sodium ions flow into the postsynaptic cell. If the depolarization caused by the sodium influx through kainic-acid-type glutamate receptor channels is large enough, it will enable the opening of a different type of glutamate receptor channel on the same cell. This glutamate receptor, known as the **N-methyl-D-aspartate** (NMDA) **receptor**, contains a channel that is also permeable to calcium. Opening of the NMDA receptor channel by glutamate is contingent on the presence of glycine and simultaneous depolarization of the cell. Depolarization releases a Mg^{2+}-ion that blocks the NMDA cation channel. The influx of calcium through NMDA-type glutamate receptor channels mediates structural changes in the synapse that facilitate activation of this synapse in the future. This is thought to underlie the physiologically observed phenomenon of **long-term potentiation** (LTP): once a neural pathway is stimulated, it becomes easier to stimulate the same pathway subsequently. LTP (and its counterpart LTD, long-term depression) may underlie learning and memory formation, and is especially well-documented in an area of the mammalian brain known as the **hippocampus**, thought to be involved in the acquisition of memory (Figure 3.19).

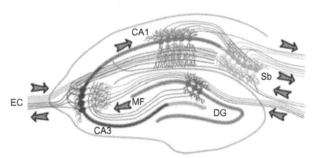

FIGURE 3.19 The hippocampal circuit. The hippocampus is an integral component of the limbic system. Information arrives from the **entorhinal cortex** (EC) to the **dentate gyrus** (DG) and CA3 **pyramidal neurons**. CA3 neurons also receive input from the dentate gyrus via **mossy fibers** (MF). CA3 neurons project to the ipsilateral and contralateral CA1 regions. CA1 neurons send axons to the **subiculum** (Sb), from where output is returned to entorhinal cortex, completing a neural loop.

The short-lived free radical gas, **nitric oxide** (NO), plays an important role as neuromodulator in LTP. NO is synthesized from the amino acid arginine by nitric oxide synthetase, an enzyme that is regulated by **calmodulin**, a ubiquitous calcium-binding protein. During LTP, calcium

Box 3.5 The case of H.M.: Everything happens for the first time

On 1 September, 1953, at the age of 27, a patient referred to as H.M. (Henry M.) was to become the most intensely-studied patient in the history of neuroscience, when he underwent bilateral resection of his medial temporal lobes in an effort to treat his frequent debilitating epilepsy attacks. The medial temporal lobe resection removed the amygdala, hippocampal gyrus, and the anterior two-thirds of the hippocampus. Following his surgery, H.M. displayed a profound memory defect, characterized by loss of short-term declarative memory. He could recall experiences that had occurred before his surgery, and could draw accurate maps of his childhood home. However, he could not acquire new memories, did not know his age, and lived in a world in which Truman was forever president and Shirley Temple the hottest star. H.M. has been studied extensively by Brenda Milner at the Montreal Neurological Institute, but, despite their 50-year-long interactions, he would experience every encounter with her as if it happened for the first time, and would not remember having met her before, whenever she momentarily left the room. Standard psychological tests showed that his IQ was above average and unchanged as a result of the operation. He had no difficulties in perception, abstract thinking, or reasoning, and seemed normal in every respect, except all events in his life since his surgery seemed to happen for the first time without any previous recollection. Subsequently, several other patients with hippocampal lesions have been studied with similar, albeit less severe, impairments. Studies on those patients showed that the hippocampus is a gateway for the acquisition of new memories but that once memories are formed and committed to long-term storage they are no longer dependent on the hippocampus. H.M. expressed his experiences in the following touching quote attributed to him: "Right now, I'm wondering, have I done or said anything amiss? You see, at this moment everything looks clear to me, but what happened just before? That's what worries me. It's like waking from a dream. I just don't remember."

Unfortunately, H.M.'s inability to consolidate new memories prevented him from realizing the great contributions he has made to understanding memory and the human brain.

flowing into the postsynaptic cell activates the production of nitric oxide, which then diffuses back to the presynaptic terminal, where it activates the production of the second messenger cyclic guanosine 3,5-monophosphate (cyclic GMP) by guanylyl cyclase. Cyclic GMP promotes "strengthening" of the synaptic connection by activation of a cyclic GMP-dependent protein kinase which phosphorylates presynaptic proteins. Thus, nitric oxide acts here as a **retrograde** neurotransmitter (Figure 3.20, p. 53).

Inhibitory Neurotransmitters

Decarboxylation of glutamate gives rise to the neurotransmitter γ-aminobutyric acid (GABA) (Figure 3.21, p. 53). GABA is the major inhibitory neurotransmitter in the mammalian brain. Binding of GABA to its receptor opens an anion channel through which chloride flows, thereby hyperpolarizing the cell, i.e. generating an inhibitory postsynaptic potential (IPSP). Barbiturates and benzodiazepines, a class of anxiolytic (anxiety-relieving) drugs, such as Valium, bind to the GABA receptor and potentiate the GABA-mediated opening of the anion channel. GABA receptors are prominent in the cerebellum, a region of the brain involved in motor control. Deficiencies in GABA in this region will result in uncontrolled movements.

Whereas GABA is the major inhibitory neurotransmitter in the brain, the amino acid glycine is the major inhibitory neurotransmitter in the spinal cord. Glycinergic interneurons mediate inhibition of contralateral and antagonistic ipsilateral muscles. This inhibition is essential for the coordination of muscle movements. Glycine receptors

are the target for the poison strychnine. GABA receptors, glycine receptors, and even nicotinic acetylcholine receptors are structurally-related molecules that evolved from an ancestral gene. Together they are members of the family of ligand-gated ion channels.

Peptide Neurotransmitters

The class of peptide neurotransmitters includes substance P, involved in mediating pain, and the enkephalins, endorphins, and **dynorphin**, endogenous peptides that block the transmission of pain and bind to the same receptors to which **analgesic** (pain-relieving) drugs, such as opium, bind (Figure 3.22, p. 54). The discovery of the enkephalins and endorphins is one of the most exciting stories in **neuropharmacology**. In 1973, Solomon H. Snyder and his colleagues, using radiolabeled ligands, identified high-affinity receptors in the brain for opium and its derivatives, drugs that had long been used to relieve pain. Since opium is not normally found in the body, the existence of endogenous substances that modulate pain was postulated. This led eventually to the discovery of the **opioid peptides**, the enkephalins, endorphins, and dynorphin, all of which are produced in the **periaqueductal gray** matter around the aqueduct in the brain, an area implicated in modulating nociception.

The spectrum of neuropeptides in the central nervous system is diverse and includes peptides previously identified as hormones, such as **cholecystokinin** (CCK), which is produced in the small intestine and promotes release of bile from the gallbladder. In some cases, different neuropeptides are present in the same synaptic terminal or are

Box 3.6   Autonomic nervous system

The **autonomic nervous system** innervates involuntary effectors, including cardiac muscle, smooth muscle of visceral organs (organs in the body cavity), and glands. The autonomic nervous system comprises the sympathetic (thoracolumbar) and the parasympathetic (craniosacral) system.

The sympathetic system prepares the body for physical exertion. Norepinephrine released at synapses between sympathetic axons and their target organs and epinephrine, released by the adrenal medulla in the blood stream, cause acceleration of the heartbeat, bronchodilation, stimulation of glycolysis, and vasoconstriction, which increases the blood pressure. In skeletal muscles sympathetic nerves release acetylcholine which causes vasodilation. Vasoconstriction in the visceral organs and vasodilation in skeletal muscles effectively diverts blood from the viscera to the muscles, preparing the body for fight or flight.

In different tissues epinephrine and norepinephrine act via distinct subtypes of adrenergic receptors, linked to different second messenger pathways. Thus, drugs can be designed that are specific for a particular adrenergic receptor and will elicit a therapeutic effect without unwanted side effects (e.g. to treat asthma drugs can be used that act as specific agonists on β_2-adrenergic receptors in the lungs and cause bronchodilation without increasing the heart rate or blood pressure, which are regulated by β_1-adrenergic receptors).

Axons of parasympathetic neurons release acetylcholine at their target organs. Acetylcholine binds here to muscarinic cholinergic receptors. In contrast to the nicotinic acetylcholine receptors, which are members of the ligand-gated ion channel family, muscarinic acetylcholine receptors are members of the GPCR family.

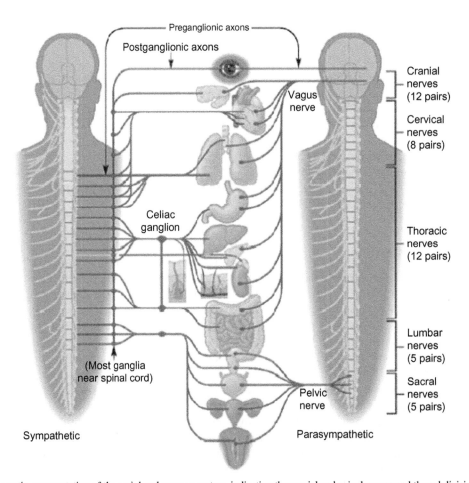

FIGURE A   Schematic representation of the peripheral nervous system, indicating the cranial and spinal nerves and the subdivisions of the autonomic nervous system. Sympathetic nerve fibers synapse on a row of sympathetic ganglia near the spinal cord, whereas postganglionic fibers innervate the target organs. Sympathetic ganglia are highly-interconnected, causing mass activation of the sympathetic system during the fight or flight response. In contrast, parasympathetic ganglia are located in or near the target organs, giving rise to long **preganglionic** and short **postganglionic** fibers. Sympathetic and parasympathetic effects on target organs are often, but not always, antagonistic. Note the extensive ramifications of the vagus nerve which mediates a large proportion of parasympathetic regulation.

Many actions of the parasympathetic nervous system are antagonistic to those of the sympathetic nervous system, and the balance between sympathetic and parasympathetic activity controls homeostasis. For example, sympathetic innervation increases the heart rate, whereas parasympathetic innervation slows it down. Sympathetic innervation causes dilation of the pupil, whereas parasympathetic innervation causes constriction of the pupil. Parasympathetic innervation also causes bronchoconstriction and vasodilation.

The autonomic nervous system is controlled by higher brain regions, especially the medulla oblongata and the hypothalamus. The limbic system also influences the autonomic nervous system, and this is the reason why visceral responses accompany certain emotional states, e.g. blushing, pallor, fainting, breaking out in a cold sweat, a racing heartbeat, and "butterflies in the stomach."

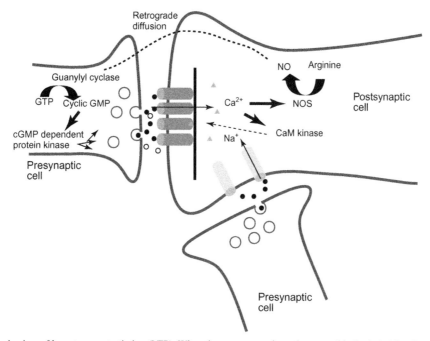

FIGURE 3.20 The mechanism of long-term potentiation (LTP). When the neurotransmitter glutamate (black circles) is released on a postsynaptic cell under conditions in which kainic acid-type receptors (light cylinders) are opened to depolarize the postsynaptic cell, magnesium (triangle) is released from NMDA-type glutamate receptors (dark cyclinders) and in the presence of the co-agonist glycine (open circles), calcium can enter the postsynaptic cell. This leads to activation of nitric oxide synthetase (NOS) and calcium–calmodulin dependent protein kinase. The latter phosphorylates postsynaptic density components (black rectangle), which anchor NMDA-type receptors to strengthen the synapse. NO diffuses to the presynaptic cell, where it activates guanylyl cyclase generating cyclic GMP, which in turn activates a cyclic GMP-dependent protein kinase.

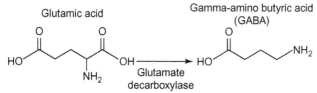

FIGURE 3.21 Decarboxylation of glutamate generates the inhibitory neurotransmitter γ-aminobutyric acid (GABA).

released together with a small-molecule neurotransmitter. They may modulate the response of postsynaptic neurons to a small-molecule neurotransmitter, or exert profound physiological and behavioral effects. Examples are the effects of CCK and **neuropeptide Y** (so named because the first and last amino acids are tyrosines, indicated by Y

in single letter amino acid code). Both neuropeptides are part of a complex neuroendocrine system that regulates food intake. Neuropeptide Y, the most abundant neuropeptide, induces hunger, whereas CCK suppresses appetite.

FMRF-amide-(Phe-Met-Arg-Phe-NH₂)-like neuropeptides are an abundant family of neuropeptides in the nervous systems of invertebrates which all share an FMRF-amide sequence at their C-termini. In the nematode *Caenorhabditis elegans*, FMRF-amide-like peptides are expressed in at least 10% of the neurons, including motor, sensory, and interneurons that are involved in behavioral and physiological functions, such as movement, feeding, defecation, and reproduction. FMRF-amide-like neuropeptides are also abundantly present in the insect brain, where they perform signaling functions during development, as well as in the adult.

Leucine enkephalin	Tyr Gly Gly Phe Leu
Methionine enkephalin	Tyr Gly Gly Phe Met
α-Endorphin	Tyr Gly Gly Phe Met Thr Ser Glu Lys Ser Gln Thr Pro Leu Val Thr
β-Endorphin	Tyr Gly Gly Phe Met Thr Ser Glu Lys Ser Gln Thr Pro Leu Val Thr Leu Phe Lys Asn Ala Ile Val Lys Asn Ala His Lys Gly Gln
Dynorphin	Tyr Gly Gly Phe Leu Arg Arg Ile Arg Pro Lys Leu Lys Trp Asp Asn Gln
Substance P	Arg Pro Lys Pro Gln Gln Phe Phe Gly Leu Met
Cholecystokinin	
octapeptide (CCK-8)	Asp Tyr Met Gly Trp Met Asp Phe
Neuropeptide Y	Tyr Pro Ser Lys Pro Asp Asn Pro Gly Glu Asp Ala Pro Ala Glu Asp Leu Ala Arg Tyr Tyr Ser Ala Leu Arg His Tyr Ile Asn Leu Ile Thr Arg Gln Arg Tyr

FIGURE 3.22 Some examples of neuropeptides. The enkephalins, endorphins, and substance P are associated with modulation of pain. Cholecystokinin and neuropeptide Y regulate hunger and satiety.

SUMMARY

The mammalian brain develops around a fluid-filled space which gives rise to the four cerebral ventricles and the central canal of the spinal cord. Concentrations of cell bodies in the adult brain appear as gray matter, whereas myelinated fiber tracts appear as white matter. Ascending sensory information is relayed to the appropriate region of the contralateral cortex via a thalamic relay (except olfactory information, which is not processed in the thalamus). Descending activity is coordinated by the motor cortex, the cerebellum and the basal ganglia, and descends via the spinal cord to direct the activity of the musculature. Both sensory perception and motor activity are mediated via the contralateral side of the brain, and the representation of sensory and motor areas in the cortex is proportional to the extent of peripheral innervation. The oldest part of the brain evolutionarily contains the limbic system, which is a neural circuit dedicated to placing sensory input in an emotional context, and to instinctive behaviors that serve basic physiological needs. The limbic system exerts control over the autonomic nervous system that regulates involuntary physiological activities, but has few connections with the cortex, the main region of the brain that mediates cognitive functions. Whereas the insect brain has developed along a different evolutionary trajectory, homologous regions can be identified, including optic lobes and antennal lobes for processing visual and olfactory information, and the central complex and mushroom bodies, which direct sensorimotor integration and associative learning, respectively. In any neural circuit, the activity of a neuron in response to external signals is determined by the summed effect of an ensemble of intracellular messages that result from the perturbation of a complex intracellular biochemical network.

STUDY QUESTIONS

1. Describe the relationships between dilatations in the rostral neural tube during early development and regions of the adult brain.
2. What is long-term potentiation and what roles do calcium and nitric oxide play in this process?
3. Describe evidence that supports the notion that the hippocampus is required for acquisition of memory, but not for long-term storage of memory.
4. Imagine that you discovered a nucleus in the hypothalamus that you suspect mediates grooming behavior in mice. What experiments would you perform to prove that neurons in this region indeed mediate this behavior?
5. A physiologist inserts an electrode into the brain of a monkey and when he passes current through this electrode, he observes that the monkey moves its left index finger. In what region of the brain is the electrode inserted?
6. There are at least three distinct neural tracts in the brain that utilize dopamine as neurotransmitter. Name them and explain the functions associated with each of them, indicating related diseases and their possible treatments.
7. What are the functions of the central complex and the mushroom bodies in the insect brain?
8. Describe the neural circuits that mediate withdrawal of the foot after accidentally stepping on a thumb tack.
9. Explain the concept of "receptive field." Could different neurons have overlapping receptive fields?
10. How would a typical ganglion cell respond to a spot of light that travels in a straight line across the center of its receptive field?

RECOMMENDED READING

Bear, M. F., Connors, B., and Paradiso, M. (2006). *Neuroscience. Exploring the Brain*, 3rd edition. Lippincott, Williams and Wilkins, Baltimore, MD.

Bennett, M. R., and Hacker, P. M. (2002). The motor system in neuroscience: a history and analysis of conceptual developments. *Prog. Neurobiol.*, **67**, 1–52.

Burt, A. (1993). *Textbook of Neuroanatomy*. Elsevier/Academic Press, New York, NY.

Chapman, R. F. (1998). *The Insects: Structure and Function*, 4th edition. Cambridge University Press, Cambridge, UK.

Macmillan, M. (2000). *An Odd Kind of Fame: Stories of Phineas Gage*. MIT Press, Cambridge, MA.

Purves, D., Fitzpatrick, D., and Augustine, G. (eds) (2004). *Neuroscience*, 3rd edition. Sinauer Associates, Inc., Sunderland, MA.

Webster, R. (ed.) (2001). *Neurotransmitters, Drugs and Brain Function*. Wiley & Sons, Inc., Hoboken, NJ.

Chapter 4

Measuring Behavior: Sources of Genetic and Environmental Variation

OVERVIEW

Behaviors represent the means by which animals interact with, and adapt to, their environments. In addition to genetic factors, large environmental effects on behavioral plasticity also contribute substantially to individual variation in behavioral phenotypes. Thus, obtaining quantitative measurements of behaviors can be challenging. It is important to define precisely the behavioral parameter that is being measured, and to identify and exclude or account for confounding behavioral elements (e.g. impairments in aggressive behavior could be due to locomotion deficits rather than aggression *per se*); to control as much as possible for environmental variation (which can be done more readily in the laboratory than in the field); and, whenever possible, to control or assess effects of genetic background. In this chapter, we will describe the caveats inherent in measuring behaviors; consider the effects alleles at a single locus may exert on the behavioral phenotype; describe how the effects of alleles at different loci can interact to affect behavior; and discuss environmental effects that may contribute to phenotypic variation.

BEHAVIORAL ASSAYS

In the Field and in the Laboratory

Behaviors can be observed in the field or in the laboratory. Field studies have the advantage in that analyses can be conducted under ecologically-relevant conditions. Some types of experimentation are possible in the field. For example, different traps baited with different pheromone blends can be used to determine which pheromones attract a particular species of moth (Figure 4.1). Investigating herbivore–host plant interactions or foraging behaviors requires field studies as a matter of necessity. Simple observations that do not require experimentation have yielded a vast amount of information about behaviors

of large mammals that could not be obtained in the laboratory, e.g. the social organization and hunting behaviors of lions, in which the male has a harem of females, who predominantly do the hunting, or rough-and-tumble play of young wolves, which shapes their social skills (Figure 4.2). Although DNA analysis from samples gathered in the field can be used to determine kinship relations during long-term studies of groups of wild animals, gaining insights into the underlying genetic and neurobiological principles that give rise to behaviors requires experimental manipulation, which in most cases is feasible only in the laboratory.

Laboratory studies are advantageous in that breeding, and thus the genetic background, and environmental conditions, can be controlled. Temperature, humidity, light/dark cycle, diet, social environment, and age can be controlled precisely. In addition, transgenic approaches and invasive experimental procedures (e.g. castration) can be applied readily in the laboratory. Whereas laboratory studies are indispensable for the study of behavioral genetics, they also face limitations. The behavioral repertoire and the range of variation observed in the field are generally broader than that observed under standard laboratory conditions. In addition, laboratory experiments on behavior are plausible for a relatively limited number of species, generally designated as **model organisms**. It is assumed that fundamental insights gained from studies on such model organisms can be applied either to related species, or more generally, across phyla. A comprehensive analysis of behavior benefits from a combination of field and laboratory studies. Insect models, such as honey bees, have been especially amenable to such approaches (see also Chapter 12).

Dissecting Behavior

Unlike morphological traits, behaviors change constantly as environmental and physiological conditions change. Males and females, more often than not, differ in their behaviors (other than, of course, self-evident differences in mating and parental care behaviors). Thus, sexes should be analyzed separately. Behaviors are shaped through continuous sensory input and appropriate adjustments

FIGURE 4.1 A commercial moth trap baited with a pheromone that attracts male gypsy moths, which are an agricultural pest. Traps like these can be used to capture moths in the wild. However, when traps are collected it is not uncommon to find a host of other insects in the trap in addition to the targeted species. Such "contamination" would not occur in a laboratory setting, but is unavoidable when experiments must be carried out in nature.

FIGURE 4.2 Rough-and-tumble play among wolf pups is an example of a socializing behavior that cannot be readily replicated in the laboratory.

of motor output, and different behaviors may be interdependent. Such complexities may confound experimental observations aimed at a particular behavior. For example, failure of a female mouse to retrieve her pups to the nest could be interpreted as an impairment of maternal behavior. However, deafness would have the same result, as pup retrieval occurs in response to ultrasonic vocalization

of the pups. Maternal behavior and aggression are often correlated, and are an example of interdependent behaviors. When analyzing aggression, one must consider that lack of aggressive behavior could be the result of impaired locomotor ability, rather than an intrinsic lack of aggression. Sleep deprivation is likely to reduce exploratory behavior, whereas starvation may increase exploratory behavior. Thus, individual behaviors are a reflection of the animal's physiological state and are integrated with its entire behavioral spectrum. To complicate matters further, mutations of individual genes are likely to affect multiple phenotypes, as similar genes can function within multiple genetic networks associated with different phenotypes. Thus, assays that quantify a specific behavior must be interpreted with caution, as other relevant behavioral phenotypes may exist that are not being investigated and that may profoundly impact the behavior under study. Characterizing the behavior of an animal is best achieved through a battery of behavioral assays. Below, we will give examples of commonly-used behavioral tests for mice, rats, and fruit flies (*Drosophila melanogaster*), the most frequently studied model organisms in behavioral genetics. Such tests will be discussed in more detail in later chapters.

Behavioral Assays in Mice and Rats

The behavioral assays described here are not exhaustive, but merely a small illustrative sampling of the many behavioral assays that exist. One of the most fundamental behavioral assays is the **open field test.** Open field behavior monitors spontaneous exploratory behavior when an animal is placed in an enclosure. Often open field behavior is video-recorded and total activity, average speed, number of activity bouts, trajectory, and other parameters can be quantified (Figure 4.3). However, the behavior can also be scored manually, for example by counting the number of times the animal transits the borders between different quadrants of the arena. Open field behavior can be used as an indication of anxiety (see also Chapter 11). Anxiety is inferred when an animal in a brightly-illuminated arena freezes, or constantly hugs the wall of the enclosure. Excessive defecation and urination during the open field test are also indications of anxiety. A frequently-used modification of the open field test to assess anxiety is the amount of time spent in a dark area compared to a bright area. Mice are nocturnal and choosing the illuminated area over the dark area is a sign of boldness, whereas the opposite behavior is interpreted as anxiety. Many modifications of open field behavior exist. For example, the time it takes for a mouse to locate a buried food pellet in an open field arena after a period of food deprivation to motivate exploratory behavior is used as a rapid, albeit crude, measure of chemosensory ability.

Spontaneous endogenous locomotor activity can also be measured by recording revolutions of a **running wheel**

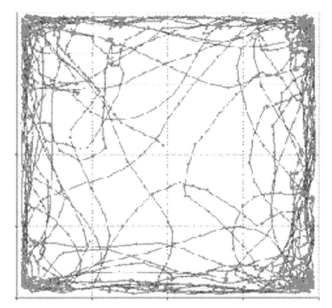

FIGURE 4.3   Record of the trajectory of a mouse in an open field. Note that most of the time is spent in the corners and along the sides of the wall, a behavior known as **thigmotaxis**. (Image from www.phenotyping.com)

FIGURE 4.4   Mice on a rotarod.

placed in the animal's cage during a fixed time period. Mice can run many kilometers a night on the running wheel. There is variation in running activity among strains, and this variation has a genetic component. In addition to serving as an indication of locomotor ability, the running wheel can also be used to monitor general circadian activity.

Another simple behavioral test is the **rotating rod** (rotarod). This test is a measure of balance (**proprioception**). The animal is placed on a slowly revolving rod, and the time it takes for the animal to fall off the rod is measured (Figure 4.4). Often the rotation of the rod is gradually accelerated. The rotarod test is useful for assessing the effects of mutations on proprioception and postural control, or to assess the effects of drugs. Mice exposed to alcohol or drugs will not be able to stay on the rod as long as untreated controls. Here, the time to falling off the rod is a measure of behavioral sensitivity to the drug exposure.

Preference tests can be employed to assess an individual's preference for a particular soluble substance (typically a drug) by providing the animal a choice of a container that holds a solution of the compound versus a water control (see also Chapter 13). Preference for the drug can be measured by calculating a preference index of the total amount consumed as $C/(W+C)$, where W is the water control and C the solution of the test compound. To separate post-ingestive physiological effects from taste preference, the number of licks of each solution can be recorded, in addition to the total amount consumed. This assay uses an instrument known as a **lickometer** (Figure 4.5).

In addition to the light/dark box test mentioned above, the **elevated plus maze** is a commonly used assay for anxiety. The apparatus is a four-armed runway in the shape of

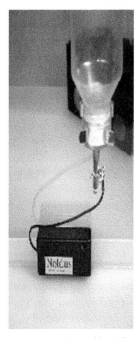

FIGURE 4.5   A lickometer designed by Noldus Informatics, Inc. The instrument detects a change in capacitance between the spout of the water bottle and a metal plate attached to an electronic box. A change in capacitance only occurs when the tongue of a rat or mouse makes contact with the spout.

a plus (+) sign suspended above the ground, two arms of which are enclosed. Exploration of the open arms is scarier than exploration of the closed arms. Avoidance of the open arms is an index of anxiety (see also Chapter 11).

A variety of learning and memory tests have been developed for mice and rats. Classically, animals are trained to navigate a maze for a food reward (Figure

FIGURE 4.6   A radial arms maze to test learning and memory. The animal is placed in the center, and receives a food reward when it remembers to enter the correct arm of the maze, based on previous experience.

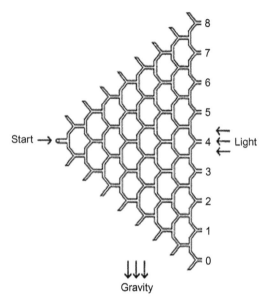

FIGURE 4.7   Diagram of a vertical maze used to measure geotactic behavior in *Drosophila*. Flies are negatively geotactic, i.e. they move upwards against the gravitational field. The example shown fractionates flies into nine collection tubes.

4.6). Learning is assessed by the number of trials it takes to reach the end of the maze without error, and memory is assessed by the performance after a period of no training. A variation of the classic maze is the **Morris water maze** in which animals are trained to use visual cues to swim toward a hidden platform in a circular pool (see also Chapter 14). Learning can also be assessed by avoidance conditioning, where an animal is exposed to an electric shock paired to a loud sound or a particular odor. This assay measures conditioned fear by observing the freezing response of the subject when the same sound or odor is presented without the shock. **Place-preference tests**, where animals are given a choice to spend time in one area versus another (e.g. an area with soiled bedding of a conspecific, or an area with an odor paired previously with an electric shock), are also frequently used behavioral assays (see also Chapters 14 and 15). Finally, various tests exist to measure aggression, the most common of which is the **resident-intruder test** where a strange animal is introduced into the cage of a resident, and the number of aggressive encounters during a fixed time period is counted.

Behavioral Assays in Flies

One of the first behavioral assays applied to *Drosophila* was the **phototaxis** assay developed by Seymour Benzer, who invented a "countercurrent distribution sorting technique" to measure the response of flies to light. Benzer constructed an apparatus which consists of several pairs of test tubes (up to 15) joined by a celluloid sleeve, and laid horizontally in a black rack. A light source is located perpendicular to the tubes. A group of flies are placed at the bottom end of the first double tube (away from the

light) and allowed to distribute freely for a minute. Then the tube is divided into two halves, and the same procedure repeated, up to fifteen times. This yields a frequency distribution of flies that made 0, 1, 2, and up to 15 positively phototactic decisions.

A similar strategy was devised by Jerry Hirsch to measure **geotaxis.** Here, flies are sorted in a mass screening maze according to their tendency to move opposite (negatively geotactic) or towards (positively geotactic) the earth's gravitational vector. The maze consists of a series of vertically bifurcating tubes, and the flies make an "up" or "down" decision at each bifurcation (Figure 4.7). With a 15-unit maze there are 16 collection vials at the end of the maze. Flies scoring 16 made 15 "up" decisions, whereas flies scoring 1 made 15 "down" decisions.

A large literature exists on courtship and mating behavior in *Drosophila*. The courtship behavior is composed of sequential actions that exchange auditory, visual, and chemosensory signals between males and females, allowing for the individual components of the behavior to be separated. First the male aligns himself with the female. Then he taps the female with his foreleg, performs a "courtship song" by vibrating one wing, extends his proboscis to lick the female's genitalia, attempts to copulate, and (if the female accepts his advances) copulates (see also Chapter 12). Courtship latency (the time to initiate courtship) and copulation latency (the time to copulate) are measures of male mating behavior.

Locomotion in flies can be measured as spontaneous open field behavior, similar to open field tests used with rodents. **Circadian activity** can be measured by placing a single fly in a narrow tube and recording movement each

FIGURE 4.8   A simple "dipstick" assay for measuring olfactory behavior. Five flies of the same sex are exposed to a scented cotton wool swab and the number of flies in the compartment remote from the odor source, marked by the dotted line on the left, is counted every five seconds. Ten consecutive counts are averaged to give a response score. Baseline activity is assessed by exposing the flies to an unscented swab.

time it breaks an infra-red light or laser beam. Locomotion can also be induced by tapping a vial containing a single fly sharply against the bench, and measuring the amount of time the fly moves during a brief period following the mechanical disturbance. This **locomotor reactivity** is in essence a **startle-response**, similar to the conditioned fear tests employed with rodents described above (see also Chapter 11).

Chemosensory behavior can be assessed using an olfactory trap assay. Olfactory traps are constructed that allow flies to enter a baited microcentrifuge tube through a tapered pipet tip; the tapered entrance makes it difficult for flies to leave once trapped. Groups of flies are placed in a petri dish containing the trap, and the number of flies trapped after 48 hours is recorded. This assay is limited by the long time it takes to conduct the sometimes small percentage of flies that enter the trap, and because flies may die during the assay period. A faster assay that relies primarily on odor avoidance employs a "dipstick" technique. Here, single sex groups of five flies are placed into an empty culture vial, followed by insertion of a cotton swab saturated with odorant. Marks on the outside of the vial divide it into two compartments. The numbers of flies in the compartment away from the odor source is recorded at five second intervals, for a total of ten measurements, and the average of these ten measurements is the avoidance score (Figure 4.8).

Olfactory avoidance assays have been recruited to study learning and memory in flies. These assays use a device reminiscent of Benzer's countercurrent distribution sorting method. Light at the top of the device encourages the flies to move into the top tubes. Flies are introduced into a rest tube with no odor, and left there for 60 seconds. The device is then slid to the first training tube, which contains an electrified grid. The flies are shocked while exposed to a conditioning odor. They are then returned to the rest tube for 60 seconds, after which they are again introduced to a tube containing an electrified grid. This time, they are exposed to a different odor without shock, and again returned to the rest tube. This cycle is repeated three times, after which the flies are given a choice between the two odors (Figure 4.9). The difference between the numbers of flies responding to each odor is a measure of learning (see also Chapter 14).

Aggressive behavior can also be assessed in fruit flies. In one assay, flies are deprived of food for a brief period to

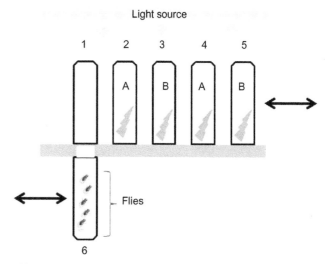

FIGURE 4.9   Diagram of the basic olfactory learning paradigm. Tube 1 is the rest tube; tube 2 is the tube in which flies are shocked in the presence of the conditioning odor. Tube 3 contains the unconditioned odor. Tubes 4 and 5 contain the conditioned and unconditioned odor, respectively. Tube 6 is the start tube. After being exposed to odor A (with shock) or odor B (without shock), flies are tested for responsiveness to each odor without electric shock. The jagged arrows in tubes 2–5 represent electric grids. (The diagram is based on the apparatus described by Quinn, W. G. et al. (1974). *Proc. Natl. Acad. Sci. USA*, **71**, 708–712).

induce aggressive behavior, and then placed in a container with a small drop of food. The number and nature of aggressive encounters (wing threat, chasing, boxing, tussling) is then recorded for a defined period (see also Chapter 12).

CONTROLLING EXPERIMENTAL VARIATION

A major challenge for quantifying behaviors is the fact that behaviors are intrinsically plastic. To minimize experimental error it is essential to conduct behavioral assays, as much as possible under standard environmental conditions. Measurements should be done each day during the same window of time to control for circadian variation, and under controlled conditions of temperature, humidity, air flow, and illumination. In instances where behaviors are subject to seasonal variation, all measurements should be made within the relevant seasonal window. Whenever it is possible, the behaviors of the sexes should be measured separately (Figure 4.10). If no statistical differences are evident between the sexes, measurements can be pooled *post hoc*.

The physiological state of the test subjects may profoundly influence their behaviors. Thus, when possible, diet and age should be standardized. Similarly, whether animals are virgin, pregnant or sexually-experienced may influence certain behaviors. For example, in *Drosophila melanogaster*, females, once mated, become less receptive to courting males. The social environment should also be controlled, as animals that are reared in isolation

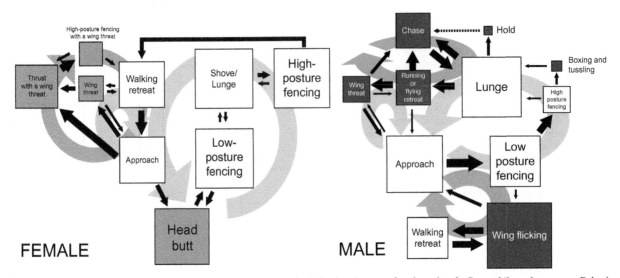

FIGURE 4.10 Ethograms that illustrate differences between aggressive behaviors between female and male *Drosophila melanogaster*. Behavioral patterns that are common between males and females are indicated in the white boxes, whereas female-specific and male-specific behaviors are highlighted in light gray and dark gray boxes, respectively. (From Nilsen, S. P. et al. (2004). *Proc. Natl. Acad. Sci.* USA, **101**, 12342–12347.)

may behave differently from animals that are reared under crowded, competitive conditions. The developmental history of the test subjects may also contribute to the behavior they display as adults. Clearly, it is easier to control all these variables in the laboratory than in the field, and even in a laboratory setting environmental variation will contribute to experimental error.

To further minimize experimental error it is important to accumulate large numbers of measurements for each genotype, and to spread behavioral measurements out over several days or weeks while randomizing measurements on different genotypes. An experimental design aimed at standardization and randomization will help to average, and thus reduce, daily experimental variation.

In addition to environmental contributions, much variation in the expression of behaviors is genetically-determined. This variation can be reduced or eliminated by controlling the genetic background, either by using inbred strains, comparing litter mates, or – in human studies – by comparing **monozygotic** and **dizygotic twins**. If the genetic background cannot be controlled, larger sample sizes will be necessary to obtain accurate average estimates of phenotypic values.

Phenotypic values are characterized by a mean and variation around the mean. The latter can be dissected in different sources that give rise to the total variation by statistical methods, collectively known as **analysis of variance** (ANOVA). From a genetics perspective, both the mean and variation of a trait value are informative. The genes that give rise to the behavioral phenotype can be classified in two categories: genes that contribute to *manifestation* of the trait; and a subset of these genes, which contribute to *variation* in the trait. An example of the former is eye pigmentation. In brown-eyed African populations, or

among blue-eyed Scandinavian people, there may be little or no variation in eye color; yet, eye color is genetically, not environmentally, determined (see also Chapter 6). Genes that contribute to the manifestation of a behavioral trait, but that do not vary in a population can be detected through mutagenesis approaches, as will be described in Chapter 9. Genes that give rise to phenotypic variation are the province of quantitative genetics, and their identification and estimates of their relative contributions to the observed phenotypic variance require statistical analyses (see Chapter 8). In the remainder of this chapter we will take a closer look at the sources of genetic and environmental variation that determine the manifestation of behaviors.

SOURCES OF VARIATION IN BEHAVIOR

Genetic Variation

How do we determine what genes affect behavior? Classical genetics relies on genetic variation to identify genes affecting any trait. One can either create variation *de novo* via mutagenesis (see Chapter 9), or rely on mutations that occurred spontaneously in nature and remain segregating in populations to give rise to naturally-occurring variation in behavior. Unbiased mutagenesis screens tell us what genes are required to produce a given behavior, and knowing these genes can give powerful insights into the cellular and neurobiological mechanisms modulating the behavior. Analysis of natural variants tells us what subset of the genes defined by mutagenesis actually varies in nature. The two approaches are thus complementary.

Whether we are examining the effects of new mutations or natural variants, we are quantifying *allelic* variation. We generally use the term **mutation** to refer to a laboratory-generated allele, and **polymorphism** to refer to an **allele** segregating in nature, although the polymorphism occurred via a spontaneous mutation. In population and quantitative genetics models, we refer to a locus affecting a particular behavior with a capital letter (A, B, C. . .), and the different alleles at that locus with subscripts. In the simplest one locus, two allele case, the alleles would be A_1 and A_2.

Types of Mutant Alleles

Naturally occurring mutations include point mutations and insertions or deletions (the latter are together often designated as "indels"). When such mutations segregate in a population they are referred to as **polymorphisms**. Point mutations are changes of one nucleotide to another and are known as **single nucleotide polymorphisms**, or SNPs (pronounced "snips"). Mutations can occur in coding regions of the DNA, as well as in noncoding regions. The most frequent point mutations are substitution of a purine for another purine (A to G or *vice versa*), or a pyrimidine for another pyrimidine (C to T or *vice versa*). Such mutations are known as **transitions**, whereas substitutions of a purine for a pyrimidine (**transversions**) occur less frequently.

The consequences of mutations depend on where they occur. Eukaryotic protein coding genes are typically split into several **exons** separated by non-coding **introns**, with non-coding 3′ and 5′ regulatory sequences. Adjacent genes may be separated by stretches of sequence with no known function, or genes may be nested in the same or opposite orientation within other genes.

There are three types of point mutations (Figure 4.11). **Silent** or **synonymous mutations** are nucleotide substitutions, often in the third position of a codon, that do not result in a change in the amino acid. However, in some cases such silent mutations may exert effects on the phenotype, if the change in the nucleotide results in a change in the stability of the structure of the mRNA. **Missense** or **nonsynonymous mutations** result in an amino acid change. The consequence of such a substitution, if any, depends on the nature of the encoded polypeptide, and is expected to be greater the more disparate the physical properties (charge or size) are of the interchanged amino acids at this position. **Nonsense mutations** result in the premature formation of a stop codon, thereby generating a truncated protein product. If this protein is completely dysfunctional, such mutations are **null mutations**, i.e. mutations that cause the complete abolishment of a functional gene product.

Given the triplet code, insertions of extra nucleotides in the DNA can cause a change in the reading frame of the coding sequence, if the total number of nucleotides that has inserted is not a multiple of three. This type of mutation is known as a **frameshift mutation**. Similarly, deletions of nucleotides can result in frameshift mutations. Insertion

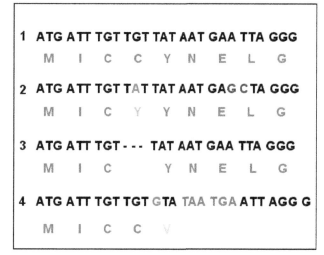

FIGURE 4.11 Examples of mutations. The original nucleotide sequence and the encoded amino acid sequence in single letter code are shown in (1). In (2) three mutations are shown, two of which are synonymous and do not result in a change in the encoded amino acid and one of which is a nonsynonymous substitution (missense). In (3) an in-frame deletion has occurred which results in a single rather than a double cysteine in the encoded peptide sequence. In (4) insertion of a single nucleotide (gray color) results in a frame shift, which changes a tyrosine into a valine and generates two consecutive stop codons (TAA and TGA) resulting in premature termination of the polypeptide and, consequently, most likely an inactive protein.

and deletion mutations occur when one or more nucleotides are added (insertions) or subtracted (deletions), and give rise to **indel** polymorphisms, or **copy number variation** (CNV). Indels can involve small numbers of bases, entire genes, gene regions, and even chromosomes. Indels can also affect mRNA splicing, and thus the formation of alternative transcripts, or expression of the gene, if they occur in the promoter.

In addition to local mutations in the genome, large-scale rearrangements in chromosomal structure can contribute to genetic diversity, including **gene duplications**, chromosomal **translocations, inversions**, and large deletions that result in the loss of one or more genes. **Inversions** occur when a segment of the genome within a chromosome is reversed, and can be polymorphic in natural populations. **Translocation**s occur when a segment of the genome is deleted from one region of the genome and reinserted in another – either on the same or a different chromosome. Translocations may involve a single DNA segment, or be reciprocal translocations of two different segments.

Mutations can further be classified as **loss-of-function** or **null mutations**, in which the mutation abolishes the function of the affected gene, and **hypomorphic mutations**. Null mutations are usually recessive. When the mutation prevents survival of an individual, the mutation is a **lethal mutation**.

Hypomorphic mutations are mutations that do not abolish the function of the gene altogether, but reduce the efficiency of transcription, mRNA stability or splicing, or effectiveness of the function of the encoded protein. Hypomorphic mutations are especially valuable for the study of behaviors, as they enable the development of a viable individual, which, after all, is a prerequisite for behavioral studies.

Gain-of-function mutations (hypermorphic mutations) arise when a mutation results in a new or abnormal function of the gene product (When a new function results from a mutation, the mutation is **neomorphic**.) The abundance and stability of the gene product can increase, and its expression pattern can change. Such mutations often give rise to dominant phenotypes. A special case of mutations are **dominant–negative mutations**, which result in altered gene products that interfere with the normal wild-type allele.

Microsatellites (also called simple sequence repeats, or SSRs) are simple sequences of two, three, or four nucleotides that are tandemly repeated 10–100 times. The number of repeats can vary substantially among individuals, leading to variation in repeat copy number.

Transposable elements (TEs) can replicate by jumping to different genomic locations, and are present in multiple locations in the genome. There are many different families of TE, based on their size, structure, and mechanism of transposition. Examples of TEs include the *Ac/Ds* elements of maize, the first transposable element system to be described by Barbara McClintock (see also Chapter 1); and the *P* element of *Drosophila*, which has been harnessed as an important transformation vector (see also Chapter 9). The genome copy number of TEs can vary, from less than 10 to over one million, such as the *Alu* TE in humans. There can be polymorphism for both the total number of transposable elements, as well as for individual insertion sites. TE insertions in coding regions can result in non-functional proteins, or proteins with novel functions. Mutations in regulatory regions could alter the timing, efficiency, or tissue-specific patterns of gene expression, or alter the number and size of transcripts by changing splice sites (see also Chapter 9).

EFFECTS OF MUTATIONS ON BEHAVIORAL PHENOTYPES

Now we need to consider how genetic variation – whether *de novo* mutations or naturally-segregating polymorphisms – affect behavioral phenotypes. In other words, what is the relationship between genetic variation and phenotypic variation? Let us start by considering the simplest case: a single **autosomal** locus, A, with two alleles (A_1 and A_2) that affect a behavioral **phenotype**. We measure the behavioral phenotype for many individuals of the three possible **genotypes** at this locus, the A_1A_1 and A_2A_2 **homozygous** genotypes, and the A_1A_2 **heterozygote**. We expect that there will be variation in the behavior within

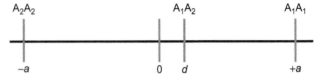

FIGURE 4.12 Diagram of genotypic effects at a single locus with two alleles that affect a behavioral phenotype. The **homozygous effect** (*a*) is one-half of the average difference in phenotype between the two homozygous genotypes. The **heterozygous effect** (*d*) is the difference between the average phenotype of heterozygous individuals and the mean phenotype of the two homozygous genotypes. By convention, the A_1 allele is considered to be the allele that increases the behavioral phenotype.

each genotypic class, because the individuals may differ for other loci affecting the behavior, and, as we will explain below, the expression of alleles affecting behavior is highly sensitive to the environment, and different individuals in the population will experience different environments. However, since this is a locus that has an effect on the behavior, we expect on average that there will be a difference in our quantitative measure of the behavioral phenotype between the two homozygous genotypes. We call this difference the **homozygous effect** of the locus, symbolized *a*. We can estimate the homozygous effect of the locus as a numerical value: $a = (A_1A_1 - A_2A_2)/2$, where the bold face genotype designates the mean behavioral phenotype of many individuals of the genotype (Figure 4.12). By convention, the A_1A_1 genotype is considered to have the highest mean value of the trait. Alleles can have a range of homozygous effects. If the effect is very large, then it can be scored as a Mendelian allele. Many induced mutations have large homozygous effects; alleles with large effects are only rarely found in natural populations. Most alleles affecting natural variation in behavior have moderate to small effects. It is relatively easy to map and clone genes if alleles have large effects, but much more difficult if alleles have smaller effects.

The mean behavioral phenotype of heterozygous individuals may or may not be distinguishable from one or both homozygotes, depending on dominance relationships at the locus with respect to the behavior. We can estimate a quantitative value of the **heterozygous effect** of the locus, symbolized *d*, as the mean value of the heterozygous genotype expressed as a deviation from the mean of the two homozygotes: $d = A_1A_2 - (A_1A_1 + A_2A_2)/2$. Because of the way we have defined the homozygous effect, the average of the homozygous effects of A_1A_1 and A_2A_2 is 0 (Figures 4.12 and 4.13). If $d = 0$, the mean value of the heterozygote genotype is exactly intermediate between the two homozygous genotypes. In this case, we say that **gene action** is **additive**. If $d \neq 0$, there is *non-additive interaction* between alleles at the locus (i.e. some form of **dominance**). If $d = -a$, the A_2 allele is completely dominant; conversely, if $d = +a$, the A_1 allele is completely dominant. If $d < 0$ but $d > -a$, the A_2 allele is partly dominant. If $d > 0$ but $d < +a$, the

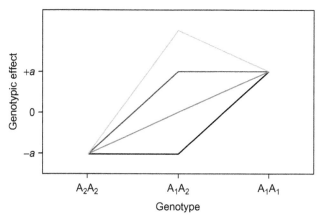

FIGURE 4.13   Graphic representation of the range of homozygous (*a*) and heterozygous (*d*) effects for two alleles at a single locus affecting a behavioral phenotype. The degree of dominance of the A_1 or A_2 allele depends on the heterozygous effect relative to the homozygous effect. If the mean phenotype of heterozygous individuals is exactly between the phenotype of individuals of the two homozygous genotypes, the alleles act additively (straight line). The A_1 (A_2) allele is dominant if the behavior of heterozygous individuals is indistinguishable from that of A_1A_1 (A_2A_2) homozygous individuals (center and bottom broken lines, respectively). Overdominance occurs if the average behavioral phenotype of the heterozygous individuals exceeds that of the A_1A_1 homozygous individuals (upper broken line).

A_1 allele is partly dominant. Finally, if $d > +a$ (or $d < -a$) then there is **over (under) dominance**. The degree of dominance is often expressed as d/a. On this scale, $d/a = 0$ indicates additive gene action, $d/a = 1$ (or -1) means the A_1 (or A_2) allele is completely dominant, $-1 < d/a < 1$ indicates partial dominance, and $-1 > d/a > 1$ indicates over (under) dominance. Dominance and **recessivity** are common for Mendelian alleles with large additive effects, and the term **co-dominant** is used to refer to any category of gene action in which the heterozygous genotype is intermediate between the two homozygous genotypes. Alleles with smaller effects have a greater spectrum of gene action. The homozygous and heterozygous effects at a single locus are collectively called the **genotypic effects** at that locus. Box 4.1 gives an example of genotypic effects at a single locus, *neuralized* (*neur*), affecting startle-induced locomotion in *Drosophila*.

If more than one segregating locus affects a behavioral phenotype, we need to consider their joint effect on the trait. If the total effect on the trait is simply the sum of the homozygous and heterozygous effects of the genotypes of the two (or more) loci affecting the trait, we say the loci combine additively. If, however, the effects of two (or more) genotypes on the trait cannot be predicted from the sum of the effects of the genotypes of each individual locus, then there is non-additive interaction, or **epistasis**, between the loci. Thus, **dominance** refers to non-linear (non-additive) interactions between alleles at the same locus, and **epistasis** refers to non-linear (non-additive) interactions between alleles at *different* loci. We have used the term *additive* to refer to both

---

Box 4.1   Computing homozygous and heterozygous effects

An example of how we can estimate homozygous and heterozygous effects for a behavior comes from studies on startle-behavior in *Drosophila melanogaster*. Yamamoto *et al.* (2008, *Proc. Natl. Acad. Sci. USA*, 105, 12393–12398) assessed the effects of over 600 *P*-element insertional mutations (see also Chapter 9) on the startle-response of *Drosophila melanogaster* following a mild mechanical stress (see also Chapter 11). In this assay, flies were subjected to a brief tap of the vial against the bench top, and the total time spent moving in the 45 seconds immediately following the mechanical disturbance was recorded for 20 male and 20 female flies of each mutant line, for the *P*-element free control line of the same homozygous genotype, and for the heterozygote containing one *P*-element mutant allele and one wild-type allele. An insertion in the *neuralized* (*neur*) gene significantly reduced the average time spent moving following the mechanical stimulus. The means of the three genotypes are given below, as are the theoretical genotypic effects:

Genotype	*neur*	*neur*/*neur*+	*neur*+
Time moving (seconds)	25.48	37.63	42.15
Genotypic effect	− *a*	*d*	+ *a*

(Note on *Drosophila* genetic nomenclature. The "plus" superscript indicates the wild-type allele. The "/" indicates two different alleles at the same locus; i.e. a heterozygote. If there is no "/," the genotype is homozygous for the indicated allele.)

From these data, we can compute the homozygous effect, *a*, $a = (42.15 - 25.48)/2 = 8.335$. The heterozygous effect, *d*, is estimated as $d = 37.63 - [(42.15 + 25.48)/2] = 3.815$. The dominance ratio, $d/a = 3.815/8.335 = 0.458$. Thus, the *neur* mutation is partially recessive; the phenotype of the heterozygote is closer to the wild-type homozygote than the mutant homozygote.

---

intra- and *inter*-locus interactions. Epistasis occurs when the additive and/or dominance effects at one locus are dependent on the genotype of a second locus affecting the trait. When considering alleles with large effects at two loci, epistasis is used to denote *masking* of effects of one locus by another, and is detected by a modification of the expected 9:3:3:1 phenotypic ratio in the F_2 of a cross of homozygous dominant mutations at two independent loci (i.e. $A_1A_1B_2B_2 \times A_2A_2B_1B_1$, where A_1 and B_1 are dominant alleles; Figure 4.14). Epistasis between alleles with small effects at two loci refers to any interaction that is not additive. Epistasis is complicated, and can involve interactions of homozygous effects at two loci ($a \times a$), heterozygous effects at two loci ($d \times d$), or homozygous and heterozygous effects at two loci ($a \times d$). Epistasis can

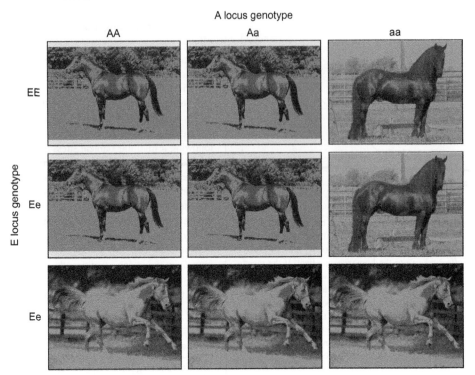

FIGURE 4.14 Some of the best examples of epistasis come from the genetics of coat color. In horses, the E (extension locus) has a dominant allele, "E," which allows the production of both red and black pigment; and a recessive allele, "e," which only allows the production of the red pigment. EE or ee horses have black pigment somewhere on their body; ee horses are chestnut with no black. A second locus, called Agouti (A), also has a dominant ("A") and recessive ("a") allele. The "A" allele restricts the black pigment to the "points" of the horse (mane, tail, forelock, lower legs). The "a" allele allows expression of the black expression on the entire body. There are thus nine possible genotypes combining these two loci: AAEE; AaEE; aaEE; AAEe; AaEe; aaEe; Aaee; Aaee; and aaee, as illustrated in the figure. All horses who are ee are chestnut, regardless of what genotype they are at the A locus. Thus, the E locus is epistatic to (masks the expression of) the A locus, and expression of the A locus depends on the genotype at the E locus. (Image of bay horse from http://www.le-cheval-bleu.com/cardoun.jpg; image of black horse from http://www.ultimatehorsesite.com/images_colors/black6.jpg; image of chestnut horse from http://www.freewebs.com/horsesarethe_best/chestnut%20horse.jpg.)

also occur between genotypes at more than two loci – with increasing difficulty in writing all possible interactions as the number of interacting loci increases. Note that dominance (an interaction between alleles at a single locus), and epistasis (an interaction between alleles at different loci) are independent; one can have dominance with no epistasis, and epistasis with no dominance (Figure 4.15). Epistasis is important, since it occurs when two loci genetically interact, and can be used to infer genetic pathways, or networks (see also Chapter 10). Box 4.2 gives an example of epistasis between two loci affecting startle-induced locomotion in *Drosophila*.

Pleiotropy refers to the effect of a gene on more than one trait. In the narrow sense, pleiotropy can be restricted to a particular allele. In a broader sense, pleiotropy refers to effects of different alleles at the same gene on different traits. For example, the *Drosophila* gene *scribble* contributes to the determination of epithelial cell polarity during development. Null (non-functional) alleles of this gene are embryonic-lethal, but a hypomorphic viable allele caused by a *P* transposable element insertion in the promoter region affects adult olfactory behavior.

ENVIRONMENTAL VARIATION

As mentioned above, behavioral traits are exquisitely sensitive to variation in the environment, including but not limited to differences in nutrition, temperature, humidity, barometric pressure, time of day, handler, lighting, age of the individual, and maternal (developmental) and social environment. Therefore, measurements of the same behavior for individuals that have the same genotype (e.g. a highly-inbred line that is homozygous for all loci, and hence not genetically variable) will not be identical, even if the individuals are reared under highly-controlled environmental conditions. As noted above, when we speak of homozygous, heterozygous, and epistatic genetic effects, we refer to the *average* behavior of individuals with the same genotype. If all individuals are genetically identical, the *variation* about the average is due to environmental effects. The higher the variance, the larger the numbers of individuals that must be measured to get an accurate estimate of the mean, as judged by its standard error (Box 4.3).

It is conventional to scale the estimates of homozygous effects by the standard deviation of the trait phenotype

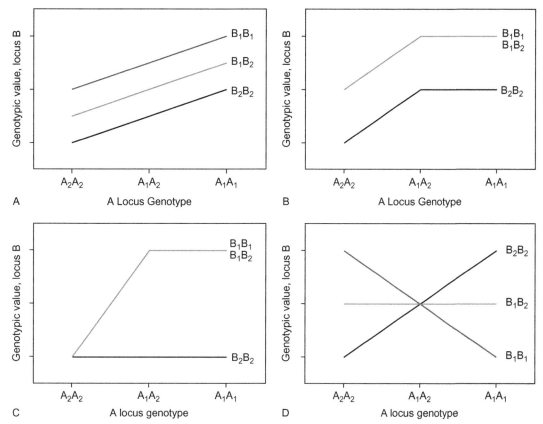

FIGURE 4.15   Graphic representations of genotypic effects at two loci (A, B), each with two alleles. The A_1 and B_1 alleles each increase the behavioral phenotype. (A) This panel represents additivity within loci (no dominance) and between loci (no epistasis). For each locus, the heterozygous effect is exactly intermediate between the two homozygous effects; and the effects of the B locus genotypes are the same regardless of the A locus genotype (and *vice versa*). (B) This panel illustrates dominance of both the A_1 and B_1 alleles; the heterozygous effect is equal to the homozygous effect of the A_1A_1 and B_1B_1 homozygotes, respectively. There is, however, additivity between loci (i.e. no epistasis). (C) This case is where the B_1 allele is dominant, and the B locus is epistatic to the A locus. (D) This panel represents a hypothetical example showing that epistasis can theoretically alter genotypic values profoundly. In this example the B_1B_1 homozygous genotype has the highest value in the A_1A_1 genetic background, and the lowest value in the A_2A_2 genetic background, but there is no dominance (the B_1B_2 genotype is intermediate in the A_1A_1 and A_2A_2 genetic backgrounds).

Box 4.2   Computing epistatic effects

Let us look again at startle-behavior in *Drosophila*, to illustrate how we can compute epistatic effects. In addition to the *neur* mutation described in Box 4.1 Yamamoto et al. ((2008). *Proc. Natl. Acad. Sci. USA*, 105, 12393–12398) found many other mutations that affected startle-induced locomotion in *Drosophila* (described in Chapter 11), including mutations in *roundabout* (*robo*) and *Semaphorin 5c* (*Sema5c*). The average times spent moving following the mechanical stimulus for homozygotes of these mutations were:

Genotype	Time moving (seconds)
robo	21.80
Sema-5c	23.88
robo+; Sema5c+	42.15
robo; Sema-5c	25.50

(The semi-colon is used to indicate alleles on different chromosomes.)

robo and *Sema5c* interact epistatically. To deduce this, we first compute the homozygous effects for each locus separately. For *robo*, a(*robo*) = (42.15 − 21.80)/2 = 10.175. For *Sema5c*, a(*Sema5c*) = (42.15 − 23.88)/2 = 9.135. If the two loci interacted additively (i.e. no epistasis) to affect startle-induced locomotion, we expect the difference between the wild-type double homozygote and the mutant double homozygote to be 2[(a(*robo*) + a(*Sema5c*))] = 38.62. Since the phenotype of the wild-type homozygous genotype is 42.15, the predicted phenotype of the *robo; Sema5c* double mutant with additive interactions between the loci is 3.53 seconds. The observed phenotype of the *robo; Sema5c* double mutant, however, is much greater (25.50 seconds); therefore, there is epistasis. The homozygous effect of the *robo* mutation is 10.175 in the *Sema5c* wild-type background, but (23.88 − 25.5)/2 = −0.81 in the *Sema5c* mutant background. The homozygous effect of the *Sema5c* mutation is 9.135 in the *robo* wild-type background, but (21.80 − 25.5)/2 = −1.85 in the *Sema5c* mutant background. This is an example of epistasis by interactions of homozygous effects at two loci ($a \times a$ epistasis).

Behavioral plasticity is very common, and thus it is very important to control the conditions under which behavior is measured, if this is possible, by ensuring individuals are reared under the same nutritional, climatic, developmental, and social conditions, and that the measurements are taken at the same time during the circadian cycle. Otherwise, the environmental variance will increase greatly, hindering our ability to detect genetic effects (i.e. σ will be increased). If control of the environment is not possible, then different environments must be included in the statistical model. We will discuss correct experimental design to account for environmental variation in Chapter 6.

GENE–ENVIRONMENT CORRELATION AND INTERACTION

A **genotype–environment correlation** can occur if better phenotypes are given better environments. For example, throughout Western history, royalty and aristocrats experienced better nutritional and sanitary conditions than commoners and, thus, were on average healthier; this had little to do with their breeding (see also Chapter 6).

Independence of genotype and environment also implies that a specific difference in environment has the same effect on different genotypes. A particular environmental deviation can be associated with a specific difference in environment, irrespective of the genotype on which it acts. When this assumption is violated, there is **genotype-by-environment interaction** (GEI). GEI can arise if specific differences in the environment have greater effects on some genotypes than others (changes in variance), or if there is a change in order of merit of a series of genotypes measured under different environments (changes in rank). GEI thus occurs if the environmental sensitivities are not the same for different genotypes. GEI is common for behavioral traits. GEI is easily visualized by plotting reaction norms for different genotypes. If they are not parallel, there is GEI (Figure 4.17; see also Chapter 6).

There are often differences between the mean value of a quantitative trait between males and females of the same genotype, called **sexual dimorphism**. A special case of GEI is variation in the amount of sexual dimorphism between different genotypes; here sex is considered as two different environments (Figure 4.18).

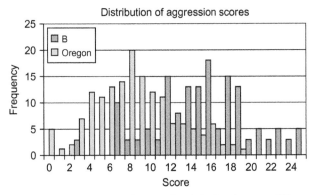

FIGURE 4.16 Measures of aggressive behavior for two different *Drosophila* strains, Canton S (B), designated B in the figure (light gray), and Oregon (dark gray). Individuals of each strain are genetically identical and reared under the same constant environmental conditions. The range of phenotypic variation in aggressive behavior within each strain is due to environmental variation; the difference in mean level of aggression between the strains is due to genetic differences between the strains. (Data from Edwards and Mackay (2009). *Genetics* (in press).)

(σ). Any effect $>3\sigma$ is large enough to segregate as a Mendelian locus; smaller effects will give overlapping distributions of the two homozygous genotypes (as in Figure 4.16). Thus, genetic effects on behavior are statements of the probabilities of exhibiting a particular behavior, not exact predictions.

The tendency for the average behavior of individuals of a given genotype to vary if the environment changes (e.g. crowded social conditions, sleep deprivation, altered nutrition) is called **behavioral plasticity**. We can visualize behavioral plasticity by plotting the mean values of the phenotypic measurement of the behavior across an environmental gradient. Such a plot is called a **reaction norm**.

Penetrance and Expressivity

Quantitative traits, such as behaviors, allow for a range of phenotypes to be expressed by a single genotype, due to environmental sensitivity, epistatic interactions with other loci, and/or GEI. In human genetics, the terms **penetrance** and **expressivity** are used to indicate that the relationship between genotype at a locus and the phenotype of the trait is not perfect. Penetrance refers to the proportion of the

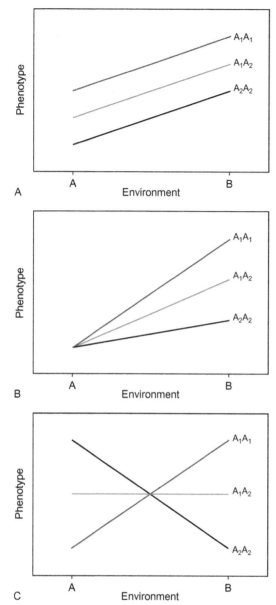

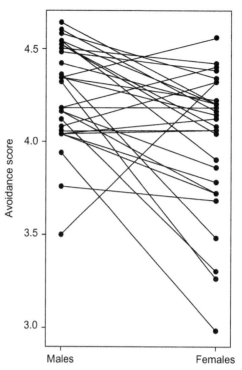

FIGURE 4.18   An example of reaction norms that illustrate sexual dimorphism in a behavioral phenotype. Variation for avoidance response to benzaldehyde among isogenic chromosome *3* substitution lines in *Drosophila*. The third chromosome of an inbred line was replaced by third chromosomes from a natural population. The male and female avoidance scores of each line are connected. Extensive crossing-over of the lines is diagnostic of genotype by sex interactions. (From Mackay et al. (1996). *Genetics*, **144**, 727–735).

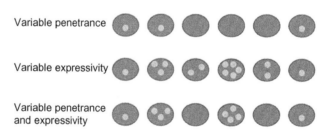

FIGURE 4.19   Diagram of penetrance and expressivity. The dark circles represent individuals with the same hypothetical homozygous genotype that affects spotting (small light gray circles). Individuals can vary for presence and absence of spots (variable penetrance), number of spots, (variable expressivity) or both.

FIGURE 4.17   Reaction norms that illustrate environmental effects on phenotype. (A) This panel illustrates phenotypic plasticity. All genotypes have higher phenotypes in environment B than environment A. However, the genotypic values of the three genotypes are the same in both environments; i.e. there is no genetic variation in phenotypic plasticity, and no GEI. The reaction norms of the genotypes are parallel. (B) This is an extreme example of GEI due to **conditional neutrality**. Here, genetic variation at the A locus is only expressed in environment B; there is no difference in phenotype attributable to this locus in environment A. The reaction norms are not parallel. There is greater variance in genotypic values at the A locus in environment B than environment A. (C) This is an extreme example of GEI due to **antagonistic pleiotropy**. The homozygous effects at the A locus are exactly opposite in the two environments, with the A_1A_1 genotype having the highest phenotype in environment B, and the A_2A_2 genotype having the highest phenotype in environment A. The heterozygote is insensitive to the change in environment in this example. The reaction norms are not parallel, but show crossing-over. Note that reaction norms can be considered across a range of environments, and are not restricted to two environments. Also, reaction norms across multiple environments can have any shape, and are not necessarily linear.

individuals with a given genotype that display the phenotype associated with that genotype. Expressivity refers to quantitative variation in expression of a phenotype among individuals of the same genotype (Figure 4.19). In both cases, the cause of the phenomenon could be epistasis (modifying loci) or environmental effects.

SUMMARY

Whether conducting studies in the field or in the laboratory, measuring behaviors is challenging, as behaviors are intrinsically plastic and can vary dramatically depending on the environment, including differences in nutrition, temperature, humidity, barometric pressure, time of day, handler, lighting, age of the individual, and maternal (developmental) and social environment. Furthermore, males and females can express the same behaviors differently. Thus, when possible, sexes should be measured separately. In nature spontaneous mutations that have arisen during the course of evolution, known as polymorphisms, give rise to allelic variation. Alleles at a single locus can affect the difference in phenotype between individuals with alternative homozygous genotypes (homozygous effects), and heterozygote genotypes (heterozygous effects). The heterozygous effect is the mean value of the heterozygous genotype, expressed as a deviation from the mean of the two homozygotes, and reflects the degree of dominance of the genotype with respect to the behavioral phenotype. Epistasis occurs when the homozygous and/or heterozygous effects at one locus are dependent on the genotype of a second locus affecting the trait. The tendency for the average behavior of individuals of a given genotype to vary if the environment changes (e.g. crowded social conditions, sleep deprivation, altered nutrition) is called behavioral plasticity. A genotype–environment correlation can occur if better phenotypes are given better environments. If the environmental sensitivities are not the same for different genotypes, genotype-by-environment interaction (GEI) is inferred. Sexual dimorphism refers to differences between the mean value of a trait between males and females of the same genotype. Sex can be considered an environment, and sex by environment interaction is a special case of GEI. Environmental sensitivity, epistatic interactions with other loci, and/or GEI are the reasons why a single genotype can give rise to a range of phenotypes. The term penetrance is used to refer to the proportion of individuals with a given genotype that display the phenotype associated with that genotype, while expressivity refers to quantitative variation in expression of a phenotype among individuals of the same genotype.

STUDY QUESTIONS

1. Discuss the advantages and disadvantages of field studies on behavior compared to studies conducted in the laboratory.
2. Describe at least three behavioral assays that use locomotion as a read-out.
3. Prepare a list of environmental conditions that can lead to changes in average behavioral phenotypes.
4. Define the concepts "homozygous effect" and "heterozygous effect." What is overdominance?
5. What is epistasis? Construct reaction norms that illustrate additivity versus epistasis between two loci, A and B, each with two alleles.
6. What is the difference between genotype–environment correlation and genotype–environment interaction?
7. What is meant by the statement that "the phenotype is not fully penetrant?" What could be the reason for incomplete penetrance?

RECOMMENDED READING

Anholt, R. R. H., and Mackay, T. F. C. (2004). Quantitative genetic analyses of complex behaviours in *Drosophila*. *Nat. Rev. Genet.*, **5**, 838–849.

Crawley, J. N. (1999). *What's wrong with my Mouse? Behavioral Phenotyping of Transgenic and Knock-Out Mice*. John Wiley & Sons, Inc, New York, NY.

Falconer, D. S., and Mackay, T. F. C. (1996). *Introduction to Quantitative Genetics*, 4th edition. Prentice Hall, Harlow, UK.

Hirsch, J., and Erlenmeyer-Kimling, L. (1962). Studies in experimental behavior genetics: IV. Chromosome analyses for geotaxis. *J. Comp. Physiol. Psychol.*, **55**, 732–739.

Konopka, R. J., and Benzer, S. (1971). Clock mutants of *Drosophila melanogaster*. *Proc. Natl. Acad. Sci. USA*, **68**, 2112–2116.

Chapter 5

Mapping Genotype to Phenotype in Populations

OVERVIEW

Behaviors are complex traits determined through the interactions of many segregating loci. In order to make predictions about the inheritance of behavioral traits, we must attempt to understand the relationship between genotype and phenotype in populations. This involves not only assessing the effects of genetic variation at a given locus on the trait phenotype, but also their allele frequencies in a population. In this chapter we will examine the rules that govern the proportions of segregating alleles at a single locus in an idealized, very large, randomly-mating population, and see how these rules are affected by inbreeding, i.e. breeding of related individuals. We will discover how the addition of multiple loci contributing to the phenotype gives rise to a bell-shaped (normal) distribution of phenotypic values, characteristic of complex traits. We will further explain how the population mean of a phenotype is affected by genotypic effects and allele frequencies at loci affecting the phenotype, in random mating and inbred populations. Finally, we will define the breeding value of an individual as the breeding quality of that individual judged by the mean value of its progeny, and explain how estimates of breeding values can provide us with information about the quality of an individual to contribute to a phenotype in its offspring, even without detailed knowledge of the genetics that underlies the inheritance of the trait.

Alleles affecting complex behaviors in natural populations have two important properties. We discussed the first important property in Chapter 4: the genetic *effects* on the trait (homozygous and heterozygous effects, epistatic effects, pleiotropic effects). The second important property is the *frequency* of the genotypes and alleles in the population.

The genotype frequency is simply the proportion of each of the three genotypes in the population. If we sample a large number of individuals (N) from a population, and determine their genotype at an autosomal locus (A) with two alleles (A_1, A_2), we will obtain a count of the number of individuals of each of the three genotypes, with N_{11} A_1A_1 individuals, N_{12} A_1A_2 individuals, and N_{22} A_2A_2 individuals (where $N = N_{11} + N_{12} + N_{22}$). The frequency of each genotype is simply the number in each genotypic class divided by the total number of individuals in the sample. Thus, the frequency of the A_1A_1 genotype $= P = N_{11}/N$; the frequency of the A_1A_2 genotype $= H = N_{12}/N$; and the frequency of the A_2A_2 genotype $= Q = N_{22}/N$. Note that $P + H + Q = 1$, because there are only three possible genotypes with two alleles (Figure 5.1).

The allele frequencies can now be computed directly from the genotype frequencies. Each A_1A_1 homozygote contains two A_1 alleles, and each A_1A_2 heterozygote contains one A_1 allele. Thus the frequency of the A_1 allele $= p = P + 1/2H$. Similarly, the frequency of the A_2 allele $= q = Q + 1/2H$. With only two alleles at a locus, $p + q = 1$, and therefore $q = 1 - p$.

GENES IN POPULATIONS: RANDOM MATING

In a large (technically, infinitely large) random mating outbred population, in which none of the evolutionary forces of mutation, migration, and natural selection occur, we can derive the relationship between allele frequencies of parents and genotype frequencies in their offspring. Random mating means random union of gametes, i.e. an A_1 gamete will pair with either an A_1 or an A_2 gamete according to

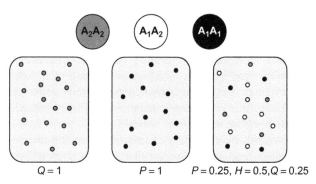

FIGURE 5.1 Populations with different genotype frequencies.

R. Anholt and T. Mackay: Principles of Behavioral Genetics
ISBN: 978-0-12-372575-2

TABLE 5.1 Random union of gametes

	Male gametes (frequency)	
	$A_1(p)$	$A_2(q)$
Female gametes (frequency) $A_1(p)$	$A_1A_1(p^2)$	$A_1A_2(pq)$
$A_2(q)$	$A_1A_2(pq)$	$A_2A_2(q^2)$

FIGURE 5.2 Godfrey Hardy (left) and Wilhelm Weinberg (right), pioneers of population genetics.

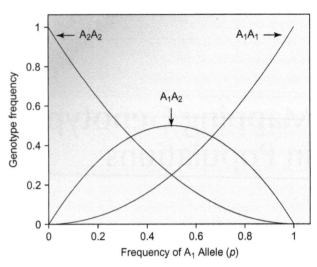

FIGURE 5.3 Genotype frequencies as a function of allele frequencies in a population that is in Hardy–Weinberg equilibrium.

their frequency, as shown in Table 5.1. Thus, with random mating in such an *ideal* population, the genotype frequencies in offspring are $A_1A_1 = p^2$; $A_1A_2 = 2pq$; $A_2A_2 = q^2$; $p^2 + 2pq + q^2 = 1$. These are known as Hardy–Weinberg equilibrium (HWE) proportions, in honor of the two scientists who independently discovered this relationship early in the last century (Figure 5.2). Thus, genotype frequencies in offspring can be *predicted* from allele frequencies in parents. Further, genotype and allele frequencies are *stable* from generation to generation in the absence of evolutionary forces that change gene frequencies, such as mutation, selection, migration, or inbreeding/genetic drift.

The term "equilibrium" usually refers to the tendency of a population to return to the same value after a perturbation. In this case, however, "equilibrium" refers to the return to the same *proportions* in a single generation of random mating after a change in allele frequency, even though the actual frequencies will not be the same. For example, assume that a population is in HWE with $p = 0.8$, and $q = 0.2$. The genotype frequencies are thus $P = p^2 = 0.64$, $H = 2pq = 0.32$, and $Q = q^2 = 0.04$. Now, let us assume that in one generation some of the A_2A_2 individuals died or failed to reproduce, changing the allele frequencies to $p = 0.9$, and $q = 0.1$. With random mating with these parental allele frequencies, the new equilibrium genotype frequencies in the progeny generation are $P = p^2 = 0.81$, $H = 2pq = 0.18$, and $Q = q^2 = 0.01$.

If we plot the frequency of each genotype for all possible allele frequencies, we see that the frequency of heterozygotes is maximal at $2pq = 0.5$, when $p = q = 0.5$ (Figure 5.3). Also, when the frequency of an allele is low,

the rare allele occurs predominantly in heterozygotes, with very few homozygotes. For example, when $q = 0.01$, the frequencies of the three genotypes are $P = p^2 = 0.9801$, $H = 2pq = 0.0198$, and $Q = q^2 = 0.0001$. The ratio of heterozygous carriers to homozygotes is 198:1.

Polymorphisms, i.e. sequence variations in the DNA segregating in natural populations, are often referred to as "common" or "rare," depending on the frequency of the less frequent allele (**minor allele frequency**, or **MAF**). By convention "common" polymorphisms are those for which the MAF is greater than 0.05, while "rare" polymorphisms are those for which the MAF is less than 0.05.

GENES IN POPULATIONS: INBREEDING

The HWE discussed above is valid for large random mating populations. However, many species are not randomly mating, but have some degree of **inbreeding**. Inbreeding means mating of related individuals. We quantify the amount of inbreeding by the **inbreeding coefficient**, F. F is the probability that two alleles at a locus are **identical by descent** (IBD); i.e. are copies of the same allele from a common ancestor (Figure 5.4). The statement that "the inbreeding coefficient of individual I is $F = 0.125$" has two equivalent meanings. First, this means that a randomly chosen locus in individual I has both alleles IBD with probability $F = 0.125$, and hence is a homozygote. Second, it means that 12.5% of the loci in individual I are IBD, and are hence homozygous.

We can determine the inbreeding coefficient of an individual if we know the pedigree. Inbreeding can only occur if both copies of an allele in individual I come from a common ancestor. Thus, we can derive the inbreeding coefficient of individual I by labeling the two alleles at a locus in the common ancestor x and y, and calculating the probability that individual I has either two copies of x or two copies of y at the locus. (We label the alleles in the ancestor

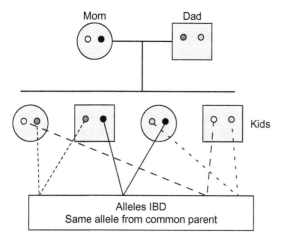

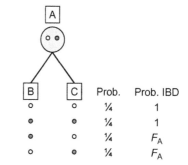

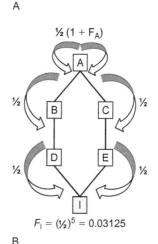

FIGURE 5.4 Identity by descent. Each of the four alleles in two parents is color-coded. Two progeny can have the same copy of an allele as the parents; such alleles are **identical by descent** (IBD).

x and y, and not A_1 and A_2 as before, because we want to distinguish between identity by state [i.e. both alleles are A_1 or A_2] and identity by descent [both alleles are A_1 or A_2, and are the exact same copy of A_1 or A_2, transmitted through the common ancestor]. Alleles can thus be identical in state but not IBD; but all IBD alleles are also identical in state.) The first step is to calculate the probability that the same allele from the common ancestor, $\underline{A}$ (underlined to distinguish this usage from a locus), is transmitted to at least two progeny in the next generation that are also ancestors of I. There are four equally probable results: both progeny of $\underline{A}$ receive allele x or both receive allele y, in which case the alleles are IBD; and one progeny receives allele x and the other allele y (there are two ways this can happen), in which case the alleles are only IBD if the common ancestor is itself inbred (F_A) (Figure 5.5). Therefore, the total probability that both offspring of $\underline{A}$ have an allele IBD is $1/2(1 + F_A)$. In each subsequent generation, the probability that the allele is transmitted from one generation to the next is $1/2$, by Mendelian segregation. Thus, if we write the path of transmission of alleles from $\underline{A}$ to I backward from one parent of I, through each of the individuals descended from the common ancestor, and forward through each of the individuals descended from the common ancestor to the other parent of I, the probability that two alleles in I are IBD (i.e. the inbreeding coefficient of I, F_I) is $(1/2)^i(1 + F_A)$, where the exponent i denotes the number of individuals I in the path (Figure 5.5). If there are multiple paths and multiple common ancestors, the total inbreeding coefficient of I is the sum of all paths.

Note that the inbreeding coefficient is relative to a population that we consider to be not inbred, which in practice means when we have no further information about the pedigree. Thus, in the absence of additional information, the inbreeding coefficient of the common ancestor is assumed to be 0. Inbreeding coefficients from mating of various relatives are given in Figure 5.6.

FIGURE 5.5 Calculation of inbreeding coefficients. (A) Derivation of the probability (prob.) that two offspring (B and C) of individual A have alleles IBD. (B) Pedigree of individual I. Each line represents a path of transmission of an allele from one generation to the next. The pedigree is simplified to only show the ancestors of I that could contribute the same copy of an allele from the common ancestor, A. To determine the inbreeding coefficient of individual I, F_I, we multiply the total probability that two alleles in I are IBD: $1/2 \times 1/2 \times 1/2(1 + F_A) \times 1/2 \times 1/2 = (1/2)^5(1 + F_A)$. Since we have no other information on individual A, we assume $F_A = 0$. F_I is thus 0.03125. Note that if we follow the path up from one parent of I, to the common ancestor, and back down to the other parent of I, we write DB$\underline{A}$CE. The number of individuals in this path is the exponent to which 1/2 was raised.

Inbreeding also occurs in populations of finite size, N. In this case, the inbreeding coefficient after one generation equals $F = \dfrac{1}{2N}$. After t generations of restricted population size, the average inbreeding coefficient of the population will be $F_t = 1 - \left(1 - \dfrac{1}{2N}\right)^t$. Thus, the rate of inbreeding is greater the smaller the population. This is because the smaller the population, the less remote is the common ancestor. We will discuss the consequences of inbreeding in finite populations further in Chapter 16.

Inbreeding is non-random mating, and thus the HWE genotype frequency proportions are no longer valid. We model the effects of inbreeding on genotype frequency by considering an initial population in HWE that is then

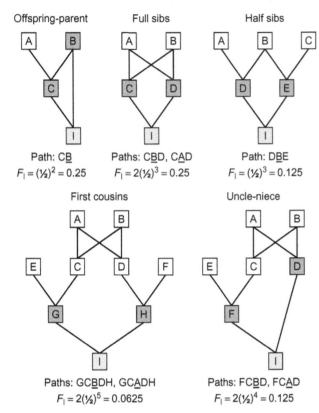

FIGURE 5.6 Derivation of inbreeding coefficients for inbred progeny (I, dark gray) of common relatives (light gray).

divided into a large number of lines, or subpopulations, each of which is inbred by amount F. For example, we could take a large outbred population and establish from it many independent lines that had one or more generations of full sibling mating, or that were maintained with the same number of parents (N) each generation. We further stipulate that each of these lines is a closed population – there is no migration from other populations. Inbreeding within each line causes an increase in homozygosity and a decrease in heterozygosity; however, in the absence of selection, there will be no systematic change in allele frequency in the collection of lines as a whole, relative to the initial outbred population. Thus, if we consider the population of inbred lines together, there will be more homozygotes and fewer heterozygotes than expected under HWE in an outbred population with the same allele frequencies. The frequency of the homozygous genotypes are each increased by pqF, so the frequency of $A_1A_1 = p^2 + pqF$, and of $A_2A_2 = q^2 + pqF$. The frequency of the A_1A_2 heterozygous genotype is correspondingly decreased by $2pqF$, and is $2pq - 2pqF$.

If inbreeding occurs each generation, F will ultimately reach 1. This happens after 10 generations of self-fertilization, and 20 generations of full sibling mating, for example. Inbreeding occurs naturally in self-fertilizing plants like *Arabidopsis thaliana* and hermaphroditic animals like

Caenorhabditis elegans. In humans, inbreeding was common among Egyptian pharaohs and European royal families. Inbreeding can also be applied deliberately in model organisms, most commonly *Drosophila* and mice. When $F = 1$, there are no heterozygotes, and all individuals are A_1A_1 or A_2A_2. This is true for a single locus, and all loci in the genome. Thus, inbreeding produces strains of genetically identical individuals. Such strains are invaluable tools for studying behavior, because different laboratories can investigate effects of the same genotype; inbred lines are also useful for gene mapping (see also Chapter 8).

QUANTITATIVE GENETIC MODEL

Most behaviors are continuously distributed in natural populations, often approximating a statistical normal distribution (Figure 5.7). Some behaviors fall into discrete classes, such as bipolar disorder or schizophrenia, where some individuals can be categorized as affected and others as not affected. Such traits are known as **threshold traits**, and are also genetically complex. Threshold traits are treated by assuming an underlying continuous liability to express the disease, with a threshold value of liability above which individuals are affected, and below which they are not. Threshold traits require different methods of analysis than continuously varying traits. For some behavioral paradigms, the quantitative score assigned to an individual is a discrete number, such as 0–15 in the *Drosophila* phototaxis assay discussed in Chapter 4. These traits are not truly continuous, and are called **meristic traits**. Meristic traits are typically treated as if they were truly continuous.

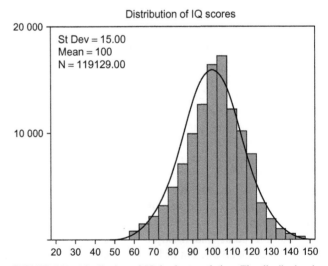

FIGURE 5.7 Distribution of IQ in the population. The distribution is normal, i.e. bell-shaped, which is characteristic for a complex trait. In the case of IQ the average is at 100. IQ is determined by performance on standardized tests (see also Chapter 14).

How can we explain continuous variation for complex behaviors when we know that alleles at each locus affecting the trait segregate according to Mendel's laws?

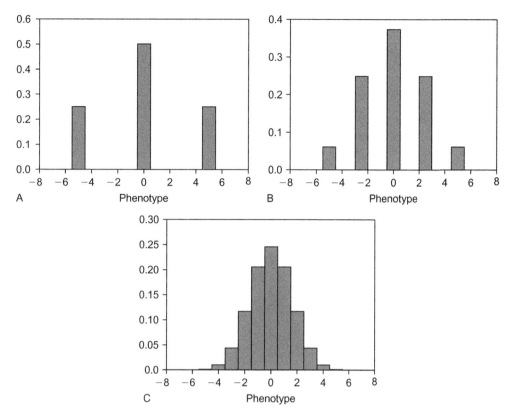

FIGURE 5.8 Relationship between genotype and phenotype for 1, 2, and 5 additive loci affecting behavior. For simplicity, we assume the genotypic effects are the same for all loci, the mean genotypic effect is the same as the mean phenotypic effect, the total range of phenotypes is constant for the different models, and that allele frequencies are 0.5 at each locus. (A) For one locus, the genotypes and phenotypes correspond exactly. The A_1A_1, A_1A_2, and A_2A_2 genotypes have effects of $a = 5$, $d = 0$, and $-a = -5$; and frequencies of p^2, $2pq$, and q^2, respectively. (B) For two loci, the $A_1A_1B_1B_1$ genotype has phenotype of 5 and a frequency of p^4, and the $A_2A_2B_2B_2$ genotype has a phenotype effect of -5 and a frequency of q^4. The genotypes $A_1A_2B_1B_1$ and $A_1A_1B_1B_2$ both have phenotypes of 2.5, which occurs in the population with a frequency of $4p^3q$; and the genotypes $A_1A_2B_2B_2$ and $A_2A_2B_1B_2$ both have phenotypes of -2.5, which occurs with a frequency of $4pq^3$. Three genotypes ($A_1A_1B_2B_2$, $A_1A_2B_1B_2$, $A_2A_2B_1B_1$) have phenotypes of 0; the frequency of this phenotype is $6p^2q^2$. Note that the frequencies of the phenotypic classes are given by the binomial expansion $(p + q)^{n-1}$, where n is the number of phenotypic classes. (C) For 5 loci, there are 11 phenotypic classes, with frequencies given by the expansion of $(p + q)^{10}$. Note that even with as few as five loci the distribution of phenotypes begins to approximate a normal distribution.

The quantitative genetic model assumes that the continuous distribution of phenotypes results from: (1) multiple polymorphic loci affecting the trait; (2) environmental variation; and (3) possible genotype–environment interactions.

To illustrate how increasing the number of loci affecting a trait leads to a continuous distribution of phenotypes, let us compare the phenotypic distributions when 1, 2, or 5 loci, each with two alleles, affect the behavior. Assume, for simplicity, that there is strict additivity ($d = 0$) for each locus; there is no epistasis so the different loci combine together additively to affect the phenotype; the genotypic effects are the same for all loci; and the allele frequencies are 0.5 at all loci. We also assume that the range of phenotypes stays the same regardless of the number of loci: as the number of loci increases, the effect of each decreases. Finally, we assume there is no environmental variation.

For one locus, the genotypes and phenotypes correspond exactly. For two loci, there are nine genotypes and five phenotypes, and the correspondence between the genotype and the phenotype is no longer exact. For five loci,

the relationships between genotypes and phenotypes are even more complex, and we can now see how a normal distribution materializes (Figure 5.8). As the number of loci increases, there is no longer a one-to-one correspondence between genotype and phenotype; multiple genotypes can give rise to the same phenotype. The distinction between genotypic classes decreases as the number of loci increases, and hence the effect of each becomes smaller.

Next, we assume that any one genotype can take on a range of phenotypic values, depending on the environment (environmental sensitivity). Superimposed environmental variation is truly continuous, and further blurs the distinction between genotypes. With genotype-by-environment interaction (GEI, see Chapter 4), the environmental sensitivities of different genotypes are not the same, further adding to the complexity of the genotype–phenotype relationship and contributing to the continuous distribution of phenotypes. The phenotypic value (P) of an individual is thus determined by the individual's unique genotype (G) at all loci affecting the behavior, the environment (E) in which the individual is reared, and any genotype-by-environment interaction (GE): $P = G + E + GE$.

We can measure the phenotypic values of many individuals in a population for one or more behaviors of interest using the behavioral assays described in Chapter 4. Based on these observations, we can provide a statistical summary of the behavior in the population in terms of the population mean, phenotypic variance (Box 4.3), and phenotypic correlations with other behaviors and complex traits. In an outbred population, we can also measure the behavior in known relatives, and statistically quantify the resemblance between relatives (see Chapter 7). If the population is composed of inbred lines, we can statistically quantify the variance within and between lines. In some circumstances, we can measure the performance of defined genotypes in different environments to obtain quantitative assessments of phenotypic plasticity and (GEI). We can also determine the genotypes of unrelated individuals and family members in terms of molecular polymorphisms.

Our first goal in understanding the genetic architecture of behaviors is to determine how much of the phenotypic variation we observe is due to genetic variation, and how much is due to environmental variation (the nature versus nurture partition) and the interaction of genes and environment. Our second goal is to understand the detailed genetic basis of behavioral variation in terms of the actual genes affecting natural variation in behavior; their homozygous, heterozygous, and epistatic effects; their pleiotropic effects on other traits; environmental plasticity and GEI; the molecular basis of genetic variation (and GEI); and the frequency of alleles that are causal to the observed variation. Using classical quantitative genetic analyses, we can determine the answer to the first question without polymorphisms as molecular markers and knowledge of individual genes (Chapters 6 and 7). However, we need molecular marker information in order to infer the details of the genetic basis of variation in behavior (Chapter 8).

How can we determine how much variation in phenotypes is due to genetic variation, and how much is due to environmental variation? Closely related individuals share genes that are IBD to a greater extent than unrelated individuals. To the extent that genes affect behavior, closely related individuals should behave more similarly to each other than to random, unrelated individuals. We can statistically quantify the degree of relationship between relatives, and figure out how this relates to the amount of genetic variance for the behavioral trait (Chapter 7). We can also deduce how the homozygous, heterozygous, and epistatic effects of individual loci affecting the trait, and their allele frequencies, contribute to the genetic variance (Chapter 6). A variance is, by definition, the mean of the squared deviations from the population mean. Therefore, in the remainder of this chapter, we will show how the population mean depends on homozygous and heterozygous effects, and allele frequencies. Since alleles, not genotypes, are

TABLE 5.2 Calculation of the population mean of a quantitative trait in an outbred population

A_2A_2		A_1A_2		A_1A_1
$-a$		0	d	$+a$

Genotype	Frequency	Effect	Frequency × Effect
A_1A_1	p^2	a	p^2a
A_1A_2	$2pq$	d	$2pqd$
A_2A_2	q^2	$-a$	$-q^2a$
		Σ	$= a(p^2 - q^2) + 2pqd$
			$= a(p - q)(p + q) + 2pqd$
	$\mu_G = \mu_P$	$=$	$a(p - q) + 2pqd$

Over several loci, assume values are additive:
$\mu_G = \mu_P = \Sigma a(p - q) + 2\Sigma pqd$

transmitted from parents to offspring in a randomly-mating population, we have to consider the effect of an allele on a behavioral trait. Thus, we also define a new concept – the breeding value of an individual – in terms of allelic effects and frequencies.

Population Mean

We can now ask, what is the expected mean value of a behavioral trait under the quantitative genetic model? Let us assume that there is no genotype–environment interaction, and that in the population as a whole the environmental fluctuations cancel out, so the mean phenotypic value (μ_P) of all individuals in the population is equal to their mean genotypic effect (μ_G). We first consider just one locus affecting the trait, and assume that the genotypes at this locus are in HWE. The mean genotypic effect (and the mean phenotypic value) of the trait attributable to this locus in a random mating population can be determined by multiplying the effects of each genotype by their expected frequency, and adding up the products of frequencies and effects over all genotypes. The population mean attributable to a single locus is thus $\mu_G = \mu_P = a(p - q) + 2pqd$. Note that when we calculate the population mean in this manner, it is expressed as a deviation from the average of homozygous effects (0) (Table 5.2). Because many loci affect the trait, the simplest model for the population mean is to assume that each locus contributes additively to the trait (i.e. no epistasis): $\mu_G = \mu_P = \Sigma a(p - q) + 2\Sigma pqd$.

Therefore, the mean of a behavioral trait may be different between populations, for two possible reasons. First, the population mean depends on allele frequency, so the contribution of any locus to the mean can range from a high of $\mu_G = +a$ if the population is **monomorphic**

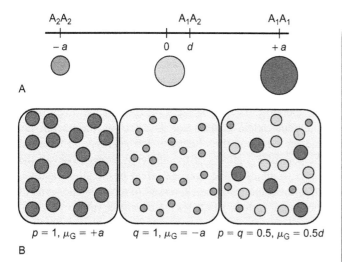

A

B

$p = 1, \mu_G = +a$ $q = 1, \mu_G = -a$ $p = q = 0.5, \mu_G = 0.5d$

FIGURE 5.9 The population mean depends on allele frequency. (A) Homozygous and heterozygous effects at a locus affecting a behavioral trait. The size of the circles is proportional to the effect on the trait. (B) Three populations with different allele frequencies. In the population to the left, the A_1 allele is fixed ($p = 1$, $q = 0$), and the population (μ_G) mean is $+a$. In the middle population, the A_2 allele is fixed ($p = 0$, $q = 1$), and the population mean is $-a$. In the population to the right, the A_1 and A_2 alleles are equally frequent ($p = q = 0.5$), and the population mean is $0.5d$.

Box 5.1 Mean of an outbred population in Hardy–Weinberg equilibrium

Recall the *Drosophila neuralized* (*neur*) locus from Chapter 4, at which alleles affect startle-induced locomotion (time spent moving after a mechanical disturbance). In Box 4.1, we determined the homozygous and heterozygous effects at this locus:

Genotype	*neur*	*neur/neur$^+$*	*neur$^+$*
Time moving (seconds)	25.48	37.63	42.15
Genotypic effects	-8.335	3.815	8.335

We can use this example to illustrate how the population mean of an outbred population in Hardy–Weinberg equilibrium (HWE) depends on allele frequency. The population mean, expressed relative to 0, the average of the two homozygous effects, is $\mu_G = a(p - q) + 2pqd$. In a population where the frequency of the *neur$^+$* allele (p) is 0.8, the population mean is $\mu_G = 8.335(0.8 - 0.2) + 2(0.8)(0.2)(3.815) = 6.2218$. However, in a population where the frequency of the *neur$^+$* allele (p) is 0.2, the population mean is $\mu_G = 8.335(0.2 - 0.8) + 2(0.2)(0.8)(3.815) = -3.7802$. This makes intuitive sense, because in the first case there are more homozygous *neur$^+$* individuals in the population, which have the highest startle responses, whereas in the second case there are more homozygous *neur* individuals in the population, which have the lowest startle responses.

for the A_1A_1 genotype ($p = 1$), to a low of $\mu_G = -a$ if the population is monomorphic for the A_2A_2 genotype ($q = 1$) (if there is no **overdominance**) (Figure 5.9). Box 5.1 illustrates different population means with changing allele frequencies at a locus affecting the startle response in *Drosophila*. Second, the values of a and d are likely to be context-dependent, and may vary if the environment changes (Figure 5.10). If populations experience different environments, the means could be different, even if the allele frequencies of loci affecting the trait are the same. For example, similar allele frequencies at loci affecting behaviors in different populations could still result in different mean values of the behavior if one population experiences a benign environment while the other is physically and socially stressed.

We can similarly derive the mean of a population with inbreeding (Table 5.3). The difference between the mean of the random mating, outbred ($\mu_G O$) and inbred ($\mu_G I$) populations is $\mu_G I = \mu_G O - 2dpqF$, or $\mu_G O - 2F\Sigma pqd$ over all loci. Thus, there will be a change of mean value on inbreeding only if $d \neq 0$. For a single locus, if $d > 0$, inbreeding will reduce the mean value of the trait. This will occur if the A_1 allele is partially or completely dominant, or if there is **overdominance** (Figure 5.11). If $d < 0$, inbreeding will increase the mean. This will occur if the A_2 allele is partially or completely dominant, or if there is **underdominance**. Box 5.2 illustrates the effect of inbreeding on the population mean at a locus affecting the startle response

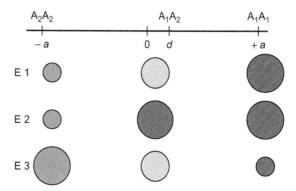

FIGURE 5.10 Homozygous and heterozygous effects can change in different environments. Homozygous and heterozygous effects at a locus affecting a behavioral trait are depicted in three different environments (E1, E2, E3). The size of the circles is proportional to the effect on the trait. In E1, the A_1A_1 genotype has the greatest effect on the trait, and the A_1 allele is partially dominant. In E2, the A_1A_1 genotype also has the largest effect on the trait, but here the A_1 allele is completely dominant. In E3, the A_2A_2 genotype has the largest effect on the trait, and the A_2 allele is partially dominant.

in *Drosophila*. With multiple loci, a decrease in the mean under inbreeding (**inbreeding depression**) requires **directional dominance**, with the sum of the dominance effects at all loci tending to be positive. Inbreeding depression is common for behavioral traits (Figure 5.12). If loci combine additively (i.e. there is no epistasis), the change in mean should be directly proportional to F, the inbreeding coefficient. When $F = 1$, the mean of an inbred line $= \Sigma a$ over all loci (ignoring epistasis).

TABLE 5.3 Calculation of the population mean of a quantitative trait in a population of inbred lines

A$_2$A$_2$			A$_1$A$_2$		A$_1$A$_1$
$-a$			0	d	$+a$

Genotype	Frequency	Effect	Frequency × effect
A$_1$A$_1$	$p^2 + pqF$	$+a$	$p^2a + apqF$
A$_1$A$_2$	$2pq - 2pqF$	d	$2pqd - 2pqdF$
A$_2$A$_2$	$q^2 + pqF$	$-a$	$-q^2a - apqF$

Mean of inbred (average of all lines) $=$
$$\mu_G I = \Sigma \text{ Frequency} \times \text{Effect} = p^2a - q^2a + 2pqd - 2pqdF$$
$$= a(p + q)(p - q) + 2pqd(1 - F)$$

Mean of outbred base $= \mu_G 0 = a(p - q) + 2pqd$
$$\mu_G I = \mu_G 0 - 2pqdF$$
$$\mu_G I = \mu_G 0 - 2F\Sigma pqd \text{ over all loci}$$

Average Effect, Breeding Value, and Dominance Deviation

The population mean does not describe the breeding properties of the population, because alleles and not entire genotypes are transmitted from generation to generation. We therefore have to derive a measure of the effect of an individual allele in terms of its effect on the phenotype, and then express the effect of each genotype in terms of the sum of the effects of the alleles it carries. We call the effect of an allele the **average effect** of the allele, and the effect of each genotype expressed in terms of the sums of the average effects of alleles the **breeding value** of a genotype. These are two key concepts in quantitative genetics, and also two of the most difficult concepts to understand.

We define the average effect of an allele as "the mean value of all progeny that received that allele from a parent, the other allele having come at random from the population, expressed as a deviation from the overall population mean." Let us work through this definition step-by-step to see how the average effect of an allele depends on the homozygous and heterozygous effects of genotypes, and allele frequencies. Imagine a pool of gametes from a random mating population. The gametes are either A$_1$ or A$_2$, and they occur with frequency p and q, respectively. Now we take an A$_1$ gamete from the pool and ask what genotypes that gamete can make. It can combine with another A$_1$ gamete and produce an A$_1$A$_1$ genotype with effect a,

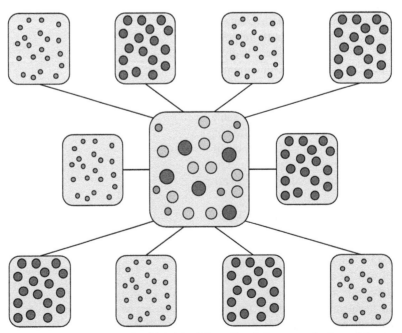

FIGURE 5.11 Schematic illustration of the derivation of multiple inbred lines ($F = 1$) from an outbred population in which the frequency of two alleles affecting a behavioral trait are $p = q = 0.5$ (center population). The shaded circles represent the three genotypes, as in Figures 5.9 and 5.10, with the size of the circles proportional to the size of the effect on the trait. With complete inbreeding, the inbred populations are either fixed for allele A$_1$ ($p = 1$, $q = 0$) or allele A$_2$ ($p = 0$, $q = 1$). The proportion of inbred populations fixed for A$_1$ and A$_2$ is equal to the allele frequencies in the outbred population, here $p = q = 0.5$. The mean of the outbred population is $0.5d$, and the mean of the population of inbred lines is 0. Thus, this is an example of inbreeding depression.

Box 5.2   Population mean with inbreeding

Consider again the *Drosophila neur* locus from Chapter 4 and its effect on startle-induced locomotion behavior. The genotypic effects of the *neur⁺*, *neur/neur⁺* and *neur* genotypes are, respectively 8.335, 3.815, and −8.335. We can use this example to illustrate the change of mean of a population of inbred lines compared to the outbred population in HWE from which they were derived. The mean of the population of inbred lines, expressed relative to 0, the average of the two homozygous effects, is $\mu_G = a(p - q) + 2pqd(1 - F)$. In a population of lines derived by one generation of full sibling mating ($F = 0.25$), where the frequency of the *neur⁺* allele (p) is 0.8, the population mean is $\mu_G = 8.335(0.8 - 0.2) + 2(0.8)(0.2)(3.815)$ $(1 - 0.25) = 5.9166$. In a similar population of inbred lines where the frequency of the *neur⁺* allele (p) is 0.2, the population mean is $\mu_G = 8.335(0.2 - 0.8) + 2(0.2)(0.8)$ $(3.815)(1 - 0.25) = -4.0854$.

In both cases, the mean of the population of inbred lines is less than the outbred population from which they were derived, that is, the *neur* locus shows inbreeding depression:

	$p = 0.8$	$p = 0.2$
Mean of outbred population	6.2218	−3.7802
Mean of inbred population ($F = 0.25$)	5.9166	−4.0854
Inbreeding depression	0.3052	0.3052

The inbreeding depression is due to the loss of $2pqF$ heterozygotes and the gain of pqF homozygotes, and the partial dominance of the *neur⁺* allele (i.e. $d > 0$).

or it can combine with an A_2 gamete and produce an A_1A_2 genotype with effect d. The probability of the first event is p and of the second is q. Thus, the mean value of all progeny produced by combining A_1 alleles at random with other alleles in the population is obtained by summing the products of frequency and effect for the two offspring genotypes, or $pa + qd$. To obtain the average effect of allele A_1, we need to subtract the population mean from this expression: $pa + qd - [a(p - q) + 2pqd]$. After some algebra, this expression simplifies to $q[a + d(q - p)]$, which is the average effect of allele A_1, symbolized α_1. Following similar logic, we can deduce the average effect of allele A_2 as $\alpha_2 = -p[a + d(q - p)]$ (Table 5.4).

As for the population mean, the average effect of an allele is not an inherent property of the allele, but depends on the particular population under consideration. Like the overall mean, the average effect of an allele depends on allele frequency. By definition, the average effect is the effect of an allele in combination with a random sample of alleles from the population. Further, the homozygous effects (a) and heterozygous effects (d) can also change, according to the particular constellation of environments to which individuals of the population are exposed.

We now define the breeding value of a genotype as the *sum of the average effects of the genes it carries*. Thus, the breeding values of individuals with genotypes A_1A_1, A_1A_2, and A_2A_2 are $2\alpha_1$, $\alpha_1 + \alpha_2$, and $2\alpha_2$, respectively. Breeding values are expressed as deviations from the population mean; therefore, we expect the breeding values of some genotypes to be positive, and some to be negative. Because the breeding value of the heterozygote is always intermediate between the breeding values of the two homozygotes, breeding value is also called the **additive genetic value**, and symbolized A.

Breeding value is a central concept in quantitative genetics, because it can be measured directly, and thus gives us a way to define the value of an individual in the absence of knowledge of the genetic details underlying the inheritance of the trait. We define the breeding value of an individual empirically in terms of the mean phenotype

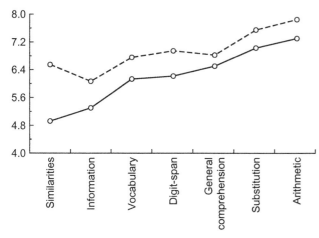

FIGURE 5.12 An example of inbreeding depression on Japanese children. The solid line shows the performance of children of first-cousin marriages on a Japanese version of the Weschler intelligence test (WISC; see also Chapter 14). The dotted line shows the performance of unrelated matched controls. Inbreeding has a deleterious effect on IQ as measured by this test. (From Schull and Neel (1965). *Science*, **150**, 332–333.)

TABLE 5.4 Calculation of the average effect of an allele

Type of gamete	Effects and frequencies of genotypes produced			Mean effect of genotypes produced	Average effect, α_i of allele
	A_1A_1	A_1A_2	A_2A_2		
	a	d	$-a$		
A_1	p	q		$pa + qd$	$\alpha_1 = q[a + d(q - p)]$
A_2		p	q	$-qa + qd$	$\alpha_2 = -p[a + d(q - p)]$

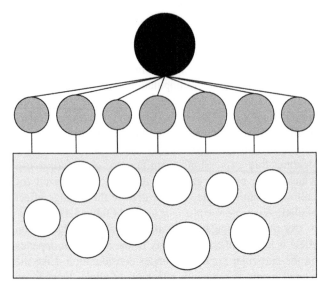

FIGURE 5.13 A schematic representation of breeding value. The black individual is mated at random to a large number of individuals (gray circles) from the same population, and the trait of interest is measured on the progeny (white circles) of these matings. The size of the circles reflects the quantitative value of the trait. The black individual has a higher phenotype than most individuals in the population, and the offspring of this individual consequently have higher values of the phenotype than average, indicating a genetic basis to the trait. We can calculate the breeding value of the black individual by determining the mean value of the phenotype of the offspring, and multiplying by 2. A real example would be to determine the breeding value of a stallion by judging the performance of his offspring out of different mares (see also Figure 5.14).

FIGURE 5.14 The legendary stallion Pion, one of the foundation sires of the Dutch Warmblood Studbook Registry. Performance testing of progeny is common in horse breeding industries in Europe designed to select only stallions with the best breeding values for breeding. Pion sired numerous Grand Prix dressage champions, as well as jumpers, before his death in January 29, 2003 at the age of 29.

of the offspring of that individual. A defined in this way can be determined experimentally. More specifically, we assume that the expected mean environmental deviation of the progeny is zero. The breeding value of individual i (A_i) is then twice the difference in the mean phenotype of its progeny from the population mean (twice because the parent only provides half the genes in the progeny, the other half coming at random from the population) (Figures 5.13, 5.14). A represents the part of the genotypic value that is transmitted to progeny. Breeding value is thus the value of the trait transmitted from parents to offspring.

Previously, we defined the genotypic effects (homozygous and heterozygous effects) of individual genotypes. Now that we are considering populations of individuals,

we need to define genotypic values of genotypes as the genotypic effects expressed as deviations from the population mean. We symbolize these genotypic values as G (Table 5.5). Note that the breeding values and the genotypic values are identical if there is no dominance ($d = 0$). The breeding value of heterozygotes is always exactly intermediate between the homozygote breeding values, but this is not true for the genotypic values, which can exhibit partial dominance, complete dominance of A_1 or A_2, and over- or under-dominance ($d \neq 0$). The difference between the genotypic values and the breeding values of a genotype is thus due to dominance, and we can split the genotypic values into two components: the breeding value (A); and the *dominance deviation* (D). Thus, $G = A + D$, and $D = G - A$ (Table 5.5). D represents the effect of putting genes together in pairs to make genotypes. The relationship between genotypic values, breeding values, and dominance deviations is illustrated in Box 5.3 for a locus affecting the startle response in *Drosophila*.

TABLE 5.5 The relationship between genotypic values (G), breeding values (A) and dominance deviations (D) of genotypes at a locus with two alleles, expressed as deviations from the population mean ($\mu_G = a(p - q) + 2dpq$)

Genotype	A_1A_1	A_1A_2	A_2A_2
G	$2q(a - pd)$	$a(q - p) + d(1 - 2pq)$	$-2p(a + qd)$
A	$2q[a + d(q - p)]$	$(q - p)[a + d(q - p)]$	$-2p[a + d(q - p)]$
D	$-2q^2d$	$2pqd$	$-2p^2d$

Box 5.3   Relationship between genotypic values, breeding values, and dominance deviations

Continuing with the example of the *Drosophila neur* locus and its effect on startle-induced locomotion behavior from Chapter 4, we can show the relationship between genotypic values (*G*), breeding values (*A*), and dominance deviations (*D*) in an outbred population in HWE, and how the values depend on allele frequency. Consider two populations, in one of which the frequency of the *neur+* allele (*p*) is 0.8, and in the other of which the frequency of *neur+* is 0.2. Recall that the genotypic effects of the *neur+*, *neur/neur+*, and *neur* genotypes are, respectively 8.335, 3.815, and −8.335. Following the definitions in the text, we obtain:

			$p = 0.8$	$p = 0.2$
Population mean	$\mu_G = a(p - q) + 2pqd$		6.2218	−3.7802
Genotypic effects (*G*)	*neur+/neur+*	$a - \mu_G$	2.1132	12.1152
	neur+/neur	$d - \mu_G$	−2.4068	7.5952
	neur/neur	$-a - \mu_G$	−14.5568	−4.5548
Average effects	α_1 (*neur+*)	$q[a + d(q - p)]$	1.2092	8.4992
	α_2 (*neur*)	$-p\,[a + d(q - p)]$	−4.8368	−2.1248
Breeding values (*A*)	*neur+/neur+*	$2\alpha_1$	2.4184	16.9984
	neur+/neur	$\alpha_1 + \alpha_2$	−3.6276	6.3744
	neur/neur	$2\alpha_2$	−9.6736	−4.2496
Dominance deviations (*D*)	*neur+/neur+*	$G - A$	−0.3052	−4.8832
	neur+/neur	$G - A$	1.2208	1.2208
	neur/neur	$G - A$	−4.8832	−0.3052

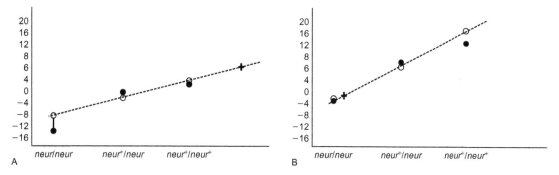

FIGURE A   Relationship between genotypic effects (solid circles), breeding values (open circles) and dominance deviations (vertical lines) on startle response for genotypes at the *neur* locus in populations in which the frequency of *neur+* is (A) 0.8 and (B) 0.2. The cross indicates the population mean.

Note that genotypic values, breeding values, and dominance deviations are different in populations with different allele frequencies. This is because we have defined these values as deviations from the population mean, and the population mean depends on allele frequency. The relationship between *G*, *A*, and *D* can be more easily visualized by plotting graphs of these values for 0, 1, and 2 *neur+* alleles. The breeding values and population mean fall on a straight line, as expected given the definition of breeding value. However, the breeding value of the *neur/neur* homozygote is better than its genotypic value, and much more so in the population in which the frequency of the *neur+* allele is higher. This is because the *neur* allele can combine with either a *neur* or *neur+* allele, and in the latter case will generate a *neur+/neur* heterozygote, which has a greater startle response than the *neur/neur* homozygote. This occurs more frequently in the population in which the frequency of *neur+* is 0.8 than the population in which the frequency of *neur+* is 0.2.

The breeding value of the *neur+/neur* heterozygote is worse than its genotypic value. This is because the *neur* allele can combine with another *neur* allele to generate a *neur/neur* homozygote, which has a lower startle response than the *neur+/neur* heterozygote.

Finally, the *neur+* homozygous genotype has a breeding value that is greater than its genotypic value. This is because the progeny of this genotype are either *neur+* homozygotes or *neur+/neur* heterozygotes. The mean of these genotypes will always be greater than the population mean, which includes the *neur* homozygotes and their low locomotor startle response.

With multiple loci, we also have to consider that non-additive interactions between loci can affect an individual's genotypic value. We can represent all possible interactions by adding an interaction deviation, I, to the decomposition of the genotypic value: $G = A + D + I$. If $I = 0$, genes are said to act additively between loci. As noted in Chapter 4, "additive" can be used in two different contexts. Referring to one locus, additive means $D = 0$ (no dominance); referring to many loci additive means $I = 0$ (no epistasis).

SUMMARY

In a large random mating outbred population, in which none of the evolutionary forces of mutation, migration, and natural selection occur, genotype frequencies in offspring can be predicted from allele frequencies in parents. The genotype frequencies in offspring are $A_1A_1 = p^2$; $A_1A_2 = 2pq$; $A_2A_2 = q^2$; $p^2 + 2pq + q^2 = 1$, and are known as Hardy–Weinberg proportions. These proportions are altered in a population that experiences inbreeding, i.e. mating between related individuals. The inbreeding coefficient F indicates the probability that two alleles at a locus are identical by descent, i.e. are copies of the same allele from a common ancestor. With inbreeding, genotype frequencies are modified such that inbred populations have more homozygotes and fewer heterozygotes than an outbred population.

Phenotypic values of most behavioral traits are continuously distributed in natural populations, often approximating a statistical normal distribution. The distinction between genotypic classes decreases as the number of loci increases, and hence the effect of each becomes smaller. In addition, sensitivity to environmental variation further blurs the distinction between genotypes. The population mean of a trait can be estimated from the effects and frequencies of alleles that contribute to the trait. The mean of a population of inbred lines can be less than the outbred population from which they were derived, that is, exhibit inbreeding depression, if there is net directional dominance at the loci affecting the trait. The average effect of an allele is defined as the mean value of all progeny that received that allele from a parent, the other allele having come at random from the population, expressed as a deviation from the overall population mean. The value of each genotype can then be expressed in terms of the sums of the average effects of alleles, and is known as the breeding value. The breeding value can be empirically estimated as the value of the trait transmitted from parents to offspring. The breeding values and genotypic values at a locus affecting a complex behavioral trait are identical if there is no dominance, but will be different in the presence of dominance. This difference is called the dominance deviation.

STUDY QUESTIONS

1. What is the Hardy–Weinberg equilibrium? Why is it important?
2. What is meant by the concept "identity by descent?"
3. What is inbreeding? Explain how inbreeding affects the proportions of homozygotes and heterozygotes in a population of inbred lines.
4. In Wagner's *Ring Cycle* the God Wotan has fathered the fraternal twins Sieglinde and Siegmund, who subsequently fall in love with each other and give rise to Siegfried. Siegfried rescues Brünhilde, another daughter of Wotan from a different mother, by rousing her from a deep sleep amid a ring of fire, to which Wotan had condemned her for disobeying him. If Siegfried and Brünhilde had a son what would be his inbreeding coefficient, assuming Wotan himself is not inbred?
5. How does continuous variation for a quantitative trait arise?
6. What determines the mean of a behavioral trait in an outbred population?
7. What determines the mean of a behavioral trait in a population of inbred lines?
8. What is inbreeding depression? Why does inbreeding depression occur?
9. What is the "average effect" of an allele?
10. What is meant by "breeding value"? How can the breeding value of an individual be measured?
11. What is the difference between homozygous and heterozygous effects and genotypic values?
12. What is meant by "dominance deviation?"
13. What is the relationship between genotypic value, breeding value, and dominance deviation?

RECOMMENDED READING

Falconer, D. S., and Mackay, T. F. C. (1996). *Introduction to Quantitative Genetics*, 4th edition. Longmans Green, Harlow, Essex, UK.

Lynch, M., and Walsh, B. (1997). *Genetics and Analysis of Quantitative Traits*. Sinauer Assoc., Inc., Sunderland, MA.

Chapter 6

Partitioning Phenotypic Variance and Heritability

OVERVIEW

Because both genetic factors and environmental factors contribute to the manifestation of behavioral phenotypes, it is important first to identify the sources of phenotypic variation and second to quantify their relative contributions to the phenotype, if we are to understand the genetic architecture of complex traits. As described in Chapter 5, genetic factors include breeding values, dominance deviations, and interactions due to epistasis. The environmental contribution to phenotypic variation can be dissected similarly into common spatial and temporal environmental factors, intrauterine maternal effects (in mammals), and residual environmental factors of unknown origin. The genetic and environmental components that contribute to variation in a behavioral phenotype can be expressed as variance components. This enables us to estimate them statistically by using analysis of variance. Of particular interest is that fraction of the phenotypic variance that is attributable to variation in genetic factors. In this chapter, we will describe how phenotypic variation can be partitioned into genetic ("nature") and environmental ("nurture") components, and how the genetic contribution to the phenotypic variance (the "heritability") should be interpreted.

We have seen in Chapter 5 that the behavioral phenotype (or indeed any complex trait phenotype) of an individual has three major components: the individual's unique genotype at all loci affecting the trait (nature, G); the particular external and social environment experienced by the individual (nurture, E); and any interaction or correlation between alleles affecting the trait and the environment (the interaction between nature and nurture, GE), $P = G + E + GE$. We further showed that the genotypic effect can be split into the breeding value, A, or additive effect of each genotype that is transmitted from parent to offspring, and the dominance (D) and interaction (I) deviations. The two latter terms reflect any non-additive effects of alleles at a single locus (D), and from combining multiple loci affecting the behavior (I). Non-additive effects are not transmitted from parent to offspring, because genotypes segregate at meiosis. Thus $P = A + D + I + E + GE$ (see Chapter 5).

All we can observe is the individual's phenotype. How can we determine the relative contribution of additive and non-additive genetic effects, the environment and any genotype-environment interaction to the behavior? The answer is that this is not possible for any single individual. However, it is possible to do this if we have a population of individuals in which we have replicate measures of the behavioral phenotype for genotypes at any locus affecting the trait. We will explain how this is achieved in the remainder of this chapter.

COMPONENTS OF VARIANCE IN RANDOM MATING POPULATIONS

We can dissect genetic and environmental contributions to the manifestation of behavioral phenotypes by screening families of related individuals. Relatives share alleles that are **identical by descent** (see also Chapter 5). Analyzing the phenotypes of related individuals can tell us the average proportionate contribution of genes and environment to individual behavior, without knowing anything about the actual underlying loci. Because we need to assess the behavior at the level of the population, we need to consider the effects of genotypes and environments on the variance of the behavior among individuals in the population. The population phenotypic variance $\sigma_P^2 = \sum \frac{(X_i - \mu)^2}{N}$ is equal to the mean of the squared deviations from the mean. That is, not everyone in the population has the exact same behavioral phenotype. The numerical value of σ_P^2 indicates how much the phenotypes are dispersed about the mean. If the variance is small, individuals have very similar behaviors; if the variance is large, the behaviors vary greatly among individuals (Figure 6.1). In the same way that an individual's phenotype can be split into the effects of genes, environments, and the gene–environment interaction, the phenotypic variance (variance of phenotypic values) of

R. Anholt and T. Mackay: Principles of Behavioral Genetics
ISBN: 978-0-12-372575-2

the population can be partitioned into similar corresponding components:

$$\sigma_P^2 = \sigma_G^2 + \sigma_E^2 + 2cov_{GE} + \sigma_{GE}^2$$

$$\sigma_P^2 = \sigma_A^2 + \sigma_D^2 + \sigma_I^2 + \sigma_E^2 + 2cov_{GE} + \sigma_{GE}^2$$

where:

σ_P^2 = phenotypic variance, variance of phenotypic values

σ_G^2 = genotypic variance, variance of genotypic values

σ_A^2 = additive genetic variance, variance of breeding values

σ_D^2 = dominance variance, variance of dominance deviations

σ_I^2 = interaction variance, variance of interaction deviations

σ_E^2 = environmental variance, variance of environmental deviations

cov_{GE} arises if there is a correlation between genotypic values and environmental deviations

σ_{GE}^2 refers to any interaction between genotype and environment

Partitioning the variance in this way now enables us to examine systematically the contributions of each of the components separately.

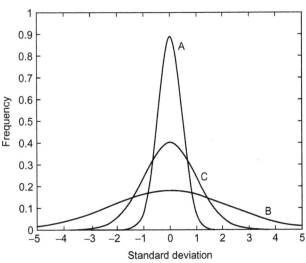

FIGURE 6.1 Three populations with different variances. The variance is smallest in population A, greatest in population B, and intermediate in population C. The x-axis is in standard deviation units (the square root of the variance) of the trait, because the standard deviation is expressed in the same units of measurement as the trait, while the variance is expressed as (measurement)^2.

Genetic Components of Variance

In Chapter 5 we showed that we can express the genotypic values, breeding values and dominance deviations for the three genotypes at a single locus as deviations from the population mean (Table 5.5). The population *variance* for

each of these terms is thus obtained by squaring the deviations from the population mean for each genotype, multiplying by the genotype frequency, and summing over all genotypes. Doing this, as shown in Table 6.1, we obtain algebraic expressions for the additive genetic variance, dominance variance, and total genetic variance for a single locus affecting a behavioral trait:

$$\sigma_A^2 = 2pq[a + d(q - p)]^2$$

$$\sigma_D^2 = (2pqd)^2$$

$$\sigma_G^2 = \sigma_A^2 + \sigma_D^2$$

Additive genetic and dominance variance depend on allele frequency (Figures 6.2, p. 86 and 6.3, p. 87). They are thus maximal when allele frequencies are intermediate, and become smaller as allele frequencies approach 0 and 1. Recessive alleles at low frequency contribute little variance, because most copies of rare recessive genes are in heterozygotes. If one allele is fixed, then $\sigma_A^2 = \sigma_D^2 = 0$. In this case, all individuals have the same genotype, so there is no genetic variance. Note that even if there is complete dominance at a locus affecting the trait, the genotype still contributes additive genetic variance (Figure 6.2B). With additive gene action, $d = 0$ and $\sigma_A^2 = 2pqa^2$ and $\sigma_D^2 = 0$. Such populations are useful for gene mapping studies, as described in Chapter 8.

Because of the dependence on allele frequency, the additive and dominance variance components, σ_A^2 and σ_D^2, for the same trait will be different in populations with different allele frequencies. Box 6.1 illustrates how genetic variance in locomotor startle response due to segregation of alleles at the *neuralized* locus changes in populations with different allele frequencies. σ_A^2 and σ_D^2 for the same trait may also be different in different populations, even if they have the same allele frequencies, if a and d, which can depend on the environment, vary between the populations. A special situation arises when populations are formed by crossing two inbred lines. Such populations have a simpler genetic architecture than randomly-mating populations, since allele frequencies at all segregating loci in such populations are equal to 0.5 (p = q = 0.5 for all loci). In this special case, $\sigma_A^2 = \frac{1}{2}a^2$ and $\sigma_D^2 = \frac{1}{4}d^2$.

If there are epistatic interactions among loci, they will give rise to epistatic variance, σ_I^2. A full partitioning of σ_I^2 would require knowing the variance contributed by all two-way, three-way, and higher order interactions, and the nature of those interactions. However, it is commonly thought that three-way and higher order interactions contribute vanishingly small amounts of variance, and can be ignored.

Environmental Components of Variance

The environmental variance, σ_E^2, is, by definition, all non-genetic variation. Some obvious sources of variation in the environment that can give rise to differences in behavioral

---

TABLE 6.1 Genetic components of variance at a single locus

The mean genotypic effects, breeding values, and dominance deviations for the three genotypes at a single locus as deviations from the population mean are:

Genotype	A_1A_1	A_1A_2	A_2A_2
Frequency	p^2	$2pq$	q^2
G	$2q(a - pd)$	$a(q - p) + d(1 - 2pq)$	$-2p(a + qd)$
A	$2q[a + d(q - p)]$	$(q - p)[a + d(q - p)]$	$-2p[a + d(q - p)]$
D	$-2q^2d$	$2pqd$	$-2p^2d$

The variance of each term is computed by squaring the values in the table, multiplying by the genotype frequency, and summing over all genotypes.

$$\sigma_A^2 = 4p^2q^2[a + d(q - p)]^2 + 2pq(q - p)^2[a + d(q - p)]^2 + 4p^2q^2[a + d(q - p)]^2$$
$$= [4p^2q^2 + 2pq(q - p)^2 + 4p^2q^2][a + d(q - p)]^2$$
$$= 2pq[2pq + (q - p)^2 + 2pq][a + d(q - p)]^2$$
$$= 2pq[2pq + q^2 - 2pq + p^2 + 2pq][a + d(q - p)]^2$$
$$= 2pq[q^2 + p^2 + 2pq][a + d(q - p)]^2$$
$$= 2pq[a + d(q - p)]^2$$

$$\sigma_D^2 = 4p^2q^4d^2 + 8p^3q^3d^2 + 4p^4q^2d^2$$
$$= 4p^2q^2d^2(q^2 + 2pq + p^2)$$
$$= (2pqd)^2$$

$$\sigma_G^2 = \sigma_A^2 + \sigma_D^2$$

---

phenotypes between individuals are differences in nutrition and temperature, and differences in social context. In mammals, differences in prenatal and postnatal maternal nutritional influences (maternal effects) can give rise to phenotypic differences between individuals, especially infants. Finally, some phenotypes, particularly physiological and behavioral phenotypes, can be difficult to measure with precision. In humans, psychological traits are assessed by a series of questions, where the answer to one question often determines the next question to be answered, so all individuals being evaluated do not even receive the same test! Blood pressure in humans can vary from minute to minute, depending on stress level and posture. Many behavioral and physiological traits also vary according to the circadian clock, making the time of day at which the measurement is taken a critical factor. Finally, most traits are affected by "intangible" changes in the environment. For example, if we measure the aggressive behavior of fruit flies of the same sex and genotype, all housed together in the same culture vials under conditions of constant temperature and humidity, there will still be substantial variation in aggressive behavior among the individuals (Figure 6.4). The environmental variation stemming from unknown causes is called **intangible variation**, and is symbolized as σ_{EW}^2.

In addition, we need to take account of environmental variation due to rearing individuals in the same, or common, environment. **Common environmental variance** (σ_{EC}^2) is defined as the variance attributable to environmental conditions that cause phenotypes of individuals reared in that environment to be more similar to each other than to individuals reared in different common environments (Figure 6.5, p. 88). The concept can be clarified by considering the main sources of common environment.

Examples of common *spatial environments* are fruit flies that are reared in the same vial, mice that are housed in the same cage, or humans in the same household. There will be differences in the environment between vials, cages, and households, and these differences in environments can affect quantitative traits, such as behaviors. Such differences give rise to common spatial environmental variance ($\sigma_{EC(S)}^2$).

Examples of common temporal environments are measurements taken for one group of individuals at one time of

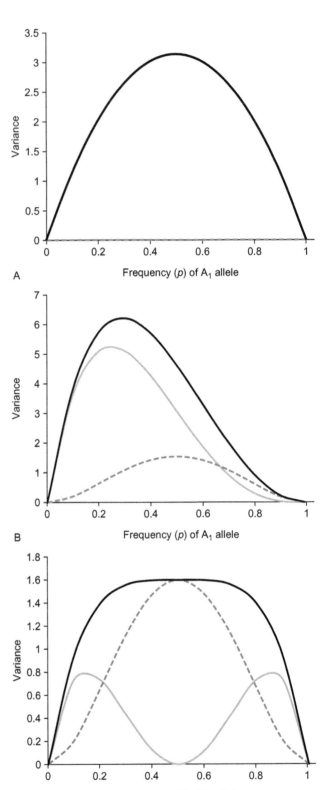

Box 6.1 Components of genetic variance at a single locus affecting a behavioral trait

Let us continue with the example of the *Drosophila neur* locus, and its effect on startle-induced locomotion behavior described in the preceding chapters. We can derive the additive genetic variance (σ_A^2), dominance variance (σ_D^2), and total genetic variance (σ_G^2) due to segregation at this locus, and show that these depend on allele frequency. Consider two populations, in one of which the frequency of the *neur*$^+$ allele (p) is 0.8, and in the other of which the frequency of *neur*$^+$ is 0.2. Recall that the genotypic effects of the *neur*$^+$, *neur/neur*$^+$, and *neur* genotypes are, respectively, 8.335, 3.815, and −8.335. Following the definitions in the text, we obtain:

		$p = 0.8$	$p = 0.2$
Additive genetic variance (σ_A^2)	$2pq[a + d(q - p)]^2$	11.6973	36.1182
Dominance variance (σ_D^2)	$(2pqd)^2$	1.4904	1.4904
Total genetic variance (σ_G^2)	$\sigma_A^2 + \sigma_D^2$	13.1877	37.6086

day, week, or year, and another group with measurements taken at a different time of day, week, or year. Differences in circadian cycle, temperature or season between these different measurement times can affect quantitative traits, and give rise to common temporal environmental variance ($\sigma_{EC(M)}^2$) in behavior.

Finally, mammalian siblings share a common **maternal environment**. Differences between mothers in pre- and postnatal nutrition can cause differences between quantitative traits of their offspring, and thus give rise to common maternal environmental variance ($\sigma_{EC(T)}^2$). In mammals, common intra-uterine maternal effects can be estimated directly by embryo transplants, and common postnatal maternal effects can be estimated by cross-fostering experiments. These effects can be quite striking, as illustrated in Figure 6.6 (p. 88). A classical experiment performed in 1966 compared the behavior of C57BL/6 and BALB/c mouse pups that were either reared by mothers of their own strain or by mothers of the other strain, in a number of tests of activity (Figure 6.6). The tests were open field activity (see also Chapter 11), a hole-in-the wall emergence test, where the time is measured at which mice would escape through an opening in an enclosure, a water-escape test, and a barrier climbing test. There were differences between strains in all tests except for the barrier climbing task.

C57BL/6 mice were more active in the open field, took longer to emerge from the hole in the wall, and required longer to escape from the water than BALB/c mice, regardless

FIGURE 6.2 Genetic variance at a single locus affecting a behavioral trait, as a function of allele frequency. The solid dark gray curve depicts the additive genetic variance; the dotted dark gray curve the dominance variance, and the solid black curve the total genetic variance. (A) This panel shows additivity ($a = 2.5, d = 0$). The total genetic variance is equal to the additive genetic variance, because the dominance variance is zero. (B) This panel shows complete dominance ($a = 2.5, d = 2.5$). (C) This panel illustrates overdominance ($a = 0, d = 2.5$).

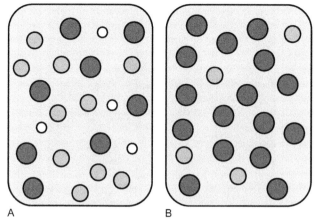

A   B

FIGURE 6.3 Schematic representation of the dependence of genetic variance on allele frequency. The circles represent three genotypes at a locus affecting a behavioral trait, with dark gray, light gray and white indicating the A_1A_1, A_1A_2, and A_2A_2 genotypes, respectively. The size of the circles is proportional to the homozygous effect ($a = 2.5$) and heterozygous effect ($d = 2.0$). (A) In this population the frequency of the A_1 allele $= p = 0.6$, and the frequency of the A_2 allele $= q = 0.4$. The additive genetic variance, $\sigma_A^2 = 2.1168$; the dominance variance, $\sigma_D^2 = 0.9216$ and the total genetic variance, $\sigma_G^2 = 3.0384$. (B) In this population the frequency of the A_1 allele $= p = 0.9$, and the frequency of the A_2 allele $= q = 0.1$. The additive genetic variance, $\sigma_A^2 = 0.1458$, the dominance variance, $\sigma_D^2 = 0.1296$, and the total genetic variance, $\sigma_G^2 = 0.2754$. Even though the homozygous and heterozygous effects are the same in both populations, the population in A is 11 × more genetically variable than the population in (B), because allele frequencies are intermediate in population (A) and more extreme in population (B).

of mode of rearing. BALB/c mice showed significant effects of cross-fostering in the open field and hole-in-the-wall tests, in the direction of the C57BL/6 performance. Conversely, C57BL/6 mice showed significant effects of cross-fostering in the water escape test, in the direction of the BALB/c performance. These cross-fostering experiments demonstrated that the environmental effect on these behavioral traits was significant and could assess the relative magnitude of the different rearing environments on these behaviors.

GENOTYPE–ENVIRONMENT CORRELATION AND INTERACTION

A correlation between genotype and environment occurs if genotypic values are not independent of environmental deviations. Such a correlation could occur if "better" phenotypes are given "better" environments. If we consider the correlation between two parameters X and Y, statistical theory defines the correlation between these parameters as $\sigma^2(X + Y) = \sigma_X^2 + \sigma_Y^2 + 2cov_{XY}$, where cov_{XY} denotes the covariance. We can apply this to genotype–environment correlation, as follows: $\sigma_P^2 = \sigma_G^2 + \sigma_E^2 + 2cov_{GE}$. cov_{GE} can occur in natural settings. A good example is provided by race horses (Figure 6.7). The average speed of race horses has not increased significantly during the last century, despite the high premium placed on stakes winners.

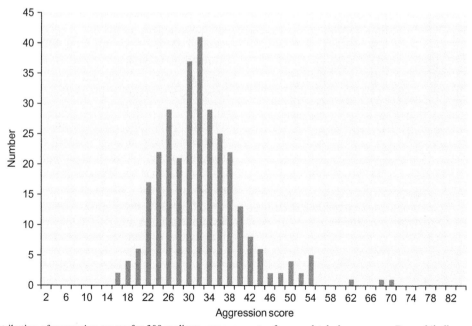

FIGURE 6.4 Distribution of aggression scores for 300 replicate measurements of a completely homozygous *Drosophila* line, reared in a constant environment. For each replicate, aggression was measured as the number of aggressive encounters (chase, wing threat, kicking, boxing) between eight flies in a two minute period. Since there is no genetic variation in a completely homozygous line, the wide range of scores is entirely due to small differences in micro-environment experienced by the flies, as every effort was made to house them under identical conditions. (Data are from Dr Charlene Couch.)

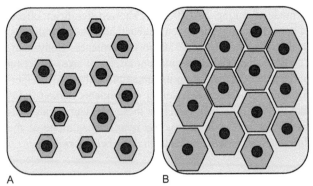

FIGURE 6.5 Schematic representation of the effects of a common environment. The circles represent the genotypic value of individuals. Since all circles are identical, this means that all individuals have the same genotype, as would be the case if they were members of a fully-inbred line. The hexagons represent the effects of the environment. Panels A and B represent the phenotypes of individuals that were reared in two different environments. There is variation in the magnitude of the environmental effect on the phenotype within each of the two environments. However, there is a much greater difference between environments, with the conditions of environment B causing a large increase in the value of the phenotype relative to environment A. The two environments could represent different cages, times of day, year of measurement, or different maternal environments.

FIGURE 6.7 Despite the premium placed on pedigree, the average speed of race horses has not increased during the last century; "nurture" rather than "nature" makes a stakes winner.

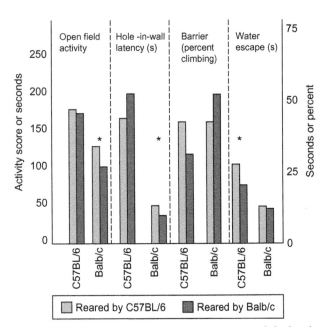

FIGURE 6.6 Effect of maternal environment on mouse behaviors in mice reared by their own dams or dams from a different strain. (Data are from Reading (1966). *J. Comp Psychol.*, *62*, 437–330).

Elite young racehorses sell for huge amounts of money, based on their pedigree and the performance of their parents. They consequently go to top trainers and are ridden by top jockeys, i.e. they are given the best possible environment for top performance.

In experimental quantitative genetics, cov_{GE} is avoided because correct experimental design stipulates that genotypes are randomized across environments. If cov_{GE} exists

and we do not know about it, the practical consequence is that it is considered part of the genotype, and σ_G^2 is increased by $2cov_{GE}$.

We have already noted that independence of genotype and environment means that a specific difference in environment has the same effect on different genotypes. If this assumption is violated, there is an interaction between genotype and environment, and $\sigma_P^2 = \sigma_G^2 + \sigma_E^2 + \sigma_{GE}^2$. Genotype environment interaction (GEI) can arise if specific differences in the environment have greater effects on some genotypes than others (changes in variance), and/or if there is a change in order of merit of a series of genotypes measured under different environments (changes in rank; see also Chapter 4). We can estimate σ_{GE}^2 if the environmental differences are large (macroenvironments); otherwise, σ_{GE}^2 is considered part of the environmental variance. A classic example of GEI for the behavior of mouse strains in an open field arena, measured in different laboratories, is given in Figure 6.8.

PARTITIONING PHENOTYPIC VARIANCE

Now that we have described the genetic and environmental components of phenotypic variance separately, we can obtain a complete description of the "nature" and "nurture"

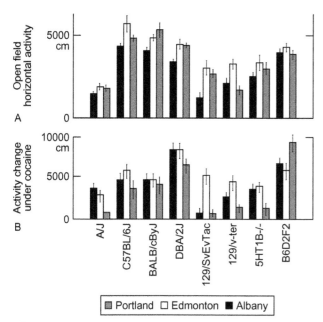

Portland □ Edmonton ■ Albany

FIGURE 6.8 Activity of eight different strains of mice in an open field arena, tested in three laboratories (Portland, Edmonton, and Albany). Mice were either untreated, or given 20 mg/kg cocaine. Despite considerable efforts of the investigators to keep conditions constant in the different environments, GEI was apparent. For example, C57BL/6J mice tested in Edmonton were more active than those tested in Albany or Portland; 129/SvEvTac mice tested in Albany were very inactive compared to their counterparts in other labs; and B6D2F2 mice were more responsive (and A/J mice insensitive) to cocaine in Portland, but not at other sites. (The figure is reproduced from Crabbe et al. (1999). *Science*, *284*, 1670–1672.)

components of behavioral traits. A full partitioning of phenotypic variance would be as follows:

$$
\begin{aligned}
\sigma_P^2 &= \sigma_G^2 &+ \sigma_E^2 \\
&= \sigma_A^2 + \sigma_D^2 + \sigma_I^2 &+ \sigma_{EC(S)}^2 + \sigma_{EC(T)}^2 + \sigma_{EC(M)}^2 \\
& \quad + 2cov_{GE} &+ \sigma_{EW}^2 + \sigma_{GE}^2 \\
&= \text{Nature} &+ \text{Nurture}
\end{aligned}
$$

The partitioning of phenotypic variance is usually expressed as the ratio of one component to the total phenotypic variance, (σ_P^2). For example, we can express the total genetic variance as a fraction of the total phenotypic variance. This is where the concept of **heritability** enters, which is explained in the next section.

HERITABILITY: THE CONCEPT

The term **heritability** is used in two contexts, known as **broad sense heritability** and **narrow sense heritability**. Broad sense heritability is indicated as H^2, and is the basic "nature-nurture" partition, which represents the degree of

Box 6.2 Heritability at a single locus affecting a behavioral trait

We now finish with the example of the *Drosophila neur* locus and its effect on startle-induced locomotion behavior by computing the broad and narrow sense heritabilities due to segregation at this locus in an outbred population, and showing how they depend on allele frequency. To do this, however, we first need an estimate of the environmental variation within each of the three genotypes – the intangible environmental variance (σ_{EW}^2). σ_{EW}^2 was determined experimentally to be 38 s^2. Using this information we can now compute the heritabilities due to this locus in two populations, in one of which the frequency of the $neur^+$ allele (p) is 0.8, and in the other of which the frequency of $neur^+$ is 0.2, as follows:

		$p = 0.8$	$p = 0.2$
Phenotypic variance (σ_P^2)	$\sigma_G^2 + \sigma_{EW}^2$	51.1877	75.6086
Narrow sense heritability (h^2)	σ_A^2/σ_P^2	0.2285	0.4777
Broad sense heritability (H^2)	σ_G^2/σ_P^2	0.2576	0.4974

genetic determination of a behavior. Thus, $H^2 = \sigma_G^2/\sigma_P^2$, or more completely:

$$
H^2 = \frac{\sigma_A^2 + \sigma_D^2 + \sigma_I^2 + 2cov_{GE}}{(\sigma_A^2 + \sigma_D^2 + \sigma_I^2 + 2cov_{GE} + \sigma_{EC(S)}^2 + \sigma_{EC(T)}^2 + \sigma_{EC(M)}^2 + \sigma_{EW}^2 + \sigma_{GE}^2)}
$$

H^2 indicates the extent to which variation in phenotypes in the population is due to variation in genotypic values of individuals. Because non-additive effects are included in this definition, H^2 describes a current population, and cannot be used to make inferences about the transmission of the behavior from parents to offspring. This is where the narrow sense heritability, designated h^2, comes in:

$$
h^2 = \frac{\sigma_A^2}{\sigma_P^2},
$$

or more completely:

$$
h^2 = \frac{\sigma_A^2}{\sigma_P^2} = \frac{\sigma_A^2}{(\sigma_A^2 + \sigma_D^2 + \sigma_I^2 + 2cov_{GE} + \sigma_{EC(S)}^2 + \sigma_{EC(T)}^2 + \sigma_{EC(M)}^2 + \sigma_{EW}^2 + \sigma_{GE}^2)}
$$

h^2 indicates the extent to which variation in phenotypes is determined by variation in effects of alleles transmitted by parents; i.e. the proportion of variation due to additive genetic variation (variation in breeding values).

Both H^2 and h^2 are ratios of a part to a whole; hence their maximum values are 1 and minimum values are 0. Heritability is a slippery concept, since it sounds like it refers to whether or not a behavior is inherited, which is not at all what it means. All behaviors are inherited, in the sense that individuals cannot exist in the absence of genes. Heritability refers to the fraction of the variance of the behavioral phenotype that is due to the variance of all genetic effects (broad sense heritability, H^2) or of individual breeding values (narrow sense heritability, h^2).

Narrow sense heritability is particularly important in quantitative genetics, because it is the proportion of the deviation of the parental phenotype from the population mean that is expected to be transmitted to the progeny, and is therefore predictive. That is, if an individual is 1 unit above average for phenotype, its breeding value is expected to be h^2 units above average. In other words, h^2 is the regression of breeding value on phenotypic value ($h^2 = b_{AP}$). The best predictor of breeding value (A) from phenotypic value is $A = h^2 (P - \mu_P)$, where μ_P is the mean population mean phenotype. Expressed in these terms, h^2 is thus the reliability of an individual's phenotypic value as an indicator of its breeding value. If h^2 is high, an individual above average in phenotype is likely to be above average genetically. In this case, phenotype is a reliable indicator of genotype. If h^2 is low, however, phenotypic differences among individuals are due more to environmental variation than to genetic differences. In this case, phenotypic merit indicates little about the true genetic merit.

To illustrate these concepts, consider the three hypothetical populations illustrated in Figure 6.9. In each scenario the total phenotypic variance is the same. However, in population A the breeding values and environmental contributions to the phenotypic value both vary among the individuals. The h^2 is the proportion of phenotypic differences among individuals in a population brought about by differences in their breeding values, and is approximately 0.4. Individuals with high breeding values also tend to have high phenotypic values, and individuals with low breeding values tend to have low phenotypic values. In population B, all phenotypic variance is caused by environmental variance, and all individuals have the same breeding values ($\sigma_A^2 = 0$). The heritability is zero; knowledge of the breeding value gives no information about the resulting value of the trait phenotype. Again, it is critical to understand that a low h^2 does not mean the character is not inherited; rather, it means that variation in breeding values contributes little to the variance in phenotypes (Figure 6.10). In population C, all phenotypic variance is caused by variance in breeding values. The environmental contribution to phenotypic variance is constant among the individuals ($\sigma_E^2 = 0$), and the heritability is 1. It is very important to note that a high h^2 does not mean the character is insensitive to the environment; rather, it means that environmental variance contributes little to the variance in phenotypes relative to the variance of breeding values (in the range of environments

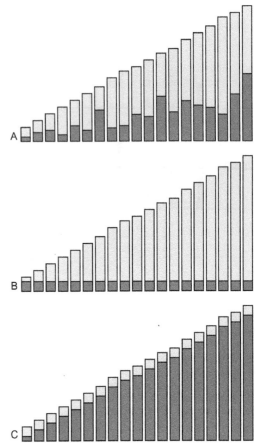

FIGURE 6.9 Three populations with different heritabilities for an additive trait. The individuals in each population are arranged from left to right in ascending order of their behavioral phenotype. The total phenotypic variance is the same in each population, as indicated by the height of the bars. The height of the dark gray bar denotes the breeding value of the individual, and the height of the light gray bar denotes the contribution the effects of the environment to individual phenotypic value. In this example there is no contribution of non-additive genetic variation (dominance and epistasis) to variation in the trait. (A) In this population individuals vary for both their breeding values and the effect of the environment. The heritability is approximately 0.40. (B) In this population all the individuals have the same breeding value, and all the phenotypic variance is due to variation in environmental effects. The heritability is zero. (C) In this population all the individuals have the same contribution to the phenotype from the environment, and all phenotypic variation is due to variation in breeding values. The heritability is one.

experienced by the population in which the measurements were taken).

Properties of h^2

It is important to emphasize that h^2 of a particular character has reference only to the population for which it was estimated, at that time in the population history, and in the environmental circumstances under which the individuals were reared. h^2 for a character can be different for different populations, because σ_A^2 depends on allele frequencies

FIGURE 6.10 There is little, if any, phenotypic variation for the blue eye color of Siamese cats; hence, heritability for this trait is essentially zero. Yet eye color is genetically, not environmentally, determined.

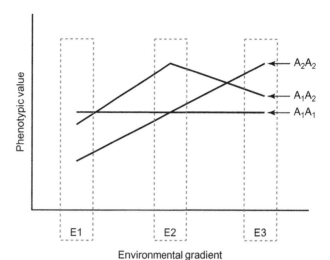

FIGURE 6.11 Norms of reaction for three genotypes at a locus affecting a behavioral trait, along an increasing environmental gradient (*x*-axis). The *y*-axis denotes mean phenotypic values. The A_1A_1 genotype is insensitive to the change of environment; the A_1A_2 genotype has the highest phenotypic value in the middle of the range of environments; and the A_2A_2 genotype shows a linear increase in phenotype across the environmental gradient. The extensive crossing of reaction norms indicates this trait shows extensive genotype by environment interaction due to this locus. Note that the genotypic effects differ in the three environments labeled E1, E2, and E3, and consequently the heritability of the trait will differ between these environments.

at each locus contributing to the trait, and these frequencies may be different in different populations. Box 6.2 shows how the heritability of startle-induced locomotion in *Drosophila* due to variation at the *neuralized* locus changes with different allele frequencies. Because h^2 is a function of allele frequency, it will be expected to change under conditions which operate to change allele frequency, such as inbreeding (see also Chapters 7 and 16).

h^2 for a character can also be different for different populations because h^2 depends on the range of environments encountered, because the denominator, σ^2_P, depends partly on the variance due to environmental differences among individuals. Increasing σ^2_E will decrease h^2, and keeping σ^2_E constant (as under defined laboratory conditions) will increase h^2. h^2 also depends on the environment in a more subtle manner, because the values of a and d at each locus determining the trait are also influenced by the environment. Consider the hypothetical norms of reaction for three genotypes shown in Figure 6.11. In the range of environments indicated by E1, A_1 has the largest phenotypic value and is nearly dominant to A_2; there is overdominance in the E2 environment; and in E3 the A_2 allele has the highest phenotypic value, and effects are close to additive. Because the values of a and d are different between these environments, estimates of heritability of the trait will be different, even though the total phenotypic variance is the same for all the populations.

It is important to realize that h^2 conveys no information about the nature of differences in mean phenotype between populations. It is a common misconception to assume that if the h^2 of a trait is high, and there is a difference in mean phenotype between populations, then the basis for the difference in mean must be genetic. The equally incorrect corollary is that if the h^2 of a trait is low, and there is a difference in mean phenotype between populations, then the basis for the difference in mean must be environmental. The

heritability of a trait within populations gives no information about the cause of differences in mean values of the trait between populations. Consider two populations in which the h^2 of a trait is high, but one population experiences a beneficial environment and the other a deleterious one (such as a mouse population on *ad lib* or restricted diets). In this case, the resulting difference in mean of the trait between the two populations is entirely environmental. Alternatively, consider two populations of highly-inbred lines, in each of which $h^2 = 0$. These populations may have different mean values of the trait, and if grown in the same environment this difference is entirely genetic.

CONTROLLING AND ESTIMATING ENVIRONMENTAL COMPONENTS OF VARIATION

How can we estimate the different genetic and environmental components of phenotypic variation? The only variance that can be easily obtained by observing values of the trait in a random mating population is σ^2_P, the variance of phenotypic values. To estimate genetic components of variance and heritability, we need measurements of the behavioral phenotype on related individuals, as discussed in Chapter 7. In organisms amenable to crossing and rearing under standardized conditions, some environmental components of variance can be eliminated by correct experimental design, or they can be estimated. For example, cov_{GE}, $\sigma^2_{EC(S)}$ and $\sigma^2_{EC(T)}$ can be eliminated if one can ensure genotypes are randomized with respect to environments. Alternatively, $\sigma^2_{EC(S)}$ and $\sigma^2_{EC(T)}$ can be estimated if replicate measurements are taken across different spatial and temporal environments (i.e. the same genotype is measured in at least two different vials or cages, or at two different time points).

The statistical technique of **analysis of variance** (ANOVA) can be used to partition environmental variation (Box 6.3). For example, if one rears a single inbred line of *Drosophila* in multiple vials, and measures a behavioral trait for several individuals from each vial, the ANOVA model $Y = \mu + V + E$ is used to analyze the data, where V refers to the variance between vials, and E to the variance within vials. The ANOVA results in estimates of the components of variance among vials and within vials. The variance among vials is due to differences in mean phenotype between vials, due to common spatial environment. This component is an estimate of $\sigma^2_{EC(S)}$. The variance within vials arises from "intangible" variation. This component thus estimates σ^2_{EW} (Figure 6.12A).

ANOVA models can also be used to account for temporal environmental variation. If one repeats the same experiment as above, but at two or more different times, the model $Y = \mu + T + V(T) + E$ is used to analyze the data, where T designates variance due to time of assay. The ANOVA gives estimates of the components of variance among times, among vials, and within vials. The variance among times is due to differences in mean phenotype between time periods, due to common temporal environment. This component thus estimates $\sigma^2_{EC(T)}$. As before, the variance among vials estimates $\sigma^2_{EC(S)}$, and the variance within vials estimates σ^2_{EW} (Figure 6.12B).

In Chapter 7, we will assume (unless otherwise specified) that variance due to genotype–environment correlation is eliminated by randomizing genotypes across environments, and that common spatial and temporal environmental variance is either eliminated by randomizing across environments, or accounted for by replication. However, pre- and postnatal maternal common environment will be assumed to be present, due to the difficulty of eliminating this component experimentally.

Box 6.3   Analysis of variance (ANOVA)

First developed by R. A. Fisher in 1918, analysis of variance (ANOVA) has become a central statistical tool in quantitative genetics. The method is an expansion of the standard Student's *t*-test, which assesses the statistical significance of differences between the mean values of two data sets relative to the variance within each. However, ANOVA allows multiple comparisons, and divides the observed variance among its contributing sources; that is, it quantifies to what extent variation attributable to each factor – for example genotype, age, sex, experimental treatment – contributes to the overall variation around the mean.

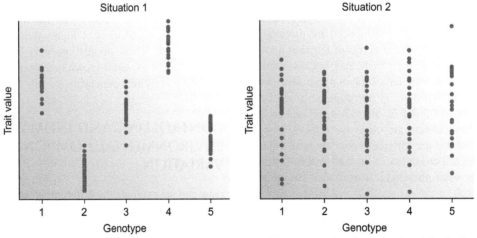

FIGURE A Illustration of the principle of ANOVA. The diagram shows two different scenarios of variation in a behavioral trait within and between five distinct genotypes. In each case we would like to know whether there is a significant difference between the five genotypes. The ANOVA model used to determine this is represented as $Y = \mu + G + E$, where Y designates the observed value for each individual, μ is the overall mean value, G is the effect of genotype, and E is the variation within each genotype. The effect of genotype will be significant if the variance between the mean values of the behavior between genotypes is greater than the variance in the behavior within genotypes. Here, the effect of genotype would be significant in situation 1, but not in situation 2. (From Anholt, R. R. H., and Mackay, T. F. C. (2004). *Nature Reviews Genetics*, 5, 838–849.)

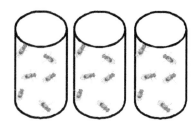

Source of variation	Variance component
Between vials	$\sigma^2_{EC(S)}$
Within vials	σ^2_{EW}

A

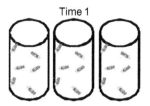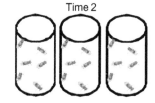

Time 1   Time 2

Source of variation	Variance component
Between times	$\sigma^2_{EC(T)}$
Between vials	$\sigma^2_{EC(S)}$
Within vials	σ^2_{EW}

B

FIGURE 6.12 Method for evaluating common environmental variance components. (A) A single inbred genotype of *Drosophila* is reared in several culture vials, and individuals from each vial are measured for a behavioral trait. Flies grown together in the same vial may behave more similarly to each other than to flies from other vials, because they share a common spatial micro-environment. This component of variation can be explicitly estimated using ANOVA. The average variance within vials is also purely environmental in origin, since the fly strain is inbred, and thus homozygous at all loci, so there is no genetic variation. (B) Here the flies are not only reared in different culture vials, but at two different times. There is thus an extra common environmental variance component to be estimated which reflects the similarity between flies reared at the same time.

SUMMARY

Since behaviors are complex traits, their manifestation (like any quantitative trait) depends on the interplay between genetic and environmental factors. The genetic and environmental contributions can be dissected by analyzing sources of phenotypic variation in a population, and partitioning these sources of variance into genetic and environmental components. Genetic components of variance consist of variance in breeding values (additive genetic variance), variance in dominance deviations, and variation in interaction deviations (due to epistasis), all of which depend on allele frequency. Environmental variance consists of spatial environmental variance, temporal environmental variance, environmental variance attributable to

maternal effects (in mammals), and "intangible" environmental variance of which the origin cannot be determined. Genotype–environment correlations of the environment experienced depend on the genotypes. Genotype–environment interaction (GEI) can arise if differences in the environment have greater effects on some genotypes than others, and/or if there is a change in rank order of a series of genotypes measured in different environments. The contribution of each source of phenotypic variation can be expressed as a fraction of the total phenotypic variance. That portion of the total phenotypic variance that is due to genetic sources of variation is known as the heritability. Broad sense heritability (H^2) indicates the extent to which variation in phenotypes in a population is due to variation in genotypic values, and includes non-additive genetic variation. Narrow sense heritability (h^2) is the portion of additive genetic variance that contributes to the total phenotypic variation, and represents the extent to which variation in phenotypes is determined by variation in effects of alleles transmitted by parents to offspring. The heritability depends on allele frequencies and the environment, and thus applies only to the population for which it was estimated, at that particular time in the population's history, and in the environmental circumstances under which the individuals were reared. Furthermore, estimates of h^2 convey no information about the nature of differences in mean phenotype between populations. In appropriately-designed experiments, common spatial and temporal environmental variance is either eliminated by randomizing across environments, or accounted for by replicate measurements. Similarly, genotype–environment correlation can be eliminated by randomizing genotypes across environments.

STUDY QUESTIONS

1. Name the components of phenotypic variance and explain what they mean.
2. What are the genetic components of phenotypic variance?
3. What factors determine the magnitude of genetic variance components?
4. Why can genetic variance components differ between populations?
5. What is meant by common environmental variation? Give an example.
6. Explain the difference between genotype–environment correlations and genotype–environment interaction. Give an example of each.
7. What is meant by the term "heritability?"
8. Explain the difference between broad sense heritability and narrow sense heritability.
9. Why is narrow sense heritability important?
10. What can heritability tell us about the genetic basis of differences between populations in their mean value of a behavioral trait?

11. Why can heritability differ between populations?
12. How can we account for common spatial and temporal variation in behavior genetic experiments?

RECOMMENDED READING

Falconer, D. S., and Mackay, T. F. C. (1996). *Introduction to Quantitative Genetics*, 4th edition. Longmans Green, Harlow, Essexb, UK.

Lynch, M., and Walsh, B. (1997). *Genetics and Analysis of Quantitative Traits*. Sinauer Assoc., Inc., Sunderland, MA.

Sokal, R. R., and Rohlf, F. J. (1995). *Biometry: The Principles and Practice of Statistics in Biological Research*, 3rd edition. W. H. Freeman and Co, New York, NY.

Estimating Heritability

OVERVIEW

Determining the fraction of phenotypic variation that is contributed by genetic factors is important for gene mapping studies, as will be described in Chapter 8. Heritability estimates are also useful for human behavioral traits, such as cognitive disorders or addiction, as they can provide insights into what extent genetic components contribute to the likelihood that the disorder is transmissible from parents to offspring. Since genetically, closely related individuals resemble each other more than unrelated individuals, studies on phenotypic resemblance between relatives are an effective way for dissecting genetic and environmental contributions to the observed phenotypic variance. In this chapter we will describe the statistical approaches that can be used to obtain heritability estimates by comparing phenotypes between parents and offspring, or by comparing phenotypes among siblings. We will also describe in detail how comparisons between monozygotic twins and dizygotic twins can yield heritability estimates for human behavioral traits. Finally, we will explain how such estimates have to be interpreted with caution in the face of several sources of bias.

PHENOTYPIC RESEMBLANCE BETWEEN RELATIVES

The **narrow sense heritability** (h^2) is an important property of a behavioral trait. It tells us what fraction of the total population variation in the trait (the phenotypic variance, σ_P^2) is due to variation in breeding values (the additive genetic variance, σ_A^2). This information is useful if we are interested in mapping the genes contributing to natural variation in behavior, since this will only be successful in a population where there is genetic variation for the trait, i.e. the h^2, is greater than zero. h^2 is also important because it predicts how closely the phenotypic measurement of an individual's behavior matches the breeding value of the individual; that is, the value of the behavior transmitted to its progeny (see Chapter 6). We will show in this chapter that h^2 is

also important in determining how similar the behavior of related individuals is, compared to random individuals from the population. We can take advantage of this relationship, and infer the value of h^2 based on a statistical quantification of the degree of resemblance among relatives.

Why does the extent to which relatives resemble each other for a behavioral trait depend on the amount of genetic variation there is for the trait, relative to the total phenotypic variation? To answer this question, we first need to understand how to quantify the phenotypic "resemblance among relatives" statistically, and then use statistical procedures to dissect the genetic and environmental components of the phenotypic variance. The intuition behind this exercise is that relatives should resemble each other for quantitative traits more than they do unrelated members of the population, because relatives share alleles. The closer the relationship, the higher the proportion of shared alleles. In addition, relatives can also share the same environment; if they do, this will increase their phenotypic resemblance over and above that expected from shared alleles.

The most common sorts of relatives considered in behavioral genetics are parents and their offspring, and siblings (including twins in human studies). Parents and offspring are examples of ancestral relatives that span generations; and siblings are examples of collateral relatives from the same generation (Figure 7.1). Our goal is to determine a statistic that estimates the phenotypic resemblance among each kind of relative, and then derive the genetic and environmental contributions to the phenotypic relationship.

Resemblance between Parents and Offspring

To quantify the resemblance between offspring and their parents, we need to obtain quantitative measurements of our behavioral trait for one or both parents, and at least one of their offspring, for many families in a population. Such a study will yield pairs of data for each of the families, namely the mean value of the trait of the offspring and the value of either one of the parents, or the average of the two parents (the mid-parent value), if both are measured. The obvious statistic to quantify these data is the regression of mean offspring phenotypic value on the parental phenotypic values (b_{OP}). The regression is $b_{OP} = (Cov_{OP})/(\sigma_P^2)$,

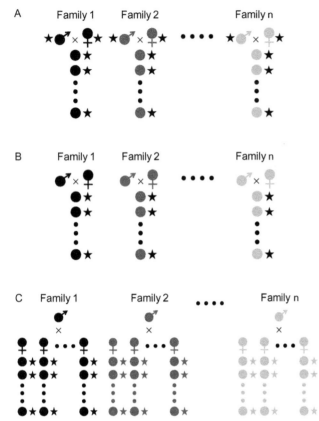

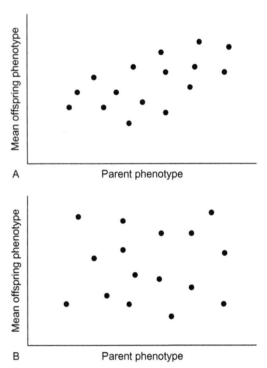

FIGURE 7.1 Common relatives used in behavioral genetics studies. ★ indicates the relatives for which measurements of the behavioral trait are obtained. (A) Parents and offspring. We obtain measurements of our trait for one or both parents, and one or more of their offspring, for *n* families in a population. (B) Full siblings. We obtain measurements of the trait for one or more offspring in each of *n* families in the population, but not for the parents. (C) Half-siblings. We obtain measurements for one or more offspring in each of *n* half-sibling families, where the common parent is mated to at least two individuals, but not the parents.

FIGURE 7.2 Cartoons showing the resemblance between offspring and their parents. (A) In this population, the heritability is high. Parents with higher than average values of the trait have, on average, offspring with higher than average values of the trait; and *vice versa* for parents with lower than average values of the trait. (B) In this population the heritability is zero. The value of the trait in offspring does not depend on the value of the trait in parents.

where Cov_{OP} is the covariance of trait phenotypes between offspring and parents, and σ_P^2 is the variance of parent phenotypes. There are three possible offspring–parent regressions: regression of mean offspring phenotype on the female parent (P_F); the male parent (P_M); or the mid-parent value ($MP = (P_F + P_M)/2$). The higher the offspring–parent regression coefficient, the greater is the amount of genetic variance for the trait, as a fraction of the total phenotypic variance (Figure 7.2).

Resemblance between Siblings

There are two common sorts of siblings used in behavioral genetic analysis. Full siblings have both parents in common. To quantify the relationship between full siblings, we need to obtain measurements of the behavioral phenotype for at least one offspring in each of many families in the population, but measurements for the parents are not needed (Figure 7.1). Half-siblings have one parent in common. This is typically the male parent in animal populations. To

quantify the relationships between half-siblings, we need to obtain measurements for at least one offspring in each of many half-sibling families, where the common parent is mated to at least two individuals. If there are more than one offspring per half-sibling family, there are two kinds of siblings to consider. Some individuals have the same mother and father, and are full siblings, while some individuals have just one parent in common, and are half-siblings (Figure 7.1).

For sibling designs, we use **analysis of variance (ANOVA)** to partition the phenotypic variance between families (σ_b^2) and within families (σ_w^2) The between group component refers to differences in the mean values of phenotypes between families. The total variance (σ_t^2) is the sum of the variances between groups and within groups ($\sigma_t^2 = \sigma_b^2 + \sigma_w^2$). We summarize the phenotypic resemblance among siblings as the intraclass correlation coefficient, *t*, which expresses the relative magnitudes of these components of phenotypic variance as $t = (\sigma_b^2)/(\sigma_b^2 + \sigma_w^2) = (\sigma_b^2)/(\sigma_t^2)$. A property of ANOVA is that any between-group variance equals the covariance of members of that group. Thus, the larger the covariance between members of a full- or half-sibling family is, the larger is the fraction of the total variance that is attributed to differences between family means. This measure of phenotypic

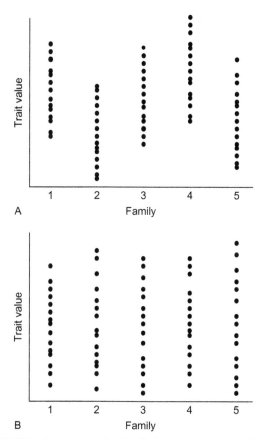

FIGURE 7.3 Cartoons showing distribution of phenotypes for five full-sibling families from: (A) a population in which the intraclass correlation between full siblings is 0.25; and (B) a population in which the intraclass correlation between full siblings is zero.

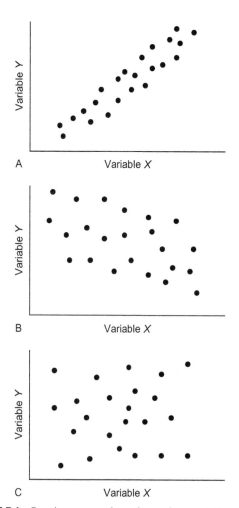

FIGURE 7.4 Covariance means the tendency of two variables, X and Y, to vary together. If the covariance is positive, high values of X are associated with high values of Y. If the covariance is negative, high values of X are associated with low values of Y. The covariance is high if knowing the value of X gives a good indication of the value of Y, low if knowing the value of X only partially indicates the value of Y, and 0 if X and Y are independent. (A) High positive covariance. (B) Low negative covariance. (C) Zero covariance.

resemblance among siblings can be considered equivalent to a measure of the *similarity* among siblings of the same family for the phenotype of a behavioral trait, compared to random members of a population, or also as the *difference* in phenotype between different families. The larger the variance between groups of siblings, the greater the difference between group means, and the more similar are the phenotypes of members of a group.

This concept is illustrated in Figure 7.3, which depicts two extreme distributions of phenotypes in full-sibling families. In both cases, the total phenotypic variation in the population is the same at $\sigma_t^2 = 9$. In panel A, the between-group variance $\sigma_b^2 = 2.25$, and the within-group variance $\sigma_w^2 = 6.75$. Here, members of a family resemble each other more closely than they do members of other families. There are differences in average phenotype between families. The interclass correlation is $t_{FS} = Cov_{FS}/\sigma_t^2 = 0.25$. Phenotypic resemblance is high within families, but not between families; thus, genetic variation is high. In panel B, however, the between-group variance $\sigma_b^2 = 0$ and the within-group variance $\sigma_w^2 = 9$. In this case, members of a family resemble each other no more than they do members of other families. There are no

significant differences in average phenotype between families; in other words, all families have the same mean value. Here, $\sigma_b^2 = 0$ and $t_{FS} = 0$, and there is no genetic variation for the trait.

For all types of relationship, it is the observed phenotypic covariance among relatives that statistically quantifies the amount of resemblance among them. Recall from statistics that covariance is a measure of the association between two variables (X and Y), and is defined as the mean product − product of means: $Cov_{XY} = \overline{XY} - \overline{X}\,\overline{Y}$. Defined in this way, the variance is the covariance of a variable with itself: $\sigma_X^2 = \overline{XX} - \overline{X}\,\overline{X} = \overline{X^2} - (\overline{X})^2 =$ mean square − square of means. If X and Y are independent (the value of one tells us nothing about the value of the other), then their covariance is zero (Figure 7.4).

GENETIC CAUSES OF COVARIANCE BETWEEN RELATIVES

The **genetic covariance** between relatives is the covariance of the genotypic values of the related individuals, or the amount by which **breeding values** and dominance deviations vary together in the two relatives being considered. Genetic covariances arise because two related individuals are more likely to share alleles than two unrelated individuals. By "sharing alleles," we mean having alleles in common between two relatives that **are identical by descent (IBD)**, such that both are copies of the same allele in a recent common ancestor (see also Chapter 5). To determine the genetic covariance among any set of relatives, we first need to determine the probabilities that the relatives have 0 (P_0), 1 (P_1), or 2 (P_2) alleles IBD.

Consider first a single parent and its offspring. To determine the probabilities that alleles are IBD, label the alleles at a single locus in one of the parents A_1 and A_2, and in the other parent A_3 and A_4. There are only four possible offspring genotypes: A_1A_3, A_1A_4, A_2A_3 and A_2A_4. Thus, all offspring have one allele IBD with either parent ($P_1 = 1$).

The offspring of a single pair of parents are full siblings. To compute the probabilities that full siblings have 0 (P_0), 1 (P_1), or 2 (P_2) alleles IBD, we make a table of the 16 possible combinations of genotypes for two full siblings, and write down the number of alleles IBD for each combination (Table 7.1). We find four genotype combinations have none and four have both alleles IBD, and eight genotype combinations have one allele IBD. All 16 combinations are equally likely, so $P_0 = 4/16 = 1/4$; $P_1 = 8/16 = 1/2$; and $P_2 = 4/16 = 1/4$.

We can similarly determine the probabilities of 0, 1, and 2 alleles that are IBD in any pair of relatives. Table 7.2 gives these probabilities for common relationships, including the ones we have just derived from first principles. The genetic covariance between relatives is a simple function of these IBD probabilities. We define $r_{XY} = 1/2\,P_1 + P_2$ and $u_{XY} = P_2$, where X and Y are the two relatives (i.e. parents and offspring, full siblings, or any relative pair). Then the genetic covariance between relative X and relative Y is:

$$Cov(G_X, G_Y) = r_{XY}\,\sigma_A^2 + u_{XY}\,\sigma_D^2 + r_{XY}^2\,\sigma_{AA}^2 + r_{XY}\,u_{XY}\,\sigma_{AD}^2 + u_{XY}^2\,\sigma_{DD}^2$$

Thus, the genetic covariance between any pair of relatives is a function of the **additive** (σ_A^2), **dominance** (σ_D^2), and two-locus **epistatic** ($\sigma_{AA}^2, \sigma_{AD}^2, \sigma_{DD}^2$) variances of the trait, and the probabilities are of having one and two loci IBD in the particular pair of relatives under consideration. r_{XY}, the coefficient of the additive genetic variation, is the correlation of breeding values between relatives X and Y. u_{XY}, the coefficient of the dominance variance, is the probability that relatives X and Y have the same genotype at a locus due to IBD, which can only occur

in an outbred population if the relatives have two parents or more distant ancestors in common.

When more than one locus contributes to the genetic variance of a behavioral trait, any two-locus, three-locus, and higher order epistatic interactions between loci contribute to the genetic covariance between relatives. The expression above can be indefinitely expanded to include all higher order interactions (i.e. $r_{XY}^3\sigma_{AAA}^2$, $r_{XY}^2 u_{XY}\sigma_{AAD}^2$, $r_{XY} u_{XY}^2\sigma_{ADD}^2$, $u_{XY}^3\sigma_{DDD}^2$, etc.). However, for simplicity we have included only the two locus interactions. This is justified, because for most relatives the coefficients of the epistatic terms are small. The largest value

TABLE 7.1 Worksheet for calculating the probability that 0, 1, or 2 alleles are IBD in two full siblings (O_1, O_2)

O_1	O_2	Alleles IBD
A_1A_3	A_1A_3	2
	A_1A_4	1
	A_2A_3	1
	A_2A_4	0
A_1A_4	A_1A_3	1
	A_1A_4	2
	A_2A_3	0
	A_2A_4	1
A_2A_3	A_1A_3	1
	A_1A_4	0
	A_2A_3	2
	A_2A_4	1
A_2A_4	A_1A_3	0
	A_1A_4	1
	A_2A_3	1
	A_2A_4	2

The parent genotypes are designated A_1A_2 and A_3A_4. The table gives the 16 possible genotypes of two full siblings, and the number of alleles IBD in each combination.

TABLE 7.2 Probabilities that 0 (P_0), 1 (P_1), and 2 (P_2) alleles are IBD in common relatives

Relationship	P_0	P_1	P_2
Identical (MZ) twins	0	0	0
Offspring–parent	0	1	0
Full siblings, fraternal (DZ) twins	1/4	1/2	1/4
Grandparent–grandchild, half-siblings, uncle–nephew	1/2	1/2	0
Great grandparent–great grandchild, cousins	3/4	1/4	0

of r_{XY}^2 is 1/4 for offspring and parents and full siblings, and u_{XY} is often zero. Thus, epistasis usually contributes relatively little to the genetic covariance among relatives. The major exception is for **monozygotic twins**, which have identical genotypes and thus also share covariance from all possible epistatic interactions. Note, however, that all covariances among relatives include a contribution from additive-by-additive interaction variance, and increasingly negligible contributions from higher order interactions between breeding values at loci affecting variation in the trait. Genetic covariances among common relatives are given in Table 7.3.

TABLE 7.3 Genetic covariances among common relatives

Relatives	Symbol	σ_A^2	σ_D^2	σ_{AA}^2	σ_{AD}^2	σ_{DD}^2
			Contribution to genetic covariance			
Identical (MZ) twins	Cov_{MZ}	1	1	1	1	1
Offspring–parent	Cov_{OP}	$\frac{1}{2}$	0	$\frac{1}{4}$	0	0
Full siblings	Cov_{FS}	$\frac{1}{2}$	$\frac{1}{4}$	$\frac{1}{4}$	$\frac{1}{8}$	$\frac{1}{16}$
Fraternal (DZ) twins	Cov_{DZ}	$\frac{1}{2}$	$\frac{1}{4}$	$\frac{1}{4}$	$\frac{1}{8}$	$\frac{1}{16}$
Half-siblings	Cov_{HS}	$\frac{1}{4}$	0	$\frac{1}{16}$	0	0
Cousins	Cov_C	$\frac{1}{8}$	0	$\frac{1}{64}$	0	0

ENVIRONMENTAL CAUSES OF THE RELATIONSHIP BETWEEN RELATIVES

We saw in Chapter 6 that common spatial, temporal, or maternal environments could cause individuals reared in the same environment to behave more similarly to each other than to individuals of the same genotype reared in a different common environment. When this occurs, we need to add additional components of phenotypic variance attributable to the spatial ($\sigma_{EC(S)}^2$), temporal ($\sigma_{EC(T)}^2$), or maternal ($\sigma_{EC(M)}^2$) common environments. Members of a family will often be reared together, and thus they will share a common environment. If the common environmental circumstances are different for each family, the variance due to common environment will cause greater similarity among members of a family, and greater differences among families, than would be expected from the proportion of alleles they share. Variance due to common environment thus increases the phenotypic covariance between relatives. For example, the propensity for alcohol addiction is greater in families where parents or siblings regularly engage in heavy drinking. Another example is religious beliefs (e.g. church-going), which are strongly influenced by members of the same family, and can differ greatly between families in the population. Common environment mainly contributes to resemblance of siblings, but maternal environment can also contribute to resemblance between mother and offspring. $\sigma_{EC(S)}^2$ and $\sigma_{EC(T)}^2$ can be eliminated, or estimated, by correct experimental design, but it is very difficult (except by cross-fostering) to eliminate or estimate $\sigma_{EC(M)}^2$ from Cov_{FS}.

Phenotypic Covariance among Relatives

It is clear from the previous sections that the phenotypic covariance among relatives is the sum of the genetic covariance and any **common environmental variance** shared by the relatives. Phenotypic covariances for common relatives are given in Table 7.4. We can now write the observed, statistical quantification of the phenotypic resemblance between relatives for a behavioral trait (i.e. the regression or intraclass correlation determined by measuring the behavior in families as described at the beginning of this

TABLE 7.4 Phenotypic covariances among common relatives

Relatives	Phenotypic covariance
Identical (MZ) twins	$\sigma_A^2 + \sigma_D^2 + \sigma_{AA}^2 + \sigma_{AD}^2 + \sigma_{DD}^2 + \sigma_{EC(MZ)}^2$
Offspring–parent	$\frac{1}{2}\sigma_A^2 + \frac{1}{4}\sigma_{AA}^2 + \sigma_{EC(OP)}^2$
Full-siblings	$\frac{1}{2}\sigma_A^2 + \frac{1}{4}\sigma_D^2 + \frac{1}{4}\sigma_{AA}^2 + \frac{1}{8}\sigma_{AD}^2 + \frac{1}{16}\sigma_{DD}^2 + \sigma_{EC(FS)}^2$
Fraternal (DZ) twins	$\frac{1}{2}\sigma_A^2 + \frac{1}{4}\sigma_D^2 + \frac{1}{4}\sigma_{AA}^2 + \frac{1}{8}\sigma_{AD}^2 + \frac{1}{16}\sigma_{DD}^2 + \sigma_{EC(DZ)}^2$
Half-siblings	$\frac{1}{4}\sigma_A^2 + \frac{1}{16}\sigma_{AA}^2 + \sigma_{EC(HS)}^2$
Cousins	$\frac{1}{8}\sigma_A^2 + \frac{1}{64}\sigma_{AA}^2 + \sigma_{EC(C)}^2$

TABLE 7.5 Relationship between observed phenotypic resemblance between relatives and the theoretical phenotypic covariance, for common relatives

Relatives	Regression (b) or correlation (t)
Identical (MZ) twins	$t_{MZ} = \dfrac{\sigma_A^2 + \sigma_D^2 + \sigma_{AA}^2 + \sigma_{AD}^2 + \sigma_{DD}^2 + \sigma_{EC(MZ)}^2}{\sigma_P^2}$
Offspring – one parent	$b_{OP} = \dfrac{\frac{1}{2}\sigma_A^2 + \frac{1}{4}\sigma_{AA}^2 + \sigma_{EC(OP)}^2}{\sigma_P^2}$
Offspring – mid-parent*	$b_{O\bar{P}} = \dfrac{\sigma_A^2 + \frac{1}{2}\sigma_{AA}^2 + 2\sigma_{EC(OP)}^2}{\sigma_P^2}$
Full-siblings	$t_{FS} = \dfrac{\frac{1}{2}\sigma_A^2 + \frac{1}{4}\sigma_D^2 + \frac{1}{4}\sigma_{AA}^2 + \frac{1}{8}\sigma_{AD}^2 + \frac{1}{16}\sigma_{DD}^2 + \sigma_E^2}{\sigma_P^2}$
Fraternal (DZ) twins	$t_{DZ} = \dfrac{\frac{1}{2}\sigma_A^2 + \frac{1}{4}\sigma_D^2 + \frac{1}{4}\sigma_{AA}^2 + \frac{1}{8}\sigma_{AD}^2 + \frac{1}{16}\sigma_{DD}^2 + \sigma_E^2}{\sigma_P^2}$
Half-siblings	$t_{HS} = \dfrac{\frac{1}{4}\sigma_A^2 + \frac{1}{16}\sigma_{AA}^2 + \sigma_{EC(HS)}^2}{\sigma_P^2}$
Cousins	$t_C = \dfrac{\frac{1}{8}\sigma_A^2 + \frac{1}{64}\sigma_{AA}^2 + \sigma_{EC(C)}^2}{\sigma_P^2}$

The resemblance between offspring and parents is $b_{OP} = Cov_{OP}/$(variance of parents). For regressions on one parent, the variance of the parents is σ_P^2. For the regression on mid-parent, the variance is of group means with n = 2. From statistics, the variance of a mean = σ^2/n, so the variance of mid-parents is $\sigma_P^2/2$.

Thus $b_{O\bar{P}} = \dfrac{\frac{1}{2}\sigma_A^2 + \frac{1}{4}\sigma_{AA}^2 + \sigma_{EC(OP)}^2}{\frac{1}{2}\sigma_P^2} = \dfrac{\sigma_A^2 + \frac{1}{2}\sigma_{AA}^2 + 2\sigma_{EC(OP)}^2}{\sigma_P^2}$

chapter) to the theoretical phenotypic covariance among relatives, written in terms of genetic and environmental variance components (Table 7.5). These expressions hint at how we can estimate variance components from observations on different sorts of relatives, and also how complications will arise from the confounding effects of common environment and epistasis. Even if we had measured the behavioral trait in all the relatives listed in Table 7.5, we would not be able to estimate all the genetic and environmental variance components, because there are more variance components than there are types of relatives.

With the exception of monozygotic twins, the greatest contribution to genetic covariance among relatives is the variance of breeding values, σ_A^2. All of the phenotypic relationships have a component that is the ratio of additive genetic variance to the total phenotypic variance, σ_A^2/σ_P^2, i.e. the narrow sense heritability, h^2. This is why we said at the beginning of this chapter that the resemblance between relatives depends on h^2. Therefore, multiplying the estimate of the regression intraclass correlation by the appropriate coefficient gives an estimate of the narrow sense heritability (Table 7.6).

It is important to realize that none of these estimates are unbiased estimates of heritability; they are all potentially confounded by environmental variance common to the sets of relatives in the analysis, and other genetic variance components (additive-by-additive epistasis, and sometimes dominance and other epistatic variance components). In experimental organisms, we can account for any common spatial or temporal environments with correct experimental design. In mammals, however, common maternal environment will always inflate the heritability estimates from mother–offspring, twin, and full-sibling families. Estimates from father–son regressions and paternal half-sibling families are generally regarded as unbiased estimates of h^2. Heritability estimates from twins and full siblings are also biased upward by the inclusion of dominance variance. Thus, it is important to recognize that most estimates of heritability are too high. How much too high depends on the particular relatives used for estimating heritability. Even the best experimental designs for estimating heritability suffer from the inclusion of epistatic variance from interactions of breeding values at loci affecting the trait. These complications do not matter

TABLE 7.6 Heritability estimates from common relatives

Relatives	Regression (b) or correlation (t)
Identical (MZ) twins	$t_{MZ} = h^2 + \dfrac{\sigma^2_D + \sigma^2_{AA} + \sigma^2_{AD} + \sigma^2_{DD} + \sigma^2_{EC(MZ)}}{\sigma^2_P}$
Offspring – one parent	$2b_{OP} = h^2 + \dfrac{\frac{1}{2}\sigma^2_{AA} + 2\sigma^2_{EC(OP)}}{\sigma^2_P}$
Offspring – mid-parent	$b_{O\bar{P}} = h^2 + \dfrac{\frac{1}{2}\sigma^2_{AA} + 2\sigma^2_{EC(OP)}}{\sigma^2_P}$
Full-siblings	$2t_{FS} = h^2 + \dfrac{\frac{1}{2}\sigma^2_D + \frac{1}{2}\sigma^2_{AA} + \frac{1}{4}\sigma^2_{AD} + \frac{1}{8}\sigma^2_{DD} + 2\sigma^2_{EC(FS)}}{\sigma^2_P}$
Fraternal (DZ) twins	$2t_{DZ} = h^2 + \dfrac{\frac{1}{2}\sigma^2_D + \frac{1}{2}\sigma^2_{AA} + \frac{1}{4}\sigma^2_{AD} + \frac{1}{8}\sigma^2_{DD} + 2\sigma^2_{EC(DZ)}}{\sigma^2_P}$
Half-siblings	$4t_{HS} = h^2 + \dfrac{\frac{1}{4}\sigma^2_{AA} + 4\sigma^2_{EC(HS)}}{\sigma^2_P}$
Cousins	$8t_C = h^2 + \dfrac{\frac{1}{8}\sigma^2_{AA} + 8\sigma^2_{EC(C)}}{\sigma^2_P}$

if most variation affecting the trait is additive genetic variation with no dominance and epistasis. However, it is wrong to assume that contributions from dominance and epistatic variance to the heritability estimates are zero just because we cannot readily estimate their relative magnitudes (see below).

It is possible to combine information from multiple types of relatives and unrelated individuals that share a common environment, to obtain estimates of all variance components. For example, the difference between b_{OP} from offspring–mid-parent regression and $2t_{FS}$ from full siblings provides an estimate of the dominance variance, σ^2_D, if we can assume $\sigma^2_{EC(OP)}$ and $\sigma^2_{EC(FS)}$ are the same, and there is no epistatic variance due to interactions of dominance deviations. Human geneticists and animal breeders typically use multiple sources of information from relatives to estimate heritability and variance components.

HERITABILITY ESTIMATES IN HUMANS

In human nuclear families, the easiest data to obtain are parent–offspring and full-sibling family measures. The latter are preferable for several reasons. Our long generation time means that measurements on parents are not always available. Further, some behavioral phenotypes change over time, or are expressed only late in life, so measurements at the same age of parents and offspring are difficult and, in most cases, impossible to obtain. In addition, the environment is not constant between parental and offspring generations. Full siblings not only share a common maternal environment ($EC(M)$), but also a common home environment (cultural, social, educational, and dietary, $EC(S)$). While the common maternal environment can decrease in importance with age, the common home environment is a very significant source of bias, increasing the covariance of full siblings and biasing estimates of heritability upwards:

$$Cov_{FS} = \frac{1}{2}\sigma^2_A + \frac{1}{4}\sigma^2_D + \frac{1}{4}\sigma^2_{AA} + \frac{1}{8}\sigma^2_{AD}$$
$$+ \frac{1}{16}\sigma^2_{DD} + \sigma^2_{EC(M)} + \sigma^2_{EC(S)};$$

$$2t_{FS} > h^2 = \frac{\sigma^2_A}{\sigma^2_P} + \frac{\left(\begin{array}{c}\frac{1}{2}\sigma^2_D + \frac{1}{2}\sigma^2_{AA} + \frac{1}{4}\sigma^2_{AD} + \frac{1}{8}\sigma^2_{DD} \\ + 2\sigma^2_{EC(M)} + 2\sigma^2_{EC(S)}\end{array}\right)}{\sigma^2_P}$$

where $(\sigma^2_A)/(\sigma^2_P)$ would be the exact estimate of the narrow sense heritability, and

$$\frac{\left(\begin{array}{c}\frac{1}{2}\sigma^2_D + \frac{1}{2}\sigma^2_{AA} + \frac{1}{4}\sigma^2_{AD} + \frac{1}{8}\sigma^2_{DD} \\ + 2\sigma^2_{EC(M)} + 2\sigma^2_{EC(S)}\end{array}\right)}{\sigma^2_P}$$

represents the biased overestimate.

In humans, twin births are not uncommon. In the United States, approximately four births per thousand are monozygotic (MZ) twins, while roughly eight births per

FIGURE 7.5 Identical twins. Notice the similar posture. Twin studies have shown that personality traits, such as shyness and engaging in risky behavior have a genetic component. (Photograph from the *Smithsonian* magazine, November 2004, Jonathan Torgovnik.)

thousand are **dizygotic** (DZ) twins. As noted in Table 7.3, MZ twins are genetically identical (Figure 7.5), while DZ twins have the same genetic covariance as full siblings. From Table 7.5:

$$t_{MZ} = \frac{\sigma_A^2 + \sigma_D^2 + \sigma_{AA}^2 + \sigma_{AD}^2 + \sigma_{DD}^2 + \sigma_{EC(MZ)}^2}{\sigma_P^2} \text{ and}$$

$$t_{DZ} = \frac{\begin{pmatrix} \frac{1}{2}\sigma_A^2 + \frac{1}{4}\sigma_D^2 + \frac{1}{4}\sigma_{AA}^2 + \frac{1}{8}\sigma_{AD}^2 \\ + \frac{1}{16}\sigma_{DD}^2 + \sigma_{EC(DZ)}^2 \end{pmatrix}}{\sigma_P^2}$$

Thus, if we are willing to assume that the common maternal and spatial environments are the same for MZ and DZ twins ($\sigma_{EC(MZ)}^2 = \sigma_{EC(DZ)}^2$), and that epistasis is negligible ($\sigma_{AA}^2 = \sigma_{AD}^2 = \sigma_{DD}^2 = 0$), subtracting the estimate of the intraclass correlation between DZ twins from the intraclass correlation between MZ twins (Table 7.5) gives $t_{MZ} - t_{DZ} = (\frac{1}{2}\sigma_A^2 + \frac{3}{4}\sigma_D^2)/\sigma_P^2$. Doubling the difference in the correlations between MZ and DZ twins thus gives $2(t_{MZ} - t_{DZ}) = h^2 + 1.5\sigma_D^2/\sigma_P^2$. Here, $1.5\sigma_D^2/\sigma_P^2$ represents the biased overestimate of the narrow sense heritability due to dominance variance. The majority of information on the relative contribution of genetic variance to the total phenotypic variance for human complex traits, including complex behavioral traits, comes from studies that compare phenotypic variation between monozygotic and dizygotic twins. Indeed, National Twin Registries have been established for this purpose in many countries (Box 7.1).

The assumption that the common environment is the same for MZ and DZ twins has been criticized. MZ twins

Box 7.1   The Swedish Twin Registry

The Swedish Twin Registry was established in 1961, and is a research resource maintained at The Karolinska Institute in Stockholm. The registry includes an old cohort, comprised of 10 945 pairs of like-sexed twins born 1886 through 1925, a middle cohort comprised of approximately 50 000 pairs of like- and unlike-sexed twins born 1926 through 1967, and a young cohort of approximately 45 000 pairs born since 1967. Questionnaire data have been collected from members of the old cohort in 1961, 1963, 1967, and 1970, and from like-sexed members (born 1926–1958) of the middle cohort in 1973. The questionnaire data include items about health, health-related behaviors, personality, physical activity, eating habits, and environmental stressors.

FIGURE A   The Swedish Twin Registry.

are often treated more similarly than DZ twins; particularly since half of DZ twin pairs are opposite sexes. Also, intrauterine effects may differ. These considerations motivate the use of adoption studies to compare correlations of MZ and DZ twins reared together and reared apart, to further tease apart common environment effects (Box 7.2). Even these studies are not perfect, since the environments of the adoptive twins are probably highly correlated, because of screening procedures used for adoptive parents. In addition, twins reared apart after birth still share a common maternal environment. Studies of unrelated individuals reared in the same household have been used to estimate the common spatial and cultural environment.

Box 7.2   Partitioning genetic and environmental variance of IQ in humans

Devlin et al. (1997, *Nature*, *388*, 468–471) performed a meta-analysis of 212 studies on human IQ. The mean correlation of MZ twins reared together was $t_{MZ} = 0.85$, and of DZ twins reared together was $t_{DZ} = 0.59$. Thus, $2(t_{MZ} - t_{DZ}) = 0.52 = h^2 + 1.5\sigma_D^2/\sigma_P^2$, assuming the same common environmental variances in MZ and DZ twin pairs, and no epistatic contribution to the genetic variance of IQ. However, the authors had data on IQ of many different sorts of relatives, including full siblings reared together and apart. This enabled them to make a full partitioning of additive and dominance variance components, common maternal environment for twins ($\sigma_{EC(MT)}^2$), common maternal environment for siblings ($\sigma_{EC(MS)}^2$), and common spatial environment ($\sigma_{EC(S)}^2$). The expected covariances ignoring any potential contribution of epistasis and observed correlations or regressions are given below.

Relationship	Raised	Expected covariance	Observed t or b
MZ twins	Together	$\sigma_A^2 + \sigma_D^2 + \sigma_{EC(MT)}^2 + \sigma_{EC(S)}^2$	0.85
MZ twins	Apart	$\sigma_A^2 + \sigma_D^2 + \sigma_{EC(MT)}^2$	0.74
DZ twins	Together	$\tfrac{1}{2}\sigma_A^2 + \tfrac{1}{4}\sigma_D^2 + \sigma_{EC(MT)}^2 + \sigma_{EC(S)}^2$	0.59
Siblings	Together	$\tfrac{1}{2}\sigma_A^2 + \tfrac{1}{4}\sigma_D^2 + \sigma_{EC(MS)}^2 + \sigma_{EC(S)}^2$	0.46
Siblings	Apart	$\tfrac{1}{2}\sigma_A^2 + \tfrac{1}{4}\sigma_D^2 + \sigma_{EC(MS)}^2$	0.24

The authors used these data to partition the phenotypic variance of IQ from the model that best fit the data. The h^2 of IQ was estimated to be about 1/3. Surprisingly, maternal effects for MZ and DZ twins were large and important.

Variance	Symbol	Estimate/σ_P^2
Additive genetic	σ_A^2	0.34
Dominance genetic	σ_D^2	0.15
Total genetic	σ_G^2	0.49
Maternal (twins)	$\sigma_{EC(MT)}^2$	0.20
Maternal (sibs)	$\sigma_{EC(MS)}^2$	0.05
Common spatial environment	$\sigma_{EC(S)}^2$	0.17
Intangible environmental	σ_{EW}^2	0.09

Precision of Heritability Estimates

If we wish to conduct an experiment to estimate components of phenotypic variance and estimate heritability, we need to ask how many families and individuals per family we need to measure in order to obtain reliable estimates. This is done by considering the precision of heritability estimates. In a nutshell, large numbers of individuals are needed to obtain precise estimates of variance components and heritability. By "precision," we mean small **standard errors** (the square root of the variance) of the estimate (Figure 7.6, Box 7.3).

For offspring–parent regressions, the standard error of the heritability is a function of the number of families (i.e. parent–offspring pairs). Given that there are usually a fixed total number of individuals that we can measure, precision is maximized when we measure as many families as possible with only one offspring per family. The total numbers of individuals (parents and offspring) that need to be measured for different standard errors (SEs) are given in Table 7.7. A rough guide to significance of an estimate is 2SE.

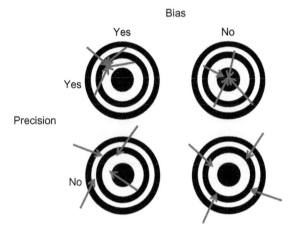

FIGURE 7.6 Precision and bias. These concepts are illustrated by the simile of shooting arrows (statistical estimates) with the goal of hitting the bull's eye of a target (the true population parameter). Precision refers to how closely the arrows (estimates) cluster to each other. High precision means low variance between estimates. Bias refers to any difference between the average of the estimates and the true population value of the parameter. No bias means that on average the estimates equal the true population value.

Thus, if SE = 0.05, heritability estimates greater than 0.1 are significantly different from zero; if SE = 0.2, only heritability estimates greater than 0.4 are significantly different from zero. The regression of offspring on the mean of both parents gives better precision than regression on the mean of a single parent, for the same total sample size (Table 7.7). The smaller the heritability, the greater the sample size required to detect it as significantly different from zero. The precision of regression methods can be increased by selection of extreme parents and/or **assortative mating** of parents. Heritability estimated from the regression of offspring means on the mean of parents selected to have either high or low values of the behavioral trait is called the **realized h^2**.

In sibling designs, the SE of the heritability depends on the heritability itself, making it difficult to work out in advance. In full-sibling designs, or in half-sibling designs where there is only one offspring per dam, precision for a given total number of offspring is maximized when the number of offspring per family (or dams per sire) = $1/t$, where t is the intraclass correlation. Precision in half-sibling designs is maximized by increasing the number of dams per sire, at the expense of the number of offspring per dam. Table 7.8 shows the sample sizes required to detect a range of h^2 values, under the optimal designs for full- and half-sibling estimates. In general, for the same total sample size estimates of heritability from full siblings are more precise than those for half siblings. However, the full-sibling design is biased by inclusion of dominance variance and possible common maternal environment: there is a trade-off between precision and bias.

HERITABILITY IN OTHER POPULATIONS

Often, the choice of whether to use offspring–parent or sibling methods will be determined by the biology of the species of interest. For species with long generation times, offspring–parent relationships may not be feasible. For organisms that cannot be bred in the laboratory, naturally occurring broods (usually full siblings) must be used. If the total number of individuals that can be measured is fixed, and there is a choice between methods, offspring–parent regression is more efficient than sibling correlations for $h^2 > 0.2$, and sibling correlations are more efficient for $h^2 < 0.2$.

Other methods for determining **broad sense heritabilities** (H^2) can be used in organisms that can be inbred to homozygosity and crossed, such as mice, *Drosophila*, and the nematode worm, *Caenorhabditis elegans*. One such method involves measuring the phenotypic variances of individuals within two parental lines that are completely inbred (P1, P2), the F_1 progeny derived from crossing these lines, and the F_2 progeny derived from crossing F_1 individuals. Because the two parental lines are not genetically variable, the phenotypic variation for the trait within each of these lines results from **intangible environmental variation**, σ_{EW}^2. The F_1 generation is also not genetically variable: all individuals are heterozygous for all loci that are fixed for alternative alleles in the two parental strains. Thus, phenotypic variation among the F_1 individuals is also due to intangible environmental variation. Therefore, the best estimate of σ_{EW}^2 is the average of the phenotypic variances of the two parental lines and the F_1. The phenotypic variance in the F_2 population is due to all genetic variance components plus intangible environmental variance. Therefore $\sigma_{PF2}^2 - \dfrac{\sigma_{PP1}^2 + \sigma_{PP2}^2 + \sigma_{PF1}^2}{3}$ estimates $\sigma_A^2 + \sigma_D^2 + \sigma_{AA}^2 + \sigma_{AD}^2 + \sigma_{DD}^2 = \sigma_G^2$, the total genetic variance. The broad sense heritability is $H^2 = \dfrac{\sigma_G^2}{\sigma_G^2 + \sigma_{EW}^2}$.

It is also possible to measure behavior of many individuals for many different completely inbred lines. These lines could be derived by inbreeding from outbred animals collected from the wild, as is common for mice and *Drosophila*. They could also be derived by inbreeding an F_2 population resulting from a cross of two inbred lines, in which case they are called a **recombinant inbred line** (RIL) population. The variation in behavior within each inbred line, σ_w^2, is σ_{EW}^2, since all variation is due to intangible environmental variation, as above. The variation between inbred lines (σ_b^2) is entirely genetic, and is composed of the variation of homozygous effects at each segregating locus, and variation from two-way, three-way, and higher-order interactions between homozygous effects. In this case, the estimate of broad sense heritability is $H^2 = (\sigma_b^2)/(\sigma_b^2 + \sigma_w^2)$. These heritability estimates describe the fraction of the total phenotypic variance that is attributable to genetic factors in particular reference populations.

Box 7.3 Hiawatha designs an experiment

by W. E. Mientka

Hiawatha, mighty hunter,
He could shoot ten arrows
   upward,
Shoot them with such strength
   and swiftness
That the last had left the
   bow-string
Ere the first to earth descended.

This was commonly regarded
As a feat of skill and cunning.
Several sarcastic spirits
Pointed out to him, however,
That it might be much more
   useful
If he sometimes hit the target.
"Why not shoot a little straighter
And employ a smaller sample?"

Hiawatha, who at college
Majored in applied statistics,
Consequently felt entitled
To instruct his fellow man in
Any subject whatsoever,
Waxed exceedingly indignant,

Talked about the law of errors,
Talked about truncated normals,
Talked of loss of information,
Talked about his lack of bias,
Pointed out that (in the long run)
Independent observations,
Even though they missed the
   target,
Had an average point of impact
Very near the spot he aimed at,
With the possible exception
of a set of measure zero.

"This," they said, "was rather
   doubtful;
Anyway it didn't matter.
What resulted in the long run:
Either he must hit the target
Much more often than at present,
Or himself would have to pay for
All the arrows he had wasted."

Hiawatha, in a temper,
Quoted parts of R.A. Fisher,
Quoted Yates and quoted Finney,

Quoted reams of Oscar
   Kempthorne,
Quoted Anderson and Bancroft
(practically *in extenso*)
Trying to impress upon them
That what actually mattered
Was to estimate the error.

Several of them admitted:
"Such a thing might have its uses;
Still," they said, "he would do
   better
If he shot a little straighter."

Hiawatha, to convince them,
Organized a shooting contest.
Laid out in the proper manner
Of designs experimental
Recommended in the textbooks,
Mainly used for tasting tea
(but sometimes used in other
   cases)
Used factorial arrangements
And the theory of Galois,
Got a nicely balanced layout
And successfully confounded
Second order interactions.

All the other tribal marksmen,
Ignorant benighted creatures
Of experimental setups,
Used their time of preparation
Putting in a lot of practice
Merely shooting at the target.
Thus it happened in the contest
That their scores were most
   impressive
With one solitary exception.
This, I hate to have to say it,
Was the score of Hiawatha,
Who as usual shot his arrows,
Shot them with great strength and
   swiftness,
Managing to be unbiased,
Not however with a salvo
Managing to hit the target.

"There!" they said to Hiawatha,
"That is what we all expected."
Hiawatha, nothing daunted,
Called for pen and called for
   paper.
But analysis of variance

Finally produced the figures
Showing beyond all
   peradventure,
Everybody else was biased.
And the variance components
Did not differ from each other's
As they did from Hiawatha's.

(This last point, it might be
   mentioned,
Would have been much more
   convincing
If he hadn't been compelled to
Estimate his own components
From experimental plots on
Which the values all were
   missing.)
Still they couldn't understand it,

So they couldn't raise objections.
(Which is what so often happens,
with analysis of variance.)
All the same his fellow tribesmen,
Ignorant benighted heathens,
Took away his bow and arrows,
Said that though my Hiawatha
Was a brilliant statistician,
He was useless as a bowman.
As for variance components,
Several of the more outspoken
Made primeval observations
Hurtful of the finer feelings
Even of the statistician.

In a corner of the forest,
Sits alone my Hiawatha
Permanently cogitating
On the normal law of errors.
Wondering in idle moments
If perhaps increased precision
Might perhaps be sometimes
   better,
Even at the cost of bias,
If one could thereby now and
   then
Register upon a target.

The above poem appeared in *American
Mathematical Monthly*, Vol. 79, Number
6 (June–July, 1972), in an article entitled, "Professor Leo Moser – Reflections
on a Visit."

TABLE 7.7 The total number of individuals required to give the indicated standard error (SE) of the heritability estimated from offspring–parent regressions

SE	Offspring – single parent	Offspring – mid-parent
0.01	80 000	60 000
0.05	3200	2400
0.10	800	600
0.15	356	267
0.20	200	150

TABLE 7.8 The total number of individuals required to give the indicated standard error (SE) of the heritability estimated from correlations between full-siblings and half-siblings

| | Full-siblings | | Half-siblings | |
SE	$h^2 = 0.25$	$h^2 = 0.50$	$h^2 = 0.25$	$h^2 = 0.50$
0.01	40 000	80 000	80 000	160 000
0.05	1600	3200	3200	6400
0.10	400	800	800	1600
0.15	178	356	356	711
0.20	100	200	200	400

ESTIMATES OF HERITABILITY FOR BEHAVIORAL TRAITS

Heritability estimates for some behavioral traits in humans, mice and *Drosophila* are given in Table 7.9. Some general features of these data are worth mentioning. First, there is evidence for genetic variation for behaviors in all three species: both nature and nurture contribute to the variation in behavior we observe in populations. Second, heritability estimates are on average much higher for behavioral traits in humans than in the model organisms. This is not surprising, because any common spatial and temporal environmental variation, leading to more similarity among relatives than dictated by genetic factors alone, was controlled and accounted for in the data used to estimate heritability in mice and flies. It is also possible that non-additive genetic variation (dominance and epistasis) inflates the heritability estimates in humans, since they were all obtained from MZ and DZ twin correlations.

The *Drosophila* data indicate that the contribution of non-additive variance to variation in behavior is not trivial. Heritabilities for copulation latency, aggression, locomotor startle response (see also Chapter 11), and ethanol knockdown time (see also Chapter 15) were all obtained from the same population using two experimental designs: realized heritability from selected parents and their offspring; and variation among completely inbred lines. The inbred lines are homozygous for either A_1A_1 or A_2A_2 at any single locus affecting variation in behavior, with homozygous effects a and $-a$, and allele frequencies p and q, respectively (see Chapters 4 and 5). We can determine the genetic variation in the population of inbred lines in the same manner we did for outbred populations in Chapter 6; express the homozygous effects of each genotype as a deviation from the population mean, square these, multiply each by the allele frequency, and add the terms for each genotype together. This gives $4pqa^2$, or $2\sigma_A^2$. Thus, the population

of inbred lines has twice the genetic variance as that of the outbred population from which it was derived. When all genetic variation is additive, the narrow sense heritability, $h^2 = (\sigma_A^2)/(\sigma_A^2 + \sigma_{EW}^2)$ and the broad sense heritability $H^2 = (2\sigma_A^2)/(2\sigma_A^2 + \sigma_{EW}^2)$, and $H^2 = (2h^2)/(1 + h^2)$ The broad sense heritabilities for *Drosophila* behaviors are all much greater than predicted from strictly additive effects; therefore dominance and/or epistasis contribute to natural genetic variation for behavior in *Drosophila*, and this is also likely to be true for other species.

SUMMARY

The narrow sense heritability is an important property of a behavioral trait, because it predicts how closely the phenotypic measurement of an individual's behavior matches the value of the behavior transmitted to its progeny, and determines the extent to which relatives resemble each other for a behavioral trait. It is also important to know that the heritability is greater than zero if we are interested in mapping the genes contributing to natural variation in behavior. We can quantify the phenotypic resemblance among relatives in an outbred population by measuring the behavior in parents and their offspring, and measuring the behavior in siblings (full-siblings, half-siblings, and monozygotic and dizygotic twins). The statistics quantifying the degree of resemblance among relatives are offspring–parent regressions and intraclass correlations of siblings. For all types of relationship, it is the observed phenotypic covariance among relatives that statistically quantifies the amount of resemblance among them.

The theoretical phenotypic covariance among relatives has two components, the genetic covariance and the environmental covariance. Genetic covariance arises because two related individuals are likely to share alleles that are copies of the same allele in a recent common ancestor

TABLE 7.9 Heritability estimates of behavioral traits

Species	Trait	Estimate	Reference	Species	Trait	Estimate	Reference
Human	Schizophrenia[a]	0.62	1		Time spent in open arm elevated plus maze[c]	0.38	5
	Schizophrenia[a]	0.70–0.85	2		Time spent in closed arm elevated plus maze[c]	0.27	5
	Alcoholism (males)[a]	0.38	1		Entries into closed arm elevated plus maze[c]	0.19	5
	Alcohol dependence[a]	0.50–0.60	2		Entries into open arm elevated plus maze[c]	0.39	5
	Criminal conviction[a]	0.58	1		Activity in open field[c]	0.34	5
	Reading disorder[a]	0.60	1		Ethanol preference drinking[b]	0.32	6
	Bipolar disorder[a]	0.60–0.85	2		Nest building[b]	0.28	7
	Major depressive disorder[a]	0.40	2	Drosophila	Copulation latency[d]	0.25	8
	Autism/Austism spectrum disorders[a]	0.90	2		Copulation latency[b]	0.07	9
	Anorexia nervosa[a]	0.55	2		Copulation latency[e]	0.25	10
	Bulimia nervosa[a]	0.60	2		Aggressive behavior[b]	0.09	11
	Panic disorder[a]	0.40–0.50	2		Aggressive behavior[e]	0.78	12
	Obsessive-compulsive disorder[a]	0.60–0.70	2		Startle response[d]	0.26	13
	Attention deficit–hyperactivity disorder[a]	0.60–0.90	2		Startle response[b]	0.16	14
Mouse	Maternal defense behavior[b]	0.40	3		Startle response[e]	0.58	10
	Voluntary wheel running[b]	0.28	4		Ethanol knock-down time[b]	0.08	15
	Startle response[c]	0.41	5		Ethanol knock-down time[e]	0.24	16
	Distance moved in open arm of elevated plus maze[c]	0.42	5		Geotaxis[b]	0.13	17
	Distance moved in closed arm of elevated plus maze[c]	0.39	5				

References
(1) McCue, and Bouchard, (1998). Annu. Rev. Neurosci., **21**, 1–24; (2) Burmeister, et al. (2008). Nat. Genet., **9**, 526–540; (3) Gammie, et al. (2006). Behav. Genet., **36**, 713–722; (4) Swallow, et al. (1998). Behav. Genet., **28**, 227–237; (5) Valdar, et al. (2006), Genetics, **174**, 959–984; (6) Belknap, et al. (1997). Behav. Genet., **27**, 55–66; (7) Lynch, (1980). Genetics, **96**, 757–765; (8) Moehring, and Mackay, (2004). Genetics, **167**, 1249–1263; (9) Mackay, et al. (2005). Proc. Natl Acad. Sci. USA, **102**(Suppl. 1), 6622–6629; (10) Ayroles, et al. (2009). Nat. Genet., **41**, 299–307; (11) Edwards, et al. (2006). PLoS Genet., **2**, e154; (12) Edwards, et al. (2009). Genome Biol., **10**, R76; (13) Jordan, et al. (2006). Genetics, **174**, 271–284; (14) Jordan, et al. (2007). Genome Biol., **8**, R172; (15) Morozova, et al. (2007). Genome Biol., **8**, R231; (16) Morozova, et al. (2009). Genetics, in press; (17) Watanabe, and Anderson, (1976). Behav. Genet., **6**, 71–86.
Superscripts indicate the method used to estimate heritability.
[a]$2(t_{MZ} - t_{DZ})$;
[b]Realized h^2;
[d]Recombinant inbred lines;
[c]Correlations among relatives (siblings, parents, grandparents);
[e]Inbred lines.

(i.e. identical by descent, IBD). The genetic covariance between any pair of relatives is a function of the additive dominance, and two-locus epistatic variances of the trait, and the probabilities of having one and two loci IBD. Environmental covariance can arise if families share common spatial, temporal, or maternal environments. Variation due to common environment will cause greater similarity among members of a family, and greater differences among families, than would be expected from the proportion of alleles they share, and thus will increase the phenotypic covariance between relatives. We can write the regressions or intraclass correlations quantifying the observed degree of resemblance among relatives to the theoretical phenotypic covariance among relatives written in terms of genetic and environmental variance components. The narrow sense heritability, h^2, is a significant component of all of the phenotypic relationships, and the estimate of h^2 is a simple multiple of the estimates of the regression and intraclass correlations. Most estimates of h^2 are too high: they are biased upward by inclusion of environmental variance common to the sets of relatives in the analysis and nonadditive genetic variance (dominance and epistasis). Much information about heritability estimates for human behavioral traits have been obtained from studies that compare phenotypic variation between monozygotic and dizygotic twins. Large numbers of individuals are needed to obtain precise estimates of heritability. The smaller the heritability, the greater the sample size required to detect it as significantly different from zero.

In model organisms that can be inbred to homozygosity, broad sense heritabilities (H^2) can be measured by crossing inbred lines. Variation for most behaviors studied has a significant genetic component. Studies on *Drosophila* have revealed major contributions of dominance and/or epistatic interactions to natural genetic variation for behavior.

STUDY QUESTIONS

1. Why are heritability estimates useful?
2. Explain how we can statistically quantify the resemblance between offspring and their parents.
3. Explain how we can statistically quantify the resemblance between siblings.
4. What causes genetic covariance between relatives?
5. What causes environmental covariance between relatives?

6. Describe three ways in which heritability can be estimated. Which of these methods would you prefer to use, and why?
7. Discuss potential sources of bias of heritability estimates. What is the consequence of these potential biases?
8. Explain the concept of "realized heritability."
9. In a population in which the heritability of a behavioral trait is 0.45, what is: (1) the correlation among full siblings; and (2) the regression of offspring on one parent? What assumptions have you made?
10. Imagine that you have conducted a study on alcohol drinking in humans by questioning fathers and their adult children about their alcohol consumption per week. You have data from over 500 families. You calculate the regression of mean alcohol intake of adult children and their fathers to be 0.35. What is you estimate of heritability of alcohol intake? Comment on the reliability of the estimate (i.e. is it precise or is it biased?)
11. Imagine that you have conducted a study on schizophrenia in humans. You measured the correlation of pairs of monozygotic twins as 0.7, and of pairs of dizygotic twins as 0.3. All twin pairs were reared together in the same household. What is your estimate of heritability of schizophrenia? How is the estimate biased?
12. Why is it preferable when estimating heritability using a single parent–offspring design to compare fathers and their children, rather than mothers and their children?

RECOMMENDED READING

Falconer, D. S., and Mackay, T. F. C. (1996). *Introduction to Quantitative Genetics*, 4th edition. Longmans Green, Harlow, Essex, UK.

Lynch, M., and Walsh, B. (1997). *Genetics and Analysis of Quantitative Traits*. Sinauer Assoc. Inc., Sunderland, MA.

Chapter 8

Quantitative Trait Locus Mapping

OVERVIEW

Understanding the genetic basis of behavior requires linking phenotypic variation to variation in the DNA sequences of genes that contribute to the behavior. To accomplish this, strains that differ in the phenotype under study can be crossed, and segregation of parental polymorphisms can be correlated with variation in the phenotype. Statistical methods are employed to detect regions of the genome in which segregation of polymorphic markers correlates with differences in phenotypic values. Such regions, known as **quantitative trait loci** (QTLs) typically contain a large number of genes, and identifying the causal genes within these QTL regions is often a formidable challenge. This challenge can be addressed either by creating large numbers of additional recombinants within the area of interest, or by testing whether candidate genes in a QTL region contain polymorphisms that are associated with phenotypic variation in an outbred population. Additional studies may be necessary to confirm conclusions drawn on gene identification from linkage and/or association studies. Despite the challenging and labor-intensive nature of these approaches, linkage studies and association analyses have been applied to virtually every behavior that has been studied to date. In this chapter we will describe the general principles that underlie linkage and association analyses for mapping QTLs, and give a few examples that illustrate these principles. Additional examples are given in other chapters, including identification of genes that underlie taste perception in people (Chapter 13), startle-induced locomotion behavior in *Drosophila* (Chapter 11), alcohol intake in people (Chapter 15), and alcohol and drug withdrawal responses in mice (Chapter 15).

We have seen in Chapters 5–7 that the underlying quantitative genetic basis of variation in behavioral traits is determined by the homozygous (a), heterozygous (d), and epistatic allelic effects and allele frequencies (p, q) of individual loci affecting the trait. We can estimate what fraction of the phenotypic variation for behavior in a population is due to variation in breeding values (the narrow

sense heritability, h^2). However, this parameter summarizes the aggregate effects of all loci affecting the trait, and gives no information about effects and frequencies of individual QTLs causing the variation.

Knowledge of the QTLs affecting naturally occurring variation in behavioral traits is important from several perspectives. First, understanding the genes and genetic networks affecting variation in behavior can give insights about the underlying cellular and neural mechanisms affecting behavior, which can lead to novel therapeutic interventions for extreme and pathological behaviors. Performing these analyses in model organisms provides candidate genes for study in humans. Finally, we need to know not only the genes, but also the actual molecular polymorphisms affecting variation in behavior if we are to understand the evolution of behavior, and why genetic variation for behavioral traits is maintained in natural populations.

Mapping QTLs affecting all quantitative traits, including complex behaviors, is complicated because we expect that the genetic variation will be due to multiple QTLs, each with small genotypic effects relative to the total phenotypic variation, and whose expression is sensitive to the environment. How then can we map the QTLs? The answer is that we can do this if they are genetically linked to visible marker loci with genotypic classes that we can unambiguously categorize. This principle was recognized by Sax early in the twentieth century, but its application was stymied by the lack of appropriate marker loci until relatively recently, when the advent of DNA sequencing technologies made resequencing individual gene regions (and even more recently, whole genomes) possible. This led to the discovery of abundant polymorphic markers that can be used in mapping studies. Today, the most frequently used markers are **single nucleotide polymorphisms** (SNPs), and **copy number variants** such as **microsatellites** and **indel** polymorphisms. There are commercially-available genotyping platforms for mapping studies in humans and mice with over 500 000 SNPs that can be scored simultaneously. In other organisms it is necessary to develop custom genotyping assays for each application – identifying markers still remains a significant effort in most QTL mapping studies.

There are two major methods for mapping QTLs. The first, **linkage mapping**, maps QTLs in families from an outbred population, or in segregating generations derived

R. Anholt and T. Mackay: Principles of Behavioral Genetics
ISBN: 978-0-12-372575-2

from crosses of inbred lines in model organisms. The second, **association mapping**, maps QTLs in a sample of individuals from an outbred population, or many inbred lines derived from an outbred population in model organisms. In both cases, we need to obtain measurements of the behavioral phenotype for all individuals in the mapping population, and determine the marker locus genotypes for all individuals in the mapping population, at all marker loci. Then we use a statistical method to determine whether there are differences in behavior between marker genotypes; if so, the QTL is linked to the marker.

LINKAGE MAPPING

The simplest linkage mapping experiments involve crosses between completely inbred lines that differ genetically for the behavior in model organisms. Such lines are homozygous for all loci, including marker loci and QTLs affecting the behavior. However, many of the loci will be homozygous for *different* molecular marker alleles and QTL alleles; it is these loci that are informative for mapping. The F_1 between two inbred lines is also genetically uniform: all individuals are heterozygous for all variable loci. However, F_1 progeny derived from backcrossing to either parent (BC progeny) or by mating F_1 individuals to each other to produce the F_2 generation are genetically variable, and can be used for mapping (Figure 8.1).

In organisms with short generation times that are amenable to inbreeding, the F_2 population can be subdivided into a large number of lines that are inbred to a point where they are all homozygous. Such recombinant inbred lines (RILs) are very useful, and have been used extensively to

map QTLs in mice, *C. elegans* and *Drosophila*. Although RILs take a long time to construct, they have the advantage that markers only need to be genotyped once, and the same lines can be phenotyped in multiple traits and in different environments. Each BC or F_2 individual has a unique genotype and phenotype.

The choice of experimental design (BC, F_2, RILs) often depends on the biology of the species for which we wish to map QTLs affecting variation in behavior. Reciprocal BCs to both parents and the F_2 design both permit estimating homozygous and heterozygous effects. It is only feasible to construct RILs in organisms with short generation times and sufficiently high reproductive rates to be able to tolerate the inevitable loss of lines during inbreeding, such as mice and *Drosophila*. We can only estimate homozygous effects of QTLs in RIL populations, but crossing each RIL to both parental lines is a good strategy for estimating homozygous and heterozygous effects, while preserving the benefits of this design.

Having constructed a linkage mapping population, we now measure the behavioral trait for all individuals (or RILs) in the population, and we determine the genotype of all individuals for many polymorphic molecular markers covering the entire genome (Figure 8.2). We then perform statistical tests to determine whether there is a difference in behavior between individuals with different molecular marker genotypes, to infer if the QTL is linked to the marker. We systematically scan the genome for QTLs by performing these tests for each marker locus in turn (**single marker analysis**), or for pairs of adjacent markers (**interval mapping**), to estimate both the chromosomal locations and genetic effects (a, d) of the QTLs throughout the entire genome.

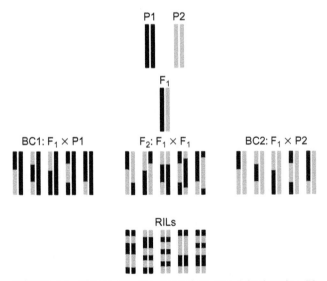

FIGURE 8.1 Linkage mapping populations. Two inbred strains, P1 and P2, are crossed to produce the F_1 generation. F_1 individuals can be crossed to P1 and P2 to produce backcross (BC) mapping populations, or to each other to produce an F_2 mapping population. Inbreeding of the F_2 by 20 generations of full-sibling mating results in a population of recombinant inbred lines (RILs).

					Genotype				
Individual	M1	M2	M3	M4	M5	M6	M7	M8	Phenotype
1	1	1	2	2	2	3	3	1	20
2	3	3	1	1	1	1	1	3	8
3	1	1	2	2	1	1	2	1	14
4	3	2	3	2	2	2	3	2	24
5	1	3	2	2	2	2	3	3	28
6	3	3	2	3	1	3	2	1	18
7	3	2	3	1	3	1	3	3	10
8	2	3	1	3	2	2	1	2	22
9	2	1	3	3	3	3	1	3	20
10	1	2	1	3	3	3	2	2	17

FIGURE 8.2 Hypothetical data from a QTL mapping experiment. Each row in the table represents an individual from an F_2 mapping population. Molecular marker genotypes (M1–M8) are determined for each individual, with 1 indicating the same genotype as parent 1; 2 indicating the same genotype as parent 2; and 3 indicating a heterozygote. In addition, each individual is evaluated for performance in a behavioral paradigm, and given a quantitative score (phenotype).

Single Marker Analysis

To illustrate the principle of QTL mapping, let us first consider an F_2 design, and test whether a QTL is linked to a single molecular marker. The distance between the QTL (locus A) and the marker (locus M) is measured in recombination units, or centimorgans (cM), c (Figure 8.3). c is the recombination fraction between M and A, and ranges from $c = 0$ (M and A are the same locus and there is no recombination between them) to $c = 0.5$ (M and A are unlinked; i.e. they are on different chromosomes or 50 cM apart on the same chromosome, so they segregate at random).

We distinguish the alleles at M and A for one homozygous parental strain (P1) by the subscript 1 and the other homozygous parental strain (P2) by the subscript 2. Thus we write the genotype of P1 as M_1A_1/M_1A_1, where the '/' indicates the two homologous chromosomes. The homozygous effect of this genotype is a. Similarly, the genotype of P2 is M_2A_2/M_2A_2, with a homozygous effect of $-a$; and the genotype of the F_1 is M_1A_1/M_2A_2, with a heterozygous effect of d. The F_2 population is produced by random union of F_1 gametes. There are four F_1 gamete types. The parental gamete types, M_1A_1 and M_2A_2, are produced when there is no recombination between the QTL locus and the marker locus. Since the probability of recombination is c, the probability of no recombination is $1 - c$. Thus, these gamete types are each produced with frequency $(1 - c)/2$.

The recombinant gamete types, M_1A_2 and M_2A_1, are the two products of a single recombination between the QTL locus and the marker locus, and are therefore each produced with frequency $c/2$.

The F_2 produced by random mating of the F_1 gives 10 possible F_2 genotypes at both the QTL and marker loci, but only three marker genotype classes (Table 8.1). We can work out the contribution of each marker genotype class to the F_2 mean for the behavioral phenotype by multiplying the frequency of each genotype by its genotypic value, then summing within marker genotype classes. We want the actual means of the behavioral phenotype for each marker class, which are obtained by dividing the contribution to the F_2 mean by the frequency of that marker class, which is the Mendelian segregation ratio of 1/4 for the homozygotes, and 1/2 for the heterozygotes (Table 8.1).

Now we can compare the mean behavioral phenotypes between individuals of different marker genotypes, and interpret the differences between marker genotype classes in terms of additive and dominance effects of the QTL (a and d), and the recombination fraction, c. Let us indicate the mean value of a behavioral trait for many individuals of a particular marker genotype by a bar above the genotype. From Table 8.1, we see that $(\overline{M_1/M_1} - \overline{M_2/M_2})/2 = a(1 - 2c)$ and $\overline{M_1/M_2} - (\overline{M_1/M_1} + \overline{M_2/M_2})/2 = d(1 - 2c)^2$. Therefore, a significant difference in the mean value of a quantitative trait between homozygous marker genotype classes indicates linkage of a QTL and the marker locus. However, estimates of a and d/a from single marker analysis are confounded with recombination frequency, and will underestimate the true values by $(1 - 2c)$. For example, if the difference between homozygous marker class means is 4, this could be due to a QTL with homozygous effect $a = 2$ at the marker locus, or a QTL with homozygous effect $a = 2.5$ located 10 cM ($c = 0.1$) from the marker locus.

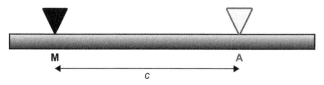

M   A

$\longleftarrow \quad c \quad \longrightarrow$

FIGURE 8.3 Single marker analysis. The figure depicts a marker locus (M) and a QTL locus (A). c is the recombination fraction between M and A.

TABLE 8.1 Single marker analysis

Genotype	Frequency	Effect	Marker class	Marker frequency	Contribution to F_2 mean	Actual mean
M_1A_1/M_1A_1	$(1 - c)^2/4$	a				
M_1A_1/M_1A_2	$c(1 - c)/2$	d	M_1/M_1	1/4	$a(1 - 2c)/4$ $+ dc(1 - c)/2$	$a(1 - 2c) + 2dc(1 - c)$
M_1A_2/M_1A_2	$c^2/4$	$-a$				
M_1A_1/M_2A_1	$c(1 - c)/2$	a				
M_1A_1/M_2A_2	$(1 - c)^2/2$	d	M_1/M_2	1/2	$d[(1 - c)^2 + c^2]/2$	$d[(1 - c)^2 + c^2]$
M_1A_2/M_2A_1	$c^2/2$	d				
M_1A_2/M_2A_2	$c(1 - c)/2$	$-a$				
M_2A_1/M_2A_1	$c^2/4$	a				
M_2A_1/M_2A_2	$c(1 - c)/2$	d	M_2/M_2	1/4	$-a(1 - 2c)/4$ $+ dc(1 - c)/2$	$-a(1 - 2c) + 2dc(1 - c)$
M_2A_2/M_2A_2	$(1 - c)^2/4$	$-a$				

The table shows the calculation of the mean of three marker locus (M) genotypes attributable to a linked QTL (A) for an F_2 mapping population.

INTERVAL MAPPING

Interval mapping analysis considers pairs of adjacent marker loci (M and N), and asks whether there is evidence for a QTL (A) affecting the trait between the marker loci. The distance between the marker loci is c, the distance between M and A is c_1, and between A and N is c_2 (Figure 8.4). To simplify the discussion, we will assume that M and N are close enough together that there are no double crossovers between them, and therefore $c = c_1 + c_2$ (technically, this assumption is one of complete **crossover interference**, which is likely when the markers are less than 10 cM apart). The genotypes of P1, P2, and F_1 are $M_1A_1N_1/M_1A_1N_1$, $M_2A_2N_2/M_2A_2N_2$, and $M_1A_1N_1/M_2A_2N_2$, respectively, with corresponding genotypic effects of a, $-a$ and d.

Interval mapping can be done with any mapping population (BC, F_2, RIL). Let us consider a BC to P1. Similar to single marker analysis, we tabulate BC genotypes, frequencies and means, assuming no double recombination, and calculate expected marker genotype means (Table 8.2). As for the single marker example, contrasts between backcross marker class means estimate the effects of the QTL. In contrast to the single marker analysis, however, the map position of the QTL relative to the flanking markers can also be estimated. From Table 8.2, we see that $\overline{M_1N_1/M_1N_1} - \overline{M_2N_2/M_1N_1} = a - d = \gamma$; and

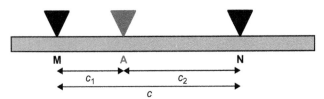

FIGURE 8.4 Interval mapping analysis. The figure displays two marker loci (M and N) and a QTL locus (A). c is the recombination fraction between M and N, and c_1 and c_2 are the recombination fractions between M and A, and N and A, respectively. We assume that M and N are sufficiently close together that no double crossovers will occur between them, so that $c = c_1 + c_2$.

$\overline{M_1N_2/M_1N_1} - \overline{M_2N_1/M_1N_1} = (a - d)(c_2 - c_1)/c = \eta$. For a backcross to one parent strain, the estimate of a is unbiased only if $d = 0$, so recessively-acting QTLs may not be detected. This problem can be overcome by backcrossing to both parental lines, or by using an F_2 design, in which case unbiased estimates of both a and d are obtained.

Using interval mapping, we can estimate c, the recombination frequency between the marker loci, directly from the observed numbers of recombinant and non-recombinant marker genotypes. Once we have this information, we can also estimate c_1 and c_2, and thus estimate the position of the QTL between the marker loci. From the two contrasts of marker class means above, we have $\eta/\gamma = (c_2 - c_1)/c = (c - 2c_1)/c$. Since we have estimates of γ, η, and c, we can solve for c_1, and then solve $c = c_1 + c_2$ to obtain the estimate of c_2.

STATISTICAL ANALYSIS

Interval mapping is preferred over single marker analysis, because the estimate of QTL effect is not confounded with the map distance of the QTL from the marker. However, this bias becomes less of a problem as the marker density increases, and single marker regression models are more flexible than interval mapping methods in terms of adding covariates (such as litter size and sex) to the models for evaluating differences between marker class means.

We expect that multiple QTLs will be responsible for genetic variation for a behavioral trait between the parental lines and the mapping population derived from them. Thus, when we are testing for the presence of a QTL at any one marker or pair of adjacent markers, there will be variation between individuals within each marker class from segregation of other QTLs affecting the trait, in addition to environmental variation. Statistical methods have been developed to test for linkage of a QTL to a marker (or pair

TABLE 8.2 Interval mapping

Genotype	Frequency	Effect	Marker class	Marker frequency	Contribution to BC mean	Actual mean
$M_1A_1N_1/M_1A_1N_1$	$(1 - c)/2$	a	M_1N_1/M_1N_1	$(1 - c)/2$	$a(1 - c)/2$	a
$M_1A_1N_2/M_1A_1N_1$	$c_2/2$	a	M_1N_2/M_1N_1	$c/2$	$(ac_2 + dc_1)/2$	$(ac_2 + dc_1)/c$
$M_1A_2N_2/M_1A_1N_1$	$c_1/2$	d				
$M_2A_1N_1/M_1A_1N_1$	$c_1/2$	a	M_2N_1/M_1N_1	$c/2$	$(ac_1 + dc_2)/2$	$(ac_1 + dc_2)/c$
$M_2A_2N_1/M_1A_1N_1$	$c_2/2$	d				
$M_2A_2N_2/M_1A_1N_1$	$(1 - c)/2$	d	M_2N_2/M_1N_1	$(1 - c)/2$	$d(1 - c)/2$	d

The table shows the calculation of the mean of four marker locus (M, N) genotypes attributable to a linked QTL (A) for a population derived from a BC of the F_1 to $M_1A_1N_1$.

of markers) that include the effects of other QTLs in the model. These methods reduce the variance within marker genotype classes, improve estimates of QTL map positions and effects, and improve the ability to detect multiple-linked QTLs over methods that do not consider multiple QTLs simultaneously.

We test many markers for linkage to a QTL in a genome scan. Therefore, we cannot use a *P*-value that would be appropriate if we had only done a single test. A *P*-value of α is the probability that a test statistic as extreme as the one observed would occur by chance, if there truly was no difference in mean value of the behavioral trait between the marker genotypes. Thus, if we do *n* independent tests, we expect $n\alpha$, not α, false positive test results. To control the false positive rate of the whole experiment, we should set the *P*-value for each independent test to α/n (a Bonferroni correction; see also Chapter 10). However, the number of independent tests will be less than the number of markers, because of linked markers. Therefore, **permutation tests** (Box 8.1) are typically used to determine appropriate experiment-wise significance levels, which account both for multiple tests and correlated (linked) markers.

The results of linkage mapping analyses are depicted graphically as a QTL plot, where the locations of the markers (in cM) are given on the *x*-axis, and the result of the statistical test is indicated on the *y*-axis, scaled so that the largest values are the most significant. We also indicate the value of the permutation threshold as a horizontal line parallel to the *x*-axis, and intersecting the *y*-axis at the appropriate value. Evidence for linkage of a QTL with markers occurs when the test for linkage generates a significance level that exceeds the permutation threshold. The best estimate of the QTL location is the position on the *x*-axis corresponding to the greatest significance level. Figure 8.5, overleaf, illustrates such a graphical representation of QTLs affecting mating behavior in *Drosophila*.

There are many freely-available statistical programs for implementing QTL mapping methods and using permutation to determine appropriate significance thresholds. Two popular software suites are QTL Cartographer (http://statgen.ncsu.edu/qtlcart/) and R-QTL (http://www.rqtl.org/).

Number of Individuals and Markers

How many individuals do we need to map QTLs? The answer to this question has two components: the number of individuals needed to *detect* a QTL; and the number required to *localize* the QTL. The power to detect a difference in mean between two marker genotypes depends on δ/σ_w, where δ is the difference in mean between homozygous marker classes ($2a$) or between homozygous and heterozygous marker classes [$(a - d)$ or $(-a - d)$], and σ_w is the standard deviation of the behavioral trait within each marker genotype class.

Box 8.1   Permutation tests

Permutation tests are used to determine significance thresholds for QTL mapping experiments, which account for the multiple statistical tests performed and linked markers. To perform a permutation, we shuffle the observed phenotypic data among the individuals of the mapping population, while keeping the genotype data for each individual the same (Figure A, panel A). We then repeat the genome scan for QTL on the permuted data, and record the highest test statistic (or lowest *P*-value) resulting from the analysis of the permuted data. We do this at least 1000 times. We then plot the distribution of test statistics (or *P*-values) from analyses of the permuted data (Figure A, panel B). This is the distribution of *P*-values under the null hypothesis of no association between markers and the trait. The permutation significance threshold, α, is the value of the test statistic or *P*-value corresponding to the most extreme α% of this distribution (Figure A, panel B).

	Genotype									Phenotype			
Ind.	M1	M2	M3	M4	M5	M6	M7	M8	O	P1	P2	P3	P4
1	1	1	2	2	2	3	3	1	20	14	24	8	20
2	3	3	1	1	1	1	1	3	8	20	22	17	28
3	1	1	2	2	1	1	2	1	14	18	20	28	10
4	3	2	3	2	2	2	3	2	24	8	18	20	8
5	1	3	2	2	2	2	3	3	28	22	10	18	17
6	3	3	2	3	1	3	2	1	18	24	20	14	22
7	3	2	3	1	3	1	3	3	10	17	14	24	20
8	2	3	1	3	2	2	1	2	22	28	17	20	18
9	2	1	3	3	3	3	1	3	20	10	8	22	24
10	1	2	1	3	3	3	2	2	17	20	28	10	14

A

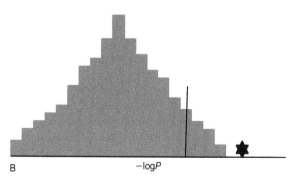

B $-\log P$

FIGURE A Permutation tests. (A) Hypothetical genotype (M1–M8) and phenotype (O) data for 10 F_2 individuals (Ind.) from a QTL mapping experiment (see Figure 8.2). The observed phenotypic data are permuted (shuffled) among the individuals (P1–P4), as indicated. (B) Distribution of permuted *P*-values, scaled as $-\log P$. The vertical line indicates the *P*-value corresponding to the 5% significance threshold. The star shows the *P*-value computed form the observed data. Since the observed *P*-value is more extreme than the 5% significance threshold, we can say the observed data are more extreme than expected by chance, after accounting for multiple tests.

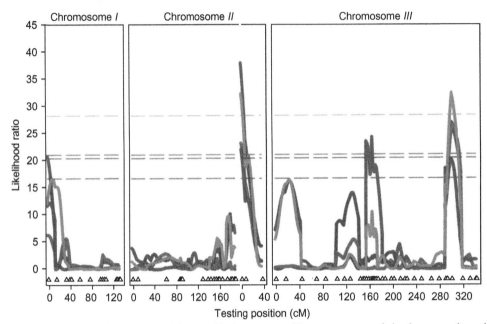

FIGURE 8.5 Genome scan for QTLs affecting *Drosophila* courtship latency, courtship occurrence, copulation latency, and copulation occurrence. Triangles on the *x*-axis denote the locations of molecular markers on the three major *D. melanogaster* chromosomes. Numbers on the *y*-axis are the results of a statistical test for the presence of a QTL in each marker interval. The horizontal lines are the empirical significance thresholds based on permutation tests. Any observed test statistic greater than these values are formally significant taking multiple tests into account. The most likely location of the QTL is the position on the *x*-axis associated with the highest test statistic. (From Moehring and Mackay (2004). *Genetics*, **167**, 1249–1263.)

Table 8.3 shows sample sizes necessary to detect QTLs with a range of values of δ/σ_w in BC and F_2 designs. Genotypes and phenotypes for large numbers (>500–1000) of individuals in the mapping populations are necessary to detect QTLs with moderate effects ($\delta/\sigma_w = 0.25$). Extremely large sample sizes are necessary to detect QTLs with small effects. An advantage of using RILs for linkage mapping of QTLs is that one can measure many individuals of the same genotype to get an accurate estimate of the mean genotypic effect of each line; this reduces σ_w and increases the power to detect QTLs relative to an F_2 or BC mapping population of the same size as the population of RILs.

Localizing a QTL depends on increased recombination, which in turn requires large numbers of individuals to ensure fine-scale recombination and increased marker density to identify recombinants. Table 8.4 shows the numbers of individuals and marker genotypes necessary to map QTLs for a range of recombination fractions (*c*). The relationship between recombination fraction and physical distance varies between species, and across the genome within species, but we can infer the scale of mapping using the *Drosophila* genome as an example. The physical size of the *Drosophila* eukaryotic genome is 120 Mb, which encodes approximately 14 000 genes. The total recombination map distance is 286 cM. Thus, a QTL localized to a 5 cM interval (*c* = 0.05) would span 2100 kb, and include on average 245 genes, while a QTL localized to a 1 cM interval would span 420 kb, and include 49 genes.

TABLE 8.3 Number of individuals required to detect a QTL with effect δ/σ_w in linkage mapping populations

δ/σ_w	n	$N(BC)$	$N(F_2)$
1	21	42	84
0.5	84	168	336
0.25	336	672	1344
0.125	1344	2688	5376
0.0625	5376	10752	21 504

The calculations were done assuming a simple model in which a t-test is used to judge the significance of the difference of two marker class means, with a false positive error rate of 0.05 and a false negative error rate of 0.1. We also assume that the QTL is completely linked to the marker (c = 0). n is the total number of individuals required per marker class, and N is the total sample size for a BC and F_2 population.

Clearly, extremely large mapping populations would be needed if we attempted to simultaneously detect QTLs and localize them to small chromosomal regions. Thus, linkage mapping of QTLs is typically an iterative procedure, where we first determine the general location (~10–20 cM intervals) of QTLs in a mapping population of several hundred to approximately 1000 individuals. We then narrow in on the regions that we know contain the QTLs, determine their location more precisely by focusing on individuals in

TABLE 8.4 Numbers of individuals (N) and markers (M) needed to localize a QTL

c	N	$M/100\,cM$	$NM/100\,cM$
0.10	29	11	319
0.05	59	21	1239
0.01	298	101	30098
0.005	598	201	120198
0.001	2994	1001	2996994

c is the recombination fraction. N is the number of individuals to detect at least one recombinant ($N = \log(0.05)/\log(1 - c)$, where 0.05 is the significance threshold). M is the number of markers needed to detect recombinants per 100 cM, and NM is the total number of marker genotypes needed per 100 cM.

which recombination has occurred between the markers flanking the QTL, and essentially repeat the whole procedure on the smaller genomic regions. This phase requires generating many more individuals to obtain the necessary recombination, and identifying molecular markers within the region of interest. These experiments are very laborious. Many QTLs affecting behavioral traits have been mapped approximately; only a few have been identified at the level of a single gene.

Caveats and Solutions

Estimates of the number of QTLs from any experiment are always minimum estimates. This is because each experiment has a limited number of samples, and the sample size sets the threshold for detection of the size of QTL effects; there will always be other loci with effects that are too small to be detected by an experiment of a particular size. In addition, the sample size determines the density of the recombination map, and the more recombination, the greater the power to separate closely-linked QTLs. Finally, linkage mapping can only detect QTLs that are segregating in the mapping population (i.e. between the two parental strains); other loci will probably be found in other strains. Thus, several strategies have been proposed and used to increase the power to detect QTLs, increase recombination, and increase genetic diversity in QTL mapping populations.

Often, the major factor limiting the sample size of a QTL mapping study is the expense of genotyping. Thus, for a fixed number of genotyping reactions, one can increase power by measuring the behavior of a large number (many thousands) of individuals, but only genotyping the high and low tails of the distribution (several hundred). Selective genotyping also gives a dense recombination map proportional to the sample size of the population for which

behavior was measured, giving more precise localization of QTLs than would be expected from the size of the population genotyped.

Recombination maps refer to a single generation of recombination. In model organisms with rapid generation intervals, it is often possible to maintain a mapping population by random mating of the BC or F_2 for many generations. In such **advanced intercross lines**, the opportunity for recombination in any region increases with time, and this expansion of the recombination map increases the precision of localizing QTLs.

Genetic diversity can be increased by starting linkage mapping populations from eight parental strains, not just two strains. This strategy has been embraced by mouse geneticists. One such mapping population is a heterogeneous stock (HS) of mice derived over 30 years ago by crossing eight inbred mouse strains. The HS population has been maintained with 40 breeding pairs per generation for over 60 generations, and has an average distance of 1.7 cM between recombinants. Thus, this population offers the increased mapping resolution of an advanced intercross population, while simultaneously increasing genetic diversity. The mouse Complex Trait Consortium has proposed the development of a novel community resource for QTL mapping: the Collaborative Cross. This population was also initiated by crosses among eight inbred mouse strains (not the same strains used to establish the HS population). The goal of this project is to develop 1000 RILs from the eight-way cross by 20 generations of full sibling inbreeding of each line. Although these lines will not be as fine-grained a mosaic as the HS population, they offer the community all the advantages of RILs.

ASSOCIATION MAPPING

Using linkage mapping to resolve QTLs to the level of individual genes responsible for variation in behavior is very challenging, due to the very large numbers of individuals necessary to obtain recombinants that implicate single genes. Association mapping utilizes historical recombination in random mating populations to identify QTLs. Association mapping is based on **linkage disequilibrium** (LD) between the QTL and a marker locus, so we need to understand LD in order to understand how association mapping works.

Linkage Disequilibrium

LD is a measure of the correlation in allele frequencies between two loci. Let the two loci be A and B, each with two alleles (A_1, A_2, B_1, B_2). The frequencies of A_1 and A_2 are p_A and q_A, and of B_1 and B_2 are p_B and q_B. If the allele frequencies at these loci are uncorrelated, the expected frequency of each of the four gamete types is the product of the allele frequencies at each locus separately (Table 8.5). The gamete types are called **haplotypes**, because it is the haploid gametes that contribute to the genotypes in each successive generation. If allele frequencies are uncorrelated, the population is in "linkage equilibrium," and $P_{11}P_{22} - P_{12}P_{21} = 0$, where P_{11}, P_{12}, P_{21}, and P_{22} are the observed frequencies of the haplotypes A_1B_1, A_1B_2, A_2B_1, and A_2B_2, respectively (Table 8.5). If allele frequencies are non-randomly associated, the gamete frequencies are not the simple product of the allele frequencies, but depart from this by amount D, where D is the **coefficient of linkage disequilibrium** (Table 8.5). With LD, $P_{11}P_{22} - P_{12}P_{21} = D$. D is a measure of how much the observed haplotype frequencies differ from their expected frequencies under random association. We can test whether or not two marker loci are in LD by comparing observed and expected haplotype frequencies using a χ^2 goodness of fit test, with one degree of freedom (Box 8.2).

The actual numerical value of D depends on allele frequencies at the two loci. The highest possible value of D is if $p_A = q_A = p_B = q_B = 0.5$, and haplotype A_1B_2 and A_2B_1 are missing (complete linkage disequilibrium); D is then 0.25 (Figure 8.6, p. 118). Since the labels of alleles are arbitrary for marker loci, we will consider only positive values of D. Because of the dependence on allele frequency, values of D estimated from data are typically scaled by the observed allele frequencies. Two common measures of LD are D' and r^2, the squared correlation of allele frequencies. D' scales the observed value of D by its theoretical maximum value, given the observed gene frequencies: $D' = D/D_{max}$. Scaled in this way, the maximum value of D' is always 1. We can determine D_{max} by noting, from Table 8.5, that $P_{12} = p_Aq_B - D \geq 0$ and $P_{21} = q_Ap_B - D \geq 0$. Thus $D \leq p_Aq_B$ and $D \leq q_Ap_B$; to satisfy both inequalities, D_{max} must be the smaller of p_Aq_B or q_Ap_B. r^2 scales the squared value of D by the product of all allele frequencies: $r^2 = D^2/p_Ap_Bq_Aq_B$. Population geneticists like r^2, because its expected value is $E(r^2) = 1/(1 + 4Nc)$ in an equilibrium

population, where N is the population size and c the recombination fraction between the two loci.

How can LD be used for QTL mapping? To understand this we need to realize two things. First, when the mutation affecting the behavioral trait first occurs in a population, it is in complete LD with all other polymorphic loci in the genome (see also Chapter 16). Second, LD does not remain complete. D declines in successive generations in a random mating population by an amount which depends on the recombination fraction, c. After t generations of random mating the disequilibrium is $D_t = D_0 (1 - c)^t$. With unlinked loci and free recombination ($c = 0.5$), the amount of disequilibrium is halved by each generation of random mating; with linked loci the disequilibrium disappears more slowly (Figure 8.7, p. 118).

Disequilibrium between pairs of loci in a large random mating population is expected to be small unless the loci are very tightly linked. Thus, we can use historical recombination in a population, rather than recent recombination in a linkage mapping population, to map QTLs (Figure 8.8, p. 118). Table 8.6 on p. 118 shows how the frequency of recombinants increases with the number of generations of random mating. Therefore, much smaller sample sizes are needed to localize QTLs compared to linkage studies.

Statistical Analysis

Association mapping can be performed for a candidate gene, for a QTL region implicated by a linkage mapping study, or genome wide. Association mapping uses single marker analysis. We sample individuals from the population, measure their behavioral phenotype, and genotype them for molecular markers. If the behavioral trait is continuously distributed, we group the data into genotypes for each marker, and perform a statistical test (usually **analysis of variance** (ANOVA)) to assess whether there is a difference between the mean of the trait between different marker genotypes. If so, the locus affecting the trait is in LD with the marker locus (Figure 8.9, p. 119). For binomial traits, we group the data according to whether individuals are affected or not affected, and perform a statistical test (G-test or χ^2 test) to determine whether there is a difference in genotype frequencies or allele frequencies between classes. If so, the locus affecting the trait is in LD with the marker locus (Figure 8.9). Association mapping experiments have the same multiple-testing problem as linkage mapping experiments. In human genome-wide association studies, association tests are performed for many hundreds of thousands of markers. Thus, significance thresholds for association studies are typically adjusted using a Bonferroni correction for the total number of tests, or derived using permutation.

We have seen that using single marker analysis in the context of linkage mapping leads to underestimates of QTL

TABLE 8.5 Observed and expected haplotype frequencies for two loci, each with two alleles, in a population in linkage equilibrium (LE), and one in linkage disequilibrium (LD)

Haplotype	Observed frequency	Expected frequency (LE)	Expected frequency (LD)
A_1B_1	P_{11}	p_Ap_B	$p_Ap_B + D$
A_1B_2	P_{12}	p_Aq_B	$p_Aq_B - D$
A_2B_1	P_{21}	q_Ap_B	$q_Ap_B - D$
A_2B_2	P_{22}	q_Aq_B	$q_Aq_B + D$

Box 8.2   Linkage disequilibrium

Let us consider data from DNA sequences of 172 alleles of the *Drosophila melanogaster* Dopa decarboxlyase (*Ddc*) gene (De Luca et al. (2003). *Nat. Genet.*, **34,** 429–433), sampled from a large outbred population. The observed numbers of haplotypes for SNPs at two polymorphic loci in the promoter region were:

Locus 1 (A)	Locus 2 (B)	Number	Frequency
C	A	40	$P_{11} = 40/172 = 0.2326$
C	G	42	$P_{12} = 42/172 = 0.2442$
G	A	14	$P_{21} = 14/172 = 0.0814$
G	G	76	$P_{22} = 76/172 = 0.4419$

Is this population in linkage equilibrium at these loci? If not, what is the coefficient of linkage disequilibrium?

To determine whether or not the population is in linkage equilibrium at these loci, we need to compute the expected frequencies and expected numbers of haplotypes. First, we need to calculate the allele frequencies at each locus. From the data given above,

we can calculate $p_A = 82/172 = 0.477$; $q_A = 90/172 = 0.523$; $p_B = 54/172 = 0.314$; and $q_B = 118/172 = 0.686$. The expected frequencies and numbers of the four haplotypes are in the table below.

It appears there are more CA and GG haplotypes, and fewer CG and GA haplotypes, than expected. To test whether this difference is statistically significant, we perform a χ^2 goodness of fit test with one degree of freedom, where $\chi_1^2 = \sum \frac{(observed - Expected)^2}{Expected}$, where observed and expected refer to the numbers of each haplotype, and the sum is over all haplotypes. $\chi_1^2 = 21.99$, $P < 0.0001$. This is a highly significant result (i.e. a χ_1^2 statistic this extreme would occur by chance if these two loci were truly in linkage equilibrium less than once in 10 000 samples). The χ^2 statistic has one degree of freedom, because we used the total sample size, as well as the allele frequencies at Locus A and Locus B to compute the expected haplotype numbers.

The numerical value of $D = P_{11}P_{22} - P_{12}P_{21} = (0.2326)(0.4419) - (0.2442)(0.0814) = 0.0829$. $D' = D/D_{max}$, where D_{max} is the smaller of $p_A q_B$ or $q_A p_B$. Here, $D_{max} = 0.1642$, and $D' = 0.505$. This is quite strong LD, 50% of its theoretical maximum. Alternatively, $r^2 = D^2/p_A p_B q_A q_B = 0.128$.

Locus 1 (A)	Locus 2 (B)	Observed number	Expected frequency	Expected number
C	A	40	$p_A p_B = (0.447)(0.314) = 0.1498$	$(0.1498)(172) = 25.76$
C	G	42	$p_A q_B = (0.477)(0.686) = 0.3272$	$(0.3272)(172) = 56.28$
G	A	14	$q_A p_B = (0.523)(0.314) = 0.1642$	$(0.1642)(172) = 28.25$
G	G	76	$q_A q_B = (0.523)(0.686) = 0.3588$	$(0.3588)(172) = 61.71$

effects if the QTL is not exactly at the marker. Similarly, association mapping underestimates QTL effects unless the molecular marker genotyped is the causal variant, sometimes called the **quantitative trait nucleotide** (QTN). Let a be the true effect of the causal QTN, and α the estimated effect. $a = [p(1 - p)/D]\alpha$, where p is the frequency of the marker, and D is the LD between the causal QTN and the marker. Since the maximum value of D and $p(1 - p)$ are both 0.25, D is always $\leq p(1 - p)$, and therefore the true effect a is always greater than or equal to the estimated effect α.

Association mapping results are usually presented graphically, on a plot where the x-axis shows the molecular markers for which genotypes were determined in linear order according to their physical position, and the y-axis indicates the result of the statistical test for association of marker genotype and trait phenotype. The y-axis is often

scaled as $-\log P$, such that high values denote the most significant associations (Figure 8.10, p. 119).

Number of Individuals and Markers

The power to detect a difference in mean between marker genotypes in association mapping analyses also depends on δ/σ_w, where δ and σ_w are the same as described above for linkage mapping analyses. Table 8.7, p. 120 shows sample sizes necessary to detect QTLs with a range of values of δ/σ_w for different allele frequencies of the causal variant. Association mapping has the same power as linkage mapping in an F_2 population for intermediate gene frequencies, but much reduced power as the frequency of the rare allele decreases. Even larger samples than indicated in this table would be necessary if the marker is in LD with the true

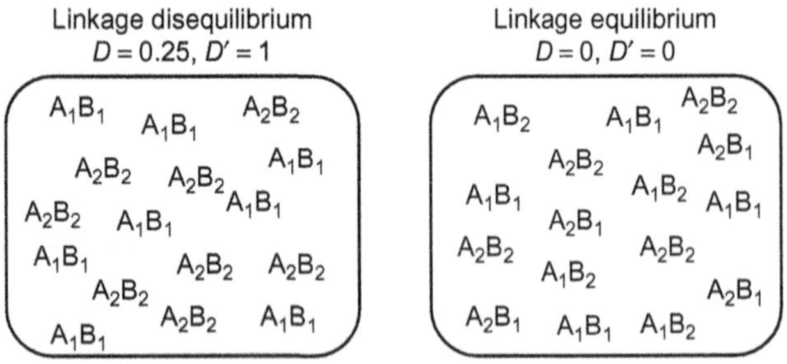

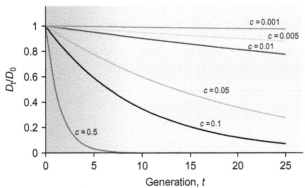

Linkage disequilibrium
$D = 0.25, D' = 1$

A₁B₁ A₁B₁ A₂B₂ A₁B₁ A₂B₂ A₂B₂ A₁B₁ A₂B₂ A₁B₁ A₁B₁ A₂B₂ A₂B₂ A₂B₂ A₂B₂ A₂B₂ A₁B₁ A₁B₁

Linkage equilibrium
$D = 0, D' = 0$

A₁B₂ A₁B₁ A₂B₂ A₂B₂ A₂B₁ A₁B₁ A₁B₂ A₁B₁ A₂B₂ A₂B₁ A₂B₂ A₁B₂ A₂B₂ A₂B₁ A₂B₁ A₁B₁ A₁B₂

FIGURE 8.6 Linkage disequilibrium. Haplotypes at two polymorphic loci, A and B, each with two alleles (A₁, A₂, B₁, B₂) at equal frequency ($p_A = q_A = p_B = q_B = 0.5$) are depicted in two populations. The population to the left is in complete LD. $P_{11} = P_{22} = 0.5$ and $P_{12} = P_{21} = 0$; haplotypes A₁B₂ and A₂B₁ are missing from the population. The value of D is its highest possible numerical value at $D = 0.25$. The population to the right is in linkage equilibrium ($D = 0$). All haplotypes are present in the frequencies expected from random association of alleles at the two loci: $P_{11} = P_{22} = P_{12} = P_{21} = 0.25$.

FIGURE 8.7 Decline of LD between two loci, each with two alleles, over time in a random mating population. All plots are for the LD at generation t (D_t) as a fraction of LD in generation 0 (D_0). The different curves are for different recombination fractions.

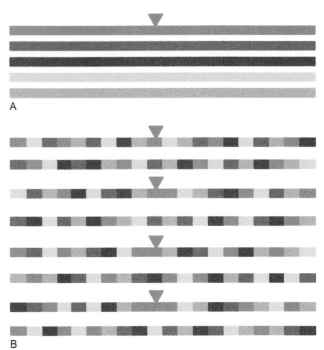

FIGURE 8.8 Historical recombination. (A) Cartoon of an ancestral population, in which a mutation (triangle) affecting a behavioral trait occurs. Each bar represents a different ancestral genotype. The mutation is initially in complete LD with all other polymorphic loci in the genome. (B) Cartoon of a modern day population, in which the frequency of the mutation is now $q = 0.5$. Recombination has shuffled the ancestral chromosomes in the modern day population. The mutation remains associated with the initial haplotype on which it arose, and is in LD with closely-linked polymorphisms.

causal variant. Thus, association mapping is a powerful method for detecting intermediate frequency variants with large effects, but not for detecting rare variants with small effects.

The higher amount of recombination in individuals from an outbred population relative to a linkage mapping population means that we need closely-spaced molecular markers to map QTLs using association mapping. "Closely-spaced," in the context of association mapping, means with respect to historical recombination, not only physical distance. Thus, understanding the scale of LD in the population or species of interest is very important in designing association mapping studies. The pattern of LD is typically visualized graphically in a plot that shows all possible pairs of molecular markers in the region of interest, with color coding of marker pairs with values of LD above a specified significance threshold and value of D' or r^2 (Figure 8.11, p. 120). The diagonal of such a plot shows adjacent pairs of markers, whereas the off-diagonal indicates

TABLE 8.6 $P_t(C)$, the expected frequency of recombinants after t generations of recombination in a random mating population

	Number of generations (*t*) of recombination			
c	5	10	50	100
0.01	0.025	0.05	0.25	0.5
0.005	0.0125	0.025	0.125	0.25
0.001	0.0025	0.005	0.025	0.05

$P_t(C) = 0.5tc$, where c is the recombination fraction.

LD between increasingly more distant markers on the physical map. The scale of LD refers to, on average, how far apart in base pairs are markers in significant LD.

The scale of LD depends on the recombination rate and population history. Recombination rates can vary dramatically across species, and within species according to chromosomal context. LD is low if recombination is high and high if recombination is low. LD also depends on population size, and is lower in large populations than in small populations.

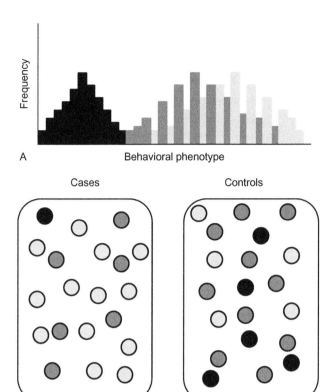

FIGURE 8.9 Association mapping for a marker genotype with two alleles. Black and light gray denote the homozygous marker genotypes, and dark gray denotes the heterozygous genotype. (A) There is a difference in mean for a continuous behavioral phenotype between marker genotypes; therefore, the marker is in LD with the locus affecting the trait. (B) There is a difference in marker allele frequency between cases and controls ($q = 0.2$ for cases and $q = 0.5$ for controls); therefore, the marker is in LD with the locus affecting the trait.

Finally, LD depends on selection (see Chapter 16), and is high for genes that have recently experienced a selective event. Thus, LD is generally quite low in *Drosophila melanogaster*, which has a large population size, but increases in regions of low recombination and near genes under recent selection. At the other end of the spectrum, LD is very high within pure breeds of domestic dogs, which have very small population sizes. Humans represent an intermediate situation, where LD is generally higher than in *Drosophila*, due to the initial small size of the human population and recent expansion of population size. However, LD in humans occurs in a distinct block-like pattern, with blocks of loci in high LD separated by recombination hot spots.

The number of markers required in an association mapping study depends on the scale and pattern of LD. If a group of markers is in high LD, we only need to genotype one of them as a proxy for all the others in the LD block. Thus, in species with large LD blocks, such as pure breeds of dogs, only a few markers may be required for QTL detection, but it will not be possible to localize QTLs very precisely by within-breed association mapping. In contrast, knowledge of all sequence variants is necessary for association mapping in species like *Drosophila*, where LD can decline very rapidly over short physical distances. Under this scenario, however, QTL localization can be quite precise, possibly even to the actual causal QTN. In humans, commercial genotyping arrays with many hundreds of thousands of markers spanning the whole genome have been developed based on **tagging SNPs** in **LD blocks**, facilitating a new era of genome-wide association studies in people. The requirement for genotyping large numbers of markers in large numbers of individuals has meant that, until recently, most association mapping studies have been for a candidate gene or candidate gene

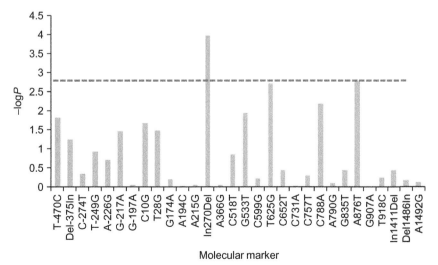

FIGURE 8.10 Association of molecular markers in *Catecholamines up* (*Catsup*) with variation in locomotor startle response in *Drosophila*. The dotted horizontal line represents the Bonferroni-corrected significance threshold. Two molecular polymorphisms at *Catsup* are in LD with a locus affecting startle response. Since there is very little LD between molecular polymorphisms in *Catsup*, and complete DNA sequences were obtained for all alleles, it is possible that these polymorphisms are QTNs causing the variation in startle response. (Data are from Carbone et al. (2006). *Curr. Biol.*, **16**, 912–919.)

TABLE 8.7 Number of individuals required to detect a QTL with effect δ/σ_w by association mapping

δ/σ_w	n	$N(q = 0.5)$	$N(q = 0.25)$	$N(q = 0.1)$
1	21	84	336	2100
0.5	84	336	1344	8400
0.25	336	1344	5376	33 600
0.125	1344	5376	21 504	134 400
0.0625	5376	21 504	86 016	537 600

The calculations were done as in Table 8.3, assuming that the QTL is completely linked to the marker (c = 0), and that δ is the difference in mean between homozygous genotypes. n is the total number of individuals required per homozygous marker class, and N is the total sample size in populations with different frequencies of the rare allele (q).

region, and used only a subset of all possible molecular polymorphisms.

Caveats and Solutions

One major disadvantage of LD mapping is that LD can also be caused if there is recent population admixture. Mixing together of populations with different values of the complex trait and different allele frequencies will give rise to LD, but such LD is spurious and can occur even between unlinked loci (Box 8.3; see also Chapter 16). However, it is possible to statistically infer whether there is population structure if one has data on genotypes at many loci throughout the genome, and then perform association tests within each group.

A second problem that has plagued association studies, especially in human populations, is that they are notoriously difficult to replicate. This is largely because many thousands of individuals are required to map QTLs with moderate or

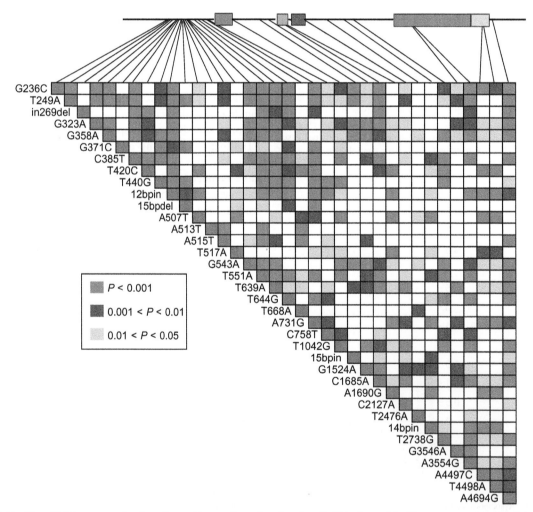

FIGURE 8.11 Plot of LD between molecular polymorphisms at *Dopa decarboxylase* (*Ddc*) in *Drosophila*. The gene structure and relative position of 36 polymorphic molecular markers in the sample of 173 alleles is given above the matrix. The matrix represents all possible combinations of markers. Pairs of markers between which LD is significant are color coded as in the inset box.

Box 8.3   Population admixture and LD

To understand how population admixture causes LD, consider two populations that are homozygous for different alleles at two loci, A and B. All individuals in population 1 are A_1B_1/A_1B_1, and all individuals in population 2 are A_2B_2/A_2B_2. If we did not know these were two separate populations and determined the haplotypes for individuals for both, we would infer the two loci were in complete LD, since $P_{11} = P_{22} = 0.5$; $P_{12} = P_{21} = 0$; and $D = 0.25$.

If individuals from these two populations were to mate at random, what would be the value of D in the following generation? To determine this, we tabulate the offspring genotypes and their frequencies, and haplotypes and their frequencies.

Genotype	Frequency	Haplotype	Frequency
$\dfrac{A_1B_1}{A_1B_1}$	0.25	A_1B_1	1
$\dfrac{A_1B_1}{A_2B_2}$	0.5	A_1B_1	$(1 - c)/2$
		A_1B_2	$c/2$
		A_2B_1	$c/2$
		A_2B_2	$(1 - c)/2$
$\dfrac{A_2B_2}{A_2B_2}$	0.25	A_2B_2	1

The haplotype frequencies in the offspring are thus:

Haplotype	Frequency
A_1B_1	$0.25 + 0.25(1 - c)$
A_1B_2	$0.25c$
A_2B_1	$0.25c$
A_2B_2	$0.25 + 0.25(1 - c)$

If A and B are unlinked, $c = 0.5$. The haplotype frequencies are thus 0.375 for A_1A_1 and A_2B_2, and 0.125 for A_1B_2 and A_2B_1, giving $D = (0.375)(0.375) - (0.125)(0.125) = 0.125$, or one-half the LD in the previous generation. For unlinked loci, D will decline rapidly after a single admixture event followed by random mating, but will persist for a longer period as c decreases (Figure 8.7).

small effects (Table 8.7), and until recently sample sizes of association studies were too small. Further, many studies did not pay attention to the need for adjusting P-values of tests for significance for multiple tests. However, it is important to reiterate that a significant association of variation in marker genotype with a behavioral phenotype does not mean that that genetic variation at the marker causes the variation in behavior; rather the marker is in LD with the true causal variant. Thus, differences in LD between the marker and the causal polymorphism between populations could lead to failure to replicate associations across populations. In the future, advances in DNA sequencing technology will enable cost-effective resequencing of candidate genes and even whole genomes for large numbers of individuals, and thus facilitate identification of polymorphisms that cause variation in behavioral traits in outbred populations.

QUANTITATIVE TRAIT LOCUS END GAME

As noted above, many thousands of QTLs have been detected that affect behavioral and other traits by linkage mapping and association mapping, but it is difficult to identify the actual genetic loci causing the genetic variation. In model organisms, high resolution mapping experiments typically begin by constructing a **congenic strain** in which the genomic region containing the QTL allele increasing the value of the trait has been transferred into the genetic background of the strain with the allele decreasing the value of the trait (or *vice versa*). This is done by successive backcrosses to one of the parental strains (the recurrent backcross parent). In each backcross generation, individuals are genotyped for molecular markers that flank the QTL region of interest, as well as markers associated with other QTLs and evenly spaced throughout the genome. Each generation, individuals are selected for the next round of backcrossing that are heterozygous for the entire QTL region, but maximally homozygous for the marker alleles of the recurrent backcross parent outside the QTL region. The final congenic strain is created by crossing individuals that are heterozygous for the QTL of interest, genotyping the progeny once more for the QTL region, and crossing individuals that are homozygous for the region containing the QTL from the non-recurrent parent together (Figure 8.12).

The next step is to cross the congenic strain to the recurrent backcross parent line, and genotype the offspring. However, in the second phase the offspring in which recombination has occurred within the QTL interval are collected, genotyped with additional markers within the QTL region to determine the **recombination breakpoints**, and phenotyped for the trait. These data are then used to map the QTL to ~2cM intervals, using the same statistical analyses described above for the initial genome scan (Figure 8.12).

In the case of *Drosophila*, geneticists can use special **deficiency stocks** to map QTLs with high resolution, without the need for further recombination. Deficiency stocks have small chromosomal regions deleted, and are maintained as viable **hemizygous** stocks against a **balancer chromosome**. Many deficiencies with overlapping breakpoints cover most of the fly genome. We can cross

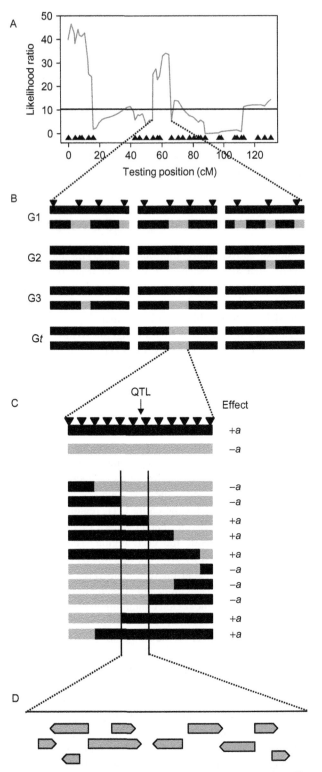

these strains and a control strain to individuals with different QTL locus genotypes, measure the behavior of interest, and infer whether a QTL affecting the difference in behavior between the parental strains resides in the deleted region (Box 8.4; see also Chapter 11). Repeating these tests using deficiencies with overlapping breakpoints spanning the QTL region can localize QTLs to sub-cM intervals with high resolution.

In non-model organisms, the only option for high resolution linkage mapping of QTLs is to increase the sample size and hunt for recombinants within the QTL region. A second option is to change the strategy from linkage mapping to association mapping, focusing on the region in which the QTL is located in a large number of individuals from an outbred population.

High resolution QTL mapping rarely succeeds in identifying individual genes affecting variation in the trait, but does pinpoint genomic regions with a small number of potential candidate genes. How can we tell which of these genes affect the variation in behavior? Supporting evidence includes potentially functional DNA polymorphisms between alternative alleles of one of the candidate genes, a difference in mRNA expression levels between homozygous genotypes, and expression of RNA or protein in tissues thought to be relevant to the trait. Associations of markers in candidate genes with the behavioral phenotype that are replicated in independent studies also constitute strong evidence that the gene affects variation in behavior. In model organisms, demonstrating that a mutation in one of the candidate genes affects the behavioral trait is strong evidence in favor of that gene (see Chapter 9). However, many mutations in candidate genes are not viable as homozygotes, and cannot therefore be assessed for behavioral traits. However, we can perform complementation tests of QTL alleles to mutations in candidate genes, even if the mutant allele is not viable as a homozygote (Box 8.4). Complementation tests to mutations are particularly useful in *Drosophila*, since mutations in over 60% of *Drosophila* genes have been generated and are available from public stock centers. Mutant complementation has also been used by Jonathan and Flint and colleagues at Oxford University, to identify a gene affecting anxiety in mice. Formal proof that a specific allelic difference causes the difference in behavior would be to replace the allele of a candidate gene in one strain with that of the other, without introducing any

FIGURE 8.12 High resolution QTL mapping. (A) A QTL region of interest is identified from a genome scan. The likelihood ratio test statistic is from a QTL mapping method that accounts for multiple QTLs. (B) An individual is chosen based on homozygosity at the focal QTL, and the fewest other QTLs, based on molecular markers (black triangles). This individual is backcrossed to the recurrent parental strain (G1). In subsequent generations, the QTL is selected for, and other QTLs selected against, based on marker genotypes (G2–G3). A congenic strain containing the focal QTL, but with minimal other genetic material from the non-recurrent strain, is constructed by breeding heterozygous individuals and selecting for the homozygous QTL region in their progeny (G*t*). (C) The congenic strain is crossed to the recurrent parent, and individuals with recombination breakpoints within the QTL region are genotyped for additional markers (black triangles), and phenotyped for the trait. The location of the QTL is inferred using the same method as the initial genome scan. (D) The final high resolution map localizes the QTL to a small region (indicated by the black horizontal line, representing the genome in bp) containing several candidate genes (indicated by the gray bars).

Box 8.4   Quantitative complementation tests

In *Drosophila*, complementation tests to deficiency (*Df*) stocks are a rapid method for mapping QTLs to sub-cM intervals. Each *Df* stock contains a deleted region of the genome and is maintained over a Balancer (*Bal*) chromosome to prevent recombination (see Chapter 9), which is marked with dominant visible mutations so it can be easily distinguished from the *Df* chromosome. The *Df/Bal* stock is crossed to both parental strains containing different QTL alleles, and the behavioral trait is evaluated for individuals of each of the four resulting genotype classes: *Df*/P1; *Df*/P2; *Bal*/P1; and *Bal*/P2. If the *Df* uncovers a QTL allele, then we expect the difference in the behavioral phenotype to be greater between the *Df*/P1 and *Df*/P2 genotypes than the *Bal*/P1 and *Bal*/P2 genotypes (Figure A, panel A). When this is true, we say the *Df fails to complement* the QTL. If we repeat these tests for overlapping deficiency stocks spanning the QTL region, we can infer the location of the QTLs from knowledge of the *Df* breakpoints, and the pattern of complementation and failure to complement (Figure A, panel B).

Complementation tests to mutations have the same logic as complementation tests to deficiencies. If the QTL alleles fail to complement a mutation in one of the positional candidate genes, then the mutation fails to complement the QTL (Figure A, panel C).

An important caveat to complementation tests to deficiencies and mutations is that failure to complement can also occur if the deficiency or mutation interacts epistatically with a different QTL affecting the behavioral trait. Thus, complementation tests are suggestive that a candidate gene region or gene corresponds to a QTL, but other studies are necessary to confirm this interpretation.

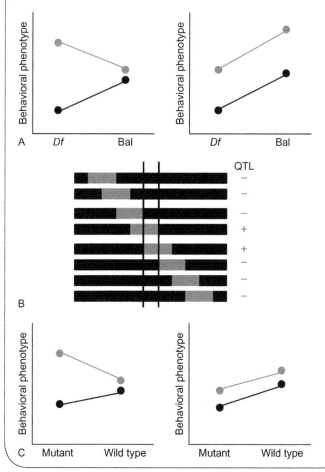

FIGURE A   Quantitative complementation tests. (A) Deficiency complementation. Two parental strains, P1 and P2 (gray and black circles) are crossed to a deficiency (*Df*) stock, in which the *Df* is maintained over a balancer (*Bal*) chromosome. If the difference in the behavioral phenotype is significantly greater between the hemizygous genotypes than the heterozygous genotypes, (P1/*Df* − P2/*Df*) > (P1/*Bal* − P2/*Bal*), and the *Df* fails to complement the QTL (left panel). If the difference in behavior between the hemizygous genotypes is not significantly different from that between the heterozygous genotypes, (P1/*Df* − P2/*Df*) = (P1/*Bal* − P2/*Bal*), and the *Df* complements the QTL (right panel). *Df* complementation tests can be improved by using *Df*s that have been generated in an isogenic background and with known breakpoints, and crossing P1 and P2 to the co-isogenic strain in which the *Df*s have been generated as the heterozygous control. (B) Illustration of overlapping *Df* strains, where the black bars represent the non-deleted chromosome and the gray bars the deleted region. If the *Df* complementation tests are repeated for each of the *Df* stocks, the pattern of complementation (−) and failure to complement (+) can be used to localize the QTL to a small genomic region, indicated by the vertical lines. (C) The logic of complementation tests to mutations is the same as that for complementation tests to deficiencies. Two parental strains, P1 and P2 (gray and black circles) are crossed to a mutant stock and a wild type control strain. If the difference in the behavioral phenotype is significantly greater between the mutant heterozygotes than the wild type heterozygotes, (P1/mutant − P2/mutant) > (P1/wild type − P2/wild type), and the mutant allele fails to complement the QTL (left panel). If the difference in behavior between the mutant heterozygotes is not significantly different from that between the wild type heterozygotes, (P1/mutant − P2/mutant) = (P1/wild type − P2/wild type), and the mutant allele complements the QTL (right panel). Complementation tests to mutations are ideally performed when the mutant allele has been generated in an isogenic background, and the co-isogenic wild type strain is used as the control.

other changes in the genetic background. Such "knock-in" experiments are not yet technically feasible in model organisms commonly used for behavioral genetic studies.

Given that identifying the genes corresponding to QTLs is so highly labor-intensive, it is not surprising that rather few QTLs affecting natural variation in behavior have been mapped to the single gene level. These studies are discussed in more detail in Chapters 11, 13, 15, and 16, in the context of particular behaviors. Most genes that have been identified to date are those with large homozygous effects. However, results of high resolution mapping of QTLs affecting behavioral traits in *Drosophila* and mice indicate that most QTLs do not correspond to genes with large homozygous effects. Instead, single QTLs that were initially detected in genome scans typically fractionate into multiple closely-linked QTLs with small effects. Further, candidate genes in QTL regions are more often than not computationally-predicted genes, or genes with known functions that have no *a priori* relationship to behavior. Recent genome-wide association studies for human behavioral disorders bolster the emerging consensus that the genetic architecture of naturally occurring variation for behavioral traits (and indeed, other quantitative traits) is typically due to large numbers of loci with individually small allelic effects. Thus, a "one gene at a time" approach to dissect the genetic basis of naturally occurring variation in behavior is neither feasible nor desirable. Future advances in understanding the genetic underpinnings of individual variation in behavior will require genome-wide analyses of genetic and transcriptional networks associated with variation in behavior (see Chapter 10). The conceptual foundation underlying more sophisticated gene mapping strategies rests on the principles outlined in this chapter.

SUMMARY

Knowledge of the QTLs affecting naturally occurring variation in behavioral traits is important, because understanding the genes and genetic networks affecting variation in behavior can give insights about the underlying cellular and neural mechanisms affecting behavior, and because we need to know the molecular polymorphisms affecting variation in behavior if we are to understand the evolution of behavior and why genetic variation for behavioral traits is maintained in natural populations. QTLs affecting complex behaviors are mapped by linkage to polymorphic molecular marker loci; if there are differences in behavior between marker genotypes, the QTL is linked to the marker. The major methods for mapping QTLs are linkage mapping and association mapping. The data required for both methods are measurements of the behavioral phenotype and analysis of marker locus genotypes for all individuals in the mapping population.

The simplest linkage mapping experiments involve crosses between completely inbred lines that differ genetically for the behavior, in model organisms. The behavioral trait and marker locus genotypes are determined for individuals resulting from a backcross of the F_1 to the parental lines, F_2 individuals, or individuals from recombinant inbred lines. Genome scans for QTLs can be performed one marker at a time (single marker analysis), or using pairs of adjacent markers (interval mapping). Estimates of the homozygous and heterozygous QTL effects are confounded with the distance of the QTL from the marker locus in single marker analyses. However, this is not a severe problem when markers are close together, and single marker analyses can readily include effects of additional covariates into the statistical model used to evaluate differences between marker class means. Interval mapping analyses jointly estimate the QTL effects and the distance of the QTL from the flanking markers. The best methods for estimating effects and location of QTLs take account of the segregation of other QTLs when evaluating marker-trait associations. Permutation tests are used to determine the threshold for declaring a QTL as significant, because multiple tests are performed in a genome scan, and the markers are linked. Large numbers of individuals are needed to detect and localize QTLs using linkage mapping. Selective genotyping and advanced intercross designs can improve the power to localize QTLs, and designs that start with more than two parental lines help address the problem of reduced genetic diversity in crosses of two inbred lines.

Association mapping uses samples of individuals from an outbred population, or many inbred lines derived from an outbred population, to map QTLs. Association mapping is thus based on a better sample of genetic diversity than linkage mapping. Association mapping uses historical recombination to identify and localize QTLs, and is based on LD between markers and loci affecting the behavior. LD, the correlation in allele frequencies between two loci, declines over time, and is expected to be small in a large random mating population unless the loci are very tightly linked. Therefore, much smaller sample sizes are needed to localize QTLs, compared to linkage studies. However, large sample sizes are still required to detect QTLs using association mapping. This design has the same power as linkage mapping for polymorphisms at intermediate allele frequencies, but has much reduced power as the frequency of the rare allele decreases. Association mapping uses single marker analysis, and underestimates QTL effects unless the molecular marker genotyped is the causal variant. Significance thresholds in association mapping experiments need to be adjusted for multiple tests. The density of markers in association mapping designs should be adjusted to the scale of LD in the population or species of interest. Very large numbers of markers are required for whole-genome association studies in most species. Association mapping studies need to take account of the possibility of recent population admixture, since this can cause spurious LD even between unlinked loci.

It is difficult to identify the actual genetic loci causing genetic variation in behavior. In linkage mapping studies of model organisms, high resolution mapping begins with constructing congenic strains, in which the genomic region containing the QTL allele of one strain is transferred into the genetic background of the other strain. The congenic strain is then crossed to the recurrent backcross parent line to generate recombinants within the QTL region, which are then used for high resolution mapping in the same manner as the initial genome scan. In *Drosophila*, complementation tests with deficiencies can be used to map QTLs to small genomic regions. In non-model organisms, high resolution QTL mapping requires increasing the sample size of the mapping population to identify recombinants within the QTL region, or to switch to association mapping of the genomic region in which the QTL is located. Strategies for determining which candidate gene(s) in an interval containing the QTL actually correspond to the QTL include: identifying potentially-functional DNA polymorphisms between alternative alleles; observing a difference in mRNA expression levels between homozygous genotypes; observing expression of RNA or protein in tissues thought to be relevant to the trait; observing failure of a mutation in the candidate gene to complement the QTL; observing that a mutant allele of the candidate gene affects the behavior; and, for association studies, independent replication of the association.

STUDY QUESTIONS

1. What is a QTL?
2. Why is QTL mapping important?
3. Explain the basic principle underlying all QTL mapping methods.
4. What are the two main methods for mapping QTLs? What are the advantages and disadvantages of each?
5. Explain the difference between single marker analysis and interval mapping. Which would you prefer to use, and why?
6. Why are large numbers of individuals needed for QTL mapping studies?
7. What statistical methods are used to deal with multiple tests in QTL mapping studies?
8. How can the power and precision of QTL mapping studies be improved?
9. What is linkage disequilibrium? Why is it important for QTL mapping?
10. Discuss strategies for identifying the actual genes corresponding to QTLs.

RECOMMENDED READING

Anholt, R. R. H., and Mackay, T. F. C. (2004). Quantitative genetic analyses of complex behaviours in *Drosophila*. *Nat. Rev. Genet.*, **5**, 838–849.

Churchill, G. A., and Doerge, R. W. (1994). Empirical threshold values for quantitative trait mapping. *Genetics*, **138**, 963–971.

Churchill, G. A. et al. (2004). The Collaborative cross, a community resource for the genetic analysis of complex traits. *Nat. Genet.*, **36**, 1133–1137.

Daly, M. J., Rioux, J. D., Schaffner, S. F., Hudson, T. J., and Lander, E. S. (2001). High- resolution haplotype structure in the human genome. *Nat. Genet.*, **29**, 229–232.

Falconer, D. S., and Mackay, T. F. C. (1996). *Introduction to Quantitative Genetics*, 4th edition. Longmans Green, Harlow, Essex, UK.

Flint, J., Valdar, W., Shifman, S., and Mott, R. (2005). Strategies for mapping and cloning quantitative trait genes in rodents. *Nat. Rev. Genet.*, **6**, 271–286.

Lander, E. S., and Botstein, D. (1986). Strategies for studying heterogeneous genetic traits in humans by using a linkage map of restriction fragment length polymorphisms. *Proc. Natl. Acad. Sci. USA*, **83**, 7353–7357.

Lynch, M., and Walsh, B. (1997). *Genetics and Analysis of Quantitative Traits*. Sinauer Assoc., Inc, Sunderland, MA.

Ryder, E. et al. (2007). The DrosDel deletion collection: a *Drosophila* genome-wide chromosomal deficiency resource. *Genetics*, **177**, 615–629.

Chapter 9

Mutagenesis and Transgenesis

OVERVIEW

Analysis of mutations has been the principal mechanism for discovering genes that mediate physiological, developmental, or behavioral processes. Whereas it is now becoming increasingly appreciated that mutations in single genes can have widespread effects on transcriptional regulation of other genes in the genome, the functional analysis of mutant alleles at single loci remains an effective starting point for uncovering mechanisms that regulate even complex traits. Mutations can occur either spontaneously, or can be induced. Mutations can result in the elimination of a gene altogether (knock-outs, null mutations), reduction of expression of the gene (hypomorphic mutations), or interference with translation or stability of mRNAs (RNA interference). Furthermore, polymorphisms that have arisen during the course of evolution can be viewed as a rich treasure trove of natural mutations, resulting in a multitude of alleles that contribute to naturally-occurring variation in behavioral phenotypes within a population. This chapter focuses primarily on the analysis of induced mutations for the study of complex behaviors. Associations between polymorphisms and phenotypic variation are discussed in Chapter 8.

THE OCCURRENCE OR INDUCTION OF MUTATIONS

Mutational analysis is an indispensable tool for the geneticist. Two types of genetic screens are often distinguished. In **forward genetic screens** mutants are isolated that produce certain phenotypes. In **reverse genetic screens** mutations are introduced in known genes to determine what phenotypes result as a consequence. Mutational screens can reach a great deal of sophistication, and can include **saturation screens**, aimed at identifying all genes that affect a certain phenotype, and **suppressor/enhancer screens**, which are used to identify alleles that modify a mutant phenotype in a **sensitized background**, i.e. a genetic background that already carries a mutation that gives rise to the phenotype

under study. In the next sections we will discuss the various strategies that can be used to obtain mutations that can give insights into genes that contribute to behaviors.

Spontaneous Mutations

Spontaneous mutations are the evolutionary substrate on which natural selection acts. Mutations can be classified as deleterious mutations, which are expected to be cleared from the gene pool over time; beneficial mutations, which are expected to accumulate in the population over time; and neutral mutations, changes in the DNA nucleotide sequence that have neither a beneficial nor a deleterious effect and, since they do not affect fitness of the species, can accumulate and account for much of the segregating DNA sequence diversity observed among individuals within a population.

Many congenital human morphological abnormalities and physiological or behavioral disorders are the result of deleterious mutations that disrupt the functions of critical genes. Recessive alleles can be carried as heterozygotes in the population, and serve as a reservoir for such deleterious mutations. In addition, alleles that function late in life after reproduction has occurred escape natural selection, and the accumulation of such genes contributes to senescence and limited lifespan. **Linkage studies** in human populations, in families, and in twins have identified a large number of mutant alleles that give rise to diseases and behavioral disorders. These include not only mutations within a single gene, but also chromosomal aberrations. One of the best characterized chromosomal abnormalities is **Down syndrome**, which results from **trisomy** of chromosome 21 and, in addition to distinct morphological features, is characterized by mental retardation (Figure 9.1).

Spontaneous mutations that affect behaviors also occur in laboratory animals. Although such spontaneous mutations are rare, their value cannot be overstated. A large number of spontaneous mutants in mice have provided important models for human neurological and behavioral disorders, including locomotor, sensory, and eating disorders.

A large number of mutant mice have been identified with locomotor disabilities. These mice perform poorly on locomotion and balance assays, such as the **rotating rod**, and show impaired spatial learning ability, as measured for

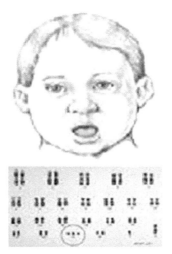

FIGURE 9.1 Down syndrome. An extra copy of chromosome 21, due to nondisjunction of the chromosome during meiosis, as shown in the **karyotype** (circled) in this figure leads to Down syndrome, characterized by mental retardation and a characteristic physical appearance with distinctive craniofacial features, including thick lips with a protruding tongue and lower lip.

example by their ability to learn the location of a hidden platform in the **Morris water maze**. One of the best-characterized mutant mouse models is the *weaver* (*wv*) mouse, which carries a guanine to adenine substitution at nucleotide position 953 of the *Girk2* gene, which encodes an inward rectifier potassium channel. This results in degeneration of neuronal populations of granule cells in the cerebellum, and progressive degeneration of mesencephalic dopaminergic neurons in the substantia nigra. Consequently, the *weaver* mouse has become the most important noninvasive animal model for **Parkinson's disease**, showing similar locomotor defects (**ataxia**) and compromised ability for spatial learning.

In addition to motor disorders, spontaneous mutations in mice have also been invaluable in studies on neurodegenerative sensory disorders. Several mutant mice have given insights into the pathogenesis of **Usher syndrome**, an autosomal recessive disorder which results in deafness and blindness, as a result of progressive degeneration of **hair cells** in the inner ear and photoreceptors in the eye. Gene-mapping studies in populations with a high prevalence of the disease, together with studies on mouse mutants, have implicated at least 11 different genes in various manifestations of Usher syndrome. The *shaker-1* mouse develops Usher syndrome-like deafness due to a mutation in the gene that encodes myosin VIIa, which is a structural component of **stereocilia** in cochlear inner hair cells. Another Usher syndrome-like phenotype is apparent in the *waltzer* mouse in a gene orthologous to the human *CDH23* gene, which encodes a member of the cadherin protein family involved in intercellular adhesion.

In addition to spontaneous mutations that have identified target genes that affect locomotor or sensory disorders, two important mice with spontaneous mutations have been

---

Box 9.1 Locomotor deficiencies in mice

In addition to the *weaver* mouse, a large number of spontaneous mouse mutants have been identified with locomotor and spatial learning defects, including the *Lurcher* and *hot foot* mutants, which affect the *Grid2* gene that encodes an AMPA-type glutamate receptor subunit (GluRδ₂). Homozygous *Lurcher* mutants are not viable, because they are defective in neonatal suckling. Heterozygotes show the typical ataxic locomotion defect and deficient spatial learning. The *hot foot* mutation results in a truncated gene product with similar phenotypes.

Several additional mouse mutants affect Purkinje cells in the cerebellum, and consequently also show defects in locomotion. In the *staggerer* mouse, the *Rora* gene, which encodes a gene product that resembles a retinoid acid nuclear receptor involved in neuronal differentiation and maturation in cerebellar Purkinje cells, has been deleted. Purkinje cells in the cerebellum coordinate sensory–motor integration, and represent the sole cerebellar output for motor coordination via GABAergic projections. In *Pcd* (*Purkinje cell degeneration*) mice the *Agtpbp1* gene, which encodes an ATP/GTP binding protein, carries a mutation that results in degeneration of cerebellar Purkinje cells, mitral cells in the olfactory bulb, thalamic neurons, and photoreceptor cells. The earliest and most prominent phenotype of *Pcd* mice, however, is impaired locomotion and impaired spatial learning.

The *reeler* mutant mouse also shows compromised performance in locomotion assays and spatial learning paradigms. Characterization of the *reeler* mutation has implicated reelin (Reln), an extracellular matrix protein, in neuronal adhesion and cell migration during development. Two spontaneous mutant alleles of the *Reln* gene have been identified, one of which consists of a deletion of the entire gene and the other of a 220 bp deletion of the open reading frame, which results in a frameshift mutation. Finally, a mutant designated as *Dystonia musculorum* (*DSt*) has a deleted *Dst* gene, which encodes dystonin, a cytoskeletal protein that organizes microfilaments and microtubules in neurons and Schwann cells. In these mice, myelination of axons is compromised and they show dramatic locomotor impairments, with most mice being able only to crawl with twisting and writhing movements. These mice are so severely impaired in locomotor ability that they cannot be tested in the Morris water maze, as they are prone to drown. (See Chapter 11 for more on locomotion).

---

identified that display abnormal feeding behaviors and develop extreme obesity. Adipose cells that store fat secrete a hormone, **leptin**, which binds to leptin receptors in the hypothalamus to inhibit further feeding. In homozygous *obese* (*ob/ob*) mice, a mutation in the leptin gene results in absence of this hormone and consequently absence of negative feedback for feeding, causing binge eating that results in obesity. In homozygous diabetic (*db/db*) mice, spontaneous mutations affect the gene encoding the leptin receptor. Thus, in contrast to the leptin deficient *ob/ob* mice, *db/db*

Box 9.2 Auditory transduction and deaf mice

Transduction in the auditory system is mediated by **hair cells** in the inner ear. These are epithelial cells that contain a **hair bundle**, which consists of a staircase array of 20–300 minute, microvilli-derived processes, called **stereocilia**, which are rigid structures that can pivot around their base. Sound waves that enter the inner ear cause mechanical displacement of the hair bundle, which results in the opening of ion channels. The opening of these channels creates a generator current, which depolarizes the cell and leads to the release of neurotransmitter from the hair cell onto synaptic nerve endings of the cochlear nerve (cranial nerve VIII) (Figure A).

The greater the intensity of the sound, the greater the displacement of stereocilia of the responding hair cells and the greater the amount of neurotransmitter that is released, resulting in a higher frequency of action potentials in the cochlear nerve. The rapid timescale of channel activation suggests that **mechanoelectrical transduction** occurs directly, without second messengers. Examination of the hair bundles by high-resolution electron microscopy reveals filamentous links between neighboring stereocilia. These **tip-links** are thought to represent the "gating springs," which open the channel on mechanical displacement. Thus, deflection of the hair bundle causes tension on the tip-links between adjacent stereocilia, which results in opening of ion channels linked to these tip-links (Figure B). When the mechanical stimulus is prolonged and the hair bundle remains deflected, the transduction apparatus adapts to the new deflected position. This adaptation process is most likely accompanied by the retensioning of the tip-links, and involves a cellular actin-myosin-like motor that moves the site of attachment of the tip-link along the stereocilium.

A large number of "deafness" genes have been identified, both in mice and humans, which have given insights into the mechanisms of auditory transduction. The *usher*, *shaker*, and *waltzer* mice are all born deaf. Mutations occur in unconventional myosins, myosin VIIA in *usher* and *shaker-1* mice, and cadherin-23 in the *waltzer* mouse.

In *shaker-2* mice the stereocilia of hair cells develop abnormally. Introduction of a normal transgene into these mice restored hearing, the first experimental example of reversion of deafness. Disorganization of stereocilia is also observed in cadherin-23-deficient *waltzer* mice.

About 85 loci and 30 deafness genes (categorized by the acronym DFN) have been identified in the human genome. Because of extensive homologies between mouse and human genes, mouse mutations have been of great use in uncovering critical components of auditory hair cells whose defects contribute to hearing loss in humans.

FIGURE A Schematic representation of an epithelial hair cell, showing the direction of displacement of the hair bundle (arrow) that causes depolarization and release of neurotransmitter from vesicles in the base of the cell where an auditory nerve fiber forms a synapse.

 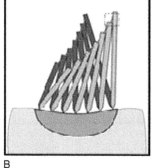 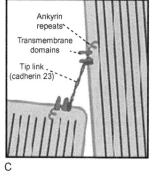

A   B   C

FIGURE B Mechanotransduction in hair cells of the inner ear. (A) Scanning electron micrograph of a hair bundle from the sacculus of the bullfrog showing the staircase array of stereocilia. (From David P. Corey.) (B) Model for mechanotransduction. Deflection of the hair bundle causes the stereocilia to pivot, placing tension on the tip links between them. (C) Ion channels attached to intracellular elastic elements (ankyrin repeats) open in response to tension on the rather rigid tip link, of which cadherin 23 is a molecular component. (Courtesy of the Theoretical and Computational Biophysics Group, Beckman Institute, University of Illinois at Urbana-Champaign.)

mice are leptin resistant. The intact hormone is produced, but cannot exert its effects due to the absence of a functional receptor. Thus, the phenotype of the *db/db* mice is the same as that of the *ob/ob* mice. In people with leptin deficiency, administering leptin can prevent or reverse obesity (Figure 9.2). It should be noted, however, that many genes contribute to risk of obesity in humans, and cases in which this can be pinned down to absence of leptin or a defect in the leptin receptor are rare.

Although spontaneous mutations in mice have provided invaluable insights in human diseases, they have one major disadvantage; they occur only rarely and the investigator has no control over the induction of the mutation. They are gratefully accepted as gifts of nature, but for effective explorations of specific behavioral phenotypes the experimental behavioral geneticist has to turn to induced mutations, which can be generated in large numbers and targeted toward the desired phenotype.

Radiation-Induced Mutations

Irradiation with ultraviolet light or X-rays introduces mutations in the DNA. X-ray irradiation has been used extensively to introduce mutations in *Drosophila* before **chemical mutagenesis**, and thereafter **transposon-mediated mutagenesis** (described below), became methods

of choice. The method was pioneered by the American geneticist H.J. Muller. X-rays induce double-stranded breaks in the DNA, which can result in large chromosomal rearrangements. This has led to the construction of **balancer chromosomes**, which remain important everyday tools in the fly geneticist's tool kit. Balancer chromosomes are chromosomes that contain multiple inversions. As a consequence, when paired with a normal homologous chromosome, vast chromosomal regions do not align during meiosis and, thus, recombination cannot occur. Together with the fact that recombination does not occur in *Drosophila* males, balancer chromosomes enable mutations to be stably propagated without fear of losing them as a result of recombination. Balancer chromosomes contain dominant visible markers, such as *Curly* (*Cy*), a marker on the second chromosome which gives rise to curly wings, or *Stubble* (*Sb*), a marker on the third chromosome that gives rise to stubbly bristles. Thus, the presence of the balancer chromosome can be inferred simply by observing the flies. In addition, balancer chromosomes carry recessive lethal markers, so that flies homozygous for the balancer are not viable. Balancer chromosomes have also recently been developed for mice, but have not found the same widespread applications as in flies.

X-ray-induced mutagenesis is no longer commonly used, as it requires expensive equipment and a specialized

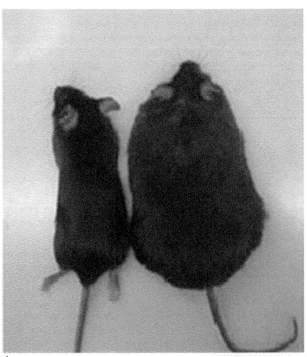

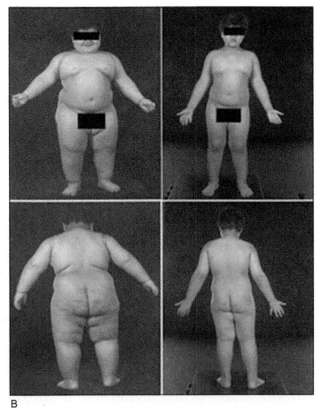

A   B

FIGURE 9.2   Leptin deficiency and obesity. Panel A shows a wild-type mouse (left) and an age-matched leptin-deficient *ob/ob* mouse (right). Panel B shows an example of congenital leptin deficiency in a young boy before (left), and after (right), treatment with leptin.

facility, presents a radiation hazard for personnel, and causes extensive mutations that are hard to map to individual genes. Chemically-induced mutagenesis has been used more widely in many model organisms, as it generates point mutations in specific genes.

Chemically-Induced Mutations

Chemically-induced mutations have employed either **base analogs** or **alkylating agents** to introduce point mutations in the DNA. Base analogs, such a 5-bromouracil and 2-aminopurine, mutate DNA only when they are incorporated in DNA during replication. In contrast, alkylating agents, such as ethyl methane sulphonate (EMS), ethyl ethane sulphonate (EES), and the ethylating agent ethyl-nitrosourea (ENU) introduce mutations both in replicating and nonreplicating DNA (Figure 9.3). These compounds are highly effective for generating point mutations in the DNA, simply by feeding them to animals. ENU acts by transferring the ethyl group to a base, usually thymine. ENU is particularly effective in spermatogonial stem cells, from which mature sperm cells are derived.

Chemical mutagenesis is an effective method for generating point mutations and null mutations, but considerable effort is required to map the mutation, and multiple mutations may occur in the same individual which have to be separated genetically, for example through repeated backcrossing. Nevertheless, chemical mutagenesis remains the best method available for introducing point mutations.

HOMOLOGOUS RECOMBINATION

The removal or inactivation of a gene in an animal was first achieved in mice by Mario Capecchi, Martin Evans, and Oliver Smithies in 1987-1989, using a process known as **homologous recombination**. The impact of this technology was recognized with the 2007 Nobel Prize. Although homologous recombination has also become possible in rats since 2003, the procedure is easier and more successful in the mouse, which remains the animal of choice for generating targeted null mutations. Mice carrying such engineered targeted mutations are referred to as

knock-out mice, since the function of the target gene has been "knocked-out."

To target a gene for homologous recombination in mouse, a plasmid or bacterial artificial chromosome carrying the gene is isolated from a DNA library, and a genetically-altered version of the gene is made that renders the gene dysfunctional. The construct is, by and large, very similar to the endogenous gene, and usually also contains a marker that allows identification of cells that contain the construct. The marker confers resistance to an antibiotic, such as neomycin, that kills cells that do not contain the genetically-engineered knock-out construct. The construct is then introduced by **electroporation** into **embryonic stem cells** that are derived from the morula stage (an early embryonic stage) of a mouse embryo and maintained in cell culture. As these cells divide, some will incorporate the genetically-engineered gene next to the endogenous gene, due to recombination. Subsequent screens can then select for cells that have lost the endogenous copy and retain the engineered one. Stem cells that have incorporated the knock-out construct in their DNA can be rapidly isolated, because they have become resistant to the antibiotic. The stem cells are then microinjected into mouse **blastocysts**, and implanted in the uterus of a female mouse. The blastocysts contain two types of stem cells, the endogenous stem cells and the stem cells that contain the knock-out construct incorporated into their DNA. Thus, the mice that are born are chimeras, since parts of their bodies are derived from the endogenous stem cells, whereas other parts result from the genetically-engineered stem cells. If the genetically-engineered stem cells are derived from a strain of mice with a coat color that is different from that of the female used for blastocyst implantation, the offspring will be a coat color mosaic. However, only those mice in which the homologous recombinant gene has been incorporated into the germ cell line will be useful. Therefore, the mice have to be bred and offspring selected that have a coat color characteristic of the strain that donated the stem cells that carry the knock-out gene (these are often mice of the 129 strain). Further breeding gives rise to homologous recombinant mice that lack the functional target gene (Figure 9.4). To identify mice that harbor the knock-out allele, and distinguish heterozygotes from homozygotes

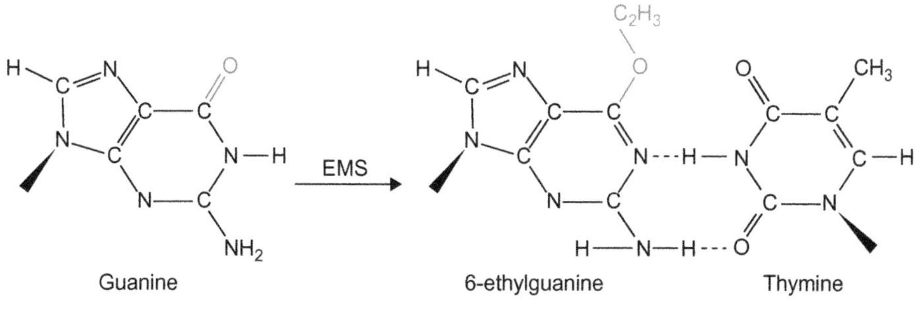

FIGURE 9.3 Alkylation by EMS attaches an ethyl group to the ketone group of guanine, eliminating a hydrogen bond for base pairing with cytosine, and inducing it to form a base pair instead with thymine. In the next replication, thymine pairs with adenine to generate a G-C to A-T change.

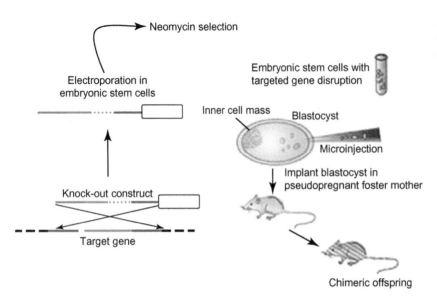

for this allele, one must obtain DNA from the mice, usually from a small tail or toe biopsy, and design PCR reactions with primers that can produce amplification products diagnostic of the intact or knock-out allele.

The entire procedure focuses on one gene at a time, is expensive, and takes at least one year from start to finish. Although knock-out mouse technology has generated a wealth of information and revolutionized mouse genetics, there are some major disadvantages and caveats that one should be aware of. First, the resulting knock-out mice may not be homozygous viable and, therefore, would be of limited use for behavioral studies. It should also be noted that genes with significant effects on behavior may complicate studies if they affect behaviors that are necessary for neonatal viability, such as suckling behavior. Second, the effect of the knock-out allele often depends on genetic background. Enhancer and suppressor effects may modulate the effect of the knock-out allele or related gene products may compensate, and these effects can be different in different genetic backgrounds. It may be necessary to **backcross** the allele into one or more different genetic backgrounds to assess its impact on the behavioral phenotype under study. Third, knocking-out a target gene may result in an unexpected and unpredicted phenotype. Whereas such phenotypes may be of interest, they may not allow assessment of the contribution of the target gene on the behavioral trait that motivated generating the mice in the first place.

A striking example of an unexpected phenotype resulting from homologous recombination is the behavior of mice in which **neuronal nitric oxide synthase** (nNOS) was deleted. Despite the central role of nitric oxide as retrograde messenger during the induction of long-term potentiation, the mice appeared overall to be normal. However, male nNOS knock-out mice showed chronic hyperaggressive behavior, to the extent that they would kill each other violently. In addition, they showed excessive sexual behaviors in the presence of females. Subsequent follow-up studies showed increased levels of serotonin in the nNOS knock-out mice, indicating that serotonin turnover may be altered as a consequence of the absence of nitric oxide. Whereas the nNOS knock-out mouse generated new insights into neural mechanisms for aggression, this fortuitous behavioral phenotype was not *a priori* predicted. Commonly, investigators are not so lucky, and unexpected phenotypes are more often than not puzzling and difficult to interpret.

Another frequently encountered problem with knock-out mice is functional redundancy. This means that removal of one gene product can be compensated for by another closely-related protein, or that an alternative pathway can be adjusted to compensate for the deleted gene, thereby obscuring any observable phenotype. In this case, double knock-outs would have to be generated to reveal the function of the deleted gene and the compensating mechanism. For example, the **dystrobrevins** (α-dystrobrevin and β-dystrobrevin) are components of a transmembrane complex that binds to dystrophin to link the cytoskeleton to extracellular proteins in many tissues, including **cerebellar Purkinje cells**. Single knock-outs of α-dystrobrevin and β-dystrobrevin have only minor behavioral defects on motor behavior, whereas the double knock-out shows severe impairments on a battery of behavioral tests (Figure 9.5).

Conditional Homologous Recombination

As the development and use of knock-out technologies in mice progressed, a growing need arose to control gene knock-out spatially and temporally. Restricting deletion of the target gene only in certain tissues and/or at certain stages would allow scientists to bypass lethality that would

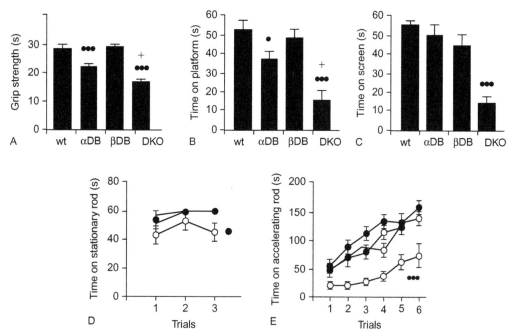

FIGURE 9.5   Defects in motor behavior in α- and β-dystrobrevin (DB) knock-out mice. Panels A-E show grip strength, time balanced on a small circular platform, time before falling from an inverted screen, time before falling from a stationary rotarod, and time before falling from an accelerating rotarod, respectively. "wt" designates the wild-type and "DKO" the double knock-out. The double knock-out mice performed significantly worse than the single knock-outs on all tests. The symbols in D and E are ■, Wild type; △, αDB; ●, βDB; ○, DKO. Significance of differences from wild-type are: * $p < 0.05$; ** $p < 0.005$; *** $p < 0.0005$. In A and B, + $p < 0.005$, significantly different from αDB. (Adapted from Grady et al. (2006). *J. Neurosci.*, **26**, 2841–2851.)

result from global abolishment of expression of an essential gene and, in many cases, would also facilitate interpretation of knock-out phenotypes. The most popular system for tissue-specific homologous recombination is the **Cre-loxP system**. Cre (an acronym for "causes recombination") is an enzyme that recombines specific DNA sequences without the need for cofactors. The target sequences for Cre-mediated recombination are known as "loxP" sites (loxP stands for "locus of crossing-over (x) in bacteriophage P1"). LoxP sites are 34 bp sequences that contain two 13 bp inverted repeats with an intervening 8 bp sequence that determines orientation. To enable Cre-mediated homologous recombination, the targeting construct contains loxP sites on either site of a critical region of the target gene. Cre recombinase will recognize these sites and, depending on their orientation, either excise (when the loxP sites are in the same orientation) or invert (when the loxP sites are in opposite orientation) the loxP flanked ("floxed") gene (Figure 9.6).

Tissue-specific homologous recombination is achieved by generating a transgenic mouse that carries a construct that encodes Cre behind a tissue-specific promoter. When such homozygous mice are crossed to homozygous mice that carry the floxed target gene, the offspring will contain one copy of the Cre recombinase, and one copy of the floxed gene. Cre recombinase will now excise the floxed gene in those cells in which it is expressed. For example, if Cre is placed under a promoter that is only active in the

nervous system, the floxed target gene will be removed from neurons, but will remain unaffected in glia and non-neural tissues.

An example of this approach is the conditional knock-out of the *Pten* gene targeted to the cerebral cortex and hippocampus via the Cre-loxP system. This gene encodes the Pten phosphatase, an enzyme that removes the phosphate from the 3′ position of phosphatidyl-inositol-3,4,5-trisphosphate and antagonizes the activity of phosphatidyl-inositol 3′ kinase, which controls a signaling pathway that regulates cell survival and proliferation. The Pten knock-out mouse is of interest, as Pten mutants show phenotypes reminiscent of **autism**. The term, **autism spectrum disorders**, or ASD, is often used to emphasize the phenotypic diversity of the various manifestations of autism (see also Chapter 14).

Abolishing *Pten* expression by conventional knock-out procedures causes embryonic lethality. However, targeting the deletion of the *Pten* gene to neuronal populations in the hippocampus and cerebral cortex enabled neuroanatomical and behavioral studies on viable adult animals. These animals showed marked **hypertrophy** of the cerebrum and hippocampus (macrocephaly) with **ectopic** dendrites and axonal tracts, abnormal **dendritic arborizations,** and increased synapses.

The mice were subjected to a battery of behavioral tests, and showed increases in locomotor activity in an

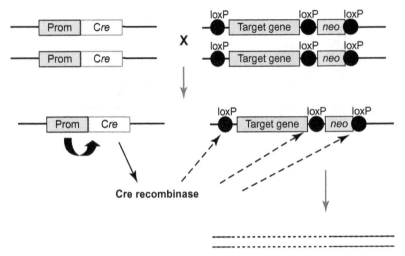

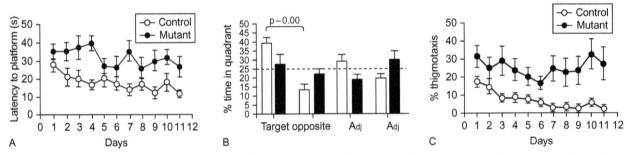

FIGURE 9.6 Conditional homologous recombination. Cre recombinase is expressed under a tissue-specific promoter (PROM) in the F1, where it recognizes loxP sites flanking a floxed target gene and a neomycin selection cassette. Homologous recombination at the loxP sites results in excision of the target gene only in those tissues in which the promoter that drives *Cre* expression is active. Note that in the example shown in this figure the neomycin resistance gene, used to select homologous recombinant stem cells, is also flanked by loxP sites, and is removed at the same time the targeted gene is knocked-out (the *neo* cassette can have undesirable effects on the phenotype).

FIGURE 9.7 Performance of *Pten* knock-out mice in the Morris water maze. A. Mutants are deficient compared to controls in learning the location of a submerged platform when tested at subsequent days. B. Whereas control mice (white bars) spend significantly more time in the quadrant of the maze in which the hidden platform is located, mutants (black bars) spend equal amounts of time searching for the hidden platform in the opposite or adjacent (Adj) quadrants. C. Mutant mice spend more time around the edge (**thigmotaxis**) of the water maze than controls similar to their open field behavior and, presumably, reflecting increased anxiety. (Modified from Kwon et al. (2006). *Neuron*, **50**, 377–388.)

open field and increased startle-responses to a loud sound. They also exhibited increased anxiety, as reflected by their reluctance to spend time in the center of an open field and increased latency to enter the light side of a light/dark box. Furthermore, they displayed learning impairment, as assessed by their poor performance in the Morris water maze (Figure 9.7). In addition, they showed abnormal social interactions with conspecifics. Their social behavioral impairments and exaggerated responses to sensory stimuli are reminiscent of human autism spectrum disorders, suggesting that disruption of the Pten/PI3K transduction pathway could contribute to some aspects of autism (see also Chapter 14).

A further modification of the Cre-loxP system designed to provide not only spatial, but also temporal, control over the removal of the target gene makes use of a Cre enzyme fused with a mutated ligand-binding domain of the estrogen receptor (ER-Cre). The fusion protein, in the absence of its ligand, remains in the cytosol, and cannot enter the nucleus. Its binding domain has relatively low affinity for endogenous estrogen, but high affinity for the estrogen analog tamoxifen. Normal circulating estrogen levels,

therefore, will not allow the fusion protein to enter the nucleus. However, when animals are fed tamoxifen, binding of tamoxifen to ER-Cre will enable translocation of the enzyme from the cytosol into the nucleus, where the *cre* recombinase can act on loxP sites. This system enables investigators to perform spatially-targeted knock-outs with temporal control that can bypass embryonic lethality. Clearly, conditional knock-outs are more complex and expensive to generate than conventional homologous recombinants, even though transgenic mice that contain Cre recombinase under a variety of different promoters are commercially available. One caveat with ER-Cre-mediated gene excision is that the efficiency is sometimes low, which may lead to incomplete excision of the floxed gene in all target cells, complicating phenotypic analysis. Furthermore, phenotypes of interest may be missed or incompletely characterized, if Cre-mediated homologous recombination is not targeted to cells that directly or indirectly contribute to the behavioral phenotype.

The **FLP-FRT system** is a variation on the Cre-loxP system, and represents another frequently-used conditional knock-out system. Here the yeast enzyme, FLP recombinase, is used instead of Cre, and recognizes FLIP recombinase target (FRT) sites in lieu of *loxP* sites. As with *loxP* sites, orientation of the FRT sequences dictates inversion or deletion events in the presence of FLP. The FLP-FRT system has been utilized to develop strategies for homologous recombination in *Drosophila*, a genetic approach previously unavailable in flies. Rong and Golic developed a method in which FLP-FRT recombination is combined with a transgenic restriction enzyme that introduces double-stranded breaks in the DNA to induce homologous recombination. An example of successful use of this method in behavioral genetics comes from the laboratory of Leslie Vosshall at the Rockefeller University, who used it to knock out the *Or83b* gene which encodes the Or83b odorant receptor in *Drosophila*. This receptor is ubiquitously expressed in olfactory sensory neurons, along with odorant receptors that are unique to each neuron (Chapter 13). Vosshall and collaborators hypothesized that *Or83b* interacts with neuron-specific odorant receptors, and facilitates their transport and insertion in the chemosensory dendritic membranes of olfactory neurons. Indeed, in flies in which the *Or83b* gene had been removed by homologous recombination, odorant receptors were not inserted in the dendritic membranes. Thus, although homologous recombination in flies has not yet found widespread use in behavioral genetics, the technology looks promising.

TRANSPOSON-MEDIATED MUTAGENESIS

Insertion of *P*-elements in the Genome of *Drosophila*

Transposon-mediated mutagenesis, also referred to as **P-element insertional mutagenesis**, is an effective experimental method for inducing mutations in *Drosophila*. Transposable elements were initially discovered by Barbara McClintock in maize as **jumping genes**. In *Drosophila*, naturally occurring transposons, *P*-elements, have inverted repeats at each end that enable insertion in the DNA when acted on by a transposase. *P*-element-derived constructs have been engineered that capitalize on these inverted repeat insertion sites. All engineered transposons contain a visible marker. This is usually a *mini-white* gene that restores eye pigmentation to flies that carry a *white* (*w*) mutation. Many *P*-elements contain **reporter constructs**, in addition to the visible marker. This is often a *lacZ* gene, which encodes β-**galactosidase**, a bacterial enzyme that can convert analogs of the sugar β-galactose into a colored product, or a fluorescent marker, such as **green fluorescent protein** (GFP). Some transposons also contain a bacterial cloning vector to enable the tagged DNA sequences to be propagated *in vitro*. After their introduction into the genome by injection into embryos, the *P*-element can

readily be moved around the genome, by crossing the original transgenic strain to a strain that provides a source of **transposase**. After allowing transposition to occur, the transposase is separated from the *P*-element in the next generation of crosses, so that following the initial transposition the *P*-element remains stable at a fixed position. Single flies are then used to generate independent lines of individuals that contain transposons at different fixed locations in the genome (Figure 9.8).

Insertion of the *P*-element in or near a gene involved in the behavior under study can disrupt the expression of this gene, and this will appear in the behavioral assay as a significant deviation from the mean phenotypic value, assessed either compared to the *P*-element-free isogenic host strain or as a statistical outlier among a large population of **co-isogenic** *P*-element insert lines (Figures 9.9 (p. 137) and 9.10 (p. 138)). One advantage of *P*-element insertional mutagenesis is the generation of hypomorphic rather than null mutations, as this allows studies of subtle phenotypic effects. Furthermore, whereas null mutations in genes that affect development might not allow the animal to grow to adulthood, hypomorphic mutations may allow development to occur normally, while having a substantial effect on adult behavior. It is important to ensure that all the generator stocks used to construct these *P*-element insertion lines are inbred. Failure to do so will make it difficult to attribute subtle behavioral effects observed in the resulting *P*-element insertion lines unambiguously to the *P*-element insertion site, as the magnitude of the effect will be of the same order as that of the background variation. Whereas, in principle, every gene in the genome could be tagged by a transposon, insertion sites are not entirely random, and there are hot spots and cold spots in the genome. Notable cold spots are gene clusters, such as genes that encode **odorant-binding proteins** or **odorant receptors**.

Characterization of *P*-element Insertion Sites and Their Phenotypic Effects

The short generation time of *Drosophila* (about 14 days), together with the mutation rate (estimated at about one mutation per genome per generation), may result in the accumulation of mutations with phenotypic effects that are unrelated to the *P*-element insertion site. It is, therefore, important to provide experimental evidence that the behavioral deviations that are observed are indeed correlated with the *P*-element insertion. This can be done in different ways. **Quantitative complementation tests** can be performed after crossing a homozygous *P*-element insert line to a balanced **deficiency stock**, to demonstrate quantitative failure of the *P*-element insert line to complement the deficiency, thereby mapping the phenotypic effect to the region of the deficiency that covers the *P*-element insertion site. Similarly, complementation tests can be performed

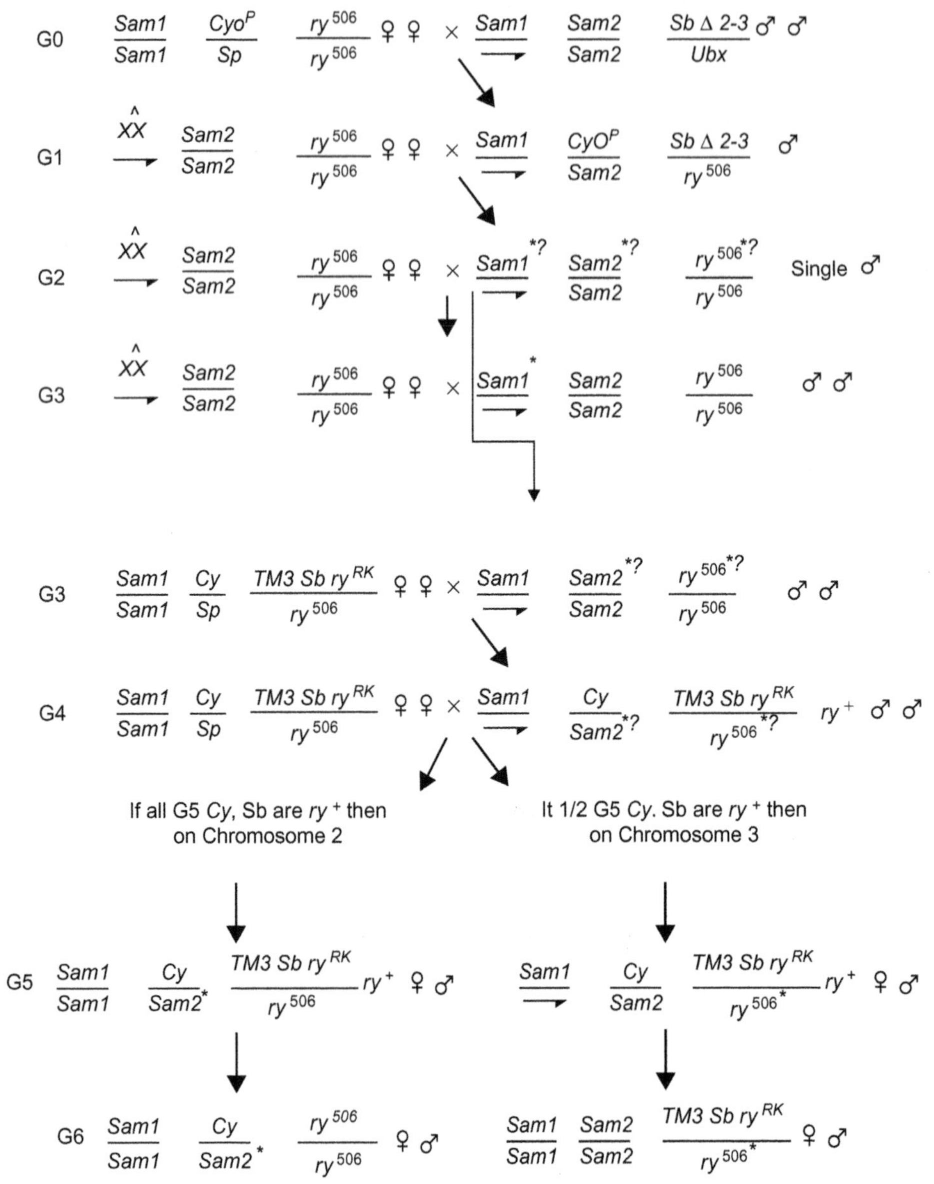

G0 $\dfrac{Sam1}{Sam1}$ $\dfrac{Cyo^P}{Sp}$ $\dfrac{ry^{506}}{ry^{506}}$ ♀♀ × $\dfrac{Sam1}{\longrightarrow}$ $\dfrac{Sam2}{Sam2}$ $\dfrac{Sb\,\Delta\,2\text{-}3}{Ubx}$ ♂♂

G1 $\dfrac{\hat{XX}}{\longrightarrow}$ $\dfrac{Sam2}{Sam2}$ $\dfrac{ry^{506}}{ry^{506}}$ ♀♀ × $\dfrac{Sam1}{\longrightarrow}$ $\dfrac{CyO^P}{Sam2}$ $\dfrac{Sb\,\Delta\,2\text{-}3}{ry^{506}}$ ♂

G2 $\dfrac{\hat{XX}}{\longrightarrow}$ $\dfrac{Sam2}{Sam2}$ $\dfrac{ry^{506}}{ry^{506}}$ ♀♀ × $\dfrac{Sam1^{*?}}{\longrightarrow}$ $\dfrac{Sam2^{*?}}{Sam2}$ $\dfrac{ry^{506\,*?}}{ry^{506}}$ Single ♂

G3 $\dfrac{\hat{XX}}{\longrightarrow}$ $\dfrac{Sam2}{Sam2}$ $\dfrac{ry^{506}}{ry^{506}}$ ♀♀ × $\dfrac{Sam1^{*}}{\longrightarrow}$ $\dfrac{Sam2}{Sam2}$ $\dfrac{ry^{506}}{ry^{506}}$ ♂♂

G3 $\dfrac{Sam1}{Sam1}$ $\dfrac{Cy}{Sp}$ $\dfrac{TM3\,Sb\,ry^{RK}}{ry^{506}}$ ♀♀ × $\dfrac{Sam1}{\longrightarrow}$ $\dfrac{Sam2^{*?}}{Sam2}$ $\dfrac{ry^{506\,*?}}{ry^{506}}$ ♂♂

G4 $\dfrac{Sam1}{Sam1}$ $\dfrac{Cy}{Sp}$ $\dfrac{TM3\,Sb\,ry^{RK}}{ry^{506}}$ ♀♀ × $\dfrac{Sam1}{\longrightarrow}$ $\dfrac{Cy}{Sam2^{*?}}$ $\dfrac{TM3\,Sb\,ry^{RK}}{ry^{506\,*?}}$ ry^{+} ♂♂

If all G5 *Cy*, Sb are *ry* $^+$ then   It 1/2 G5 *Cy*. Sb are *ry* $^+$ then
on Chromosome 2   on Chromosome 3

G5 $\dfrac{Sam1}{Sam1}$ $\dfrac{Cy}{Sam2^{*}}$ $\dfrac{TM3\,Sb\,ry^{RK}}{ry^{506}}ry^{+}$ ♀♂ $\dfrac{Sam1}{\longrightarrow}$ $\dfrac{Cy}{Sam2}$ $\dfrac{TM3\,Sb\,ry^{RK}}{ry^{506\,*}}ry^{+}$ ♀♂

G6 $\dfrac{Sam1}{Sam1}$ $\dfrac{Cy}{Sam2^{*}}$ $\dfrac{ry^{506}}{ry^{506}}$ ♀♂ $\dfrac{Sam1}{Sam1}$ $\dfrac{Sam2}{Sam2}$ $\dfrac{TM3\,Sb\,ry^{RK}}{ry^{506\,*}}$ ♀♂

FIGURE 9.8 Crossing scheme for mobilization and insertion of a *P*-element in an isogenic *Samarkand* (*Sam*) genetic background. Females who are homozygous for the *rosy* eye color mutation (the ry^{506} allele), and contain a *P*-element with an intact *ry* allele that compensates for the *rosy* mutation on the second chromosome, are crossed to males that contain the *Δ2-3* transposase, which in the F1 can move the *P*-element from its original position around the genome, causing random insertions. The *Δ2-3* transposase is removed during a subsequent cross, and single males are collected as founders for individual co-isogenic *P*-element insertion lines. Presence of the *P*-element can be monitored by a normal eye color in a homozygous ry^{506} background. Balancer chromosomes containing the *Curly* (*Cy* marker) or *Stubble* (*Sb*) marker for the second or third chromosome, respectively, are used to determine on which chromosome the *P*-element has inserted. *P*-element excision to demonstrate phenotypic reversion of a mutant phenotype is achieved via a similar crossing scheme.

with an allelic series of existing mutant stocks, if available. In addition, the *P*-element can be mobilized by crossing the *P*-element insert line to a line that provides a source of transposase. Precise excision of the *P*-element should result in reversion of the behavioral phenotype. Depending on where the *P*-element has inserted, excisions can either be precise or imprecise. In the latter case, *P*-element excision may have caused a deletion in the flanking DNA region, possibly creating a null mutation in the affected gene, or a small segment (usually the ends) of the *P*-element may remain in the genome. Deletions created by *P*-element mobilization can either be small and, if they occur in noncoding sequences, may not interfere with the phenotype, or they can be large and cause a local deletion with functional consequences that can be informative. Remnants of *P*-elements may interfere with gene expression, similar to

the original insertion, but in many cases may allow reversion to the normal behavioral phenotype (Figure 9.11, p. 138).

P-element insertion sites were traditionally determined cytologically by *in situ* hybridization to larval polytene chromosomes. However, with the availability of the complete *D. melanogaster* genome sequence since 2000, it is easy to determine precise *P*-element locations by identifying flanking sequences of the host DNA. In some cases, when the *P*-element construct contains a cloning vector, such as pBluescript (e.g. P[*lArB*]), genomic DNA can be isolated, digested with a **restriction enzyme** that cuts in the cloning site of the vector and the host DNA, ligated and transformed immediately into bacteria for cloning and sequencing (Figure 9.12, p. 138). The average size of the insert corresponds to the frequency of occurrence of the restriction site for the **endonuclease** used in the DNA (for *HindIII*, this

Box 9.3 The genome of *Drosophila melanogaster*

The genetic material of *Drosophila* is organized as three major chromosomes (Figure A). The X-chromosome (and in males also a Y-chromosome) is the sex chromosome, and there are two autosomes. In addition, flies carry a small fourth chromosome that is – like the Y-chromosome – mostly heterochromatic with few functional genes. The chromosomes are divided in 100 numbered cytological subdivisions (plus divisions 101–102 on the fourth chromosome), 20 on the X-chromosome, and 20 each on the left and right arms of the second and third chromosomes. Each cytological band is further subdivided in bands that are designated with letters A–F, and each of those subdivisions can be further subdivided numerically, e.g. 48D2. The locations of genes are frequently indicated by their cytological band locations, which have in some cases provided a useful system for nomenclature (for example, the gene which encodes odorant receptor Or83b is located on cytological location 83 on the right arm of the third chromosome, and is the second odorant receptor gene in that cytological band, the first one being *Or83a*). The telomeres and centromere (indicated by the black circles in the figure) are mostly heterochromatic and, like the fourth chromosome refractory to recombination.

Figure B shows the FM7 balancer chromosome, which is derived from the X-chromosome but contains large chromosomal rearrangements that effectively suppress recombination. The availability of balancer chromosomes, and the fact that recombination does not occur in males, allows mutations to be stably propagated from one generation to the next.

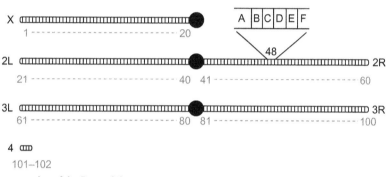

FIGURE A   Schematic representation of the *Drosophila* genome.

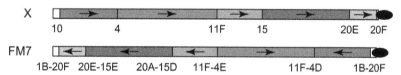

FIGURE B   Schematic representation of a balancer chromosome.

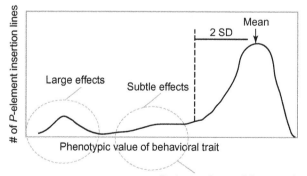

FIGURE 9.9   Schematic representation of the distribution of phenotypic values of a behavioral trait among a collection of co-isogenic *P*-element insertion lines. SD, standard deviation.

is about 2500 bp). A currently more popular way to identify flanking host DNA sequences is by **inverse PCR**. This approach can be used for *P*-element constructs that do not contain a cloning vector (e.g. *PlacW*). In this case a frequent cutting restriction enzyme (often *MboI*) is used to fragment the genomic DNA and release the *P*-element with a flanking fragment attached. This construct can now be ligated and primers against each end of the *P*-element construct can be used to amplify the host DNA fragment, contained inside the ligated circularized construct. The PCR fragment can then be cloned and sequenced (Figure 9.13, overleaf).

Whereas, in principle, identification of the DNA sequences adjacent to the *P*-element insertion sites may identify the affected gene immediately, in reality there is often ambiguity, which needs to be resolved by further

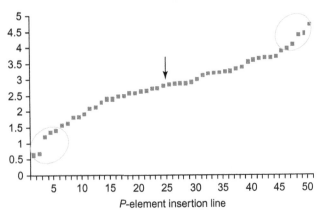

FIGURE 9.10 Frequency distribution histogram of phenotypic values of *P*-element insertion lines. The diagram shows a typical experiment in which a phenotype with quantitative measurements ranging from 0–5 was measured for 50 co-isogenic *P*-element insertion lines. Note the variation in scores around the population mean, indicated by the arrow. The circles denote *P*-element insertion lines with substantially deviant phenotypic values from the mean. Deviations from the population mean can be (and should be) statistically determined at the $P < 0.05$ (lenient criterion) to the $P < 0.01$ or $P < 0.001$ (stringent criteria) level to identify transposon-tagged candidate genes for further studies.

FIGURE 9.11 Assessment of the presence of a *P*-element based on eye color. The visible marker for this *P[GawB]*-insertion line is the *mini-white* (*w*) gene. Flies without *P*-element insertions have white eyes (far right). Presence of a single *P*-element insertion in a heterozygote turns the eye color orange (center), whereas flies homozygous for the *P*-element will show fully-compensated red eye pigmentation (left).

experimentation. In the most straightforward scenario, the transposon has inserted between the promoter and the transcription initiation site of a gene, or in an intron of a gene with no additional genes in the vicinity. However, *P*-elements can insert in regions between two genes, in regions with nested genes, or in regions remote from the most nearby predicted coding sequence. Whereas the nearest gene is usually the best bet for the transposon-disrupted effect, in some instances inserted *P*-elements can exert effects on genes located a significant distance away, or they may exert effects on more than one gene. Furthermore, the candidate gene may generate multiple transcripts, and not all of these transcripts may be equally affected by the transposon (Figure 9.14). These effects usually do not result in complete gene silencing, but rather in reduction of expression. In principle increased expression, although less common, can also occur. Finally, effects of *P*-element mutations can vary, depending on the developmental stage (embryos, larvae, pupae, and adults), as promoter and enhancer elements that are differentially-activated at

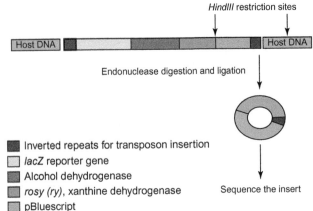

FIGURE 9.12 Identification of *P*-element flanking sequences by plasmid rescue. The diagram shows *P[lArB]*, a member of an early generation of transposable element constructs, which contains a *lacZ* reporter gene, xanthine dehydrogenase, a visible marker which gives rise to the *rosy* eye phenotype, the alcohol dehydrogenase gene, and the cloning vector pBluescript. The latter allows flanking sequences of the host DNA to be cloned directly into a plasmid vector that can be grown in bacteria following digestion with a restriction endonuclease and ligation.

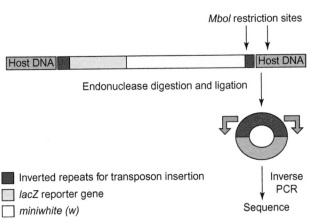

FIGURE 9.13 Identification of *P*-element flanking sites by inverse PCR. The diagram shows the transposon *P[GawB]*, which contains a *lacZ* reporter gene and a *mini-white* gene which complements the *white* mutation of the host genetic background. Following restriction endonuclease digestion and ligation, the flanking sequence of the host DNA can be identified by inverse PCR using primers that are directed from the *P*-element towards the host flanking sequence (arrows).

different developmental stages may be affected differently by the *P*-element insertion.

Evidence to Consolidate Candidate Genes

To provide evidence that the candidate gene tagged by the transposon is indeed the gene responsible for the affected phenotype, the following experiments should be performed:

1. Demonstration that expression of the candidate gene is altered in the mutant, as compared to the co-isogenic control, either by **Northern hybridization**, PCR, *in situ*

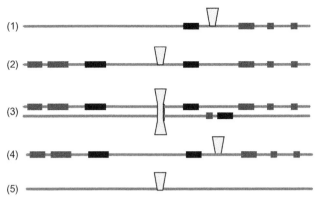

FIGURE 9.14 *P*-element insertion sites near candidate genes. In case (1) the *P*-element (inverted triangle) has inserted between the promoter region (black) and the transcription initiation site of the candidate gene (exons are indicated as gray boxes). In case (2) the *P*-element has inserted in between two genes. In case (3) the *P*-element has inserted in between two genes and a small nested gene encoded by the opposite DNA strand is nested within one of the candidate genes. In case (4) the *P*-element has inserted between the promoter region and the transcription initiation site of a candidate gene, but an effect on expression of an adjacent gene cannot be a *priori* excluded. In case (5) the *P*-element has inserted in a region of DNA without a candidate gene in the immediate vicinity. In addition, *P*-elements can insert in introns and exons, they can affect expression of alternative transcripts, and their effects may vary with developmental stage.

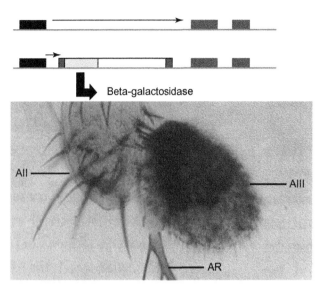

FIGURE 9.15 Enhancer trap. The diagram shows a gene with a promoter/enhancer region (black box) and two exons (gray boxes). A *P[GawB]*-element with a *miniwhite* gene (white box) and a *lacZ* reporter gene (light gray box) has inserted in the region between the promoter and the first exon. The promoter/enhancer elements of the target gene now drive expression of β-galactosidase. The lower panel shows *lacZ* staining in the third antennal segment, the main olfactory organ, driven by a reporter gene in a *P[lArB]* transposon, which has inserted near *pinocchio*, a gene that encodes a protein thought to be required for removal of foreign hydrophobic compounds, including odorants, from the fluid that surrounds the chemosensory dendrites in the antenna (the **perilymph**). AII and AIII designate the second and third antennal segments, respectively, and AR designates the **arista**, a feather-shaped mechanosensory and thermosensory appendage.

hybridization, **Western blotting**, or **immunohistochemistry**. Often slight reductions in gene expression result in profoundly-aberrant behavioral phenotypes. Such reductions may not be detectable quantitatively by most of the procedures enumerated above. **Real-time quantitative PCR** is here the most sensitive procedure. If alterations in gene expression are observed, it is imperative to show that *P*-element excision and phenotypic reversion correlates with restoration of wild-type expression levels.

2. Correlation of expression of a reporter gene contained in the *P*-element in the mutant with expression of the candidate transcript in the wild-type control. If the *P*-element has inserted near the promoter or in an intron of the affected gene, the promoter can drive a reporter gene (usually *lacZ*) in the same cells in which the candidate gene is normally expressed. The use of transposon tagging to characterize expression patterns of tagged genes is known as **enhancer trap** (Figure 9.15). The expression pattern of β-galactosidase can now be compared with immunohistochemical expression patterns or *in situ* hybridization expression patterns of the candidate gene product. If the patterns match they will provide supportive evidence. However, if there is no precise match, one cannot exclude the transposon-tagged gene as the candidate gene responsible for the phenotype, because complex promoter-enhancer regulation may give rise to different expression patterns; sometimes *P*-elements that are inserted only a few bases apart may show different reporter gene expression patterns. Furthermore, if alternative transcripts exist, expression of only one of these transcripts may match the reporter gene expression pattern. In addition, when immunohistochemistry is used to detect protein expression patterns, post-translational modifications or protein-protein interactions can interfere with antibody detection. Finally, differential stability and/or transcript abundance of messages for *lacZ* or the native gene may give rise to qualitatively different expression patterns.

3. Phenotypic rescue. In this case, a transgenic construct in which expression of the native gene is driven by a promoter, often a heat-shock protein (*Hsp70*) promoter, is introduced into the mutant background, and its introduction should compensate for the transposon-mediated disruption of its counterpart in the *P*-element insert line, thereby restoring the initially-observed behavioral phenotype. This experiment is generally considered the gold standard of proof that the *P*-element tagged gene is indeed the gene responsible for the observed mutant phenotype. Since insertion of the transgenic construct in the host DNA may itself cause gene disruption, several independent transgenic lines must be constructed to control for positional effects. In addition, when analyzing subtle phenotypic effects, it is important to

preserve the genetic background while constructing these transgenic lines. In theory, possible complications might arise from overexpression of the transgene, or from expression of the transgene in cells in which the native gene is usually not expressed. In reality, these issues seldom arise. However, dosage effects, i.e. over-expression of a transgene in cells in which the normal gene product is expressed at a critical concentration, may complicate phenotypic rescue experiments, and could result in some instances in exacerbation of the mutant phenotype rather than restoration of the wild-type phenotype.

An example of the creative use of *P*-element insertional mutagenesis in *Drosophila melanogaster* is the discovery by Dean Smith and his colleagues of an odorant-binding protein that mediates responses to 11-*cis*-vaccenyl acetate, a pheromone that mediates aggregation behavior and has been implicated as an "anti-aphrodisiac" for flies. Odorant-binding proteins are soluble proteins that are secreted in the aqueous medium, the periplymph, which surrounds the chemosensory dendrites of olfactory sensory neurons in sensilla of the **third antennal segments** and **maxillary palps**, the principal olfactory organs of the fly. A *P*-element construct that contains a *tau-lacZ* reporter gene was introduced in the genome, and lines were established that contained insertions at different cytological locations. The tau protein is a microtubule-associated protein that directs expression of β-galactosidase encoded by the *lacZ* gene to axons of neurons. When the *P*-element inserts in or near the promoter or enhancer elements of a gene, this promoter will drive the expression of the reporter gene in the same cells in which the disrupted gene is normally expressed.

Smith and colleagues identified lines that express *lacZ* in the antenna. In one line the reporter gene was expressed in **trichoid sensilla**, known to respond to 11-*cis*-vaccenyl acetate. Next, a deletion mutation was introduced by imprecise excision of the *P*-element. The affected gene encoded an odorant-binding protein, and in an olfactory trap assay mutant flies responded to high concentrations of ethanol and short chain alcohols that were repellent to wild-type flies. The gene encoding this odorant-binding protein was, therefore, initially called "*lush*," although the designation *Obp76a* is more appropriate, as it conforms to the conventional nomenclature of *Obp* genes and indicates its location at cytological band 76 on the third chromosome (Figure 9.16).

Subsequent electrophysiological and behavioral experiments showed that *lush* mutants did not respond to 11-*cis*-vaccenyl acetate, suggesting that the real function of *lush* in nature might be pheromone recognition. Normal electrophysiological and behavioral responses could be restored, either by delivering recombinant LUSH protein to trichoid sensilla into the recording pipette along with the pheromone, or by introducing an *Obp76a* transgene into the mutant background (Figure 9.17).

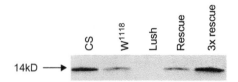

FIGURE 9.16 Expression of LUSH protein. A 14 kd immunoreactive polypeptide is seen in control antennae (*CS* and *w^{1118}*), but is absent in the *lush* deletion mutant. Expression is restored in transgenic lines that contain either 2 (rescue) or 6 (3× rescue) copies of the transgene. (Modified from Kim et al. (1998). *Genetics*, **150**, 711–721.)

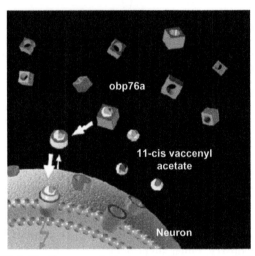

FIGURE 9.17 Transport of 11-*cis*-vaccenyl acetate by obp76a to a pheromone receptor on the chemosensory dendritic membrane of an olfactory sensory neuron in *Drosophila*. The odorant-binding protein binds to the pheromone, enabling it to be transported to membrane-associated pheromone receptors on the dendritic membranes of chemosensory neurons. However, the concept of a mere transport function for this odorant binding protein may need to be revisited, since a mutant of LUSH could constitutively activate olfactory sensilla in the absence of pheromone. This finding suggests that the pheromone induces a conformational change in LUSH upon binding, but that, once the odorant-binding protein is activated, LUSH itself can activate the odorant receptor, without further dependence on the presence of the pheromone. (Courtesy of Dr Dean Smith.)

Another example of how *P*-element insertions can provide insights into the genetic basis of behavior comes from work by Marla Sokolowski and colleagues, who discovered a naturally-occurring behavioral polymorphism in feeding behavior in *Drosophila* larvae. They observed that some larvae will travel extensively over their food source while feeding, whereas others remain mostly in one spot. These phenotypes were classified as "rovers" and "sitters," respectively, and could be attributed to a naturally-occurring polymorphism in the "*foraging*" (*for*) gene (Figure 9.18).

The *for* gene generates three alternative transcripts which encode a **cyclic GMP-dependent protein kinase**. *P*-element insertions in the *for* gene in a rover genetic background converted the rover phenotype into a sitter phenotype (Table 9.1). The rover phenotype could

TABLE 9.1 Subtle differences in enzymatic activity of cyclic GMP-dependent protein kinase (PKG), encoded by the foraging gene have profound effects on the behavioral phenotype

Genotype	Behavior	PKG kinase activity (pmol/min/mg protein)	
for R/for R	rover	12.6 ± 0.26	naturally occuring alleles
for S/for S	sitter	11.4 ± 0.28	
for S_1/for S_1	sitter	8.8 ± 0.26	sitter mutations on a rover background
for S_2/for S_2	sitter	9.4 ± 0.22	

Data are from Osborne et al. (1997). Science, **277**, 834-836.

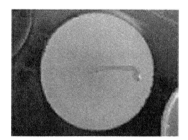

Sitter

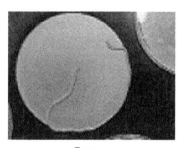

Rover
D. melanogaster

FIGURE 9.18 Examples of the trajectories of sitters (top) and rovers (bottom) during feeding.

TABLE 9.2 P-element excision and phenotypic reversion of a food search phenotype. A P-element introduced into the foraging gene in a rover background gave rise to a sitter phenotype. Excision of this P-element restored the rover phenotype

Strain	Foraging behavior on yeast	General locomotion on agar
P-element insertion line	4.3 ± 0.38 (39)	12.0 ± 0.66 (29)
P-element excision line	9.1 ± 0.55 (40)	9.4 ± 0.65 (29)

Data are from Osborne et al. (1997). Science, **277**, 834-836.

be rescued after excision of the *P*-element, or by overexpressing a *for* transgene in the mutant background. When enzymatic activities of rovers and sitters were measured biochemically in adult heads, the difference in activity of the cyclic GMP-dependent protein kinase measured in adult heads was only about 10%, illustrating the substantial influence hypomorphic *P*-element mutations can have on behavior, despite only small differences on enzyme activity (Table 9.2). Note, however, that in this case enzymatic activity was measured in adult heads, while the behavior was assessed with larvae. It remains therefore possible, and perhaps likely, that differences in enzyme activity are more pronounced and more critical at the larval stage.

Of course, many genes contribute to foraging behavior and the cyclic GMP-dependent protein kinase will affect multiple traits, but it is the feeding behavior that appears to be exquisitely sensitive to subtle differences in enzyme activity. It is of interest to note that expression of the ortholog of the *Drosophila for* gene in the honeybee *Apis mellifera*, designated *Amfor*, increases sharply at exactly the time when bees make the behavioral transition from nurses in the hive to foragers.

THE *GAL4-UAS* BINARY EXPRESSION SYSTEM IN *DROSOPHILA*

P-element insertional mutagenesis in *Drosophila* and reporter gene knock-in technology in mice enable the identification of neural pathways that drive behaviors by monitoring or controlling gene expression in specific neural circuits. More sophisticated systems also allow temporal control and analysis of gene expression.

The most popular system for the analysis of gene expression in *Drosophila* is an extension of the **enhancer trap** concept, which uses the **GAL4-UAS** binary expression system, developed by Andrea Brand and Norbert Perrimon at Harvard University. GAL4 is a yeast transcription factor not normally found in higher eukaryotes, which

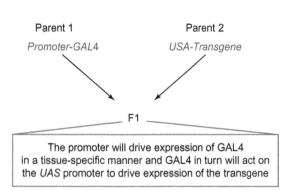

Transactivation via the binary *GAL4-UAS* expression system.

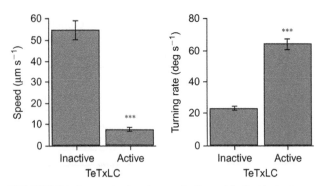

FIGURE 9.20 Locomotor impairments in *Drosophila* first instar larvae that express tetanus toxin light chain (TeTxLC) in the peripheral nervous system. An inactive form of the toxin is used as a control. (Modified from Suster, M. L., and Bate, M. (2002). *Nature*, **416**, 174–178).

binds to a promoter known as the **Upstream Activator Sequence** (UAS). Homozygous transgenic flies can be constructed that contain a construct in which a cell-specific promoter is cloned in front of the *GAL4* gene. When these flies are crossed with homozygous transgenic flies that contain a transgene of interest (e.g. a *lacZ* or *GFP* reporter gene) cloned behind UAS sequences, the heterozygous offspring will express GAL4 in specific cells, and in these cells transactivation via the UAS promoter will turn on the transgene (Figure 9.19). Using batteries of cell-specific neural promoters driving GAL4 allows expression of UAS-transgenes in different neural circuits. One caveat to these experiments, which is often ignored, is that existing GAL4 driver stocks are in different genetic backgrounds. This can complicate some analyses.

The *GAL4-UAS* system presents a simple method for analyzing neural pathways that drive behaviors. The following experiments can be performed using this system:

1. Promoter analysis. Promoters that regulate expression of the gene of interest can be cloned upstream of a *GAL4* coding sequence, and the resulting construct can be introduced into flies. Using a *UAS-lacZ* or *UAS-GFP* reporter construct will now allow visualization of the expression pattern expected of the original gene of interest. Cloning cell-specific promoters upstream of *GAL4* will enable visualization of neural projections by using a membrane-bound form of GFP (**CD8-GFP**), or a microtubule-associated form of GFP or *lacZ* (**tau-GFP** or **tau-*lacZ***). Synaptic contacts can be visualized using a **synaptobrevin-GFP** construct. Derivatives of GFP with different fluorescence spectra are available for double-labeling.

2. Inactivation or deletion of specific neural populations. Apoptotic genes (that is, genes that mediate **programmed cell death**), such as *hid* or *reaper*, or toxins, such as ricin or diphtheria toxin subunit A, can be expressed in a cell-specific manner using the *GAL4-UAS* system. This results in genetic ablation (destruction) of

neurons. A drawback of this experimental design is that the removal of neurons may lead to functional changes in neighboring neurons, introducing an aspect of uncertainty in the interpretation of results. Consequently, most experimental designs prefer to inactivate neurons rather than ablate them. This can be done by driving tetanus toxin in specific neurons. Tetanus toxin inhibits neurotransmitter release at synapses, by enzymatically cleaving the synaptic vesicle associated protein, synaptobrevin (Figure 9.20).

The ***shibire*** gene encodes a dynamin GTPase essential for synaptic vesicle recycling to provide an available pool of synaptic vesicles for neurotransmitter release. Temperature-sensitive *shibire* alleles disrupt synaptic function. These mutants have become popular, as they provide temporal control, since neuronal inactivation can be initiated by manipulating the temperature.

3. Driving expression of any transgene in specific neurons. This can be useful for misexpression studies (e.g. odorant receptors). A *Drosophila* mutant in which odorant receptors Or22a and Or22b have been deleted has silent neurons that normally express these receptors. The Or22a promoter can now be used to drive exogenous odorant receptors in these neurons via the GAL4/UAS system (see also Chapter 13).

Some temporal control over the expression of transgenes can be obtained by placing the *GAL4* gene behind a heat shock (*Hsp70*) promoter. A caveat here is that the *Hsp70* promoter may give some leaky expression. In recent years, a repressor of GAL4, **GAL80**, has been used for high-resolution analysis of neural connectivity. By introducing a temperature sensitive *GAL80* construct into a *GAL4-UAS-GFP* system, expression of the reporter gene in most neurons can be inhibited, allowing the visualization and mapping of neural projections of single cells in complex pathways. Temperature-sensitive control of GAL80 can provide temporal control over GAL4-dependent expression of transgenes, by allowing GAL4-mediated expression to occur only at temperatures at which GAL80 expression is inactive. This system (termed TARGET, an acronym for temporal and regional gene expression) uses the conventional *GAL4-UAS* system together with a temperature-sensitive GAL80 molecule, which represses GAL4 transcriptional activity at permissive temperatures. A second similar system, termed Gene-Switch, is based on a GAL4-progesterone receptor chimera that is hormone-inducible (Figure 9.21).

Tracing of neural projections through the introduction of genetically-engineered reporter constructs has also been applied successfully in mice. The best illustrations of the use of reporter gene constructs in delineating neural pathways that mediate behavior in both mice and *Drosophila* are studies on olfaction, described in Chapter 13.

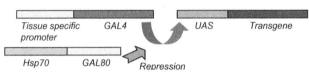

FIGURE 9.21 Schematic representation of the TARGET system for temporal control of *GAL4-UAS*-mediated transgene expression.

RNA INTERFERENCE

In the early 1990s, plant scientists in The Netherlands and the United States introduced extra copies of a gene that encodes an enzyme for flower pigmentation in petunias, with the goal of obtaining flowers with a deeper purple color. However, not only did the transgenic plants fail to grow purple flowers, their flowers were in fact less colored, and in many cases completely white. Further examination showed that introduction of the transgenes had resulted in turning off expression of both the transgene and the endogenous gene. In 1998, two Nobel prize-winning US scientists, Andrew Fire and Craig Mello, injected double-stranded RNA in the nematode *Caenorhabditis elegans*, and noticed a similar potent gene-silencing effect, which they termed **RNA interference** (RNAi).

The discovery of RNA interference heralded a new era in cell biology and molecular genetics. RNA interference appeared to be a highly-specialized process using distinct intracellular molecular mechanisms. When double-stranded RNA (dsRNA) is introduced into a cell it is processed by a ribonuclease (RNase) III enzyme called **Dicer** into **small interfering RNAs** (siRNAs); these 21–25-mer fragments subsequently direct cleavage of homologous mRNA via an **RNA-induced silencing complex** (RISC). The degree of complementarity between the RNA-silencing molecule and its cognate target determines the fate of the mRNA: blocked translation or immediate destruction. Thus, dsRNA gives rise to a number of small interfering RNAs (siRNAs) that direct RNAses to destroy transcripts complementary to the siRNAs (Figure 9.22).

RNA interference provided a new tool for the bag of tricks of molecular geneticists to silence gene expression. Because introduction of double-stranded RNA does not always completely eliminate gene expression, but rather reduces it substantially, the term "gene knock-down" instead of "knock-out" is often used. RNAi technology has been used extensively for studies in *C. elegans* and *Drosophila*. Worms will conveniently take up RNAi constructs by feeding on *E. coli* bacteria that carry such constructs. The RNAi will be taken up via the gut and distributed among the nematode's cells. However, delivery of RNAi into the 302 neurons of *C. elegans* has proven to be more difficult than delivery to other cell types. A useful feature of RNAi gene-silencing in *C. elegans* is that, once introduced, RNAi constructs in worms generally appear to be heritable. The mechanism for this phenomenon remains mysterious.

RNAi constructs can also be introduced in *Drosophila*, by designing dsRNA encoding transgenes which are incorporated in the DNA. These constructs contain tail-to-tail cloned fragments of the target gene that, when transcribed, form double-stranded hairpin structures or symmetric *UAS* elements that can direct transcription of the target gene in sense and antisense directions (Figure 9.23). RNAi-mediated knock-down of gene expression has been variable in flies, and not nearly as uniformly effective as in *C. elegans*. This can often be attributed to positional effects of the RNAi encoding transgene. Barry Dickson and collaborators at the Research Institute of Molecular Pathology in Vienna, however, have undertaken an ambitious and promising project to introduce RNAi encoding constructs for every gene in the *Drosophila* genome at a defined neutral site in

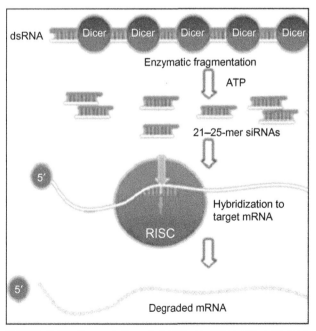

FIGURE 9.22 Fragmentation of dsRNA by Dicer and subsequent siRNA-targeted degradation of a specific mRNA.

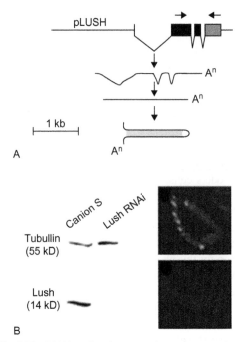

FIGURE 9.23 RNAi-mediated suppression of expression of the *Drosophila* LUSH odorant-binding protein. A. A segment of *lush* genomic DNA is ligated to a *lush* cDNA sequence in the opposite orientation and, when driven under the *lush* promoter, the resulting transcript will be spliced to an mRNA that folds into a double-stranded hairpin (boxes represent exons). B. Expression of the double-stranded *lush* RNAi obliterates expression of the LUSH protein, as shown by absence of immunoreactivity with an antibody on a Western blot (left) and absence of immunofluorescence in an antenna of a transgenic animal expressing *lush* RNAi (bottom right), compared to the control (top right), where the red fluorescence localizes LUSH to sensilla in the third antennal segment. (Modified from Kalidas, S., and Smith, D. P. (2002). *Neuron*, **33**, 177–184).

the *Drosophila* genome using the sequence specific integrase of bacteriophage PhiC31. This living RNAi library of *Drosophila* lines will be a valuable research tool for studies on behavioral genetics. Transient RNAi administration by injection into the hemolymph has been successful in many insects, other than *Drosophila*.

Why do cells have a machinery dedicated to the degradation of dsRNA? The replication cycle of many viruses involves a dsRNA stage, and it has been suggested that the RNAi interference mechanism is a cellular defense mechanism aimed at eliminating viruses. However, it has become increasingly clear that small endogenous dsRNAs are encoded and transcribed in the genome. These are known as **microRNAs**, and form double-stranded structures that can regulate the stability and translation of target genes via the RISC pathway.

Modified oligonucleotides that contain a six-membered morpholine ring instead of a ribose sugar and modified phosphorodiamidate bonds, in the nucleic acid backbone have also been used as gene-silencing tools. These antisense oligonucleotides, known as **morpholino antisense oligonucleotides**, are not degraded in the cell and are

delivered by microinjection into eggs. They are used primarily to disrupt genes associated with early development in the clawed frog, *Xenopus laevis*, and zebrafish, *Danio rerio*. They target the translation initiation sites or splice sites of the target mRNA, and result in potent suppression of translation. Their effects are diluted as cells divide, but the large developmental defects that result from the use of these morpholino antisense oligonucleotides usually result in lethality before animals reach adulthood. For these reasons, morpholino oligonucleotides have not been used for studies on behavior.

SUMMARY

Many congenital physiological and behavioral disorders are the result of spontaneous mutations that segregate in the population. Mouse models that carry spontaneous mutations have also been identified, and have served as valuable models for human diseases. However, as the occurrence of spontaneous mutations cannot be predicted or controlled, induced mutations have provided the principal means for genetic studies of behavior. Chemical mutagenesis, X-ray radiation, homologous recombination, transposon tagging, and RNA interference can be used to reduce or eliminate specific gene expression. Studies of behavior, however, require the development of a viable individual that can be assayed for behavioral impairments. Null mutations, induced either chemically or through homologous recombination, often result in lethality in early or later developmental stages. This problem can be circumvented by conditional knock-out approaches in mice, or the generation of hypomorphic mutations through *P*-element insertional mutagenesis in flies. Investigators should keep in mind that the manifestation of mutations may depend on the genetic context, and be modified by genotype-by-environment interactions. Furthermore, null mutations, including those arising from homologous recombination, may encounter complications due to functional redundancy, unpredicted phenotypes due to our limited knowledge of the dynamics of the metabolic and genetic networks in which the target gene and its product operate, pleiotropic phenotypes that are hard to disentangle, and lack of viability or fertility. Conditional knock-out strategies using the Cre-loxP or FLP–FRT systems can circumvent some, but not necessarily all, of these potential limitations. Similarly, whereas *P*-element insertional mutagenesis can rapidly identify flanking sequences of the host DNA, further experimentation is always required to confirm that the mutant phenotype is indeed due to the transposon, and not to an unrelated mutation in the genome, and that disruption of the candidate gene and not a nearby or nested gene is indeed responsible for the observed phenotype. Whereas the introduction of mutations is generally straightforward, their characterization requires careful and extensive experimentation. The *GAL4-UAS* binary expression system enables the tissue-specific expression of transgenes to identify cells that express a particular gene, to map or inactivate neural projections, or to mediate rescue of mutant phenotypes through the introduction of a wild-type transgene. Temperature-sensitive regulators provide reversibility, and more complex systems can be designed to enable temporal control of transgene expression via this system.

STUDY QUESTIONS

1. A scientist has identified a new gene of previously-unknown function which is suspected to play a critical role in regulating feeding behavior in mice. To assess the effect of this gene on feeding behavior, the scientist generates a knock-out mouse in which the gene has been deleted. What are the possible problems he or she may encounter with this approach?

2. A *Drosophila* geneticist identifies a *P*-element insertion line that shows aberrant locomotor behavior. The *P*-element has inserted exactly between two genes that are located 2 kb apart in opposite orientations. What experiments could be performed to establish which gene – or whether perhaps both genes – are causal to the observed behavioral deficit?

3. What are the advantages of *P*-element insertional mutagenesis over chemically-induced mutations for the identification of genes associated with behaviors?

4. What are the advantages and disadvantages of the Cre-loxP conditional knock-out system compared to conventional knock-out strategies in studies of behaviors?

5. Why can alleles that carry spontaneous deleterious mutations persist in a population?

6. Explain the nature and use of balancer chromosomes in *Drosophila*.

7. An investigator in California has generated a knock-out mouse that is defective in learning behavior as assessed by the Morris water maze. When shipped to the East Coast and tested in a similar water maze, the mouse appeared normal. Explain.

8. The East Coast laboratory mentioned in the previous question backcrossed the same knock-out allele for several generations into mice of a different strain. Now, when tested in the Morris water maze, the mice showed a profound learning deficit, even worse than initially observed by the Californian researcher who generated the original knock-out. Explain.

9. Compare the different actions of RNAi, morpholino antisense oligonucleotides, and ENU on inducing mutant phenotypes. Why are morpholino antisense oligonucleotides only of limited use for studies on the genetics of behavior?

10. *P*-element screens on behavioral traits in *Drosophila* performed to date typically identify a sizable fraction (e.g. ~4%) of all *P*-element insertion lines that affect a particular behavior. What is the biological implication of this observation?

11. What factors limit the resolution of detection of significant phenotypic effects on behavior during a *P*-element screen?

12. Describe the *GAL4-UAS* binary expression system and its applications in *Drosophila*.

13. Why is chemical mutagenesis still widely used?

RECOMMENDED READING

Bellen, H. J., O'Kane, C. J., Wilson, C., Grossniklaus, U., Pearson, R. K., and Gehring, W. J. (1989). *P*-element-mediated enhancer detection: a versatile method to study development in Drosophila. *Genes Dev.*, **3**, 1288–1300.

Bellen, H. J., Levis, R. W., Liao, G., He, Y., Carlson, J. W., Tsang, G., Evans-Holm, M., Hiesinger, P. R., Schulze, K. L., Rubin, G. M., Hoskins, R. A., and Spradling, A. C. (2004). The BDGP gene disruption project: single transposon insertions associated with 40% of *Drosophila* genes. *Genetics*, **167**, 761–781.

Brand, A. H., and Perrimon, N. (1993). Targeted gene expression as a means of altering cell fates and generating dominant phenotypes. *Development*, **118**, 401–415.

Branda, C. S., and Dymecki, S. M. (2004). Talking about a revolution: the impact of site-specific recombinases on genetic analyses in mice. *Dev. Cell*, **6**, 7–28.

Crabbe, J. C., Wahlsten, D., and Dudek, B. C. (1999). Genetics of mouse behavior: Interactions with laboratory environment. *Science*, **284**, 1670–1672.

Crawley, J. N. (2000). *What's Wrong with My Mouse? Behavioral Phenotyping of Transgenic and Knockout Mice*. John Wiley & Sons, Hoboken, NJ.

Duffy, J. B. (2002). GAL4 system in *Drosophila*: a fly geneticist's swiss army knife. *Genesis*, **34**, 1–15.

Filipowicz, W. (2005). RNAi: the nuts and bolts of the RISC machine. *Cell*, **122**, 17–20.

Greenspan, R. J. (1997). *Fly Pushing: the Theory and Practice of Drosophila Genetics*. Cold Spring Harbor Laboratory Press, Cold Spring Harbor, NY.

King, M. C., and Wilson, A. C. (1975). Evolution at two levels in humans and chimpanzees. *Science*, **188**, 107–116.

Kwon, C. H., Luikart, B. W., Powell, C. M., Zhou, J., Matheny, S. A., Zhang, W., Li, Y., Baker, S. J., and Parada, L. F. (2006). Pten regulates neuronal arborization and social interaction in mice. *Neuron*, **50**, 377–388.

Mello, C. C., and Conte, D., Jr. (2004). Revealing the world of RNA interference. *Nature*, **431**, 338–342.

Nelson, R. J., Demas, G. E., Huang, P. L., Fishman, M. C., Dawson, V. L., Dawson, T. M., and Snyder, S. H. (1995). Behavioural abnormalities in male mice lacking neuronal nitric oxide synthase. *Nature*, **378**, 383–386.

Osborne, K. A., Robichon, A., Burgess, E., Butland, S., Shaw, R. A., Coulthard, A., Pereira, H. S., Greenspan, R. J., and Sokolowski, M. B. (1997). Natural behavior polymorphism due to a cGMP-dependent protein kinase of *Drosophila*. *Science*, **277**, 834–836.

Papaioannou, V. E., and Behringer, R. R. (2004). *Mouse Phenotypes: A Handbook of Mutation Analysis*. Cold Spring Harbor Laboratory Press, Cold Spring Harbor, NY.

Rong, Y. S., and Golic, K. G. (2000). Gene targeting by homologous recombination in Drosophila. *Science*, **288**, 2013–2018.

Rubin, G. M., and Spradling, A. C. (1982). Genetic transformation of *Drosophila* with transposable element vectors. *Science*, **218**, 348–353.

Wang, J., and Barr, M. M. (2005). RNA interference in *Caenorhabditis elegans*. *Methods Enzymol.*, **392**, 36–55.

Xu, P., Atkinson, R., Jones, D. N., and Smith, D. P. (2005). *Drosophila* OBP LUSH is required for activity of pheromone-sensitive neurons. *Neuron*, **45**, 193–200.

Chapter 10

Genomics Approaches in Behavioral Genetics

OVERVIEW

The term "genome" was first used by H. Winkler in 1920, by merging the words GENes and chromosOMEs to designate an individual's complete set of chromosomes and their genes. The term "genomics" was coined by Thomas H. Roderick in 1986, during the planning stages of sequencing the human genome. This term gained wide acceptance, but its exact meaning still remains fluid. To some scientists genomics encompasses no more than whole-genome sequencing, others consider QTL mapping and association studies part of the genomics concept, whereas yet other investigators would use the term systems biology as a virtual synonym to emphasize the integration of the study of gene expression with the biology of the whole organism. In its broadest definition, genomics indicates the study of the evolution, structure, expression, and function of whole genomes. It represents in essence an extension of classical genetics, empowered by technological advances that enable the sequencing of whole genomes and the analysis of whole-genome transcriptional profiles. Annotated whole-genome sequences of many organisms, including *Drosophila*, mouse, rat, human, chimpanzee, honeybee, zebrafish, and many others have become available at an accelerated rate. Comparisons of sequenced genomes from closely-related species allow new insights into gene function and evolution. Large-scale analysis of protein expression profiles, or "proteomics," and global assessment of levels of cellular metabolites, or "metabolomics," are also rapidly developing areas of study. Sophisticated statistical methods for data mining the vast amount of information delivered by these procedures have developed into the field of "bioinformatics," which has become an integral part of modern genetics. These developments have had a major impact on behavioral genetics, as it has moved the field from single-gene studies toward a comprehensive analysis of genetic networks that mediate behaviors. In this chapter we will not discuss technological details of whole-genome sequencing and sequence assembly, but rather focus on an overview of major genomic technologies that can be used to study whole-genome transcriptional regulation, and thus give insights into the genetic architectures of behaviors.

DETECTING LARGE-SCALE GENE EXPRESSION

In 1973, Edwin Southern invented a blotting technique in which **restriction endonuclease** fragmented DNA could be fractionated by **electrophoresis**, transferred and bound to a nitrocellulose membrane, and hybridized to a labeled DNA probe to identify fragments in the DNA that contained sequences corresponding to the probe. Whereas many applications of the "**Southern blot**," named after its inventor, have been overtaken by PCR technology, Southern blotting is still used to determine gene copy number, for example in transgenic mice. Soon after the invention of the Southern blot, a variation of this approach was developed in which RNA is fractionated by electrophoresis, instead of DNA. This technique allows the identification of the number and sizes of specific RNA messages, as well as their tissue-specific expression patterns and, in a playful parody on Southern's name, has become known as the "**Northern blot**." With the advent of genomic technologies **expression microarrays** have appeared on the scene. Expression microarrays are used to assess global gene transcription patterns, and are sometimes described as quantitative Northern blots on a genomic scale (although detection of sizes of transcripts and reliable detection of alternative transcripts require actual Northern blots). Whereas a number of different microarray technologies are now available, they can be generally divided into **cDNA expression microarrays**, and **high-density oligonucleotide microarrays**. These arrays will be described below, along with a new generation of expression profiling techniques made possible by improved DNA sequencing methods.

Box 10.1   Serial analysis of gene expression (SAGE)

Serial analysis of gene expression (SAGE) is a method that can be used to analyze whole-genome gene expression levels, without the need to know *a priori* the genomic sequence. In this method, RNA is extracted and mRNA is bound via its polyA tail to oligo-dT bound to magnetic beads. While bound to the beads the RNA is reverse-transcribed to make complementary DNA, which is then treated with restriction enzymes to generate small fragments ("tags") that contain sequences diagnostic of the identities of the original messages. Many tags are now linked together at the cut restriction sites to form long concatamers, which are then sequenced and analyzed. A sequence from a single concatamer provides information about the presence of many mRNA species, and the frequencies with which tags are encountered in the population of sequenced concatamers provide a quantitative measure of their abundance (Figure A).

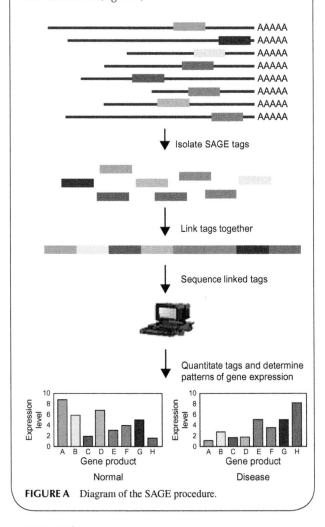

FIGURE A   Diagram of the SAGE procedure.

cDNA Microarrays

cDNA expression microarrays are constructed by isolating hundreds or thousands of PCR-amplified cDNAs, either from RNA templates or from coding regions in genomic DNA. Small nanoliter droplets of the amplified

Box 10.2   Genome-wide association studies (GWAS)

Genome-wide association studies (GWAS) use high-throughput genotyping methods to survey polymorphisms across the entire human genome, and to determine whether particular sequence variants are more frequently associated with a disease phenotype than with a normal control. Such studies have been successfully applied to identifying genes associated with risk for macular degeneration, Crohn's disease, diabetes, coronary heart disease, and bipolar disorder, and, no doubt, will be applied in future studies to addictive behaviors.

Although GWAS provides an attractive alternative over candidate gene approaches, in that it is an unbiased screen, there are a number of technical and conceptual limitations. First, surveying large numbers of single nucleotide polymorphisms (SNPs) across the entire genome results in a substantial multiple-testing problem. Thus, associations will have to exceed an exceptionally stringent statistical threshold in order to be considered significant. This means that large numbers of individuals, on the order of tens of thousands, are needed to provide sufficient statistical power, and even then only SNPs with relatively large effects can be resolved. The requirement for such large numbers of affected individuals requires extraordinary recruitment and phenotyping efforts. Such herculean endeavors have been undertaken by several large consortia, including Perlegen, Inc. in the United States, and the DECODE consortium which draws on the ethnically-homogeneous Icelandic population, where ancestries are meticulously documented.

Analyses of historical recombination in the human genome have shown that recombination often occurs at "hot spots," interspersed with large blocks of **linkage disequilibrium** (see also Chapter 8). The advantage of these large linkage blocks is that one needs to use fewer SNPs to survey the genome, as often a single SNP can characterize population variation of an entire large region. The drawback, however, is that once such a **tagging SNP** is associated with a phenotype it is difficult to determine the true causal variants among genes within the linkage block. Furthermore, migration and ethnic diversity can lead to spurious associations (see also Chapter 16), and necessitate corrections for population stratification that may reduce overall statistical power. Consequently, only one or a few associated SNPs have been identified in most genome-wide association studies. Such SNPs generally account for only a minor fraction of the phenotypic variation of a complex phenotype. Thus, when one considers that multiple genes, as well as environmental factors, determine the manifestation of a complex trait, it is difficult to assess how to interpret a person's overall risk based on only one or a few SNPs, and how to advise health care management for such an individual. Despite these challenges, the number of GWAS studies has expanded exponentially, and the popularity and power of such studies is likely to increase as experimental and statistical tools become ever-more sophisticated.

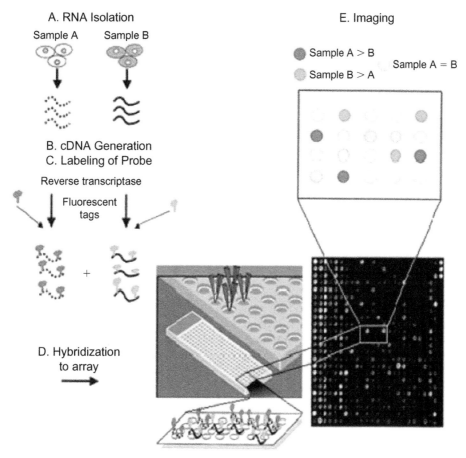

FIGURE 10.1   Hybridization to spotted cDNA expression microarrays.

fragments are "spotted" on glass slides with the use of a robotic arrayer, and the DNA is covalently bound to the surface. The slides are now hybridized to cDNA samples made separately from RNAs that the investigator wishes to compare. Sequences from RNA obtained from one sample are copied to cDNA and labeled with a green fluorescent dye (Cy3), whereas RNA sequences from the other sample are copied to cDNA and labeled with a red fluorescent dye (Cy5). If RNA species between the two samples (e.g. a control and experimental sample) are equally represented, they will hybridize to their cognate DNAs on the glass slide to the same extent, and give rise to a yellow signal when the slide is scanned with a fluorescence scanner. However, up- or down-regulation of a particular transcript in either sample will give rise to a greater intensity of the red or green fluorescent signal (Figure 10.1). When constructing cDNA expression microarrays, it is essential to verify the sequence of each amplification product, to generate fairly uniform sizes of amplicons, to eliminate repetitive sequences from the array, to have even spots on the glass slide without dust or dirt particles, and to control for dye effects by **dye switching**, that is, repeating the experiment with the same samples, but labeled with the reciprocal dyes (Figure 10.2). Because of the many steps involved, possible small variations in the quality of the arrays, and the

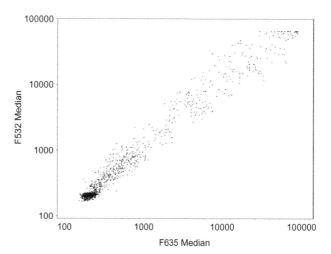

FIGURE 10.2   Scatter plot showing the correlation between fluorescent intensities of Cy3 (*x*-axis) and Cy5 (*y*-axis) for a simple expression microarray with 192 cDNAs. Points that deviate significantly from the diagonal would represent probes with dye effects. The correlation coefficient r = 0.97 indicates minimal dye effects for this particular microarray experiment.

competitive nature of the assay, a large number of independent replicates is generally needed to obtain acceptable statistical significance. The advantage of cDNA arrays is that the method can be applied to any organism, including those whose genomes have not yet been sequenced, and that each cDNA, when amplified from mRNA, is known to be an expressed transcript. In addition, once the PCR fragments are available, large numbers of arrays can be readily printed at reasonable cost. These arrays are easily customized for specific experimental purposes, as the cDNAs represented on the arrays can be readily modified. One drawback of spotted arrays is the relatively large amounts of RNA required, making it difficult to detect rare messages, or analyze expression patterns in small tissues. Furthermore, cDNA arrays often do not provide complete coverage of the genome. Despite these limitations, spotted microarrays have provided a wealth of important information.

One of the first and most elegant studies on behavior using cDNA microarrays was an analysis of transcriptional profiles that correlate with behavioral plasticity in the honey bee, *Apis mellifera*. As indicated in Chapter 12, honey bees are social insects that progress through several behavioral stages during their adult development. Young honey bees begin their careers as nurses, and through a progression of different tasks within the hive ultimately mature into foragers. Prior to completion of the honey bee genome sequence, cDNA microarrays were constructed that represent about 40% of the bee genome. Transcriptional profiles were compared between 5- to 9-day-old nurses, and 28- to 32-day-old foragers. In addition, developmental effects could be controlled by establishing **single cohort** colonies, in which precocious foragers develop that are the same age as nurses. Analyses of transcriptional profiles and cluster analysis revealed ensembles of genes that are uniquely expressed in each behavioral stage, and generated gene expression profiles diagnostic of each of these behavioral stages (Figure 10.3). Whereas these genes included transcription factors, it is of interest to note that most of these genes would not be *a priori* predicted as behavioral genes (e.g. carbonic anhydrase, trehalose-6-phosphate synthase). This finding clearly showed that a diverse spectrum of genes, including so-called **housekeeping genes** have to work together as functional ensembles to enable the expression of behavior.

High-Density Oligonucleotide Microarrays

High density oligonucleotide microarrays ("GeneChips") were pioneered commercially by the Affymetrix Company, and the original Affymetrix Gene Chips consist of synthetic 25-mer oligonucleotides which are designed based on the available genomic sequence (Figure 10.4). Total RNA extracted from the tissue of interest is converted to cDNA, and labeled with a fluorescent tag which is then hybridized to the arrays. Hybridization intensities are measured in a scanner for quantification.

In contrast to spotted cDNA microarrays, high-density oligonucleotide microarrays are not based on probe competition, which facilitates statistical analysis. Less RNA is required to obtain strong hybridization signals (as little as 6 micrograms of total RNA can be used), all

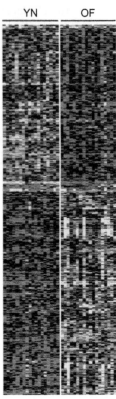

FIGURE 10.3 Clustering analysis of differential gene expression during different behavioral stages of honey bees, young nurses (YN), and old foragers (OF). Columns represent different honey bee brains and each small, colored square represents a gene that is either up-regulated or down-regulated when the two samples are compared. Clustering analysis is helpful in attempting to classify genes with altered regulation in common functional categories or cellular pathways, and can be achieved by a number of different statistical methods (*k* clustering, hierarchical clustering, principal component analysis, self-organizing maps), all of which have the same objective, grouping together genes that behave similarly in terms of their expression under different conditions. (Adapted from Whitfield, C. W. et al. (2003). *Science*, **302**, 296–299.)

known genomic sequences can, in principle, be detected, and the reproducibility among GeneChips is generally very high, because during the manufacturing process each lot of arrays is synthesized as a large slab, from which many identical GeneChips are generated thereby reducing variation in the quality of the microarrays. The only drawback of Affymetrix GeneChips is that they are far more expensive to purchase than printing cDNA microarrays, and their composition can only be altered at great expense through commercial custom synthesis.

Whereas Affymetrix GeneChips initially were designed only to cover coding sequences of known genes, a new generation of arrays has been designed in which oligonucleotide probes cover the entire genome, including non-coding sequences, with a tight spacing that ensures complete genome coverage. Such **tiling arrays** enable detection of alternative transcripts and transcripts of genes that have not been annotated, as well as small **microRNAs**.

A similar platform for whole-genome expression analysis that has gained increasingly wider use has been

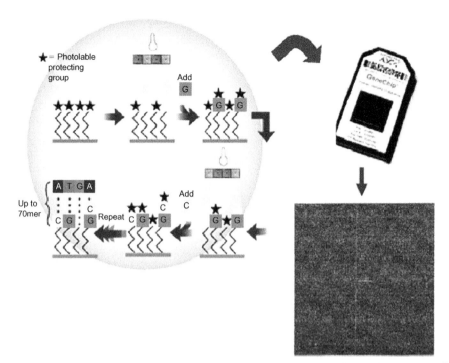

FIGURE 10.4   Affymetrix high-density oligonucleotide expression microarrays. The arrays are manufactured by a process called **photolithography**. Nucleotides are attached to a matrix and prevented from reacting with the next activated nucleotide by a photolabile blocking group, which can be removed by UV illumination. The light beam can be masked so that only desired nucleotides are de-blocked, whereas others retain the blocking group. Unblocked nucleotides can now react with another nucleotide, and by repeating the process and adjusting the mask at each iteration oligonucleotides are built, as illustrated by the diagram to the left. Whereas up to 70-mers can be readily generated by this process, Affymetrix GeneChips use 25-mers. The chip itself, shown in the upper right, is 1.28 × 1.28 cm and covered by an incubation compartment for hybridization. An example of a scanned image from a GeneChip is shown in the lower right.

developed by Illumina, Inc., and uses 50-mer or 70-mer oligonucleotides bound to beads. Random sequences provide a control for estimating noise, and the increased length of the oligonucleotides and built-in redundancy contribute to the high sensitivity of this system, which provides an efficient platform not only for high-throughput expression analysis, but also for **SNP genotyping**.

The first behavioral analyses using Affymetrix GeneChips studied changes in whole transcript profiles that correlate with **circadian activity** in *Drosophila* (further described in Chapter 11). Circadian activity arises as a result of oscillatory regulation of the Period transcription factor. The *period* gene product combines with the *timeless* gene to form an active transcription factor, and cyclic activity, arises when one of the PER/TIM activated genes, *Clock*, inhibits the expression of *per* and *tim*. Since TIM is degraded by light, the circadian cycle can be entrained to the light/day cycle (Figure 10.5; see also Chapter 11).

Using high-density oligonucleotide microarrays, Rosbash and colleagues at Brandeis University identified 134 cycling genes in wild-type flies, which included not only known members of the circadian clock, but also a large number of genes not previously known to cycle. These genes encoded detoxification enzymes, ligand carrier proteins, neuropeptide modulators, proteins involved in cuticle formation, genes involved in immune defense, and a diverse array of miscellaneous enzymes, as well as

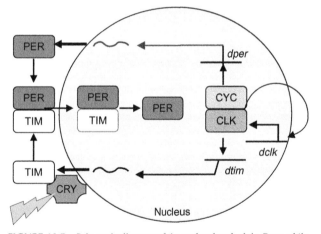

FIGURE 10.5   Schematic diagram of the molecular clock in *Drosophila*. The circadian genes *Cycle* and *Clock* encode transcription factors (CYK and CLK) that regulate the synthesis of the transcription factors Period (PER) and Timeless (TIM). Dimerization of TIM and PER is required for these transcription factors to form a functional complex that can enter the nucleus and inhibit the activity of CYC and CLK, thereby contributing to a negative feedback loop, which gives rise to circadian oscillations in the cellular concentrations of PER and TIM. Absorption of light by **cryptochrome** (CRY) entrains the circadian clock to the day/night cycle by causing the destruction of TIM. (Courtesy of Dr Guy Bloch.)

predicted proteins of unknown function. A larger group of 267 genes with altered transcriptional regulation was identified when *Clk* mutants were analyzed. Such *Clk*-regulated genes included unexpected co-regulated genes with 17 genes

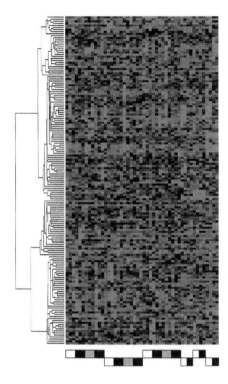

per cluster

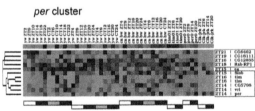

FIGURE 10.6 Hierarchical clustering analysis of circadian genes in *Drosophila*. Columns represent experimental time points and rows represent genes. The color intensity represents the magnitude of differences in gene expression. Data were collected from wild-type flies in three 12-point time courses of two days each (first 36 columns), and from three arrhythmic mutants (*period*, *timeless*, and *Clk*) in four-point time courses spanning one day each (last 12 columns). The light–dark schedule is indicated by the bars below the **clustergrams**, with white and black blocks corresponding to light and dark periods, respectively, gray indicating periods of subjective light under the free-running condition of constant darkness. The tree on the left of the clustergrams indicates the pairwise similarity relationships between the clustered expression patterns. The clustergram below shows a close-up of a segment of the overall clustergram, highlighting genes that group with the *period* cluster. (Modified from Claridge-Chang, A. et al. (2001). *Neuron*, **32**, 657–671.)

encoding antimicrobial peptides and 9 encoding pheromone or odorant-binding proteins, indicating that the *Clk* mutation has widespread direct and indirect pleiotropic effects throughout the transcriptome (Figure 10.6).

Similar results were obtained simultaneously and independently by Michael Young's group at the Rockefeller University, who identified at least 158 cycling genes, including genes implicated in learning and memory, synaptic function, vision, olfaction, locomotion, detoxification, and metabolism.

Did the two studies obtain the same results? Some trans-regulated genes were identified in both studies, including known circadian genes such as *timeless*, but there were substantial discrepancies between the transcriptional profiles. This could in part be due to different methods of analysis, but a more important factor in this case may be genetic background differences between the fly strains used, and different environmental conditions in which they were reared. Such observations are reminiscent of the difficulties in reconciling results from behavioral experiments on mice in different laboratories, mentioned in Chapter 4. It should also be noted that in both studies sexes were pooled, precluding detection of possible sexual dimorphism in circadian gene expression. Nonetheless, each of these studies showed that networks of genes under circadian control overlap with networks of genes that mediate many biological processes, including different behaviors. Perhaps the most important lesson to be learned from these experiments is the importance of controlling the time of day at which samples are collected for analysis when studying behaviors, as transcript levels may vary considerably as a consequence of the biological clock, and may give rise to erroneous interpretations of transcriptional profiles, if experimental samples and their controls are not collected contemporaneously.

PARALLEL SEQUENCING AND TRANSCRIPTIONAL ANALYSIS

During the last several years, a new generation of highly-efficient DNA sequencing techniques has emerged, which rely on simultaneous **parallel sequencing** of many copies of DNA fragments. In each case, fragmented genomic DNA is amplified by PCR and the amplification products are immobilized to a solid matrix, where they subsequently serve as templates for DNA synthesis using methods that allow the identification of each sequentially-added nucleotide (Figure 10.7). These sequencing procedures attain good precision, because of the large number of overlapping fragments that are amplified of the same DNA segments in parallel. Parallel sequencing is a revolutionary technical advance which is enabling rapid sequencing of whole genomes at reasonable cost.

Because RNA can be reverse-transcribed into cDNA, these methods can also be applied to detect RNA sequences. Moreover, the greater the abundance of a particular transcript, the greater is its representation among the DNA fragments that are synthesized using amplicons as templates. Thus, transcript abundance can be readily quantified. This methodology ensures that all transcripts are represented, and can also readily detect and quantify alternative transcripts and splice variants. Such methods offer a digital read-out of transcripts. These "digital transcriptional profiling" methods represent the next generation of procedures for whole-genome transcriptional analyses. Although their application does not depend on the availability of a

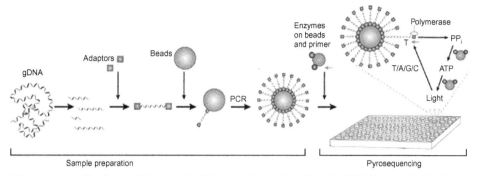

FIGURE 10.7 Parallel sequencing. The diagram illustrates the 454 sequencing method. Here, the DNA is sheared and oligonucleotide adaptors are attached. Each fragment is then attached to a bead, and the beads are PCR-amplified to generate multiple copies of the same DNA sequence on each bead. Sequencing is then performed in parallel on each DNA fragment as shown (the DNA fragment has been artificially-elongated in the figure). Nucleotide incorporation is detected by a photochemical reaction at each step. The cycle is iteratively repeated for each of the four bases. Currently, sequences as long as 250 bp can be read, and this is likely to increase in the future. In addition to 454 sequencing, other technologies such SOLiD and SOLEXA sequencing, are commonly used. These methods are conceptually similar to 454 sequencing technology in that they all employ PCR-amplification and parallel sequencing. Sequences obtained by these procedures are about 25–40 bp, and this is likely to increase in the future. (Figure modified from Medini, D. et al. (2008). *Nat. Rev.Microbiol.*, **6**, 419–430.)

reference sequence (as do traditional high density oligonucleotide microarrays), availability of a reference sequence is important for interpretation of the transcript profiles.

PLASTICITY OF TRANSCRIPTIONAL PROFILES

The dependence of gene expression patterns on circadian rhythm is only one example of the **plasticity** of whole-genome transcriptional profiles. Transcriptional profiles are highly sensitive to changes in the physical and social environment, to genetic background and sex, and to the developmental history of the organism. To add to the complexity, fluctuations in gene expression as a consequence of such environmental sensitivity will differ across the transcriptome. Thus, whereas it is reasonable to expect overlap among transcriptional profiles obtained from experiments on the same system executed in different laboratories, perfect reproducibility should not be expected. This is an important notion, as it is nonintuitive and goes against the grain of traditional scientific practice, which requires exact reproducibility of observations to render them "believable." Here, the goal is to understand the nature and extent of variation in gene expression profiles under different environmental conditions to gain insights into the mechanisms by which the genome drives the expression of behaviors, continuously adapting to environmental change.

We should view each behavior as a continuum of phenotypes, and consider the phenotypic variation itself a component of the phenotype. Again, accepting large variation as part of the phenotype runs against the traditional scientific experimental approach of attempting to minimize variation as much as possible to increase better reproducibility. Variation in nature, however, is real, especially when considering behaviors, and to the quantitative

geneticist this variation in the manifestation of the phenotype provides a treasure trove of information. Phenotypic variation enables gene-mapping studies and association analyses that can identify the genetic sources of this variation. These genes which harbor polymorphisms that cause variation in behaviors represent the substrate for natural selection and evolutionary change.

Pleiotropy

A particular gene can perform a function that affects different phenotypes. For example, the extensively-studied *period* mutant in Drosophila is best known for its arrhythmic activity, but also affects a range of other behavioral phenotypes, including courtship song and eclosion (hatching) time. Similarly, the Drosophila mutant, *amnesiac*, which encodes a neuropeptide hormone initially implicated in memory formation during conditioned learning, also affects chemically-reinforced conditioned courtship behavior, and sensitivity to ethanol exposure. Such multifunctional phenotypic manifestations arising from a single gene are described as **pleiotropy**. Pleiotropy is not an exception, but is a widespread phenomenon. For example, as mentioned in Chapter 9, natural polymorphisms in the *foraging* gene contribute to different food search strategies, in which flies either feed locally ("sitters") or migrate extensively on the food source ("rovers"). The *foraging* gene encodes a cyclic GMP-dependent protein kinase, an enzyme that clearly has widespread pleiotropic effects.

The preponderance of pleiotropy is often not recognized, as most behavioral studies focus only on one phenotype. However, when effects of mutations are assessed side-by-side on multiple behavioral phenotypes, it immediately becomes clear that pleiotropy is the rule. Indeed, expression microarray experiments on Drosophila lines selected for high and low geotactic responses, and transcriptional profiling of olfactory mutants, have identified functional categories of genes that might not *a priori* have been expected to be candidate genes for these behavioral traits. **Covariance** with other traits during selection experiments, pleiotropy, and overlap of genetic networks that drive modules of complex behaviors render it essential to associate each candidate gene directly with the phenotype under study, to implicate the action of its gene product in the manifestation of the phenotype. In *Drosophila* this can be achieved through genetic experiments using available mutant stocks for known genes.

Pleiotropic effects can vary among individuals, and different allelic polymorphisms can affect different phenotypes associated with a pleiotropic gene. An example is the *Drosophila* gene *Catecholamines up* (*Catsup*), which encodes a negative regulator of **tyrosine hydroxylase**, the rate-limiting enzyme in the catecholamine biosynthetic pathway, which is central to the expression of many quantitative traits, including behaviors. Different polymorphisms in the *Catsup* gene are associated with phenotypic variation in locomotion, lifespan, and sensory bristle number.

In the case of the *foraging* gene, subtle differences in the activity of cyclic GMP-dependent protein kinase, which do not give rise to overtly abnormal phenotypes, are sufficient to transform sitters into rovers. Polymorphisms in the *foraging* gene will result not only in variation in transcriptional networks, but may also have an even greater impact on the **proteome**, since subtle changes in the activity of this kinase may generate concomitant variations in the proteome that, through extensive posttranslational modifications, will amplify and extend transcriptional effects. Aligning functional networks of the proteome with networks at the level of the **transcriptome** is, indeed, a principal challenge for the future.

Effects of Sex

The vast majority of studies on the genetics of behavior have ignored sex as a factor that might influence gene expression. In most studies individuals of both sexes are mixed in behavioral assays and in their subsequent analyses, except when sex-specific behaviors are analyzed, such as courtship and mating behavior.

The importance of **sexual dimorphism** in the genetic architecture of behavior was illustrated in *Drosophila* for olfactory avoidance behavior in Chapter 4 (Figure 4.17), when measurements of olfactory avoidance behavior to repellent odorants were made on males and females separately, using **chromosome substitution lines**, in which either *X* or *third* chromosomes from a natural population were **introgressed** into an otherwise **isogenic** background. These studies showed that the genetic correlation of phenotypic values between the sexes was surprisingly low (Figure 4.17). Thus, the genetic architecture that orchestrates this behavior is different in males and females, and selection can operate differently in males and females. Since, under this condition, no genetic blueprint can be optimal for both sexes, the sex environment by genome interaction promotes the preservation of heterozygosity in the population for this important survival behavior.

Expression microarray studies have demonstrated that sexual dimorphism is also prevalent at the level of the transcriptome. Furthermore, mutations or polymorphisms in a sex-biased gene expressed only in males or females can affect the expression of other transcripts that are not *a priori* sex-biased, but form part of an interactive genetic ensemble. Thus, sex environment by genome interactions can exert substantial evolutionary constraints, by influencing ensembles of genes differentially in males and females. It should be readily apparent from these considerations that it is always essential to analyze sexes separately when using expression microarrays to explore the genetic architecture of behavior.

ANALYSIS OF WHOLE-GENOME EXPRESSION MICROARRAY

The vast amount of data that emerges from whole-genome transcriptional profiling studies poses challenges for analysis and interpretation. Although the most appropriate analysis of an expression microarray study will depend on its experimental design, there are four steps that must always be taken: (1) normalization; (2) defining statistical significance threshold; (3) interpretation; and (4) validation.

Normalization

There are two sources of error that contribute to variation among replicate microarrays: biological variation; and experimental variation. **Biological variation** refers to variation in transcript abundance that is inherent among independent replicate RNA extracts, whereas **experimental variation** is the variation that one would observe if the same extract were to be split in two, and both samples were processed side-by-side. Experimental error arises from technical differences in processing the samples during probe labeling and hybridization. Experimental error can result in differences in overall signal intensity of hybridization signals across arrays. Failure to correct for such global intensity differences among replicate samples will give the impression that the biological variation among replicates is larger than it truly is. Thus, it is important to assess whether overall hybridization intensities across arrays are significantly different, and if so, to normalize the data sets using a computer program. Such normalization is most effective if applied only to those probes that give detectable hybridization signals.

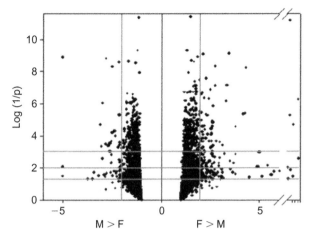

FIGURE 10.8 Example of a Volcano plot, where fold differences in whole genome transcript abundance between male (M) and female (F) *Drosophila melanogaster* on the x-axis are compared to *P*-values along the y-axis. The horizontal lines from top to bottom represent levels of statistical significance at $P < 0.001$, $P < 0.01$ and $P < 0.05$, respectively. The vertical lines indicate twofold differences in expression levels. Many transcripts with statistically-significant differences of expression at $P < 0.001$ would be missed, if one would rely on the twofold difference criterion (all points between the vertical lines and the upper horizontal line), whereas false positives would also be detected if relying only on the twofold cut-off criterion (points outside the vertical lines and below the horizontal lines, depending on the *P*-value). (From Anholt et al. (2003). *Nat. Genet.*, **35**, 180–184.)

DEFINING STATISTICAL SIGNIFICANCE THRESHOLD

When expression microarrays first appeared on the scene, many investigators used arbitrary criteria for deciding what would constitute a significant change in transcript abundance compared to control. Often, two-fold increases or decreases in expression levels were considered significant. However, for tightly-regulated genes, differences as little as 10% might be highly significant, whereas other genes may fluctuate normally more than two-fold in levels of expression (Figure 10.8). Thus, using arbitrary cut-off criteria was found to be fraught with an unacceptable number of false positives and false negatives, and led many investigators to question the validity of expression microarrays, unless changes in transcript levels were independently verified by Northern blots or PCR. In reality, however, verifying hundreds and often thousands of transcripts by quantitative PCR or Northern blots is impractical, and would negate the very use of expression microarrays.

It is now generally accepted that analysis of expression microarrays should be based on rigorous statistical analyses rather than arbitrary estimates. Analyses of variance are generally used to identify statistically-significant differences in transcript abundance among samples. What do we mean by "statistically-significant?" A confidence level of $P < 0.05$ means that there is a 5% likelihood that the observed result could have occurred by chance. Whereas in many molecular

biological experiments, where a single treated sample is compared to an untreated control, a *P*-value of <0.05 might be considered evidence for a significant difference between the samples, the situation is different when assessing expression levels among thousands of genes. Consider sampling blindly one marble from a bag that contains 95 white marbles and 5 red ones. The chance of picking a red marble the first time is only 5%. However, if we grab into the bag repeatedly, the chances of picking a red marble increase dramatically. This simple example illustrates the multiple-testing problem inherent in whole-genome expression and association studies. If we analyze an array that represents 10 000 genes, we are conducting 10 000 tests, and we must adjust our criterion for statistical significance accordingly. At $P < 0.05$, we would expect $0.05 \times 10\,000 = 500$ false positives. An appropriately corrected threshold for statistical significance in this case would be $P < 0.05/10\,000 = P < 5\,E - 06$. This simple calculation is known as the **Bonferroni correction**, named after the twentieth-century Italian mathematician Carlo Emilio Bonferroni.

Whereas the strict Bonferroni correction is mathematically correct, it is often considered too conservative, and there is concern that it may lead to discarding too many true positive observations. An alternative method to correct for multiple testing, known as **permutation analysis**, was developed by Gary Churchill and Rebecca Doerge. Here, the data set is scrambled ("permuted"), and the number of times that original data within the data set associate by chance is recorded, effectively providing a *P*-value for this single permutation trial. The process of permutation can be repeated 1000 times to obtain a distribution of the resulting 1000 randomly-associated *P*-values. The 5% tail cut-off of this distribution will indicate the corrected *P*-value for multiple testing at which the probability of identifying a real positive has 95% certainty. Permutation analysis is an attractive method for establishing significance thresholds, as it provides a multiple-testing correction that is customized for the particular data set under study.

A frequently-used procedure to correct for multiple testing is determination of a **false discovery rate** (FDR). Imagine that we are analyzing 10 000 probe sets, i.e. we are conducting 10 000 tests. A *P* value of 0.05 would indicate that there is a 5% chance that a significant difference in expression level is observed by chance. This means that, out of the 10 000 tests, 500 will be false positives (this is the Bonferroni calculation, mentioned earlier). Now, let us assume that we observe 5000 probe sets which show significant differences in expression at $P < 0.05$. In this case, the false discovery rate (FDR) is calculated as the number of significant differences expected by chance, divided by the number of significant differences observed, at a given *P* value, or in our example FDR = 500/5000 = 0.1. Unlike the *P*-value, which is the number of false positives expected when truly nothing is significant, the false discovery rate controls the proportion of false positives among all

Box 10.3   Reference designs and loop designs

In the design of expression microarray experiments, an experimental sample of interest is generally compared to an appropriate control. Comparing the same number of experimental replicates with reference replicates allows for a balanced pairwise experimental design that can readily be analyzed by standard ANOVA methods. Such pairwise experimental designs are preferable when the number of microarrays is relatively small and the same reference control can be compared to a large number of different experimental treatments, e.g. treatments of a defined genotype with a number of different drugs, compared to an untreated control.

However, standard reference designs become logistically more complex when a large number of different experimental samples are compared, each of which requires a separate reference control, e.g. comparing gene expression profiles of many different inbred strains under different environmental conditions, where each strain would have a control for each environmental condition to which it is exposed. One can argue that, in this case, half of the hybridization resources, reagents and microarrays, are used merely to produce reference controls which themselves are not of inherent biological interest.

As an alternative to the reference design, **loop designs** have been developed. In a loop design two conditions are compared via a chain of other conditions. By comparing expression profiles of each sample to all other samples in the experimental design transcripts that are upregulated or downregulated under one condition compared to all other

conditions can be identified, thereby removing the need for a reference sample. The use of loop designs limits the number of microarrays needed for analysis and is especially useful when biological material is limited (Figure A).

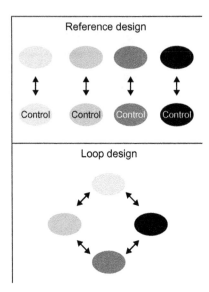

FIGURE A   Diagram of a reference design and a loop design for the analysis of expression microarray data.

---

terms declared significant. Estimating the false positive discovery rate is more in line with the process of hypothesis-generating experimental discovery than the standard reliance on conventional *P*-values. An advantage of using an FDR to assess statistical significance is that it is an easier, more user-friendly method, and less computationally intense than permutation analysis.

Interpretation

The final output of a microarray experiment is a list of up-regulated or down-regulated genes. Making sense of this list and ascribing biological significance to the observed pattern of altered transcriptional regulation can be a major challenge. First, the output of the microarray experiment needs to be considered within the context of the original experimental design. For example, RNA samples obtained from two genetically-unrelated individuals or RNA samples obtained from related individuals, but extracted under different environmental conditions, will show vastly different transcriptional profiles. The genes that are differentially-expressed among such samples will be rather meaningless. Meaningful comparisons and interpretations can be made only if experiments are appropriately controlled. For example, one can compare same-sex treated individuals to untreated controls that have the same

genetic background, were age-matched and reared in the same environment. Similarly, the effects of a mutation on genome-wide transcript abundance can be assessed if a genetically-identical control without the mutation is available. More complex comparisons are also possible, for example transcriptional profiles of artificially selected individuals for high or low phenotypic values of a behavioral trait derived from a common base population can be contrasted. In situations in which precise controls are not readily available, one can compare consistent transcriptional differences among groups. This type of scenario requires large sample sizes, to render consistent transcriptional differences among groups visible by randomizing the biological noise among individuals within each group. Thus, the golden rule that can be applied to experimental design of microarray experiments is to "either standardize or randomize" (Figure 10.9).

Let us now consider a hypothetical example in which we analyze genome-wide effects on transcript abundance that arise from a mutation in a gene that affects a behavioral phenotype, and encodes a predicted transcript of unknown function. Let us imagine that we observe 50 genes that show statistically-significant altered RNA levels in the mutant compared to a control with the same genetic background. If the vast majority of these genes encode proteolytic enzymes, it is reasonable to conclude from the

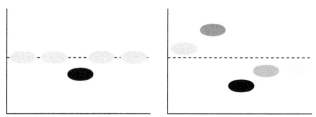

FIGURE 10.9 A diagrammatic representation of the "standardize or randomize" concept. The gray ovals in the left panel represent genetically identical individuals that serve as controls. The Y axis represents phenotypic values and the mean value among individuals is indicated by the horizontal line. The black oval represents a mutant individual with a lower phenotypic value than the mean. The right panel shows reference individuals that are not genetically identical (shades of gray) and the horizontal line represents the randomized average phenotypic value, which is compared to the phenotypic value of the black mutant individual. Note that larger sample sizes are needed when randomizing (right panel) than when standardizing (left panel) to detect statistically significant differences between mutant and wild type due to the much larger standard error.

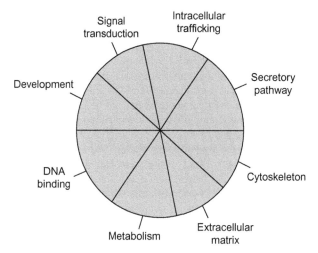

FIGURE 10.10 Example of gene ontology categories typically observed in a microarray experiment. This hypothetical example highlights eight biological function categories and the relative proportions they represent in the data set that correspond to the slices of the pie chart.

co-regulated transcriptional context that the mutated gene likely plays a role in regulating proteolysis. However, we should be aware that this is only a hypothesis, and that conclusive verification of the function of the mutated transcript as a regulator of proteolysis requires further experimentation. This simple example demonstrates that the principal utility of the lists of genes that we obtain from expression microarray experiments is to serve as hypotheses-generating tools. In other words, the list of genes with altered transcript abundance is not the end-point, but rather an intermediate stage in the overall experimental approach.

The example above is an oversimplification. Often numerous genes show altered transcript abundance and identifying a functional context, i.e. what processes are represented by these transcripts, is not straightforward. The first step in analyzing the list of differentially-expressed genes is to place the transcripts into categories based on their molecular function, the cellular process they participate in, or their subcellular localization. Such categories are known as **gene ontologies** (Figure 10.10). Although placing transcripts in gene ontologies might seem intuitively simple, in reality it is complicated because of pervasive pleiotropy. Any given gene could be assigned to multiple gene ontologies; for example, one can imagine that a gene product which binds to the cytoskeleton and vesicle-associated proteins to promote exocytosis could be classified in gene ontologies of ligand-binding, protein transport, cytoskeletal organization, cellular regulation, and secretory pathway. It may be hard to decide which description is the most relevant in the particular experimental context. This problem of pleiotropy becomes amplified when hundreds or sometimes thousands of transcripts are analyzed at the same time. To complicate matters further, transcripts that show the greatest change in abundance in terms of statistical significance compared to the control may not necessarily be the most important contributors to the behavioral phenotype under study. Similarly,

transcripts that fall just short of the statistical threshold for significance may actually be of critical importance. A large number of analytical tools have been developed to improve gene ontology analyses, and to place genes with altered transcript abundance in interconnected pathways or networks to help investigators see "the forest among the trees." One program that has become popular is named DAVID, an acronym for Database for Annotation, Visualization, and Integrated Discovery. Suites of bioinformatics programs for genomic data analysis are freely-accessible via the Internet, and many can be accessed through the National Center for Biotechnology Information website (http://www.ncbi.nlm.nih.gov/).

Validation

Assigning statistical significance to expression microarray data remains a matter of probability: in other words, we can minimize the identification of false positives or false negatives, but we can never completely eliminate them. Similarly, the problem of pleiotropy raises the issue to what extent co-regulated genes are relevant to the phenotype under study. For example, consider the genome-wide transcriptional effect of a mutation in a gene that is essential for olfactory behavior. If this gene is also important for aggression, then some genes that mediate aggressive behavior without affecting olfaction may also undergo altered transcriptional regulation. If we only phenotype the study subjects for chemosensation, we might be tempted to erroneously conclude from the transcriptional profile that these co-regulated genes contribute to olfaction, without realizing that the observed transcriptional changes are the result of overlapping genetic networks that underlie two different phenotypes.

Box 10.4   Genomic databases

The National Center for Biotechnology Information (NCBI) maintains databases and develops user-friendly bioinformatics tools that can be readily accessed with its search engine "Entrez." NCBI maintains GenBank, the universal repository for all sequence information, which can be searched via the BLAST program, and PubMed, a searchable repository for virtually all biomedical research publications. Microarray data are made freely available via the Gene Expression Omnibus (GEO) database. Most major scientific journals require authors to deposit their sequence data and microarray data in GenBank or GEO, respectively, as a precondition for publication. Some other databases curated by NCBI are listed in the table below.

Databases curated by the National Center for Biotechnology Information

OMIM	online Mendelian Inheritance in Man
OMIA	online Mendelian Inheritance in Animals
Nucleotide	sequence database (includes GenBank)
Protein	sequence database
Genome	whole genome
Structure	three-dimensional macromolecular structures
Taxonomy	organisms in GenBank
SNP	single nucleotide polymorphism
Gene	gene-centered information
HomoloGene	eukaryotic homology groups

PubChem compound	unique small molecule chemical structures
PubChem substance	deposited chemical substance records
Genome project	genome project information
Gap	genotype and phenotype
UniGene	gene-oriented clusters of transcript sequences
CDD	conserved protein domain database
3D domains	domains from Entrez Structure
UniSTS	markers and mapping data
PopSet	population study data sets
GEO profiles	expression and molecular abundance profiles
GEO DataSets	experimental sets of GEO data
Cancer chromosomes	cytogenetic databases
PubChem BioAssay	bioactivity screens of chemical substances
GENSAT	gene expression atlas of mouse central nervous system
Probe	sequence-specific reagents

Altered transcription of key genes can be verified independently by **quantitative PCR**, whereas functional tests are needed to assess to what extent co-regulated transcripts are associated with the phenotype under study. The investigator can assess whether mutations in co-regulated genes also affect the phenotype, or, if such mutations are homozygous lethal, genetic tests for epistasis can be conducted. Such manipulations are easily performed in simple model genetic organisms such as *Drosophila melanogaster* or *C. elegans*, but are more difficult in mammalian systems. Ultimately, one would wish to correlate quantitatively protein expression patterns with transcriptional profiles and phenotypic values. Whole-genome protein expression, or **proteomics**, is a field with rapidly-developing technology, and proteomics applications to behavioral phenotypes are likely to become more generally available in the near future.

We must also remember that the transcriptional profile we observe with expression microarrays provides only a static snapshot of a dynamic process. Regulation of gene expression continuously adapts to physiological and environmental changes. Developing techniques to monitor the dynamics of whole-genome transcriptional regulation in real-time may seem a hopelessly daunting endeavour today, but could conceivably become reality in the future.

IDENTIFYING GENETIC NETWORKS

Olfactory Avoidance Behavior in *Drosophila* as a Case Study

Let us now put the above information into practice, by considering genomic studies on a classical behavior, **chemosensation** in *Drosophila*. One example of how quantitative genetic approaches, together with whole-genome transcriptional profiling, can identify ensembles of genes that mediate behavior comes from studies on the

Box 10.5   Proteomics

Transcript levels determined by expression microarrays do not necessarily provide a precise indication of the amount of active gene product produced. A transcript may be produced in abundance, but could be degraded rapidly or translated inefficiently. Furthermore, many transcripts undergo alternative splicing, and give rise to polypeptides that are structurally and functionally different. In addition, posttranslational modifications, such as phosphorylation, that can activate or inactive proteins, cannot be predicted from estimates of mRNA abundance. The extent to which changes in the transcriptome influence cellular function will also depend on the lifetime of the encoded proteins. The quantification of whole-genome protein expression levels has developed into a field known as **proteomics**. To identify all the proteins in a cell or tissue, they are extracted and separated by two-dimensional gel electrophoresis. This is a two-step process in which polypeptides are first fractionated either by two-dimensional gel electrophoresis or by high performance liquid chromatography (HPLC). Two-dimensional gel electrophoresis separates the polypeptides first based on their isoelectric points, and subsequently by size. To identify proteins of interest, individual spots are cut out of the gel, extracted, and cleaved into peptides by trypsin. Alternatively, proteins can be trypsinized immediately and then subjected to HPLC for separation. The tryptic fragments are then subjected to mass spectrometry for identification. One commonly used procedure is known as **MALDI-TOF**, an acronym for matrix assisted laser desorption ionization-time of flight spectrometry, but other highly sensitive mass spectrometric applications are commonly employed. The most commonly used procedure is known as **MALDI-TOF**, an acronym for matrix-assisted laser desorption ionization-time of flight spectrometry. The peptides are deposited onto a matrix and ionized with a laser beam. A voltage difference is then generated that shoots the peptides from the matrix towards a detector. The higher the mass of the peptide, the longer the flight time. At the same time, an electrostatic reflector deflects the beam and helps separate the peptides. The deflection and the flight time enable a precise determination of the molecular masses of the peptides, from which their chemical compositions can be deduced and their identities established. A variety of techniques are also available to analyze protein-protein interactions, including **protein microarrays** and **affinity chromatography** coupled with mass spectrometry, to name but a few. As with the analysis of expression microarrays, sophisticated computational methods have been developed to enable proteomics technology.

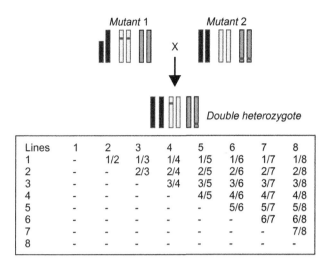

Lines	1	2	3	4	5	6	7	8
1	-	1/2	1/3	1/4	1/5	1/6	1/7	1/8
2		-	2/3	2/4	2/5	2/6	2/7	2/8
3			-	3/4	3/5	3/6	3/7	3/8
4				-	4/5	4/6	4/7	4/8
5					-	5/6	5/7	5/8
6						-	6/7	6/8
7							-	7/8
8								-

FIGURE 10.11   The **half-diallel cross** design, illustrated for experiments on *Drosophila melanogaster*. The three major chromosomes, *X, second,* and *third,* are indicated in shades of gray, with the horizontal bar in the chromosome indicating a *P*-element insertion that generates a mutation affecting the behavioral phenotype under study. Crossing two homozygotes that are co-isogenic gives rise to a double heterozygote. In the case of eight co-isogenic *P*-element insertion lines, all possible double heterozygotes can be constructed as shown in the table. A complete diallel cross is obtained if reciprocal parental crosses for each hybrid were to be included in the experimental design, which would double the size of the experiment.

olfactory avoidance response in *Drosophila*. Transposable elements were introduced into an **isogenic** host strain to generate a population of stable *P*-element insertion lines, which were measured for olfactory avoidance behavior to a standard repellent odorant, benzaldehyde. Lines in which the transposon had disrupted a gene that contributes to the behavior were identified as mutants and designated "*smell-impaired (smi)*." Because all mutants differed only in the insertion site of the *P*-element, but were otherwise in an identical genetic background, it was possible to separate

dominance effects from **epistasis**. To achieve this, non-linear **enhancer** and **suppressor** effects among these genes were analyzed by generating double heterozygotes (Figure 10.11). In this experimental scenario (i.e. all lines are in the same genetic background) one can determine the average dominance effect contributed by a given mutation in combination with all other mutations. When this is done for all possible mutations in all possible combinations, one can estimate the average dominance effect contributed by each parent for a given F1, and predict its phenotypic value. If the observed phenotypic value is statistically-significantly different from the predicted value, epistasis is inferred, with the direction of the deviation of the observed value from the predicted value indicating enhancer (more mutant than expected) or suppressor (less mutant than expected) effects. This analysis revealed extensive enhancer and suppressor effects among 12 independently-isolated "*smell-impaired (smi)*" genes, which allowed the construction of an epistasis diagram that showed a network of interactions among these loci (Figure 10.12). It should be noted that the extent of epistasis observed among this small number of independently-isolated mutations was unexpected, and far greater than would have been *a priori* predicted. A similar strategy has been employed to place genes that mediate **geotaxis** into an epistatic network (Figure 10.13).

Whereas the construction of double heterozygotes in a **half-diallel** design, as described above, can reveal small epistatic ensembles, this approach has two limitations: (1) it can only be used if a collection of mutants is available that have been generated in the same carefully-controlled genetic background; and (2) since each mutant is crossed to all other mutants, the number of crosses needed to evaluate *n* mutations is $n(n - 1)/2$; thus, the experimental design grows exponentially with expansion of the collection of mutants, and quickly becomes unfeasible in terms of the quantity of data that would need to be collected. However, with the advent of **functional genomics**, expression microarray studies can be used to extend functional gene ensembles. In our example, five of these co-isogenic *smi* lines, along with the *P*-element-free host strain, were analyzed side-by-side. RNA was extracted from male and female heads separately in duplicate from these 5 *smi* lines plus the *P*-element free control, and labeled probes were hybridized to high-density oligonucleotide microarrays

from Affymetrix, representing the *Drosophila* genome. These transcriptional profiles showed that the expression of many genes was up-or down-regulated in response to one or more *smi* mutations (Figures 10.14 and 10.15). Thus, single mutations with subtle effects in a defined genetic background precipitate ripple effects through the transcriptome that can profoundly shift the landscape of epistatic interactions. This has far-reaching implications, as it implies that polymorphisms with subtle effects on gene function should also be expected to elicit far-reaching perturbations in genetic networks. This means that any two wild-type strains will be different for many transcripts, and underscores the importance of controlling the genetic background when analyzing complex traits, such as behaviors.

Because of pervasive pleiotropy, it is important to ascertain to what extent transcripts that show altered expression on the microarray do in fact contribute to the behavior. **Quantitative complementation tests**, in which crosses were generated between these *smi* lines and mutant stocks, showed that about 67% of such co-regulated genes show epistasis at the level of phenotype. These observations indicate that behaviors are driven by complex epistatic networks of pleiotropic genes, and that biological specificity with respect to an individual behavioral phenotype is an emergent property of such complex networks.

These studies were important, in that they supported the hypothesis that complex behaviors are modular and are composed of overlapping and interacting genetic ensembles (Figures 10.16 and 10.17, p. 162). For example, courtship and mating behavior in *Drosophila* requires interactive gene ensembles that enable mate recognition, determination of the mating status of the partner (mated or virgin), orientation, locomotion, wing vibration, response to pheromones, tactile behaviors, and copulation, and this composite behavior is also affected by the circadian clock. Thus, perturbations in a single gene will have a ripple effect throughout a dynamic network, which may spill over to other overlapping genetic networks and affect related phenotypes. Moreover, we can predict that changes in transcriptional profiles during development, or as a result of genome-environment interactions, will drive correlated changes in the transcriptome that result in either subtle or dramatic modifications of the phenotype, i.e. altered expression of the behavior.

The realization that genes interact as functional ensembles raises a number of central questions regarding the genetic basis of behaviors. First, what is the extent of epistasis of each gene in a network, how plastic are these interactions, and how much variation in connectivity is there among genes within the same network? Second, how do genetic networks that drive behavior change during development? Third, how are genetic networks modified through interactions with the environment, including social interactions (**genome-by-genome interactions**) and sex-dependent effects (genome by sex environment

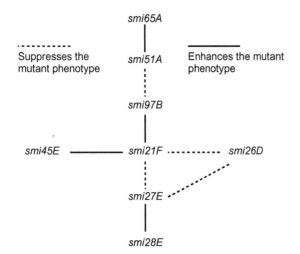

FIGURE 10.12 Diagram of epistatic interactions among eight *Drosophila melanogaster* genes implicated in olfactory avoidance behavior. The transposon-tagged mutant alleles are co-isogenic and designated *smi* (*smell impaired*) followed by the cytological location of the *P*-element insertion. Solid lines indicate enhancer effects and dotted lines indicate suppressor effects. (Modified from Fedorowicz, G. M. et al. (1998). *Genetics*, **148**, 1885–1891.)

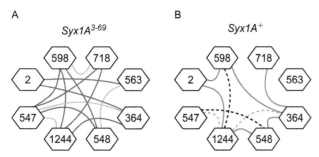

FIGURE 10.13 An epistasis diagram for **negative geotaxis behavior** among eight *Drosophila melanogaster* transposon insertion lines in a genetic background that contains a mutation in the gene that encodes the synaptic protein syntaxin (panel A), compared to a control in which the syntaxin encoding gene is intact (panel B). The diagrams demonstrate not only extensive epistasis, but also the critical dependence of epistasis on genetic context, as the interaction patterns differ substantially between panels A and B. (Modified from van Swinderen, B. and Greenspan, R. J. (2005). *Genetics*, **169**, 2151–2163.)

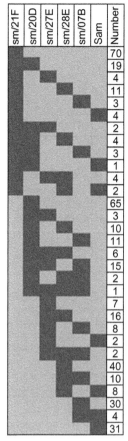

FIGURE 10.14 Overlapping transcriptional profiles among five *smi* lines and their *P*-element-free control (*Sam*). Numbers indicate the number of genes with altered transcriptional regulation in the genetic backgrounds indicated by the black boxes. For example, the expression of 70 transcripts is altered in the *smi21F* genetic background only, 19 are altered in both the *smi21F* and *smi26D* genetic backgrounds, and so forth. This analysis provides a statistically-defined quantitative appraisal of the number of co-regulated genes in each genetic background, as well as the overlap of genes with altered transcript levels in multiple co-isogenic backgrounds with different *P*-element insertions that affect the same phenotype, in this case olfactory avoidance behavior. (From Anholt et al. (2003). *Nat. Genet.*, **35**, 180-184.)

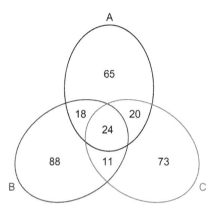

FIGURE 10.15 The numbers of genes with altered expression that are unique or common in different genetic backgrounds can also be represented as a Venn diagram, a hypothetical example of which is illustrated here, where A, B, and C represent three conditions, and the numbers indicate the number of up-regulated or down-regulated genes in the different conditions. For example, 24 genes show altered RNA levels in all three conditions, 18 in condition A and B, and 73 are unique to condition C. Venn diagrams are only useful if the number of overlapping compartments is no more than three.

interactions)? Polymorphisms in the genome provide the molecular basis for evolutionary change, as they may give rise to subtle alterations in epistatic networks. Thus, a fourth emergent question is how behaviors evolve as a consequence of evolutionary change in the context of genetic networks. The description of genetic networks elaborated above implies that evolutionary change is a consequence of the plasticity of genetic networks driven by polymorphisms in constituent genes. This notion might not be entirely surprising, as selection on pleiotropic epistatic networks rather than individual genes implies co-evolution of sometimes disparate traits with overlapping genetic architectures, which is exactly what is observed during artificial selection experiments. The notion of genetic networks also implies that environmental effects on any one component of the network will ripple through a larger sector of the genome. Such environmental effects can result from social (genome-by-genome) interactions, the sex environment, as discussed above, and effects of the physical environment on gene expression (Figure 10.18).

Microarray expression studies performed on *Drosophila* thus far have relied on RNA from whole flies or whole fly heads, with the tacit – and most likely incorrect – assumption that the observed transcriptional profiles are homogeneous throughout the fly's central nervous system. It is, however, more than likely that the observed transcriptional profiles are composites that are summed over different heterogeneous neuronal subpopulations. This concern applies not only to *Drosophila*, but to other organisms as well. A major challenge for the emerging field of quantitative neurogenomics is to develop methods to analyze transcriptional profiles in single neurons or in homogeneous neuronal populations. Placing cell-specific genetic networks in the context of neural connectivity is the next frontier in our understanding of neural and genetic principles that drive behaviors.

SUMMARY

Expression microarrays can be categorized as spotted cDNA microarrays and high-density oligonucleotide microarrays. cDNA arrays can be applied to any organism, large numbers of arrays can be readily printed at reasonable cost, and are easily customized for specific experimental purposes. In contrast to cDNA microarrays, high-density oligonucleotide based expression platforms are not based on fluorescent dye competition, require generally less RNA, enable whole-genome coverage for organisms with sequenced genomes, are highly reproducible, but

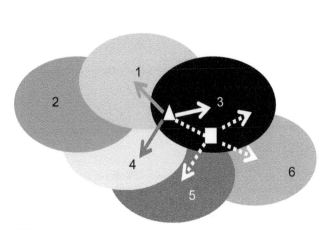

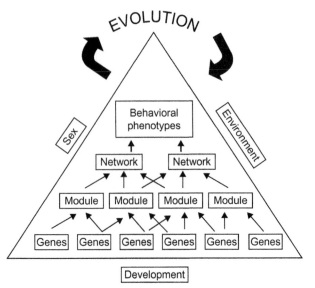

FIGURE 10.16 Schematic diagram of composite genetic modules. Transcripts are considered to be members of a module if under different conditions their expression levels are statistically more closely correlated to one another than to other genes in the genome. Each of the six ovals represents an ensemble of genes that contribute to one of six phenotypes. Because of pleiotropy, the modules overlap so that the gene represented by the triangle affects phenotypes 1, 3 and 4, whereas the gene represented by the square affects phenotypes 3, 5, and 6. The arrows indicate possible epistatic interactions of each gene within their respective modules. Note that both genes affect phenotype 3 associated with the black oval and could epistatically interact with regard to this phenotype (Modified from Anholt, R.R.H. (2004) *Bioessays* **26:** 1299-1306).

FIGURE 10.17 Diagram of the genetic architecture of behavior. Genes that contribute to the phenotype form modular ensembles that can be assembled into networks that give rise to behavioral phenotypes. These networks are plastic, and their structures depend on sex and developmental history. Genetic interactions are also influenced by the physical and social environment. Polymorphisms in genes associated with these networks that contribute to phenotypic variation provide the substrate for natural selection and evolutionary change. (From Anholt, R. R. H. (2004). *Bioessays*, **26**, 1299–1306).

Box 10.6   Quantitative complementation tests for epistasis

Quantitative complementation tests for epistasis assess whether genes, for which mutants are available, exert a non-additive effect on the phenotype under study when crossed as double heterozygotes with a mutation in a candidate target gene. The test is designed to separate dominance effects from epistasis by controlling for genetic background effects, and is most readily applied in *Drosophila*, where it was first introduced by Trudy Mackay at North Carolina State University as a variation on the principle of quantitative tests for complementation to deficiencies.

Let us consider a mutation in a gene on the second chromosome that affects olfactory behavior in *Drosophila* and is designated *smell impaired* (*smi*), and imagine that transcriptional profiling identifies a gene *X* on the third chromosome that shows altered expression in the *smi* genetic background. To find out whether the new gene *X* affects olfactory behavior, we can design a quantitative complementation test for epistasis if a mutant allele of gene *X* is available. Let us assume that such a mutant (*m*) is indeed available as a heterozygote over a balancer chromosome (*Bal*), and that the *smi* mutation is in the inbred Samarkand (*Sam*) genetic background. We can now generate heterozygotes in which the mutant allele is introduced as a double heterozygote with *smi* or as a single heterozygote in the *Sam* host strain (*m/smi* versus *m/Sam*). We can also generate *Bal/smi* and *Bal/Sam* to control for genetic background.

We can now measure phenotypic values (P) for all the crosses, analyze the data by ANOVA, and construct **reaction norms** by connecting the values for the mutant and the balancer in the *smi* and *Sam* backgrounds, respectively. Possible outcomes from such an experiment are shown in Figure A.

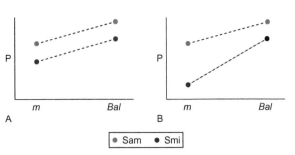

FIGURE A Quantitative complementation tests. If the reaction norms in the *Sam* and *smi* genetic backgrounds are parallel, as in panel A, there is no epistasis between *m* and *smi*; this is interpreted as complementation. However, if the reaction norms fan or cross, there is failure to complement (panel B). In this case, there is epistasis between *m* and *smi*. Panel B shows an example of where the phenotypic effect of *smi* in the presence of the *m* allele is exacerbated (an enhancer effect) over and beyond additive effects, which are represented by the parallel shift in the reaction norms in panel A.

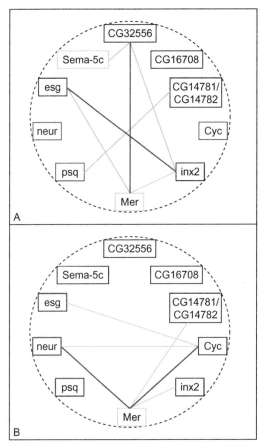

FIGURE 10.18 Environmental plasticity of epistatic networks. The diagrams show epistatic networks derived from a half-diallel cross analysis among ten co-isogenic *Drosophila* lines, with transposon insertions at the indicated genes that give rise to aberrant olfactory avoidance behavior. Enhancer and suppressor effects are here designated by the thick and thin lines, respectively. Epistatic interactions in panel A were detected when the concentration of the repellant odorant benzaldehyde was 0.1% (v/v) and in panel B the odorant concentration was threefold higher. Note that the manifestation of epistatic interactions shifts dramatically when the environmental stimulus condition is changed. Only interactions between the *Mer* (*Merlin*) and *inx2* (*innexin 2*) alleles were stable under both conditions. (Modified from Sambandan, D. et al. (2006). *Genetics*, **174**, 1349–1363.)

Analysis of expression microarrays should be based on statistical rather than arbitrary criteria. Analyses of variance are generally used to identify statistically-significant differences in transcript abundance among samples. The criterion for statistical significance should be based on correction for multiple testing, either by Bonferroni correction, permutation analysis, or estimation of the false discovery rate. The resulting transcripts that are declared statistically-significant can be classified in gene ontology categories. This analysis can be complicated because of widespread pleiotropy. The resulting classification of genes with altered transcript abundance generally serves as a hypothesis-generating tool for further investigations, rather than as an absolute experimental end-point.

Altered transcription of key genes can be verified independently by quantitative PCR or Northern blots, whereas functional tests are needed to assess to what extent co-regulated transcripts are associated with the phenotype under study. Such functional studies can be based on the use of mutations in co-regulated genes, gene silencing (e.g. by RNAi), or quantitative complementation tests for epistasis.

When a set of co-isogenic mutants is available, the phenotypic values of transheterozygotes constructed according to a half-diallel crossing design can be measured, and the contributions of dominance effects and epistasis separated to identify epistatic networks among genes that contribute to the same behavioral phenotype. Quantitative genetic studies, together with whole-genome transcriptional profiling, have demonstrated that behavioral phenotypes arise from dynamic epistatic networks of pleiotropic genes. Behaviors have modular genetic architectures that are composed of overlapping and interacting genetic ensembles. Perturbations in a single gene will have a ripple effect throughout a dynamic network, which may spill over to other overlapping genetic networks and affect related phenotypes. Thus, polymorphisms with subtle effects on gene function should also be expected to elicit far-reaching perturbations in genetic networks. This means that any two wild-type strains will be different for many transcripts, and underscores the importance of controlling the genetic background when analyzing complex traits, such as behaviors. Evolutionary change is a consequence of the plasticity of genetic networks driven by polymorphisms in constituent genes.

It should be noted that transcript levels determined by expression microarrays do not necessarily provide a precise indication of the amount of active protein produced. The quantification of whole-genome protein expression levels has developed into a field known as proteomics. Correlating the transcriptome with the proteome, and achieving this with single cell spatial and temporal resolution are ambitious future goals for behavioral neurogenomics.

are generally far more expensive than cDNA microarrays, and can only be customized commercially.

Transcriptional profiles are highly sensitive to changes in the physical and social environment, to genetic background and sex, and to the developmental history of the organism. These factors, along with the effect of circadian time, should be considered when designing an appropriately controlled microarray experiment. There are two sources of error that contribute to variation among replicate microarrays: biological variation and experimental variation. Normalization can be used to reduce experimental variation.

STUDY QUESTIONS

1. What are important considerations when designing an expression microarray experiment?

2. Imagine that you have available a large number of cDNAs that cover about 60% of the cockroach genome. How would you design a microarray experiment to find out which of these genes represented by these cDNAs change expression levels after female cockroaches have mated? Describe how you would analyze the data.

3. What are the reasons why transcript abundance levels measured with expression microarrays are often poorly-correlated with levels and activities of the corresponding proteins?

4. Explain why it is possible to assess epistasis using transheterozygotes generated via a diallel cross design of mutants only when the mutant backgrounds are co-isogenic.

5. What are the limitations on building epistatic networks using the diallel cross design?

6. Discuss how pleiotropy affects gene ontology assignments.

7. What is the multiple-testing problem and what statistical methods can be used to address it?

8. Explain the method of serial analysis of gene expression (SAGE).

9. What is proteomics?

10. Briefly outline the steps one must take to analyze expression microarray data from high-density oligonucleotide microarrays.

11. How can expression microarrays be used to gain insights into the function of a gene that encodes a transcript of unknown function, given that a mutant is available?

12. What lessons about the genetic architecture of behavior can we learn from the analysis of olfactory behavior in *Drosophila*?

13. Discuss the concept "either standardize or randomize" as it pertains to the design of expression microarray experiments.

14. What future technical developments would be desirable to enhance the power of whole-genome transcriptional profiling to study the genetics of behavior?

RECOMMENDED READING

Anholt, R. R. H., and Mackay, T. F. C. (2004). Genetic analysis of complex behaviors in *Drosophila*. *Nat. Rev. Genet.*, **5**, 838–849.

Churchill, G. A. (2004). Using ANOVA to analyze microarray data. *Biotechniques*, **37**, 173–175.

Claridge-Chang, A., Wijnen, H., Naef, F., Boothroyd, C., Rajewsky, N., and Young, M. W. (2001). Circadian regulation of gene expression systems in the *Drosophila* head. *Neuron*, **32**, 657–671.

Dennis, G., Jr., Sherman, B. T., Hosack, D. A., Yang, J., Gao, W., Lane, H. C., and Lempicki, R. A. (2003). DAVID: Database for annotation, visualization, and integrated discovery. *Genome Biol.*, **4**, P3.

Doerge, R. W., and Churchill, G. A. (1996). Permutation tests for multiple loci affecting a quantitative character. *Genetics*, **142**, 285–294.

Fedorowicz, G. M., Fry, J. D., Anholt, R. R. H., and Mackay, T. F. C. (1998). Epistatic interactions between *smell-impaired* loci in *Drosophila melanogaster*. *Genetics*, **148**, 1885–1891.

Gibson, G., and Muse, S. (2004). *A Primer of Genome Science*, 2nd edition. Sinauer Inc., Sunderland, MA.

Grozinger, C. M., Sharabash, N. M., Whitfield, C. W., and Robinson, G. E. (2003). Pheromone-mediated gene expression in the honey bee brain. *Proc. Natl. Acad. Sci. USA*, **100**, 14519–14525.

Markel, S., and León, D. (2003). *Sequence Analysis in a Nutshell: A Guide to Tools*. O'Reilly Media, Inc., Sebastopol, CA.

McDonald, M. J., and Rosbash, M. (2001). Microarray analysis and organization of circadian gene expression in *Drosophila*. *Cell*, **107**, 567–578.

Twyman, R. M., and Primrose, S. B. (2003). *Principles of Genome Analysis*. Blackwell Publishing, Oxford, UK.

Chapter 11

Neurogenetics of Activity and Sleep

OVERVIEW

Locomotion is an integral component of most behaviors. The majority of behavioral assays use locomotion as a measurement. We can distinguish two types of locomotion: spontaneous exploratory activity ("open field" behavior), which gathers information about the environment; and locomotion in response to sensory input ("locomotor reactivity"). The latter includes startle-responses, escape responses, responses to visual, chemosensory, acoustic or mechanical cues, and movement toward food sources or prospective mates. Animals go through alternating periods of activity and rest, with predictable periods of high and low activity during the day–night cycle. Such cyclic activity is known as a "circadian rhythm," and is controlled by an endogenous biological clock. The molecular components of biological clocks have been well-characterized. Circadian rhythms are driven by transcription factors that regulate each others' abundance in an oscillatory manner during the day and night. The neurons that regulate the biological clock receive input from the visual system, which allows the endogenous rhythm to be synchronized to the light–dark cycle, a process known as "entrainment." In addition to, and in coordination with, the circadian rhythm, animals show prolonged periods of rest. Once thought of as a unique mammalian phenomenon, sleep is now recognized in birds, fish, and insects. Sleep is a restorative physiological state. The precise function of

sleep remains to be elucidated. Prolonged sleep deprivation results in death. The importance of neurological and neuropsychiatric diseases that impact activity and locomotion, as well as the prevalence of sleep disorders, have resulted in a sustained interest in understanding the genetic, molecular, and neural factors that determine activity, sensory-motor integration, and sleep.

LOCOMOTION

Open Field Behavior

We can measure the spontaneous activity of a subject simply by monitoring its movement in an observation arena. This experimental design is known as **open field behavior**, and has been applied extensively to studies on rodents. Open field behaviors are usually monitored by video recording. The total amount of time the subject moves, its speed of movement, and the locations visited in the open field are some of the parameters that can be readily quantified. When animals are introduced into an unfamiliar environment, they usually start their exploratory behavior by moving closely along the walls of the arena (**thigmotaxis**), where they feel protected, before venturing into the center, where they feel more vulnerable (Figure 11.1).

The behavior of an animal in the open field paradigm is influenced by its physiological or mental state. For example, food deprivation motivates exploratory behavior in the open field. The time it takes for a hungry animal to locate

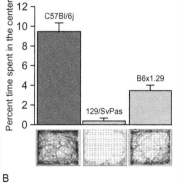

FIGURE 11.1 Open field behavior. Panel A shows a mouse displaying thigmotaxis in an open field arena. Panel B illustrates that time spent in the center (an indication of confidence versus anxiety) varies considerably among the three different strains of mice shown. Corresponding traces that record the movement of the animals as monitored by a video recorder are shown below the bar graphs.

R. Anholt and T. Mackay: Principles of Behavioral Genetics
ISBN: 978-0-12-372575-2

a buried food reward in an open field design is sometimes used to assess its ability to smell. Open field behavior can also be used to assess anxiety. In this case, the open field is divided into a dark area separated from an illuminated area. Reluctance of an animal to venture out of a dark box into an illuminated open field is a measure of anxiety. Thus, we can assess the extent of anxiety by recording the time spent in each segment of the arena, and the number of forays from the dark into the light. A variation on this experimental paradigm is the elevated plus-maze, in which animals have a choice to remain in the covered arms of the maze or venture out in the perpendicularly-oriented open arms; here, anxious animals are reluctant to venture out along the suspended open arms of the maze (Figure 11.2). Thus, in addition to their applications in monitoring spontaneous exploratory behavior, the open field arena, light–dark box transitions, and elevated plus-maze provide behavioral assays for animal models of anxiety and depression (see also Chapter 4).

Despite a vast number of studies on open field behavior, relatively little is known about the genetic basis for this complex trait. As will be described in greater detail in Chapter 12, Caspi and colleagues reported that different alleles of the serotonin transporter gene were associated with anxiety in people, but that this association was conditional on previous early adverse experiences. Indeed, mice in which the serotonin transporter had been deleted by homologous recombination showed altered behaviors in the open field and elevated plus-maze, consistent with increased anxiety. Such single candidate gene approaches, however, are limited as they tend to recapitulate previously-developed concepts ("looking for the lost key under the street light"), and often cannot be placed readily in a broader biological context.

Whole-genome studies on open field behavior have primarily utilized gene-mapping approaches (see Chapter 8). Different strains of mice or rats behave differently in open field tests, and this variation has a significant genetic component. In addition to locomotion in the open field, mice can be selectively bred for increased or decreased wheel-running behavior. Mice that run hundreds of kilometers

per night in a treadmill can be created by artificial selection, along with mice that barely engage in wheel-running. Wheel-running can be used to measure circadian activity in mice, as they are primarily active at night and show less activity during the day. The substantial heritability for spontaneous locomotion, either in the open field or on a running-wheel, is a favorable scenario for the identification of QTLs that harbor genes that contribute to phenotypic variation for these traits. Such QTL studies are performed by crossing strains that differ in phenotypic values for these behaviors, or strains artificially-selected for high or low performance in the open field or on the running wheel. Such studies have led to the identification of many QTLs, but few have actually identified candidate genes within QTL regions. One exception is a study by Jonathan Flint and colleagues at Oxford University. Flint showed first that when different assays for anxiety are compared (for example, open field, elevated plus-maze, light–dark box), different QTLs are identified. This important observation illustrates the multi-faceted nature of "anxiety," and the need for multiple behavioral assays to model this complex trait. Flint and his colleagues persevered in the dissection of one QTL, and found that it fractionated in three distinct QTL regions. Flint was the first to apply Trudy Mackay's **quantitative complementation test** (see Chapters 8 and 10) developed for *Drosophila* to a mouse model and identified *Rgs2*, which encodes a regulator of G-protein signaling, as a quantitative trait gene for anxiety.

Open field locomotion has also been studied in flies. In *Drosophila*, locomotion is directed by the central complex of the brain, which includes the **ellipsoid body** and the **protocerebral bridge** (Chapter 3). The **mushroom bodies** are associated with **sensory–motor integration**, and are important in directing spontaneous activity as moving flies constantly experience sensory input. Ablation of the mushroom bodies, or silencing of neurons in the γ lobes of the mushroom bodies by targeted expression of *shibire* (see Chapter 9), reduces the tendency of flies to avoid the center of the arena. Studies on flies illustrate that it is not easy to completely separate pure endogenous locomotion from locomotion in response to a stimulus, as animals integrate sensory input even in the open field in the absence of a deliberate external stimulus. A prime example of an extensively-studied locomotor reflex is the **startle-response**, which we will discuss below.

Locomotor Reactivity

Startle-responses (also known as **locomotor reactivity**) are critical for survival, as they prompt individuals to move away rapidly from a sudden, potentially-harmful environmental disturbance. Such responses include escape responses to avoid predators. Startle-responses are behavioral reflexes that depend on rapid sensory motor integration. The neurobiological mechanisms that underlie

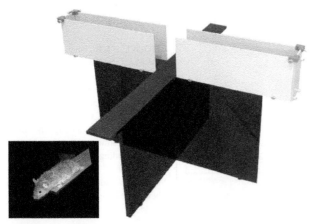

FIGURE 11.2 The elevated plus-maze.

startle-responses have been well studied, but little is known about their genetic underpinnings.

Startle-responses can be readily studied in *Drosophila* by measuring how many seconds a fly moves in response to a sudden mechanical disturbance. Two *Drosophila* strains that differ greatly in their startle-response were used to construct recombinant inbred lines to identify QTL regions that harbor genes that contribute to variation in startle behavior between the parental lines. These regions were narrowed further by **deficiency complementation mapping**, to reduce the number of candidate genes.

Among candidate genes contained within these refined QTL regions were genes associated with the synthesis of bioamines, known to be important for locomotion. One of these candidate genes encodes DOPA decarboxylase (*Ddc*), which converts DOPA into dopamine (see also Chapter 3). Indeed, when alleles of *Ddc* were analyzed in a natural population of *Drosophila*, single nucleotide polymorphisms in this gene were found to be associated with variations in startle-induced locomotor reactivity (Figure 11.3). A further study identified a network of genes associated with early development of the nervous system that

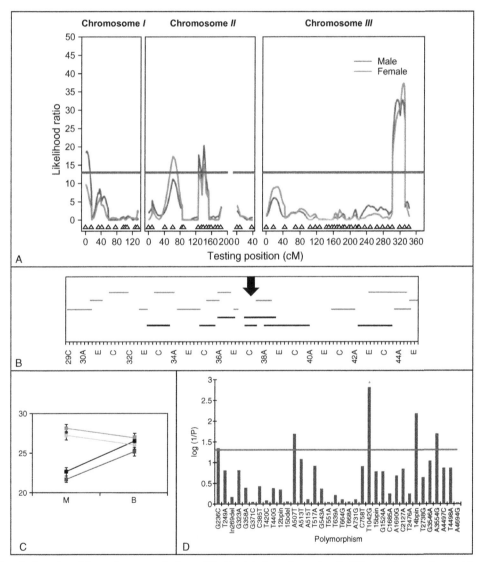

FIGURE 11.3   Genetic dissection of the startle-response in *Drosophila*. Panel A shows the results of a QTL-mapping experiment with recombinant inbred lines that differ in startle-response behavior. Genetic distances across the three major chromosomes of *Drosophila* are indicated on the x-axis in centimorgans (cM), and the positions of polymorphic *roo* transposable element markers are indicated as triangles. The horizontal lines show the threshold for statistical significance. Peaks above this threshold indicate QTL regions. Deficiency complementation testing can narrow QTL regions down, as shown in the example of panel B, where lines under the arrow indicate deficiencies that fail to complement. Quantitative complementation tests can then be conducted with candidate genes within a refined QTL region where an existing mutant allele (M) or a balancer chromosome (B) as a control is placed over each of the parental lines used to generate the recombinant inbred population (Panel C). The gene tested in this case is *Ddc* (its approximate location is indicated by the arrow in Panel B), and shows failure to complement since the differences between the parental lines over the mutants is far greater than those over the balancers for both males and females. Panel D shows polymorphisms in the *Ddc* gene that exceed the multiple-testing threshold for association (horizontal line). (Adapted from Jordan, K. et al. (2006). *Genetics*, **174**, 271–284. See also Chapter 8.)

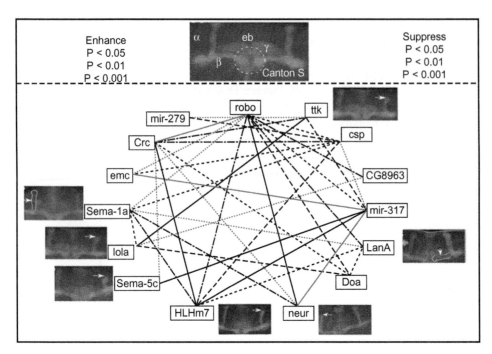

FIGURE 11.4 Diagram of **epistatic** interactions (suppressor and enhancer effects) between 15 independently-isolated *P*-element insertion lines of *Drosophila melanogaster* that affect startle-induced locomotion. The α, β, and γ lobes of the mushroom bodies, and the ellipsoid body (eb, dotted circle) are visualized by an antibody against fasciclin 2 in the control *Canton S* strain (upper panel, above the dashed line). Asymmetrically-shortened, or absence of, mushroom body lobes is seen in some of the brains of the mutants, and is indicated by an arrow or outline of the missing part (*ttk, LanA, neur, HLHm7, Sema-5c, lola, Sema-1a*). (Modified from Yamamoto, A. et al. (2008). *Proc. Natl. Acad. Sci. USA*, **105**, 12393–12398.)

mediates this behavior. These genes were identified by *P*-**element insertional mutagenesis** (Chapter 9) as mutants that displayed a reduced locomotor response to the mechanical stimulus. Examination of the brains of several of these mutants showed abnormalities in the mushroom bodies and the central complex (Figure 11.4). The startle-response in *Drosophila* and, specifically, a *P*-element insertion at the *neuralized* locus has been used in Chapters 4–6 to illustrate estimation of quantitative genetic parameters for behaviors.

Startle-responses can be suppressed by preceding the sudden startle stimulus with a similar low intensity stimulus (for example, a low intensity sound that precedes a loud sound will suppress the acoustic startle-response normally evoked when the loud sound disrupts a period of silence). This suppression is known as **prepulse inhibition**, and is observed readily in the mouse model (Figure 11.5). Deficits in prepulse inhibition (i.e. a strong startle-response despite the preceding low intensity stimulus) are thought to reflect deficits in sensory-motor processing, and are considered a characteristic indicator of **schizophrenia**, both in rodent models and in people. For this reason the prepulse inhibition paradigm is used extensively in studies on rodent models of schizophrenia. One example of how startle-responses in a model system can provide insights in a human behavioral disorder comes from studies by Takeo Yoshikawa and colleagues from the RIKEN Brain Science Institute in Saitama, Japan. These investigators conducted a QTL mapping study to identify genes that mediate prepulse inhibition of the **acoustic startle-response** in mice. They identified a promising candidate gene on chromosome 10, Fabp7 (fatty acid binding protein 7). *Fabp7* is

associated with the NMDA receptor and expressed in **glia**. It encodes a protein that binds the **essential fatty acid**, docosahexaenoic acid. To confirm the involvement of this gene in prepulse inhibition, the investigators generated knock-out mice that lack the *Fabp7* gene. These mice were indeed deficient in prepulse inhibition (Figure 11.6). Concordantly, human studies showed that levels of transcript abundance of *Fabp7* are up-regulated in the brains of schizophrenics. Furthermore, association studies indicate that alleles of *Fabp7* may be one risk factor for schizophrenia, at least in males. Effects of other gene knock-outs and drugs on prepulse inhibition in mice have identified additional gene products associated with the regulation of startle behavior. These include, perhaps not surprisingly, signaling components associated with dopamine signaling pathways. In addition to schizophrenia, prepulse inhibition is also deficient in patients who suffer from locomotion disorders, such as **Parkinson's disease** and **Huntington's disease**.

GENETICS OF HUMAN LOCOMOTION DISORDERS

There are many locomotion disorders associated with human neurological, neuropsychiatric, or neurodegenerative diseases. Some of the best studied neurodegenerative diseases, Huntington's disease and Parkinson's disease, are diagnosed foremost on the basis of characteristic motor impairments. We will limit our discussion of human locomotion disorders to these two diseases, because they provide contrasting examples of a disease that appears to be

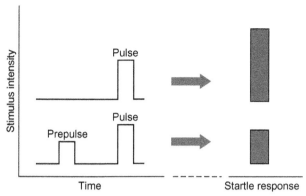

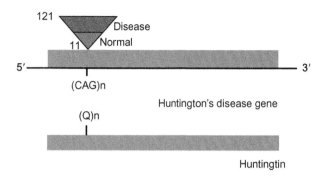

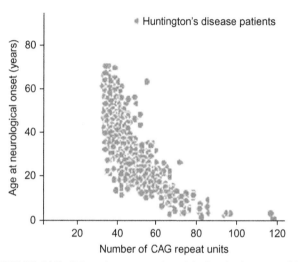

FIGURE 11.5   Diagram of prepulse inhibition, which shows that a preceding lower intensity stimulus can attenuate the startle-response to a subsequent stimulus of greater intensity.

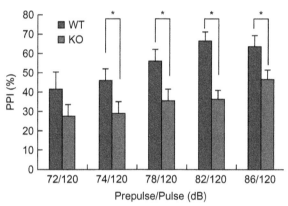

FIGURE 11.6   Prepulse inhibition (PPI, %) of an acoustic startle-response between wild-type (WT, n = 9, six males and three females) and Fabp7 knock-out mice (KO, n = 12, six males and six females). The values represent mean ± standard error (*, $p < 0.05$). (Modified from Watanabe, A. et al. (2007). *PLoS Biology*, **5**, e297.)

FIGURE 11.7   Schematic representation of the huntingtin gene and its product, and the correlation of the number of CAG repeats with the onset of Huntington's disease (Q is single letter amino acid code for glutamine). (Courtesy of the Gladstone Institute.)

entirely genetically determined (Huntington's disease), and one in which environmental effects may influence genetic predisposition (Parkinson's disease). Both diseases show similarities. Their onset occurs at an advanced age, and appears to be due to the accumulation of misfolded proteins, which form aggregates that cause a cellular stress response that culminates in neuronal cell death.

Huntington's Disease

Huntington's disease is a rare late-onset progressive disease with increasing cognitive deficits and motor impairments that are characterized by random uncontrolled jerky movements (**chorea**). The disease is named after the physician George Huntington, who first described it in 1872. It is an autosomal dominant disease. This means that one mutant allele is sufficient to cause the disease. Huntington's disease results in neuronal cell death in distinct areas of the brain, primarily the **caudate nucleus** and **putamen**, both of which

are involved in locomotor control, and to a lesser extent the frontal and temporal cortices. The culprit of this disease is a protein, named **huntingtin**. Although a role in transcription for huntingtin has been proposed, the precise function of this protein remains unclear. It contains a series of glutamine residues, encoded by a repeated CAG codon in its corresponding gene. Normally, there are 37 or fewer glutamines in this polyglutamine tract. However, CAG repeats are greatly expanded in alleles that cause Huntington's disease. This leads to the formation of long glutamine tracts in the huntingtin protein. In patients with Huntington's disease these tracts can contain more than 150 glutamines, and the lengths of these polyglutamine tracts correlate with age of onset of symptoms (Figure 11.7). The disease is entirely determined genetically. Individuals who carry a single risk allele will develop Huntington's disease at some point in their life.

Protein folding in the endoplasmic reticulum is facilitated by helper proteins, known as **chaperones**. In the event a protein is not folded properly, it is modified by the attachment of an ubiquitin moiety which labels the protein for proteolytic destruction by a specialized complex of proteases, known as the **proteasome**. Huntingtin has long polyglutamine tracts which cannot fold properly, and which also cannot be efficiently degraded. As a result, mutant protein fragments accumulate. These are translocated to the nucleus, where they form aggregates. In the neurons in which the pathogenic huntingtin protein is expressed, the long polyglutamine tracts interfere with normal transcription of DNA. In addition, axonal transport appears to be impaired in neurons that express the disease form of huntingtin. However, the pathogenic mechanisms by which aggregates of polyglutamine-containing fragments of huntingtin lead to neurodegeneration are not completely understood.

Because Huntington's disease results from a single mutant gene product, the disease can be modeled readily by expressing mutant huntingtin transgenes in mice or flies. Mice expressing huntingtin with long polyglutamine tracts in their brains show neurodegenerative symptoms similar to those observed in people. In flies, transgenic expression of huntingtin in the eyes causes degeneration of photoreceptor cells, and the severity of this phenotype correlates with the number of polyglutamine repeats (Figure 11.8). These models are being used both to gain insights in the pathogenic mechanisms of mutant huntingtin proteins, and to find ways in which the manifestation of the disease can be ameliorated, either by interfering with the expression of the mutant huntingtin gene (for example, by **RNAi interference**), by developing drugs that interfere with the formation of protein aggregates, or by modifying the misfolded huntingtin protein to render it more amenable to protein degradation.

FIGURE 11.9   Illustration of Parkinson's disease from the 1886 book *A Manual of Diseases of the Nervous System* by Sir William Richard Gowers.

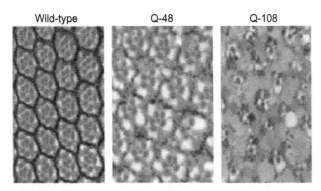

FIGURE 11.8   Overexpression of huntingtin transgenes in the *Drosophila* eye causes neurodegeneration. The wild-type eye shows the typical geometric array of ommatidia, with seven visible photoreceptor cells. Degeneration of this structure is evident when huntingtin with 48 glutamine repeats (Q-48) is expressed in the eye, and is worse when huntingtin with 108 glutamines is expressed (Q-108). (Modified from Kazantsev, A. et al. (2002). *Nat. Genet.*, **30**, 367–376.)

Parkinson's Disease

Parkinson's disease is the second-most common neurodegenerative disorder after Alzheimer's disease (see also Chapter 3). It was first described by James Parkinson in 1817 in his essay on the "shaking palsy." He described the symptoms as an "involuntary tremulus motion, with lessened muscular power, in parts not in action and even when supported, with a propensity to bend the trunk forward, and to pass from walking to a running pace, the essence of intellect being unaffected." The disease is characterized by resting tremor, slowness of movement, rigidity of the extremities and neck, and stooped posture (Figure 11.9). In some patients these motor disorders are accompanied by dementia. The symptoms of Parkinson's disease result from loss of dopaminergic cells in the substantia nigra. These dopaminergic cells project to the striatum, and form part of a motor circuit that regulates activity in the ventral nucleus of the thalamus, which in turn sends excitatory input to the motor cortex (Chapter 3). A reduction in dopaminergic activity leads to inhibition of this circuit, and hence to the manifestation of the locomotor disorders observed in Parkinson's disease.

In contrast to Huntington's disease, both genetic and environmental factors determine susceptibility to Parkinson's disease. There are several indications that genetic susceptibility to environmental toxins may determine disease risk. First, the incidence of Parkinson's disease varies geographically, and the number of cases is higher in rural communities with high exposure to agricultural chemicals, such as pesticides, either in the air or in drinking water from ground wells. Second, Parkinson's disease-like symptoms have not been reported prior to the industrial revolution (this may, of course, in part be due to less well-established diagnosis). Third, Parkinson's disease can be induced by a potent neurotoxin, MPTP (1-methyl 4-phenyl 1,2,3,6-tetrahydropyridine), which kills dopaminergic neurons in the substantia nigra. This compound has been used to induce Parkinson's disease-like symptoms in experimental animals. The neurotoxic effect of MPTP was discovered by chance; it is a by-product of the illicit manufacture of MPPP (1-methyl-4-phenyl-4-propionoxypiperidine), a synthetic opioid that resembles heroin. Individuals who made and used this synthetic by-product developed Parkinson's disease. One popular hypothesis is that the neurotoxic effects of these environmental toxins may arise from the formation of reactive oxygen species that cause oxidative damage to nigrostriatal dopaminergic neurons.

In addition to environmental factors, several genes have been identified with mutations that can predispose one to Parkinson's disease. The loss of dopaminergic neurons in the substantia nigra is accompanied by the accumulation of the protein α-synuclein, encoded by the PARK1 gene on chromosome 4q21. When α-synuclein accumulates in dopaminergic neurons, it forms filamentous aggregates, known as Lewy bodies. Mutations in α-synuclein result in congenital early-onset Parkinson's disease. In addition to the PARK1 gene, at least 14 other genes have been implicated in Parkinson's disease. The best characterized of these genes that accounts for the majority of inherited Parkinson's disease cases is PARK2, which encodes a protein called parkin. Note that, in contrast to the Mendelian inheritance of Huntington's disease, Parkinson's disease fits the description of a quantitative trait, as it results from the interactions of many genes which are sensitive to the environment.

As in the case of Huntington's disease, transgenic mice and flies have been instrumental in investigating the etiology of Parkinson's disease. Mice or flies that express human α-synuclein in dopaminergic neurons show symptoms that recapitulate the essential features of Parkinson's disease, including the formation of Lewy bodies, progressive loss of dopaminergic neurons, and locomotor dysfunction (Figure 11.10). The effect of phosphorylation of α-synuclein on neurodegeneration has been evaluated by mutating critical phosphorylation sites of the transgenic protein expressed in flies. These experiments showed that phosphorylation of α-synuclein aggravated neurotoxicity. At the same time, non-phosphorylated α-synuclein is sequestered more effectively in Lewy bodies. Thus, Lewy bodies may, in fact, serve as a mechanism that protects neurons from the toxicity of accumulated α-synuclein.

The normal function of α-synuclein is not known. Parkin, however, is an ubiquitin protein ligase. As mentioned earlier, ubiquitination tags proteins for degradation, and directs them to the proteasome. Parkin may help facilitate the removal of misfolded proteins. One substrate of parkin is Pael-R (Parkin-associated endothelin receptor-like receptor), which has been implicated as a regulator of dopamine metabolism. Unfolded Pael-R accumulates in the endoplasmic reticulum, and may lead to neuronal cell death. Parkin can protect against the effects of unfolded Pael-R, presumably by facilitating its proteolytic removal.

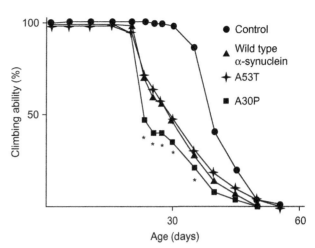

FIGURE 11.10 Locomotor impairment in transgenic flies that express α-synuclein. When flies are tapped to the bottom of a culture vial, they will respond by climbing up the wall of the vial. Flies, in which human α-synuclein is expressed in the brain using the *GAL4-UAS* binary expression system (see Chapter 9) under a **pan-neuronal** *elav* driver (i.e. a promoter of the *elav* gene that drives expression in all neurons), show a markedly more rapid decline in climbing ability with age than their wild-type controls. This behavioral defect is accompanied by loss of dopaminergic neurons. Overexpression of the A53T and A30P mutants of α-synuclein, which have been linked to familial Parkinson's disease, has the same effect. (From Feany, M. B., and Bender W. W. (2000). *Nature*, **404**, 394–398.)

All three gene products, α-synuclein, parkin, and Pael-R, are found in Lewy bodies.

The picture that emerges suggests that accumulation of misfolded proteins is a common hallmark of late-onset neurodegenerative diseases (Figure 11.11, overleaf). These diseases vary in their incidence and in the interplay between genetic and environmental risk factors, and can show a wide array of behavioral disorders, including compromised cognitive functions. When essential neurons that regulate motor circuits are affected, such neurodegenerative symptoms will include distinct locomotor defects.

CIRCADIAN RHYTHMS

Spontaneous activity is not constant throughout the day and night, but cycles according to the Earth's rotation. Every 24 hours the earth completes a rotation around its axis. This rotation generates alternating periods of light and darkness to which organisms have adapted with changes in behavior and physiology entrained to the light–dark cycle. Virtually all eukaryotic organisms, including plants, have developed endogenous clocks that regulate their activities in a diurnal manner which persists even in the absence of external cues and is about, but not exactly, matched to the 24-hour day.

The earliest observations of diurnal rhythmicity are attributed to the Greek geographer Androsthenes, who 2300 years ago accompanied Alexander the Great and observed daily

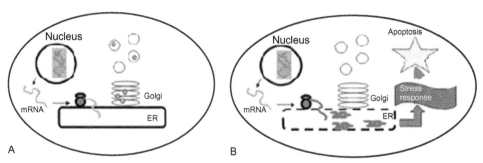

FIGURE 11.11 Diagrammatic representation of the endoplasmic reticulum (ER) stress response that leads to neurodegeneration. Panel A shows the normal synthesis of a nascent polypeptide chain on the ribosome in the endoplasmic reticulum (ER) and transport of the folded protein through the Golgi apparatus to its membrane or extracellular destination. Panel B depicts the accumulation of misfolded, aggregated proteins in the cisternae of the endoplasmic reticulum, where they elicit a cellular stress response that leads to apoptosis (cell death).

leaf movements of the tamarind tree on the island of Tylos (now Bahrein). The term **circadian**, which incorporates the Latin words "circa" (about) and "dia" (day), was coined by Franz Halberg in 1959, and heralded the beginning of the field of **chronobiology**, the study of biological clocks.

Synchronization of the endogenous clock with the external light–dark cycle is known as **entrainment**. When such synchronization is absent, for example in constant light or darkness, the persisting rhythm is said to be "free-running." The stimulus that initiates the entrainment of the circadian clock is known by the term **Zeitgeber** (which in German means "time-setter") and the progression of circadian time is often indicated as **Zeitgeber time**, based on the onset of entrainment, rather than as absolute time of day. Exposure to a flash of light prior to dawn will advance the phase of the clock, while a pulse of light delivered after dusk will delay it. In addition to their persistence in constant light or darkness, **circadian rhythms** are also resilient to changes in ambient temperature, a property known as **temperature compensation**.

Why did biological clocks evolve? One speculation implies protection from ultraviolet radiation as the driving force for the early evolution of biological clocks in Cyanobacteria and fungi, such as *Neurospora crassa*, which sporulates in a circadian rhythmicity. By timing cellular replication at periods when ultraviolet radiation is least intense, DNA damage would be minimized. For higher eukaryotes, including people, it is clear that circadian rhythms coordinate physiological parameters, such as the sleep–wake cycle, body temperature, feeding, and thirst. Certain medications are best taken at bed time, as they are most effective during sleep. For many animals the length of day serves as the most reliable environmental predictor for seasonal migration, hibernation, and reproduction.

What are the cellular mechanisms that comprise the biological clock? As mentioned in Chapter 1, in 1971 Seymour Benzer's group isolated the first circadian mutants in *Drosophila melanogaster*, which showed abnormal circadian rhythms. The gene responsible for these abnormal rhythms was named *period* (*per*), and encodes

a transcription factor. Subsequent studies identified additional circadian mutants. This resulted in the discovery of additional components of the clock, both in *Drosophila* and in mice. The clock mechanisms in flies and mammals turned out to be remarkably similar (see also Chapter 10). Biological clocks consist of interlocking negative feedback loops mediated by transcriptional regulators that undergo periodic synthesis and degradation. The clock mechanism critically depends on the intracellular transport and turn-over of the products encoded by the *period* and *timeless* genes, PER and TIM. The genes encoding these transcription factors are themselves regulated by the *Clock* and *Cycle* genes. The PER and TIM transcription factors are synthesized from mRNAs in the cytoplasm, where they form dimers. Their cytoplasmic levels start to rise at the beginning of the night. As their protein levels peak in the middle of the night, PER and TIM are translocated into the nucleus, where they shut down the transcription of their own mRNAs by inhibiting the expression of *Clock* and *Cycle* (Figure 11.12). At dawn, light activates the photopigment cryptochrome (CRY), which binds to TIM and tags it for phosphorylation followed by proteolytic degradation. Without TIM, PER also becomes susceptible to phosphorylation and degradation. Following the demise of TIM and PER, the inhibition of the expression of *Clock* and *Cycle* is lifted, and the cycle can recommence. Originally discovered by Amita Sehgal and her colleagues at the University of Pennsylvania, the *timeless* gene and its light-induced degradation embody the molecular switch that entrains the endogenous clock to the external light–dark cycle.

In the *Drosophila* brain, the PER and TIM proteins are expressed in about 75 neurons on each side of the brain, as well as in glia and photoreceptors. The two main groups of neurons are organized in lateral and dorsal clusters. The lateral neurons are absolutely essential for the expression of circadian activity. Thus, these neurons could be considered the fly equivalent of the mammalian suprachiasmatic nucleus (SCN) (Figure 11.13).

The mammalian molecular clock shows remarkable evolutionary similarities to the *Drosophila* molecular clock.

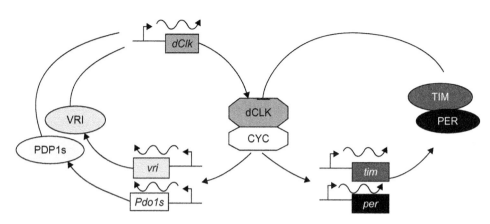

FIGURE 11.12 Two interlocking feedback loops comprise the *Drosophila* clock. The *per* and *tim* genes are negative regulators of their own transcription via the CLK/CYC transcription factors. The *Clk* gene is regulated by the *per/tim* loop via the positive and negative actions of the transcription factors PDP1ε and VRI, respectively. (From Rosato, E. et al. (2006). *Eur. J. Hum. Genet.*, **14**, 729–738.)

However, more components have been identified in the mammalian molecular clock, and their interactions are more complex. In the mammalian clock the transcriptional activators CLOCK and BMAL1 (the functional equivalent of Cycle) dimerize to activate expression of the PER and CRY genes (PER1, PER2, and PER3; CRY1 and CRY2), which form heterodimers to suppress expression of CLOCK and BMAL1, thus completing the feedback loop (Figure 11.14, overleaf).

The importance of circadian clocks to human health is dramatically evident in the Arctic Circle, where there is a high incidence of **seasonal affective disorder** (SAD), also known as **winter depression**. Individuals who are susceptible to SAD experience normal mental health during the summer, but can experience deep depression during the continuous darkness of the Arctic winter. Phototherapy is used to treat SAD. Dysfunctional clocks can also lead to irregular sleep–wake patterns, free-running sleep–wake cycles, in which individuals wake up a little later each consecutive day, or changes in sleep phase. **Delayed sleep phase syndrome** (DSPS) is a condition in which individuals experience peak alertness at night, and find it difficult to wake up in the morning. **Advanced sleep phase syndrome** (ASPS) is the reverse condition, where individuals find it difficult to stay awake in the early evening, and wake up well before dawn. Examples of mutations that result in congenital ASPS have been identified in the phosphorylation domain of the human PER gene and in the casein kinase 1 (*CK1δ*) gene, which phosphorylates this site (Figure 11.15, p. 175). People who suffer from these circadian disorders get sufficient sleep, but are at a social disadvantage, as it is difficult for them to function in a nine-to-five work environment.

SLEEP

Characteristics of Sleep

Entrainment of activity to the light–dark cycle is often most evident as alternating periods of wakefulness and sleep. Sleep is a universal feature of animals, usually manifest as a prolonged period of inactivity. People spend about

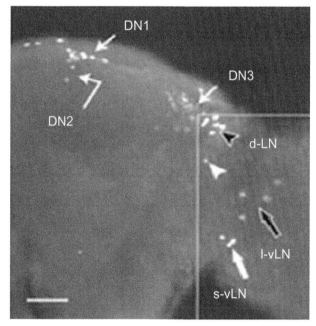

FIGURE 11.13 *Drosophila* clock neurons visualized by immunohistochemical staining with an antibody against the Period protein. Per immunostaining was performed at Zeitgeber time 0. Three groups of dorsal neurons (DN) and three groups of lateral neurons (LN) express *per*. Among the latter, a group of ventral lateral neurons, the l-vLN neurons, arborize in the optic lobe and may transmit light information to the circadian system by releasing a peptide named "pigment dispersing factor." The s-vLN neurons appear to be important for the onset of morning activity, whereas the d-LN neurons regulate circadian activity at the onset of dusk. The central brain is to the left and the optic lobe to the right. Scale bar = 50 μm. (Modified from Grima, B. et al. (2004). *Nature*, **431**, 869–873.)

one third of their lives sleeping. In humans, sleep can be defined behaviorally as a suspension of consciousness, and electrophysiologically by altered brain activity, recorded as electrical waves of an electro-encephalogram (EEG). Although the significance of sleep remains enigmatic, the restorative benefits of sleep are widely appreciated and recent studies have provided evidence that synaptic modifications occur during sleep. Animals deprived of sleep for several weeks will lose weight despite increased feeding, be unable to control body temperature, and, ultimately, die.

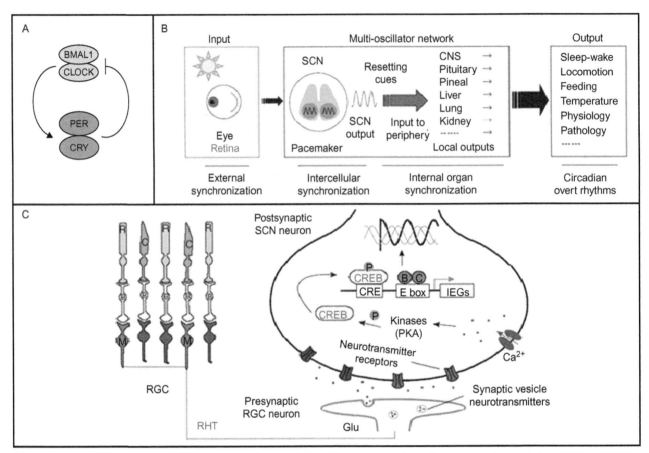

FIGURE 11.14 Properties of the mammalian circadian clock. A. clock activity arises from the cyclic activation of transcription of PER and CRY by BMAL1/CLOCK heterodimers, the expression of which is in turn inhibited by PER/CRY, giving rise to oscillations in the synthesis of these transcriptional regulators. B. Input from the eye synchronizes pacemaker activity in the suprachiasmatic nucleus (SCN), and entrains the circadian rhythm to the light–dark cycle. Output from the SCN to the periphery drives cyclic physiological and behavioral activities. C. Control of the SCN by the retinal hypothalamic pathway (rHT) leads to synchronization of firing activity in neurons of the SCN via a mechanism that involves the cyclic AMP response element binding (CREB) transcription (see also Chapter 14). Presynaptic neurons release the neurotransmitter glutamate (Glu) on the postsynaptic SCN neurons, which elicits calcium influx, followed by a phosphorylation cascade which results in phosphorylation and activation of CREB and transcription of BMAL1 (B) and CLOCK (C). The E-BOX designates regulatory elements and IEGs denote immediate early genes (R, rods; C, cones; M, melanopsin-containing neurons; RGC, retinal ganglion cells). (Adapted from Liu, A. C. et al. (2007). *Nat. Chem. Biol.*, **3**, 630–639.)

Sleep deprivation is compensated by subsequent increased periods of sleep, a phenomenon referred to as sleep homeostasis. The requirements for sleep change during our lifetime, from about 16 hours per day for infants to about 7 to 8 hours per day in adulthood. Sleep in the elderly is reduced, and tends to become fragmented.

Sleep is entrained to the circadian light–dark cycle. Sleep patterns occur in four stages (I–IV), each of which has a corresponding characteristic EEG pattern, as well as **rapid eye movement** (REM) **sleep**. During the first hour of sleep, there are transitions from stage I to stage IV; stages III and IV are characterized by high-voltage, low-frequency EEG patterns, and are known as **slow-wave sleep** or **deep sleep**. There is a drop in blood pressure, the rate of breathing slows, and body temperature is reduced. During this phase, voluntary muscles are not inhibited; thus, tossing and turning or sleep-walking can occur. Following a period of slow-wave sleep, a transition occurs to REM (rapid eye

movement) sleep. During this stage, there are rapid eye movements, and physiological characteristics (e.g. blood pressure) and the EEG pattern are remarkably similar to the awake state. This is the sleep stage during which dreams occur. During REM sleep large muscles are paralyzed and the body is immobile. Respiration and heart rate increase, and penile erections can occur. As the night progresses, increasingly longer periods of REM sleep alternate with deeper stages of sleep (Figure 11.16).

Whereas most animals sleep lying down; cows, horses, and sheep can sleep standing. However, they only experience REM sleep while lying down. Aquatic mammals, whales and dolphins, are conscious breathers and, therefore, must always be awake. In these animals, at any given time only one of the cerebral hemispheres is asleep, while the other remains awake (Figure 11.17).

The initiation of sleep is governed by the **reticular activating system** of the medulla, including a cholinergic

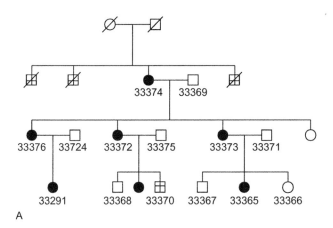

A

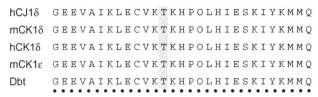

hCJ1δ	G E E V A I K L E C V K T K H P O L H I E S K I Y K M M Q
mCK1δ	G E E V A I K L E C V K T K H P O L H I E S K I Y K M M Q
hCK1δ	G E E V A I K L E C V K T K H P O L H I E S K I Y K M M Q
mCK1ε	G E E V A I K L E C V K T K H P O L H I E S K I Y K M M Q
Dbt	G E E V A I K L E C V K T K H P O L H I E S K I Y K M M Q

B

FIGURE 11.15 Familial ASPS. Panel A shows a pedigree with a segregating mutation that causes ASPS. Circles and squares represent women and men, respectively. Filled symbols show affected individuals, and open symbols show unaffected individuals. The male marked with a cross is "probably affected." Diagonal lines across symbols indicate deceased family members. Panel B shows alignments for *Drosophila* (Dbt), mouse (m), and human (h) CKIδ and CKIε proteins in single-letter amino acid code. These proteins phosphorylate, among others, PER. The T44A mutation, which causes ASPS, is highlighted. (From Xu, Y. et al. (2005). *Nature*, **434**, 640–644.)

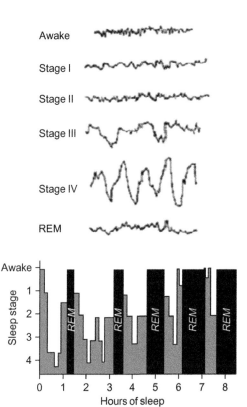

FIGURE 11.16 Progression of a typical 8-hour human sleep cycle. In the awake-state, the EEG shows characteristic **alpha waves**. The appearance of the EEG changes into **theta waves** in stages 1 and 2. Slow **delta waves** appear in stage 3, and are dominant in stage 4, slow-wave sleep. Note that during REM sleep, the EEG is similar to that observed in the awake-state.

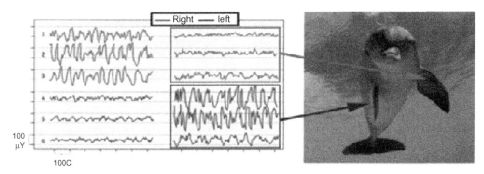

FIGURE 11.17 EEGs recorded from a resting dolphin. The left side (horizontal fin) shows a typical awake wave pattern in the right hemisphere of the brain, while the right side (vertical fin) shows a typical wave pattern characteristic of sleep in the left hemisphere of the brain. These wave patterns indicating sleep and wakefulness alternate over time in the left and right hemispheres. (Modified from Tononi, G., and Cirelli, C. (2005). *Sonno*. In: *Fisiologia Medica*, Ed. Ermes (in Italian).)

projection that extends widely throughout the cortex, as well as adrenergic projections that originate in the **locus coeruleus**, and serotonergic projections that originate from the **dorsal raphe nuclei**. Projections from the **pontine reticular formation** to the **superior colliculus**, a brain structure that controls eye movements, are thought to elicit the rapid eye movements during REM sleep.

Sleep disorders are common, and are thought to affect 80% of people at least once during their lifetime. Common sleep disorders include **insomnia**; **sleep apnea**, which is due to partial obstruction of the upper airway that makes breathing difficult and results in fragmented sleep; and **restless legs syndrome**, a congenital disorder with unpleasant tingling sensations in the legs and an urge to

Box 11.1   The interpretation of dreams

People spend about six years of their lives dreaming (about two hours each night). Dreams occur mostly during REM sleep. Many hypotheses about the significance of dreams have been proposed, ranging from divinations that reveal messages from God or prophecies about the future, to reflections of the subconscious. Sigmund Freud and Carl Jung regarded dreams as an interaction between the unconscious and the conscious. Freud argued that dreams represent the symbolic expression of frustrated desires that had been relegated to the subconscious, and he used the interpretation of dreams for psychoanalysis to uncover these desires (Figure A). The Freudian interpretation, however, has mostly been abandoned in favor of more modern theories, many of which regard dreaming as a vehicle for consolidating recent experiences into memories. In 1976, James Allan Hobson and Robert McCarley proposed that dreams are caused by sensory experiences which are fabricated by the cortex as a means of interpreting random signals that originate in the pons during REM sleep. Others, however, have challenged the theory that dreams originate from the brainstem and the pons. The neural mechanisms that evoke dreams remain mysterious.

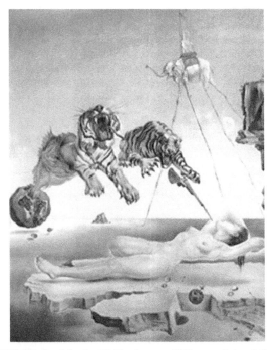

FIGURE A Salvador Dali's famous 1944 surrealist painting *"Dream caused by the flight of a bee around a pomegranate a second before awakening"* was influenced, according to the painter, by Sigmund Freud's theories about dreams. The painting shows a pomegranate which transforms into a fish that releases two tigers behind a bayonet, piercing the biceps of the woman (Dali's wife, Gala) floating above the rock, in an allegory of a bee sting. The scene anticipates the moment of the woman's abrupt awakening from her otherwise peaceful dream. The painting illustrates the symbolic nature of dreams with randomly-assorted sequences of events that seem capricious to the conscious mind.

move the legs constantly. This disease can be relieved by drugs that inhibit the release of dopamine. Finally, **narcolepsy** is a disorder characterized by sudden REM sleep attacks, which will be discussed in greater detail later in this chapter. Studies on the genetics of sleep have benefited from studies on model organisms, including *Drosophila*, mice, zebrafish, and dogs with congenital narcolepsy.

Drosophila and the Genetics of Sleep

While sleep has long been recognized as a universal feature, at least among mammals and birds, it was not recognized in invertebrates until recently, when *Drosophila* geneticists studying circadian behavior showed that periods of inactivity during the circadian day displayed features that resemble sleep. Characteristic features of such rest periods included increased arousal threshold and sleep homeostasis, i.e. compensatory resting following sleep deprivation. Sleep in flies can be monitored with infra-red activity monitors, in which movement is registered every time the fly breaks the beam. Sleep is then quantified as the total amount of time spent sleeping, and as the number and duration of sleep bouts, either during the daytime or the nighttime (Figure 11.18). Joan Hendricks and colleagues in the laboratory of Amita Sehgal at the University of Pennsylvania, and Paul Shaw and coworkers at the Neurosciences Institute in La Jolla (California), performed a detailed analysis of sleep homeostasis in *Drosophila*, and showed that flies exhibit all the major characteristics of sleep rebound after sleep deprivation that are observed in people. In addition, periods of inactivity in flies were age-dependent, similar to human sleep; young flies sleep more than older flies. The powerful genetic resources available for *Drosophila* quickly led to the identification of monoamine catabolism, and cyclic AMP and CREB signaling in the regulation of sleep in *Drosophila*, mechanisms also implicated in mammalian sleep regulation.

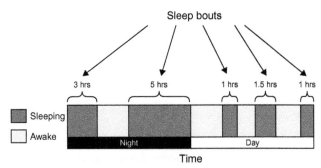

FIGURE 11.18 Parameters of sleep. Sleep can be quantified as daytime sleep and nighttime sleep. The figure shows five sleep bouts, which are longer during the night than during the day. Total sleep can be determined by adding the durations of the sleep bouts during a 24-hour period, which amounts to 11.5 hours in the example shown in the diagram. Merging of sleep bouts will lead to longer periods of uninterrupted sleep. This is known as **sleep consolidation**.

Although sleep can operate independently from the circadian clock, circadian rhythms and sleep patterns in *Drosophila* are interrelated. Many genes with altered transcriptional regulation in spontaneously awake, sleep-deprived, or sleeping flies, also cycle according to the fly's circadian rhythm. Loss-of-function mutants of the circadian clock gene *cycle* showed an abnormally large homeostatic sleep rebound response, and died after 10 hours of sleep deprivation.

One of the first sleep mutants in *Drosophila* was identified by Chiara Cirelli and Giulio Tononi at the University of Wisconsin. This mutant, named *minisleep* (*mns*), exhibited three times shorter sleep periods than normal flies. It has normal sleep rebound, and is not impaired in performance in behavioral assays such as geotaxis and heat avoidance following sleep-deprivation, as are wild-type flies. However, *mns* flies have a shorter lifespan. The *minisleep* phenotype results from a point mutation in the *Shaker* gene, which encodes a voltage-gated potassium channel (Figure 11.19). Another mutant, *fumin* (*fmn*), also shows abnormally high levels of activity and reduced periods of sleep. These mutants, however, are different from the *mns* mutant, in that they are deficient in their sleep-rebound response, and have a normal lifespan. The *fmn* mutation was mapped to the dopamine transporter gene, implicating a function for dopamine in the regulation of sleep and arousal. In addition, mutations in the GABA receptor have implicated a role for GABA in regulating onset and maintenance of sleep.

Indrani Ganguly-Fitzgerald and colleagues from the Neurosciences Institute in La Jolla showed that the duration of sleep is influenced by social experience and memory formation. Flies that were reared in social groups had longer bouts of daytime sleep than flies reared individually (Figure 11.20). Males subjected to unsuccessful courtship attempts with recently-mated females form a long-term memory of this experience, and show subsequent reduced courtship with receptive virgin females. Such males also showed increased sleep, again demonstrating that memories of social experience influence sleep patterns. When dopaminergic neuronal pathways were disrupted by targeted expression of tetanus toxin, or the apoptotic gene *reaper*, which results in programmed cell death, this socially-induced modulation of sleep was abolished. Mutations in components of the cyclic AMP-dependent CREB activation pathway that is associated with long-term memory formation (Chapter 14) also interfered with socially-induced modulation of sleep, which appears to be one regulatory mechanism that controls transitions from activity to rest in flies. These experiments provide support for the hypothesis that memory consolidation of daytime experiences may be one of the functions of sleep.

Bruno van Swinderen and Ralph Greenspan at the Neurosciences Institute in La Jolla (California) showed that inhibitors of dopamine synthesis promote sleep in *Drosophila*, whereas administration of metamphetamine suppresses sleep and promotes arousal, similar to its effects in vertebrates. Furthermore, Rob Jackson and colleagues showed that an increase in extracellular dopamine may promote waking. As mentioned previously (Chapter 3), arousal in vertebrates is also regulated by bioamines released from neurons that are located in the

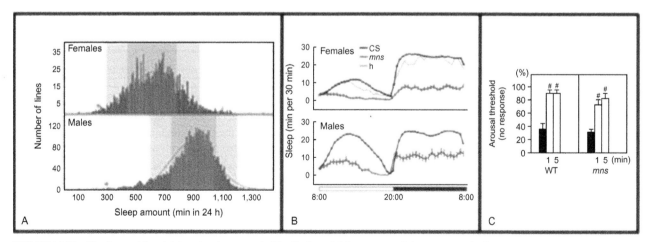

FIGURE 11.19   The *Drosophila minisleep* (*mns*) mutant. A. Distribution of daily amounts of sleep amounts in about 9000 mutant lines of *Drosophila melanogaster* generated by chemical (EMS, see Chapter 9) or *P*-element insertional mutagenesis. In almost all lines female flies sleep less than male flies. The shaded areas indicate values within one (dark shaded area) and two standard deviations (light shaded area) from the mean. Red asterisks indicate *mns* flies. The *mns* mutant has a mutation in the *Shaker* gene that encodes a voltage-gated potassium channel. These mutants sleep substantially less than their wild-type controls. B. Daily time-course (30 minute intervals) of the amount of sleep in wild-type Canton-S (CS) and *mns* flies. The white and black bars under the *x*-axis indicate light and dark periods, respectively. C. Duration (in minutes) and number of sleep and waking episodes during 24 hours of baseline recording in wild-type and *mns* flies; *mns* flies have fewer sleep bouts and spend more time awake (asterisk indicates $P < 0.05$, *t*-test). (Modified from Cirelli, C. et al. (2005). *Nature*, **434**, 1087–1092.)

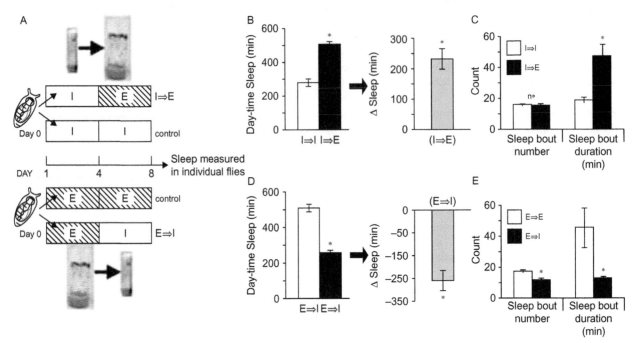

FIGURE 11.20 Effect of social environment on sleep in *Drosophila*. Panel A shows the experimental design in which flies were either reared in isolation (I) or in a socially-enriched environment (E) with 30 other flies at a 1:1 male:female ratio. When flies are switched from an impoverished to an enriched environment (I→E; panel A, top), their daytime sleep increases (panel B), because sleep bouts last longer (panel C). Conversely, when flies are transferred from a socially-enriched to an impoverished environment (E→I, panel A, bottom), they sleep less (panel D), with fewer sleep bouts of shorter duration (panel E). (From Ganguly-Fitzgerald, I. et al. (2006). *Science*, **313**, 1775–1781.)

brainstem and project widely throughout the cortex. In *Drosophila* the mushroom bodies have been implicated in the regulation of sleep. Chemical ablation of the mushroom bodies results in reduced sleep. Similarly, overexpression of cAMP-dependent protein kinase with the *GAL4-UAS* binary expression system targeted to the mushroom bodies also reduced sleep. Ralph Greenspan and colleagues discovered yet another mechanism for the regulation and maintenance of sleep in *Drosophila*. This mechanism activates a signal transduction cascade that leads to the phosphorylation of transcriptional regulators mediated via the **epidermal growth factor** (EGF) **receptor** and the **extracellular signal regulated kinase** (ERK) **pathway** (see also Chapter 14). Activation of the EGF receptor in turn is controlled by a G-protein-coupled receptor, encoded by the *rhomboid* gene, which is expressed in the **pars intercerebralis**, a brain region that separates the hemispheres of the fly brain. The *rhomboid* receptor is activated by a peptide ligand, known as "Spitz." Increased expression of *rhomboid* or *Spitz* in transgenic flies, via the *GAL4-UAS* binary expression system, or under the control of a heat shock promoter (Chapter 9), results in induction of sleep. Conversely, reduced *rhomboid* signaling decreases sleep. Changes in sleep phenotype during these experimental manipulations were accompanied by changes in the state of phosphorylation of ERK (Figure 11.21).

The various mechanisms that control the onset of sleep reveal disparate parts of a complex puzzle, of which a clear picture has yet to emerge. Chiara Cirelli and colleagues showed that knock-out mice in which the *Kcna2* gene has been deleted have reduced sleep. This gene encodes a potassium channel which is homologous to the *Shaker* channel in *Drosophila*, in which mutations give rise to the *minisleep* phenotype described earlier. Whereas molecular homologies and functional parallels have been documented extensively between the *Drosophila* and the vertebrate circadian clocks, molecular comparisons between mechanisms that regulate sleep in flies and mammals remain to be further evaluated. What sets the phenomenon of sleep apart from the well-understood circadian clock mechanism, however, is its enticing relationship to arousal, and by extension the complex concept of consciousness, the holy grail of neurobiological research.

Narcolepsy and the Hypocretin–Orexin System

Significant insights in the neural control of sleep in the mammalian brain came from genetic studies on narcolepsy. Narcolepsy is a sleep disorder that affects males and females equally, with an incidence of about 0.02% to

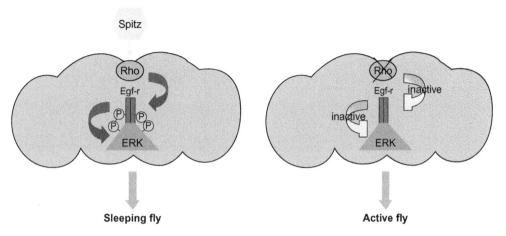

FIGURE 11.21 Control of sleep in *Drosophila* by the rhomboid signaling pathway (Egf-r, EGF receptor, P) designates phosphate groups on the EGF receptor due to autophosphorylation and on ERK. The black cross through Rho indicates experimental inactivation of the rhomboid receptor or absence of its ligand, Spitz.

Box 11.2 Are fruit flies conscious?

The nature of consciousness has been a hotly-debated topic by philosophers and neurobiologists alike. The ability to focus attention on an object or activity can be viewed as a component of consciousness. Bruno van Swinderen and Ralph Greenspan at the Neurosciences Institute in La Jolla (California) asked whether flies were able to focus their attention on relevant objects. They suspended flies, *Drosophila melanogaster*, in the center of a cylindrical arena which could rotate around the fly. They then presented a rotating object, such as a 10°-wide dark vertical bar, around the suspended animal. They positioned electrodes in the left optic lobe and medial protocerebrum of the suspended fly, and recorded **local field potentials** from its brain in response to the rotating visual stimulus. When they analyzed the pattern of electrical activity from the fly's brain, they found brain waves in the 20–30 Hz range that appeared coincident with the presentation of the rotating image. When they matched the rotation frequency of the image with a synchronous puff of banana odor, the 20–30 Hz response to the image increased. Furthermore, the average 20–30 Hz response to a rotating bar was attenuated when the fly was asleep in overnight experiments. Thus, the 20–30 Hz wave represents the physiological manifestation of attention to salient objects. To what extent this tells us whether flies are conscious or not is an issue for further debate (van Swinderen, B., and Greenspan, R. J. (2003). *Nat. Neurosci.*, **6**, 579–586).

0.18% in the Caucasian population. Individuals who are afflicted with this disorder find it difficult to function socially, since they experience unusual daytime sleepiness, and can manifest uncontrolled periods of REM sleep. Such transitions from wakefulness to REM sleep can be triggered by emotional experiences, most commonly laughing, but also by anger, surprise, excitement, or fear.

Thus, a person suffering from narcolepsy might laugh at a joke and immediately become **cataplectic**, and descend into REM sleep.

Genetic studies on mice and dogs have provided insights in the pathology of narcolepsy, and in the neural control of sleep. These studies have implicated two neuropeptides, **hypocretins** (also known as **orexins**) and their receptors in the regulation of sleep. These two peptides (orexin A and B, or hypocretin-1 and -2) are produced by a small group of cells in the lateral hypothalamus. They show about 50% sequence identity, and are produced by cleavage of a single precursor protein. They are highly-conserved peptides, thought to have arisen early in vertebrate evolution. The name orexin, coined by Masashi Yanagisawa at the University of Texas Southwestern Medical Center in Dallas, reflected the appetite-stimulating effects of these peptides. When Yanagisawa and his colleagues deleted the *orexin* gene in mice through homologous recombination, they observed to their surprise a phenotype that was remarkably similar to narcolepsy. Behavioral studies showed that homozygous knock-out mice had bouts of reduced open field activity during the night, when mice are generally most active. Electroencephalogram recordings from these mice showed that the time spent in REM sleep, and the duration of REM sleep periods, were significantly higher in the mutants than in the wild-type controls. Narcoleptic phenotypes could also be elicited in rats by expressing transgenic polyglutamine-containing ataxin-3 in orexin-producing neurons in the lateral hypothalamus (Figure 11.22). Like huntingtin, ataxin-3 is a protein which causes neurodegeneration when it contains an expanded polyglutamine tract. It is a de-ubiquitinating enzyme that is responsible for spinocerebellar ataxia type-3, a neurodegenerative disorder accompanied by impaired locomotion.

At the same time that the group of Yanagisawa in Dallas developed the mouse narcolepsy model, Emmanuel

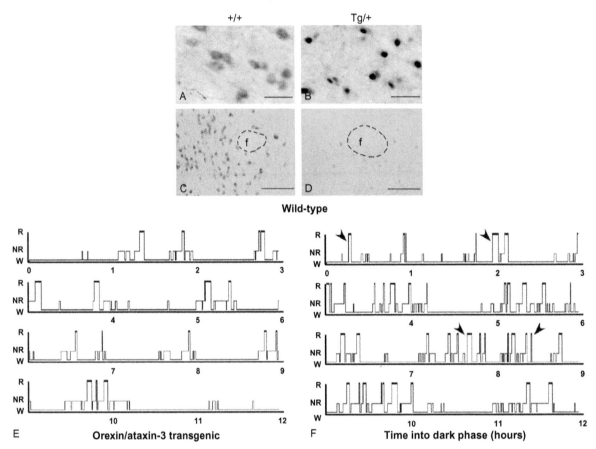

FIGURE 11.22   Elimination of orexin–hypocretin neurons in the rat brain by targeted transgenic expression of polyglutamine-containing ataxin-3 in orexin–hypocretin producing neurons. Panel A shows staining of orexin neurons in the lateral hypothalamus and Panel B shows co-localization of orexin staining with the ataxin-3 transgene (black nuclei) only in orexin neurons. After four weeks, orexin neurons are still present in the wild-type animal (panel C), but have been eliminated in animals expressing the ataxin-3 transgene (Panel D). Panels E and F record sleep–wakefulness transitions in a wild-type (E) and a transgenic rat (F), respectively. The transgenic animal shows more rapid cycling between the awake and the sleep states. It also exhibits direct transitions from wakefulness to REM sleep (indicated by arrowheads), a pattern indicative of narcolepsy (f, fornix; scale bar 40 μm in panels A and B; 200 μm in panels C and D; W, wakefulness; NR, non-REM sleep; R, REM sleep). (Modified from Beuckmann, C. T. et al. (2004). *J. Neurosci.*, **24**, 4469–4477.)

Mignot and his colleagues at Stanford University studied narcolepsy in dogs. Labrador retrievers and Doberman pinschers were the first breeds in which fully-penetrant autosomal recessive transmission of canine narcolepsy had been observed. The locus responsible for this disorder was named *canarc-1*. REM periods during daytime naps and nighttime sleep are considered the hallmark of narcolepsy, and are also observed in narcoleptic dogs. Excitement, induced either by food or play, can trigger these narcoleptic episodes. Mignot and his colleagues established a colony of Doberman pinschers in which the locus responsible for narcolepsy segregates. They conducted a painstaking series of linkage studies, followed by positional candidate gene cloning, and identified the *Hcrtr2* gene, which encodes hypocretin receptor 2, as the causal gene for narcolepsy at the *canarc-1* locus. A mutation in the *Hcrtr2* gene was identified as the culprit for the narcolepsy phenotype. Thus, canine narcolepsy is

a fully-penetrant monogenic disorder. However, human narcolepsy, except for familial cases, is thought to be a multigenic condition with sensitivity to the environment (i.e. a quantitative trait).

Long-term sleep-deprivation in rodents dramatically increases food intake and energy metabolism (catabolism), which ultimately results in death. The orexin–hypocretin system appears to be an important link between sleep, feeding behavior, and energy expenditure. Orexin increases the craving for food, and its secretion is activated by **ghrelin**, a hormone secreted by the stomach that promotes food intake, and inhibited by **leptin**, a hormone produced by fat cells that limits food intake (see also Chapter 9). Administration of orexin A (hypocretin-1) into the brain promotes wakefulness, increases body temperature, and locomotion, and consequently increases energy expenditure.

The neurons in the lateral hypothalamus that produce orexins–hypocretins project to the reticular formation

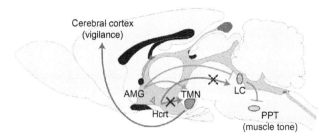

FIGURE 11.23 Simplified model for the control of arousal by hypocretin–orexin. During the normal waking state hypocretin-expressing (Hcrt) neurons in the lateral hypothalamus activate histaminergic neurons in the tuberomammillary nucleus (TMN) of the hypothalamus (which express hypocretin-2 receptors), and the noradrenergic locus coeruleus (LC) neurons in the brainstem (which express hypocretin-1 receptors). The TMN neurons activate the cerebral cortex to maintain vigilance, and the LC neurons inhibit the pedunculopontine tegmental nucleus (PPT) in the brainstem which suppresses muscle tone. Emotional events activate the limbic system, including the amygdala (AMG), which leads to increased firing of neurons in the TMN, and reduced activation of the LC. In narcoleptic dogs with hypocretin-2 receptor deficiency, there is no hypocretin signaling through the TMN (black X in the diagram), and a reduced hypocretin tone in the LC (stippled X). As a result, limbic input sometimes overcomes the excitatory inputs to the LC, leading to activation of muscle tone suppression while the TMN remains active. The result is consciousness with no muscle tone, or **cataplexy**. Where the hypocretin signal itself is deficient, as in human narcolepsy or mouse hypocretin knock-outs, limbic inputs have greater effects on the LC, and cataplexy is more prevalent. (From Sutcliffe, J. G., and de Lecea, L. (2004). *Nat. Med.*, **10**, 673–674.)

in the brainstem, the locus coeruleus, and the dorsal raphe nuclei; these regions play a role in the onset of sleep (Figure 11.23). In addition to these major nuclei implicated in sleep regulation, hypocretin-containing neurons also project to the cortex, the **basal forebrain**, the **nucleus accumbens**, and the **amygdala**. The receptors for hypocretins are G-protein-coupled receptors and are localized at the projection sites of the hypocretin neurons. Hypocretins act directly on axon terminals and modulate the release of GABA, the major inhibitory neurotransmitter, and glutamate, the major excitatory neurotransmitter. It is of interest to note that GABA has also been implicated in regulating onset and maintenance of sleep in *Drosophila*. The neuroanatomical organization of the hypocretin–orexin system suggests that hypocretins, monoamines, and acetylcholine act together in regulating sleep–wake transitions.

SUMMARY

Locomotion is a component of most behaviors. Spontaneous activity of rodents, measured as open field behavior or variations thereof, such as the elevated plus-maze or the light–dark box, can be used as an indication of emotional state, anxiety, or depression. QTL mapping studies have identified chromosomal regions associated with variation

in spontaneous activity, but to date only few genes have been identified in such regions.

Activity can also be measured in response to environmental stimuli. In this paradigm, impairment of prepulse inhibition of the startle-response serves as a model for schizophrenia. Ensembles of genes that mediate spontaneous locomotion are likely to be different from, but overlap with, those that mediate other behaviors that contain a locomotor component. Among genes involved with locomotion are genes associated with dopaminergic neurotransmission.

Genetic defects which affect neural circuitry that directs locomotion lead to locomotor impairments. The most extensively studied neurodegenerative locomotor diseases in people are Parkinson's disease, in which the function of dopaminergic neurons of the nigrostriatal pathway is compromised, and Huntington's disease, which is due to the expansion of glutamine tracts in the huntingtin protein. Misfolded aggregated proteins that cause cellular stress and lead to cell death are likely part of the pathogenic mechanisms for each disorder. Whereas Huntington's disease is a Mendelian dominant autosomal inherited disorder, Parkinson's disease is genetically more complex, and can have a neurotoxic environmental component.

The activity of animals is governed by an endogenous biological clock, and this circadian rhythm is entrained to the light–dark cycle. The mammalian clock is contained in the suprachiasmatic nucleus of the hypothalamus, whereas the clock in *Drosophila* is comprised of specific clusters of lateral and dorsal neurons in the brain. The activity of clock neurons is modulated by interlocking negative feedback loops, mediated by transcriptional regulators that undergo periodic synthesis and degradation. The cellular mechanism of the biological clock has been conserved in evolution, so that the clock mechanism of mammals is remarkably similar to that of *Drosophila*.

Sleep is an altered physiological restorative state that is often, but not always, coordinated with the circadian clock. Like mammals, periods of inactivity in flies meet the criteria of sleep, based on increased arousal threshold, homeostasis, and age-dependence. Genetic studies in *Drosophila* have identified several gene products that are important for the regulation of sleep, including the *Shaker* voltage-gated potassium channel, a dopamine transporter gene, the GABA receptor, and the *rhomboid* receptor with its ligand, *Spitz*. Insights in the control of mammalian sleep have come from studies on narcolepsy in mice and dogs, which led to the identification of neuropeptides (hypocretins–orexins) and their receptors that contribute to the regulation of sleep. In *Drosophila*, sleep appears to involve mushroom bodies and the pars intercerebralis of the brain. In mammals, the onset of sleep is regulated by activation of neuronal projections that originate in the brainstem.

Box 11.3   The suprachiasmatic nucleus

The biological clock in mammals resides in the **suprachias-matic nucleus** (SCN) of the hypothalamus, which is located immediately above the optic chiasma on either side of the third ventricle. Michael Menaker and colleagues at the University of Virginia performed early electrophysiological studies on the activity of the SCN in golden hamsters. The rate at which neurons of the SCN fire action potentials varies during the 24-hour period, with the greatest activity occurring at midday and the lowest firing rate at night. In constant darkness this endogenous rhythm is slightly longer than 24 hours (in humans it is on average 24 hours and 11 minutes). The SCN receives direct input from the retina via the **retinohypothalamic tract**. **Retinal ganglion cells**, whose axons form this projection, express a specialized photopigment, **melanopsin**. Light activation induces entrainment in the SCN, i.e. synchronization of its spontaneous circadian firing activity to the day–night cycle. The SCN interacts with many brain regions, and orchestrates the rhythmic release of several neuropeptides, hormones, and neurotransmitters via the pituitary gland and the autonomic nervous system. Among hormones that are released in a circadian manner are cortisol and the hormone **melatonin**, derived from serotonin and produced by the **pineal gland**, a small endocrine gland in the brain located rostrodorsal to the **superior colliculus**. Activation of the retinohypothalamic pathway during the day inhibits the SCN and, hence, secretion of melatonin from the pineal gland (Figure A). Thus, melatonin secretion is rhythmic, peaking at night and declining during

the day. In some species melatonin inhibits sexual maturation. Lengthening of the days with gradually less melatonin production thus enables sexual maturation during the appropriate breeding season. Destruction of the SCN results in loss of circadian rhythm. Patricia DeCoursey showed that ground squirrels with lesions in the SCN appear out of their burrows at inappropriate times, and are therefore more susceptible to predation. Rats with SCN lesions are "free-running;" they sleep the same total amount, but at random times and for random lengths at a time. Peripheral tissues also show circadian activity in their metabolism and physiology. These circadian activities are referred to as **slave oscillators**, as they are controlled and synchronized by the master oscillator, the SCN.

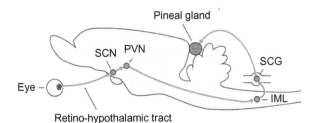

FIGURE A   The projections by which light inhibits the secretion of melatonin from the pineal gland (SCN, suprachiasmatic nucleus; PVN, paraventricular nucleus; IML, intermediolateral cell column; SCG, superior cervical ganglion). (From Takahashi, J. (1994). *Curr. Biol.*, **4**, 165–168.).

STUDY QUESTIONS

1. Define the term open field behavior. Would you expect the ensemble of genes that mediates open field behavior to be similar or different from that which enables startle behavior?

2. Which activity assays are used in mouse models of human psychiatric disorders?

3. What is prepulse inhibition?

4. What is the neural pathway affected in Parkinson's disease, and what is the characteristic locomotor phenotype of a patient with Parkinson's disease?

5. Compare and contrast the genetic basis of Huntington's disease and Parkinson's disease.

6. Explain the term entrainment. What is the molecular mechanism for entrainment in the *Drosophila* clock?

7. What would be the expected phenotype of a *timeless* null mutant? Would this phenotype be different from a *period* null mutant?

8. What would be evolutionary advantages of entrainment to the light–dark cycle?

9. Which structure houses the biological clock in the mammalian brain?

10. What are the characteristic features of rest periods in *Drosophila* that justify their designation as sleep?

11. What parameters of sleep can be quantified in flies? What does the sleep pattern of the *minisleep* mutant look like?

12. What happens to sleep in flies when *rhomboid* is overexpressed in the pars intercerebralis of the brain?

13. What are the differences between REM sleep and slow-wave sleep?

14. What is narcolepsy, and how have studies on narcolepsy contributed to our understanding of sleep?

15. What is the importance of circadian rhythms for the experimental design of studies on behavior?

RECOMMENDED READING

Allada, R., Emery, P., Takahashi, J. S., and Rosbash, M. (2001). Stopping time: the genetics of fly and mouse circadian clocks. *Annu. Rev. Neurosci.*, **24**, 1091–1119.

Dunlap, J. C., Loros, J. J., and DeCoursey, P. J. (2004). *Chronobiology: Biological Timekeeping.* Sinauer Associates, Sunderland, MA.

Flint, J., Valdar, W., Shifman, S., and Mott, R. (2005). Strategies for mapping and cloning quantitative trait genes in rodents. *Nat. Rev. Genet.*, **6**, 271–286.

Foltenyi, K., Greenspan, R. J., and Newport, J. W. (2007). Activation of EGFR and ERK by rhomboid signaling regulates the consolidation and maintenance of sleep in *Drosophila*. *Nat. Neurosci.*, **10**, 1160–1167.

Ganguly-Fitzgerald, I., Donlea, J., and Shaw, P. J. (2006). Waking experience affects sleep need in *Drosophila*. *Science*, **313**, 1775–1781.

Harbison, S. T., Mackay, T. F. C., and Anholt, R. R. H. (2009). Understanding the neurogenetics of sleep: Progress from *Drosophila*. *Trend Genet*, **25**, 262–269.

Hendricks, J. C., Finn, S. M., Panckeri, K. A., Chavkin, J., Williams, J. A., Sehgal, A., and Pack, A. I. (2000). Rest in *Drosophila* is a sleep-like state. *Neuron*, **25**, 129–138.

Jordan, K. W., Morgan, T. J., and Mackay, T. F. C. (2006). Quantitative trait loci for locomotor behavior in *Drosophila melanogaster*. *Genetics*, **174**, 271–284.

Lin, L., Faraco, J., Li, R., Kadotani, H., Rogers, W., Lin, X., Qiu, X., de Jong, P. J., Nishino, S., and Mignot, E. (1999). The sleep disorder canine narcolepsy is caused by a mutation in the hypocretin (orexin) receptor 2 gene. *Cell*, **98**, 365–376.

Liu, A. C., Lewis, W. G., and Kay, S. A. (2007). Mammalian circadian signaling networks and therapeutic targets. *Nat. Chem. Biol.*, **3**, 630–639.

Mackay, T. F. C., and Anholt, R. R. H. (2006). Of flies and man: *Drosophila* as a model for human complex traits. *Annu. Rev. Genomics Hum. Genet.*, **7**, 339–367.

Pace-Schott, E. F., and Hobson, J. A. (2002). The neurobiology of sleep: genetics, cellular physiology and subcortical networks. *Nat. Rev. Neurosci.*, **3**, 591–605.

Shaw, P. J., Cirelli, C., Greenspan, R. J., and Tononi, G. (2000). Correlates of sleep and waking in *Drosophila melanogaster*. *Science*, **287**, 1834–1837.

Siegel, J. M. (2008). Do all animals sleep? *Trends Neurosci.*, **31**, 208–213.

Watanabe, A., Toyota, T., Owada, Y., Hayashi, T., Iwayama, Y., Matsumata, M., Ishitsuka, Y., Nakaya, A., Maekawa, M., Ohnishi, T., Arai, R., Sakurai, K., Yamada, K., Kondo, H., Hashimoto, K., Osumi, N., and Yoshikawa, T. (2007). *Fabp7* maps to a quantitative trait locus for a schizophrenia endophenotype. *PLoS Biol.*, **5**, e297.

Williams, J. A., and Sehgal, A. (2001). Molecular components of the circadian system in *Drosophila*. *Annu. Rev. Physiol.*, **63**, 729–755.

Willis-Owen, S. A., and Flint, J. (2006). The genetic basis of emotional behaviour in mice. *Eur. J. Hum. Genet.*, **14**, 721–728.

Genetics of Social Interactions

OVERVIEW

The social environment of an animal can have a profound and determining effect on its behavior. Many social behaviors are clearly genetically-determined, but their manifestations are often dependent on the context of the society, i.e. the presence of conspecifics (members of the same species), predators, or prey. The genetic programs that determine these behaviors are dynamic, and the roles of the individual within its community can change as a function of development, or as a consequence of changes in environmental conditions. Social behaviors that benefit the community can benefit the individual's fitness by providing the opportunity to transfer its genes to its offspring and promote their survival. Examples are wide ranging, from social organization in bee hives and ant colonies, to sentry behavior in meerkats. Social behaviors may require the individual to perform a seemingly altruistic and therefore likely costly, function for its community, but survival of the community as a whole ultimately benefits the individual and enhances its fitness. The establishment of a social hierarchy as a consequence of aggressive encounters that determine the place individuals occupy in the dominance hierarchy is also a form of social behavior. Dominant animals typically have a greater chance to mate and transmit their genes to the next generation. Aggressive encounters can also lead to stable societies, in which energy no longer has to be devoted to fights once a "pecking order" has been established. In this chapter we will examine social interactions, how the expression of genetic programs orchestrates social behaviors, and how these genetic programs can respond to changes in the social and physical environment.

SOCIAL ENVIRONMENT AND THE GENES–BRAIN–BEHAVIOR PARADIGM

The social environment and its effect on gene expression present a special case of genotype-by-environment interaction.

Sometimes social interactions are referred to as **indirect genetic effects**, based on the notion that the genotype of one individual shapes the phenotype of another individual through social interactions. Such interactions often occur between members of the same species (conspecifics), but the term indirect genetic effects has also been applied to modification of behavioral phenotypes between members of different species, for example adaptive behaviors resulting from predator–prey interactions.

Expression of behavior is under the control of the nervous system, but brain function itself is determined through complex and plastic gene expression patterns. Thus, the "genes–brain–behavior" paradigm is the fundamental concept that describes an organism's response to the environment, including its social environment (Figure 12.1). However, this concept is perhaps a little too simplistic. Behavior itself is constantly modified by sensory input, and such sensory input affects the activity of the nervous system as well as gene expression in the brain. Thus, the genes–brain–behavior principle is a two-way interaction.

Modifications of gene expression in the brain as a consequence of social interactions are often under the control of the endocrine system. Hormonal regulation that impacts behavior in vertebrates revolves around the pituitary–gonadal, pituitary–adrenal, and pituitary–thyroid axes, all of which are under the control of regulatory hormones released by the hypothalamus. The pituitary can be viewed as a regulator of the **peripheral endocrine system**, but is itself under the control of the hypothalamus. In addition, hormones released by the posterior pituitary, **oxytocin** and **vasopressin** (also known

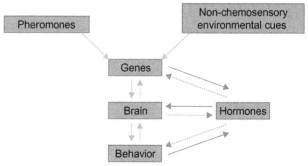

FIGURE 12.1 Diagram of the genes–brain–behavior principle.

as antidiuretic hormone, ADH), are produced in the paraventricular and supraoptic nuclei in the hypothalamus and released from axons of neurosecretory cells that terminate in the posterior pituitary. These hormones can also be released at other sites within the brain, where they perform **neuromodulatory** functions that impact social behavior.

The nervous system controls the endocrine system, but the endocrine system provides feedback information to the nervous system. On the one hand, the production of hormones requires gene regulation, but the effects exerted by hormones on their target tissues also depend on regulation of genetic programs. In addition to chemical communication within the organism that modulates the **gene-brain-behavior axis**, external chemical signals, which are known as **pheromones** and are produced by conspecifics, can trigger profound effects on the physiology and behavior of the recipient.

The literature on social interactions is immense, and cannot possibly be covered in its entirety in this book. We will, therefore, highlight only selected examples that focus on model systems and illustrate fundamental principles of the genetics of social interactions.

SOCIAL COOPERATION AND FITNESS

In light of the classic concept of "survival of the fittest," it may seem paradoxical that individuals would cooperate to form social structures rather than fending for themselves. It is, therefore, fair to ask why it would be advantageous for individuals to link their fates to those of other members of the group. The answer is simple: social interactions ultimately benefit the individual's fitness. We use the concept of **fitness** here to indicate the perpetuation of the individual's genetic signature in its offspring. Altruistic behavior that is costly to an individual may provide indirect fitness benefits by increasing the number of copies of its genes in the population through the reproductive success of closely-related relatives. When asked whether he would give his life to save his brother, the geneticist J.B.S. Haldane responded: "No, but I would to save two brothers or eight cousins."

The type of altruistic behavior that benefits close relatives is known as **kin selection**. Other forms of altruistic behavior may be advantageous if an individual is likely in the future to be the beneficiary of reciprocal altruism. Individuals that assist each other reciprocally in times of need enhance their survival chances. Social interactions in which both parties benefit are known as **mutualism**. It is, ultimately, the net fitness benefit that determines whether it is advantageous for an individual to be solitary, or to be part of a social group (Figure 12.2).

Solitary or social behaviors are not always hardwired; the decision to be social or solitary can often be modulated by changes in environmental conditions, including the social environment (an example we will describe later

FIGURE 12.2 Altruistic behavior in meerkats. Members of a meerkat community take turns standing guard to alert the community of the approach of predators, while others forage and burrow.

in this chapter is solitary or social feeding in nematodes). Such environmental effects provide striking examples of genotype-by-environment interactions, which allow individuals to switch to alternative lifestyles in different environments to optimize their fitness.

COURTSHIP AND MATE SELECTION

From an evolutionary perspective reproductive behavior is the most important social behavior. Partner selection, **courtship**, mating, **affiliative behavior** (time spent with the mate), and parenting, are instrumental for the dispersion of an individual's alleles through the population, and for optimizing the chances for survival of offspring that harbor them. These behaviors, ultimately, provide the mechanism for evolutionary change, through changes in allele frequencies.

Courtship rituals enable females to evaluate the quality of competing males. Such courtship displays can be visual (e.g. bright plumage in peacocks, display of brightly-colored dewlaps in lizards), vocal (e.g. birdsong, frog calls), or can consist of intricate performances (e.g. the construction of intricate bowers and dancing displays by bowerbirds) (Figure 12.3). The advantages of such overt displays in terms of mating benefits are countered by the inherent risk for increased predation that results from increased visibility and exposure during courtship. We will briefly examine two of the best-studied forms of courtship: birdsong and courtship in *Drosophila*. With the advent of new genomic resources, the genetic underpinnings of birdsong are on a trajectory of exciting discoveries in the near

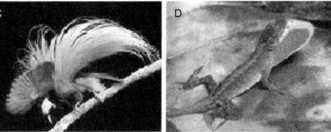

FIGURE 12.3   Visual displays that may signal reproductive quality. The bright plumage of the peacock (A) can intimidate rival males and is used for courtship; bowerbirds (B) build intricate structures and perform dances to attract prospective mating partners; (C) birds of paradise have exceptionally brilliant plumage; (D) the extended dewlap of a lizard is used as a courtship signal for potential mates, and as an aggressive display in territorial fights.

Box 12.1   Frog cocktail parties and mouse serenades

Although only a few genetic models are available to study auditory communication during courtship, vocal communication during courtship is a widespread phenomenon. Frogs have been a popular focus for studies on phonotaxis (movement toward sound), as male frogs call to attract females. For example, the wetland tungara frog has evolved an elaborate call that consists of "whines" and "clucks." Surgical modification of the vocal apparatus of the frog that eliminates the clucks reduces its attractiveness to females. When large groups of frogs call simultaneously the result is a deafening chorus, as individual males attempt to attract females while also signaling their territorial boundaries to other males. Females have an uncanny ability to distinguish individual calls among this cacophony of sound. How do they achieve this remarkable acoustic feat? This problem is known as the **cocktail party effect**, as it resembles the background noise of chattering guests at a cocktail party that makes it difficult to dissect individual snippets of conversation. Understanding how the frogs solve this acoustic resolution problem may provide insights in how to make more effective hearing aids.

A surprise observation by Holy and colleagues found that male mice also sing during courtship. In the presence of females, male mice produce ultrasonic sounds that show temporal regularity and a syllable structure not unlike that seen in songbirds (Figure A). Whether the songs of mice offer a

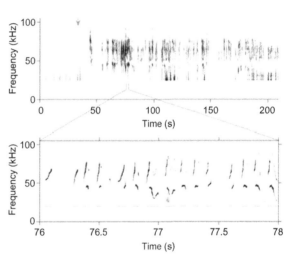

FIGURE A   Ultrasound vocalization by male mice after olfactory exploration of female urine. A male mouse was presented with a cotton swab dipped in female urine (arrow) and the frequency of ultrasonic vocalization was recorded. The 2 s expanded trace on the bottom shows individual syllables reminiscent of birdsong. (From Holy, T. E., and Guo, Z. (2005). *PLoS Biol.*, **3**, e386.).

competitive advantage for mating during courtship remains to be established, and requires studies on mice in the wild.

---

future. In contrast, genetic studies on courtship and mating in *Drosophila* have a long history that has documented many genes contributing to these behaviors.

Birdsong

Male songbirds sing to establish territories, and to attract mating partners. Development of the song requires vocal learning, a phenomenon found in many avian species (parrots, hummingbirds, and more than 4500 species of songbirds), but in only three groups of mammals: humans, bats, and cetaceans (whales and dolphins).

Pioneering studies by the laboratories of Nottebohm and Konishi identified the neural circuits in the brain that

are involved in mediating birdsong. Development of birdsong occurs in two stages: first, a period of listening to a tutor song forms an auditory memory; then activation of motor pathways results in vocalization. These processes occur during a critical period early in postnatal life. Early vocalization consists of a crude song that is gradually perfected, a process referred to as **song crystallization**.

Characteristics of birdsong resemble the acquisition of human language. Similar to birdsong, babbling gradually develops into speech during a critical neonatal period, and the acquisition of language is dependent on auditory input. In fact, development of birdsong is the only animal model for development of human speech, and also serves as a valuable model for neural plasticity during learning.

Box 12.2 The genetics of human language

While we are on the verge of learning a great deal about the genetics of birdsong, hardly anything is known about the genetic underpinnings of human speech. The evolution of language is unique to the human lineage, and is the foundation for human culture and social organization. What are the genes that were essential for the evolution of language? A significant discovery came in the early 1990s when a large family, known as the KE family, was discovered. Affected members of this family had a language disorder. They had difficulty speaking, and were virtually unintelligible. Brain imaging studies showed deficits in motor areas for speech production. The phenotype segregated in a dominant Mendelian manner, and the gene responsible was discovered to be a transcription factor of the forkhead class of transcriptional regulators, named FOXP2.

A single G to A nucleotide mutation in exon 14 of the *FOXP2* gene changed an arginine to a histidine. All affected individuals of the KE family had an allele that encodes a histidine at this position, whereas all non-affected individuals have two normal alleles with an arginine.

Interestingly, the *FOXP2* gene is one of the most highly-conserved genes, with only two amino acid differences in the human lineage between chimpanzees and humans. Phylogenetic analysis of the genomic regions surrounding the *FOXP2* gene reveals a signature of a **selective sweep**; that means **positive selection** has occurred favoring the rapid spread of a beneficial allele through the population. Genes closely-linked to this allele **hitchhiked** along generating an area of lower nucleotide diversity than expected (see also Chapter 16). The mutations in the human lineage occurred between 10 000

and 100 000 years ago, presumably coincident with the evolution of language. There are only three amino acid differences between mouse and human *FOXP2* (Figure A). Mice in which the *FOXP2* homolog was knocked-out are deficient in the ultrasonic vocalization that pups elicit from their mother when they are removed from the nest.

The discovery of the *FOXP2* gene was hailed by the popular press as the discovery of the gene for language. This is a significant exaggeration. The *FOXP2* protein is a transcriptional regulator and, thus, presumably affects expression of a wide range of genes necessary for development of the neural circuitry that enables the production of intelligible speech. Thus, *FOXP2* is only one member of a complex genetic network that enables language to develop. Nevertheless, identifying its gene targets and their interactions may bring us a step closer to understanding the genetics of human language.

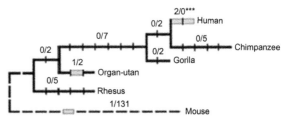

FIGURE A Nucleotide substitutions (nonsynonymous/total) mapped on a phylogeny of primates. Vertical bars represent nucleotide changes. Grey boxes indicate amino-acid changes. (Modified from Enard, W. et al. (2002). *Nature*, **418**, 869–872.).

Birds reared in isolation or birds that have been deafened fail to develop normal song. Furthermore, the development of brain nuclei that is essential for the generation of song is, at least in many nontropical species, dependent on **testosterone**. These nuclei become larger as hormone levels increase when days lengthen during the spring and summer. Because early neural development is dependent on environmental conditions, such as nutritional status, Steve Nowicki and colleagues from Duke University postulated that the quality of the crystallized song reflects the health, and, by inference, the genetic fitness, of the male.

Bird songs are species-specific, and birds are able to recognize songs from their own species (Figure 12.4). Most studies on birdsong have focused on two species, the canary and the zebra finch. What are the genetic programs that enable the development of birdsong? David Clayton and colleagues at the University of Illinois identified a transcription factor in the brains of zebra finches that is rapidly induced in auditory regions (e.g. the caudomedial neostriatum) when birds hear a conspecific song for the first time. The transcription of this **immediate early gene**, named "*zenk*," (also known as "*egr1*," i.e. "early growth response 1") correlates closely with the initiation

of auditory memory formation of a conspecific song, and is transient. This rapid transient expression of *zenk* has been termed the **genomic action potential**. What targets are regulated by *zenk* and how are these genetic programs influenced by physiological or environmental conditions? To answer these questions, a consortium of investigators embarked on the complete sequencing of the zebra finch genome. This opens the way for the use of expression microarrays – and, ultimately, proteomics approaches – to examine how transcriptional and protein networks are modified during exposure to a new song, auditory memory formation, and subsequent song generation and crystallization. The neural circuit that mediates song production is sexually dimorphic. Males and females differ, in that males must produce song to court females, but females must form a memory representation of the song that enables them to recognize and judge the male's performance. Thus, another question that begs scrutiny is whether males and females differ in the transcriptional response to auditory stimulation by conspecific songs and, if so, how? In the future access to increasingly expanding genomic resources (e.g. www.Songbirdgenome.org) is likely to provide answers to many of these questions.

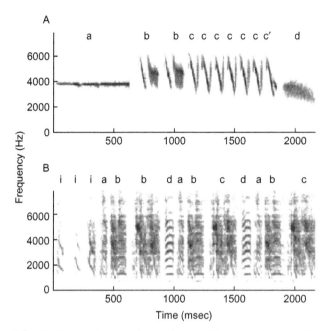

FIGURE 12.4 Sonograms of songs from a white-crowned sparrow (A) and a zebra finch (B). The white-crowned sparrow's song begins with a whistle (a) followed by repeated trills (b,c) and buzzes (d), whereas the zebra finch's song starts with introductory syllables (i), followed by a sequence of syllables (indicated by lowercase letters), that can be either simple or more complex. The syllables are composed of notes and assembled into song motifs. (From Doupe, A. J., and Kuhl, P. K. (1999). *Annu. Rev. Neurosci.*, **22**, 567–631.)

Courtship and Mating in *Drosophila*

The fruit fly, *Drosophila*, is the most extensively-studied invertebrate model system in terms of the genetics of court-ship and mating behaviors. In *Drosophila* these behaviors are multimodal and involve visual, auditory, and chemo-sensory signals. During the courtship ritual the male orients towards a female, follows her and taps her abdomen with his forelegs. He then performs a courtship song by rap-idly vibrating one of his wings. This is followed by genital licking and attempts at intromission. The female can either reject the male by walking away, flicking her wings, kick-ing with her legs or extruding her ovipositor, or she can allow copulation to occur (Figure 12.5, overleaf). Whereas the courtship ritual is characteristic for each *Drosophila* species, there is considerable variation among species. For example, acoustic characteristics, duration, and repertoire of the courtship song vary extensively among different *Drosophila* species, and females recognize the difference and respond only to males of their species.

The perception of the sound waves that are produced by the wing vibrations is mediated via an acoustic organ located in the antenna. The antenna consists of three seg-ments: the most proximal segment is the **scape**; the mid-dle segment, which houses mechanosensory neurons of **Johnson's organ** is the **pedicel**; and the third antennal seg-ment, also known as the **funiculus**, is the main olfactory organ of the fly. A feather-like structure, the **arista** arises

from the base of the third antennal segment, and communi-cates with the mechanosensory neurons of Johnson's organ. Sound vibrations cause reverberations of the arista that are transmitted to the mechanosensory neurons (Figure 12.6, overleaf).

Mutations that impair the acoustic organ and disrupt the perception of courtship song, such as *aristaless* (*al*) and *thread* (*th*), interfere with normal courtship behavior. A num-ber of neurodevelopmental genes, such as *atonal* (*ato*) and *beethoven* (*btv*), affect the structure of Johnson's organ, and thus impact auditory communication during courtship. A large number of genes alter the courtship song itself. One of the first such genes to be discovered was the circadian clock gene *period* (*per*) (see also Chapters 1 and 11). Genes in the **sex determination** pathway (*transformer* (*tra*), *doublesex* (*dsx*) and *fruitless* (*fru*)) also affect courtship song (Figure 12.7, overleaf), as do mutations that affect the wing muscu-lature (*croaker* (*cro*) and *ariadne* (*ari*)) necessary to perform the song.

Sexual dimorphism in the brain is controlled down-stream from *tra* by the *fruitless* (*fru*) gene. The *fruitless* gene exhibits sex-specific alternative splicing, and alterna-tively-spliced products are expressed in sexually dimorphic neural circuits in the brain. The *fru* mRNAs are transcribed in both males and females, but the Fru protein is only trans-lated in the male brain, and is essential for the expression of normal courtship behavior. When the male-spliced prod-uct is expressed in females, they behave as males, and *vice versa*, not only in terms of courtship behavior, but also in aggressive behavior. Male *fru* null mutants form characteris-tic rows of males attempting to mate with one another, indi-cating that sex discrimination has been abolished.

In addition to visual and auditory cues, **chemosensory signals** play a role in courtship. Whereas many insects, such as moths, release volatile mixtures of pheromones to attract mates from long distances, the best-characterized pheromones in *Drosophila* are **contact pheromones** that are incorporated in the cuticle. These contact pheromones are nonvolatile, long chain hydrocarbons, and their com-position is different in males and females. The cuticular hydrocarbons produced by female flies contain two double bonds, and 27 or 29 carbons (*cis*,*cis*-7,11-heptacosadiene and *cis*,*cis*-7,11-nonacosadiene). Small amounts of these two compounds together can elicit vigorous male courtship behavior. In contrast, male flies synthesize cuticular hydro-carbons that are 23 and 25 carbons long, with only a single double bond (*cis*-7-tricosene and *cis*-7-pentacosene). The *cis*-7-tricosene acts as an antiaphrodisiac, which can inhibit male excitation, and is transferred onto the female's cuticle during mating, which helps make mated females temporar-ily less attractive to males (Figure 12.8, p. 191).

The importance of cuticular pheromones in mediating sex discrimination and mate recognition was demonstrated elegantly by Jean-François Ferveur and his colleagues at the University of Burgundy in Dijon (France). They made transgenic flies in which expression of the *transformer* gene, a component of the sex determination pathway, was driven in **oenocytes**. These are the cells that produce the pheromone blend. Consequently, Ferveur's transgenic males produced female-characteristic pheromones. Such transgenic males themselves exhibit normal courtship behavior towards females, but they elicit homosexual courtship from other males. These experiments vividly demonstrate the importance of cuticular chemosensory signals in recognition of reproductive partners.

Further insights in pheromonal communication during courtship in *Drosophila* came from the observation that male forelegs contain about 20 male-specific taste sensilla

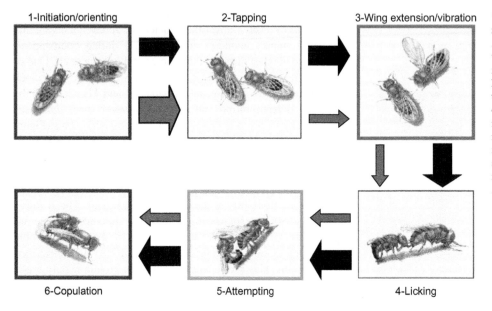

FIGURE 12.5 The courtship ritual in *Drosophila melanogaster*. The female ultimately decides whether to mate with the courting male or not. She can walk away, kick the male off, or allow copulation to occur. Whereas the sequence of events during the courtship ritual is often described as stereotypic, individual variations occur, with males sometimes short-cutting or skipping elements of the mating ritual. (Modified from Bray, S., and Amrein, H. (2003). *Neuron*, **39**, 1019–1029).

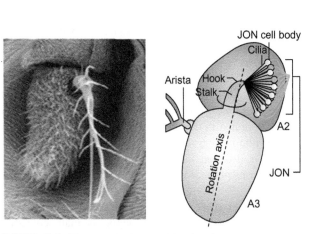

FIGURE 12.6 The antenna of *Drosophila*. The left panel shows the third antennal segment, with olfactory sensilla and the arista protruding from its base. On the right is a schematic representation of Johnston's organ (JON). Vibrations that impinge on the arista on antennal segment 3 (A3) cause the associated stalk and hook to vibrate rotationally, thereby stimulating neurons (scolopedia) of Johnston's organ within segment 2 (A2), which communicate with the auditory nerve. Movement of the arista stretches the scolopedia in one direction, while relaxing them in the opposite direction. Johnston's organ is functionally subdivided. The scolopedia in the proximal region of Johnston's organ are sensitive to wind and gravity, whereas the more distal scolopedia nearer the third antennal segment are sensitive to sound vibrations. (Diagram on the right is modified from Christensen, A. P., and Corey, D. P. (2007). *Nat. Rev. Neurosci.*, **8**, 510–521.)

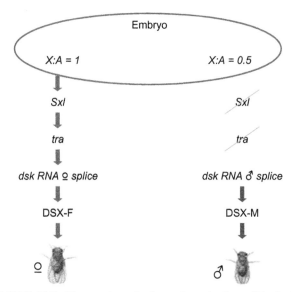

FIGURE 12.7 The sex determination pathway in *Drosophila*. In the presence of two X chromosomes in a diploid cell, the *Sex lethal* (*Sxl*) gene is active and drives expression of *transformer* (*tra*). The *tra* gene product results in splicing of the *doublesex* (*dsx*) mRNA to generate a protein that directs female development, while suppressing development of male structures. If the diploid cell has only one X-chromosome, *Sxl* and *tra* are not activated and a *dsx* splice variant results that causes male development and suppresses female development. A parallel pathway that involves *tra* results in sex-specific splicing of the *fruitless* transcript, described in the text.

that express a particular **gustatory receptor**, Gr68a (see also Chapter 13). Neurons that express Gr68a could be inactivated by targeted expression of tetanus toxin. In addition, the expression of the *Gr68a* gene could be knocked-down by targeted expression of an interfering RNAi. In either case, courtship behavior was disrupted early in the courtship ritual, with reduced wing vibration that normally follows tapping of the female by the male with its **tarsi**, or forelegs. It appears, therefore, that Gr68a is a male-specific gustatory receptor essential for the recognition of at least one of the female cuticular pheromones (Figure 12.9).

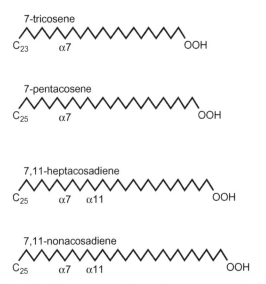

FIGURE 12.8   Male (top two) and female (bottom two) cuticular hydrocarbons of *Drosophila melanogaster*.

An excellent example of the power of transgenic approaches in *Drosophila* comes from elegant studies that have used homologous recombination to gain insights into pheromonal control of mating behavior in *Drosophila*. Barry Dickson and colleagues in Vienna generated homologous recombinant flies in which they replaced the open reading frame of the gene that encodes the *Drosophila* Or67d odorant receptor with GAL4. Homozygous null mutant males that lack *Or67d* inappropriately court other males, whereas mutant females are less receptive to courtship. Dickson demonstrated that these effects are mediated via a male-specific pheromone, 11-*cis*-vaccenyl acetate. Because the *GAL4* construct that replaces the *Or67d* open reading frame is located at exactly the same chromosomal location, the *Or67d* promoter will faithfully drive expression of GAL4 in cells that would normally express the odorant receptor. This enabled Dickson and his colleagues to use a green fluorescent protein reporter gene to identify olfactory neurons in trichoid sensilla in the third antennal segment as the cells that normally express *Or67d*. These sensilla also express the LUSH odorant-binding protein that interacts with 11-*cis*-vaccenyl acetate, mentioned in Chapter 9. In addition, Dickson and his colleagues could restore sensitivity to 11-*cis*-vaccenyl acetate by introducing a UAS-Or67d transgene in the homologous recombinant background. Moreover, instead of *Or67d*, expression of moth pheromone receptors could be directed to the neurons that normally express *Or67d*. In such transgenic flies, olfactory sensory neurons in the trichoid sensilla became responsive to the corresponding moth pheromones, and flies displayed behavioral responses to these moth pheromones that mimicked the normal response to 11-*cis*-vaccenyl acetate. Their work

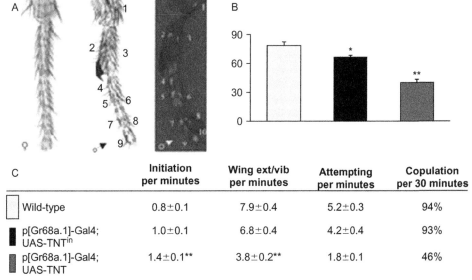

	Initiation per minutes	Wing ext/vib per minutes	Attempting per minutes	Copulation per 30 minutes
Wild-type	0.8±0.1	7.9±0.4	5.2±0.3	94%
p[Gr68a.1]-Gal4; UAS-TNT^in	1.0±0.1	6.8±0.4	4.2±0.4	93%
p[Gr68a.1]-Gal4; UAS-TNT	1.4±0.1**	3.8±0.2**	1.8±0.1	46%

FIGURE 12.9   The Gr68a gustatory receptor is essential for courtship and mating in *Drosophila*. Panel A shows the expression of the *Gr68a* gene in cells of the forelegs of male, but not female, flies by driving a green fluorescent protein (GFP) reporter gene under the *Gr68a* gene promoter. When this promoter drives tetanus toxin, courtship is significantly reduced, as evident from a reduced percentage of males courting females (panel B, dark gray bar) compared to wild-type flies (light gray bar) or flies expressing an inactive form of tetanus toxin (black bar). Panel C shows that inactivation of the *Gr68a* expressing cells by transgenic expression of tetanus toxin reduces courtship by interfering with the tapping stage of the courtship ritual. (See also Figure 12.5, modified from Bray, S., and Amrein, H. (2003). *Neuron*, **39**, 1019–1029.)

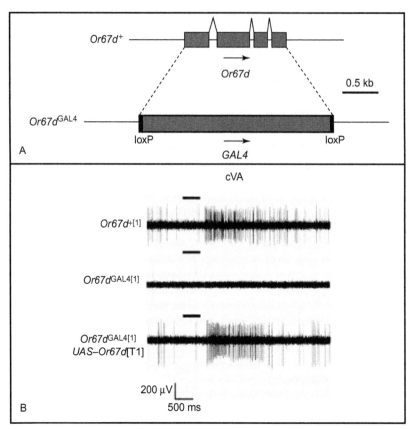

FIGURE 12.10   Replacement of the *Or67d* coding region with GAL4 by transposon-mediated homologous recombination (A). Mutant flies no longer respond electrophysiologically to application of 11-*cis*-vaccenyl acetate (cVA) to the trichoid sensilla (middle trace in panel B), but their responses can be restored when a *UAS-Or67d* transgene is introduced (lower trace, panel B). When green fluorescent protein (GFP) is expressed via the *Or67d* promoter in these homologous recombinant flies, olfactory neurons in trichoid sensilla in the antenna and the glomerulus to which their axons project in the antennal lobe can be visualized. (Adapted from Kurtovic, A. *et al.* (2007). *Nature*, **446**, 542–546.)

showed that activation of a single class of olfactory sensory neurons is both necessary and sufficient to mediate behavioral responses to 11-*cis*-vaccenyl acetate (Figure 12.10).

Studies on *Drosophila* have shown that courtship behavior involves complex genetic programs, and consists of modular "behavioral units" (such as locomotion, song production, copulation) that each depend on a complex genetic architecture. The interchange between and the controlled temporal expression of genes that govern each behavioral modality is essential to enable successful courtship and mating. Modifications of these behaviors are expected to accompany speciation, as alterations in species-specific courtship rituals may erect barriers for reproduction; thus, changes in courtship and mating behaviors may contribute to **prezygotic isolation**.

Postmating Control of Reproductive Behavior in *Drosophila*

In *Drosophila melanogaster*, mating causes profound physiological changes in the female, including a decrease in remating (mated females are less receptive and less attractive to other males), stimulation of feeding and egg production, and a reduction in lifespan. These effects are mediated by polypeptides that are produced in the male's accessory glands and transferred with the sperm into the female's reproductive system (Figure 12.11). Because they gain access to the circulation, these peptides exert their effects not only locally in the reproductive tract, but also via the peripheral and central nervous system. These chemical signals, which are transmitted in the seminal fluid, are known as **accessory gland proteins** or **Acps**. They comprise a group of about 80 different proteins that have been studied extensively by Mariana Wolfner and her colleagues at Cornell University. The functions of most Acps are not known, but several have been characterized in detail. This has been achieved by ablation of the accessory glands through the targeted transgenic expression of diphtheria toxin subunit A. The effects on females mated to ablated males and intact males could then be compared. Another experimental approach consists of assessing the physiological consequences of constitutive expression of male accessory gland proteins in transgenic female flies. Finally, RNAi-mediated knock-down of the expression of specific Acps has provided information about their functions.

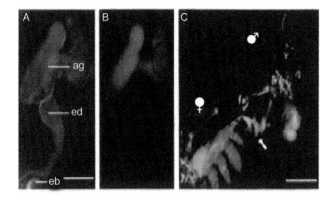

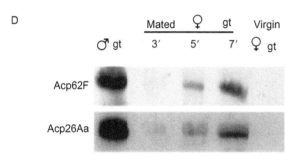

FIGURE 12.11 Transfer of male accessory gland proteins into the female reproductive tract during mating. Panel A shows a Nomarski image of the male reproductive tract with the accessory gland (ag), the ejaculatory duct (ed), and the ejaculatory bulb (eb). This male produces Acp26Aa labeled with green fluorescent protein (GFP), which can be seen in the accessory gland (panel B). Panel C shows the transfer of Acp26Aa-GFP from the male into the female reproductive tract of two flies pulled apart 10 minutes after the start of copulation. Panel D shows Western blots probed with antibodies against Acp26Aa and Acp62F. Both antibodies show heavy staining in the male genital tract (gt) and in mated females at 3, 5, or 7 minutes after the start of mating. There is no staining in virgin females. (Modified from Lung, O., and Wolfner, M. F. (1999). *Insect Biochem. Mol. Biol.*, **29**, 1043–1052.)

Acps serve specialized functions, all designed to promote the inseminating male's reproductive success. For example, Acp36DE is required for effective sperm storage, while Acp26Aa (also called **ovulin**) stimulates release of oocytes from the ovary (the number behind the Acp designation indicates the cytological location of the *Acp* gene on the chromosome). Transgenic females that express Acp26Aa ovulate at elevated rates. Thus, Acp proteins ensure that females only invest the energy to generate oocytes and release eggs when there is sperm available to fertilize them.

Since sperm can be stored in the spermatheca of the female reproductive tract, multiple matings give rise to **sperm competition**. It is, therefore, advantageous for males to render females less receptive to subsequent matings. Some seminal peptides, Acp70A (also known as the "sex peptide") and Dup99B, induce rejection behavior towards courting males in mated females. This reduced receptivity persists for up to 11 days, during which period the female uses most of the stored sperm to lay eggs. The sex peptide also stimulates feeding behavior in mated females. In addition, seminal fluid contains antimicrobial peptides, which may help protect eggs after oviposition, contributing further to the male's reproductive success. The sex peptide mediates these profound physiological effects by binding to G-protein-coupled receptors on neurons that innervate the uterus and lower oviduct and project to the abdominal ganglia, which contain neural circuits that control egg laying, and the suboesophagial ganglion, which is associated with perception of the male courtship song. This neural circuit represents a switch that controls the transition of the female from a premating to a postmating state.

During mating about 4000–6000 sperm cells are transferred, and about 25% of these are stored in the female's spermatheca and seminal receptacle. This allows a female to produce progeny for 14 days after a single mating, enabling a single fly to produce about 600–800 offspring from one mating. It is possible that the large energy investment required to produce eggs contributes to the shortened lifespan of mated females.

AFFILIATIVE BEHAVIOR

Fitness considerations must take into account not only the production of offspring, but also their survival. Nurturing offspring either by parents or closely-related kin will facilitate their survival and the further propagation of the parental alleles.

Most mammals, including humans, are polygamous and mate promiscuously. Only about 3% of mammalian species are monogamous and faithful to a single partner during their lifetime. Promiscuous mating affords many mating opportunities with any available partner, whereas monogamy ensures that a mating partner, once established, is always available without strong competition pressure from other males. The central pathways in the brain that contain the neurohypophyseal peptides oxytocin and vasopressin have been implicated in a number of social behaviors, including sexual behavior, maternal behavior, affiliation, social memory, territorial behavior, and aggression (Figure 12.12). Significant insights in the genetic and neuroendocrine control of affiliative behavior have come from studies on closely-related species of microtine rodents (voles).

The prairie vole (*Microtus ochrogaster*) is highly social, and forms lasting pair bonds after mating. Both parents take care of the offspring, and the young prairie voles remain in the parental nest for several weeks beyond weaning. In contrast, the montane vole (*Microtus montanus*), is asocial, and breeds promiscuously. Breeding partners do not form a pair bond after mating. In addition, males do not participate in providing parental care, and the females abandon their offspring in the second or third postnatal week (Figure 12.13). What are the neural and genetic mechanisms that determine these vastly different lifestyles among related species?

Many studies have implicated the hormones vasopressin and oxytocin in the control of monogamous behavior, and receptors for vasopressin (**V1a receptors**) and oxytocin have been identified in the brain. The distribution of receptors for these hormones is strikingly different in prairie voles and montane voles. Autoradiographic ligand-binding studies show that prairie vole brains express high levels of V1a receptors in the **ventral pallidum** and the **diagonal band**, whereas montane voles have high V1a receptor expression in the **lateral septum**. Furthermore, compared to montane voles, prairie voles have increased levels of oxytocin receptors in the prelimbic cortex and the **nucleus accumbens**, regions that are involved in the mesolimbic dopaminergic reward pathway (Figure 12.14).

Are different expression patterns of oxytocin and vasopressin receptors causal of the differences in affiliative behavior? A conclusive affirmative answer to this question came from elegant transgenic studies in mice. Molecular studies suggested that differences in the promoter of the

---

Arginine vasotocin (AVT)
Cys-Tyr-Ile -Gln-Asn-Cys-Pro-Arg-Gly

Arginine vasopressin (AVP)
Cys-Tyr-Phe -Gln-Asn-Cys-Pro-Arg-Gly

- Fishes
  - Killifish: stimulates spawning behavior
  - Plainfin midshipman: influences vocalization
  - White perch: increases 'attending' behavior
  - females: inhibits sexual rejection, stimulates phonotaxis, oviposition
- Reptiles: stimulates nest digging, oviposition
- Birds: stimulates sexual activity, influences aggression
- Mammals: stimulates monogamous and parental behaviors in males, territorial behaviors
- Amphibians
  - males: stimulates clasping and/or calling

FIGURE 12.12   Behavioral effects of arginine vasotocin and its closely-related counterpart arginine vasopressin.

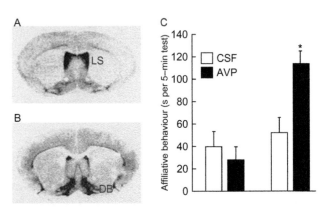

FIGURE 12.14   Distribution of vasopressin 1a (V1a) receptors and behavioral responses to arginine-vasopressin (AVP) in montane and prairie voles. Note the binding of radiolabeled AVP in the lateral septum (LS) of the montane vole (A) but not the prairie vole (B), and in the diagonal band (DB) of the prairie vole (B) but not the montane vole (A). Panel C shows that male prairie, but not montane, voles exhibit elevated affiliative behavior (time spent with mate) after direct administration of AVP into the brain (CSF indicates cerebrospinal fluid). (From Young et al. (1999). *Nature*, **400**, 766–768.)

The prairie vole (*Microtus ochrogaster*)	The montane vole (*Microtus montanus*)
Highly social	Solitary
Monogamous	Promiscuous
Forms lasting pair bonds after mating	Does not form a pair bond after mating
Both parents take care of the offspring	Males do not participate in parental care
Young prairie voles remain in the parental nest for several weeks beyond weaning	Females abandon their offspring in the second or third post-natal week

FIGURE 12.13   Social differences between monogamous prairie voles and polygamous montane voles.

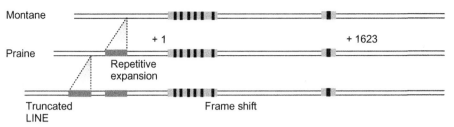

FIGURE 12.15   Microsatellite repeats in the promoter region of the vasopressin 1a receptor gene. Black boxes represent exons, gray boxes bordering exons represent introns. (From Hammock, E. A., and Young, L. J. (2005). *Science*, **308**, 1630–1634.)

vasopressin1a receptor gene might contribute to species differences in expression of this receptor. A V1a receptor transgene from the prairie vole was introduced into mice, and was expressed in these transgenic mice in a pattern that recapitulated its expression in the prairie vole. When arginine–vasopressin was injected into the ventricles of the brains of these transgenic mice, they exhibited increased affiliative behavior. These results demonstrated that the pattern of V1a-receptor gene expression in the brain is functionally associated with social behaviors.

The DNA sequence of the vasopressin gene and its protein sequence are highly-conserved. In contrast, while conserved at the protein level, V1a receptors show diversity at the level of gene structure. The 5′ flanking region of the V1a receptor gene contains a microsatellite region that differs significantly between monogamous and promiscuous vole species. A 428 bp insert is present in monogamous voles (*Microtus ochrogaster* and *M. pinetorum*), and absent in polygamous voles (*M. montanus* and *M. pennsylvanicus*) (Figure 12.15). Introduction of a V1a receptor transgene containing this 428bp insertion by viral transfer in the brains of polygamous montane voles altered the expression pattern of the V1a receptor, rendering it similar to the expression pattern seen in the brains of prairie voles. Moreover, such transgenic animals underwent a concordant switch in affiliative behavior and mating habits. This is a remarkable result, as it shows that a single insertion/deletion polymorphism in a regulatory region of a key gene can result in dramatic changes in the manifestation of a complex behavior (Figure 12.16). This is an important discovery, as it questions the often-held intuitive assumption that the evolution of complex behaviors requires the gradual accumulation of a vast number of mutations.

It is clear that the oxytocin and vasopressin systems do not act in isolation, but are an integral part of a more complex hormonal network that regulates social behavior. Female mice in which the genes for either oxytocin, or the α- or β-estrogen receptor, had been deleted by homologous recombination showed similar defects in social behaviors, including impaired social recognition and unusual anxiety in social situations. Estrogen acting through the β-estrogen receptor regulates the production of oxytocin in the paraventricular nucleus of the hypothalamus, whereas estrogen

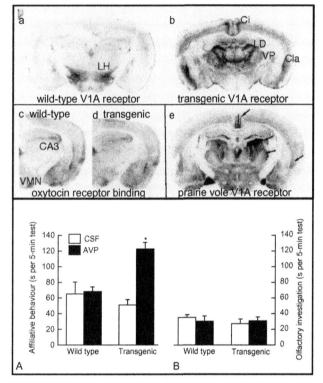

FIGURE 12.16   Expression of a prairie vole vasopressin (V1a) receptor in transgenic mice recapitulates the expression pattern of the V1a receptor in the prairie vole (upper panel, lower right), whereas there is no difference in the binding of oxytocin to its receptor between wild-type and transgenic mice. Binding of radioactive peptides is visualized by autoradiography, and V1a receptor-binding is seen in the cingulate cortex (Ci), laterodorsal (LD), and ventroposterior (VP) thalamic nuclei, claustrum (Cla) and several other regions of the prairie vole and transgenic mouse brains. The lower panel shows that intracerebral administration of arginine vasopressin (AVP) in the brains of transgenic mice induces affiliative behavior not seen with intracerebral infusion of cerebrospinal fluid (CSF) as a control and measured as time spent with a conspecific partner (A). There is no difference in investigative olfactory behavior (B). (From Young, L. J. et al. (1999). *Nature*, **400**, 766–768.)

acting through the α-estrogen receptor drives expression of oxytocin receptors in the amygdala (Figure 12.17, overleaf). It appears that the balance of this hormonal system influences the expression of appropriate social behaviors between aggressive or affiliative.

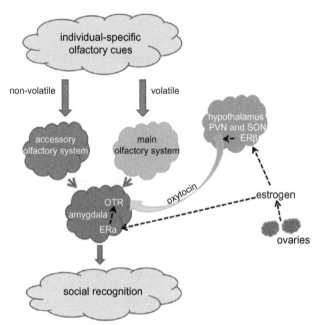

FIGURE 12.17 The role of estrogen in social recognition by regulating the oxytocin system. Studies with estrogen receptor knock-out mice indicate that ovarian estrogens regulate oxytocin (OT) secretion from the **paraventricular nucleus** (PVN) and the **supraoptic nucleus** (SON) of the hypothalamus through binding to the β-estrogen receptor (ER-β). Oxytocin reaches the amygdala through axonal projections of the PVN neurons, where estrogens regulate the expression of oxytocin receptors through binding to the α-estrogen receptor (ER-α). Activation of oxytocin receptors (OTR) is thought to facilitate social recognition. (Diagram based on Choleris, E. et al. (2003). *Proc. Natl. Acad. Sci. USA*, **100**, 6192–6197.)

AGGRESSION AND THE ESTABLISHMENT OF SOCIAL HIERARCHIES

Aggressive encounters are important for establishing social hierarchies. The establishment of such social orders will prevent future fights in the allocation of limited resources. Access to the most desirable habitats, food resources, and reproductive partners will confer the greatest degree of genetic fitness. Whereas normal levels of aggression are advantageous, excessive aggression may lead to neglect of other survival activities, such as foraging and parental care. Thus, from an evolutionary perspective, aggressive behavior is expected to be under stabilizing selection. In the following sections we will describe how crustaceans, mice, and *Drosophila* have served as model systems to investigate the neural and genetic underpinnings of aggression.

Crustaceans: A Model for Aggression

Lobsters have been used as model systems to dissect neural circuits that mediate aggression. Lobsters, in the wild, fight to establish dominance and to defend their shelters. Such aggressive encounters can be brief, or can escalate to full-blown fights. The loser retreats as the winner establishes dominance. The social hierarchy that is established as a result of such encounters is stable. Losers will not challenge an animal again that previously defeated them.

Edward Kravitz and his colleagues at Harvard University showed that serotonin and octopamine are central mediators in the neural circuitry that directs the establishment of social dominance relationships. High levels of serotonin in lobsters are associated with aggression and dominance, whereas high levels of octopamine are associated with subordinate status. The behavior of lobsters can be profoundly influenced by pharmacological treatments that interfere with serotonin metabolism. When subordinate animals are treated with fluoxetine (Prozac), which promotes serotonergic neurotransmission by inhibiting serotonin reuptake, they persist in agonistic encounters with dominant opponents that previously defeated them. This unusual behavior is not observed in animals that have not been treated.

Whereas lobsters provide a simple model system for studying neural mechanisms of aggression, they are not amenable to genetic manipulations. To understand the genetic architecture that determines aggressive behavior, mice and *Drosophila* have become the model systems of choice.

Aggression in Mice

Serotonin appears to be a central modulator of aggression across phyla. However, in contrast to crustaceans where high levels of serotonin are associated with increased aggression, in vertebrates low levels of serotonin predispose to elevated aggression. Both in mice and in humans, serotonin suppresses aggressive behavior, whereas dopaminergic and noradrenergic pathways facilitate aggression. The effects of serotonin on aggression are mediated via the 5-HT$_{1A}$ and 5-HT$_{1B}$ receptors. Male mice that lack 5-HT$_{1B}$ receptors are hyperaggressive. In contrast, deletion of the 5-HT$_{1A}$ receptor through homologous recombination results in reduced levels of aggression. Similarly, increasing serotonin levels through deletion of the serotonin reuptake transporter results in reduced levels of aggression. The neural pathways that have been implicated in aggressive behavior in mice involve the amygdala, prefrontal cortex, nucleus accumbens, ventral striatum, anterior cingulate gyrus, and hypothalamus.

Several other neurotransmitters have also been implicated in aggressive behavior in mice, including nitric oxide and GABA (see also Chapter 3). The laboratory of Randall Nelson at Johns Hopkins University reported that homologous recombinant mice that lack the neuronal nitric oxide synthase (nNOS) gene show extremely violent behavior, killing their cage mates. These nNOS knock-out mice also show reductions in the levels of serotonin, again consistent with serotonin as a key player in mediating aggressive behavior.

Box 12.3   The Bruce effect and conspecific recognition

In 1959, H. M. Bruce discovered that when a pregnant female mouse is exposed to a strange male she will abort the pregnancy. This pregnancy block effect does not require the physical presence of the male; it is also elicited when the pregnant female is exposed to the male's urine or soiled bedding. This led to the hypothesis that the Bruce effect is likely mediated by pheromones that provide a chemical signature of individual identity. Kunio Yamazaki at the Monell Chemical Senses Center reported that mice can discriminate individuals from a different strain, based on allelic differences in genes of the **major histocompatibility complex** (MHC), long-recognized to encode proteins that serve as molecular identity tags that regulate immune rejection of transplanted tissues. Yamazaki's experiments did not gain general acceptance until many years later, when Frank Zufall and Trese Leinders-Zufall at the University of Maryland showed that purified peptides isolated from male mouse urine and derived from the MHC can mediate the pregnancy block effect. Such peptides are likely proteolytic by-products that are generated during normal turnover of MHC proteins and end up in small amounts in the urine. The pregnancy block effects are highly peptide-specific, and mediated via the class of **V2R pheromone receptors** in the vomeronasal organ (Figure A). In addition, olfactory sensory neurons in the main olfactory epithelium also contribute to individual recognition signaled by these peptides. Since the pregnant female also has to perceive sex differences, other signals and neural mechanisms are also likely to be involved in kin recognition.

What are the benefits of pregnancy block? It has been suggested that the foreign male benefits, since mice can go into estrus within 1–4 days following termination of the pregnancy, thereby affording a mating opportunity for the male. It has also been suggested that it may be advantageous for the female to have offspring with the new male, as he might otherwise kill her original offspring and energy invested in completing the pregnancy would be wasted. In this scenario, pregnancy block might have evolved as a female counterstrategy to infanticide by the unrelated male. Note, however, that these hypotheses are speculative, and that the Bruce effect has been observed exclusively in the laboratory. It is not known how commonly pregnancy block occurs in nature.

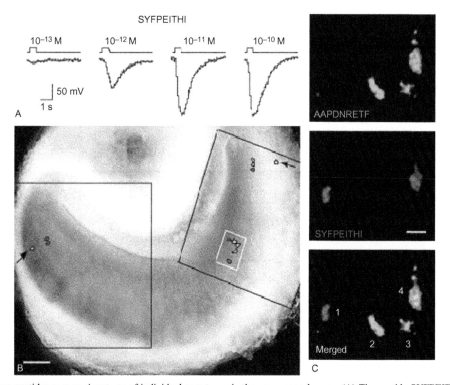

FIGURE A   Urinary peptides convey signatures of individual genotype via the vomeronasal organ. (A) The peptide SYFPEITHI activates vomeronasal sensory neurons with high sensitivity, as evident from summed field potentials evoked by brief pulses of increasing concentrations of ligand. (B) Peptide-induced activation of vomeronasal neurons in tissue slices of the vomeronasal organ visualized by calcium imaging for peptide AAPDNRETF (10^{-12} M) and SYFPEITHI (10^{-12} M). Cells responding to both peptides are shown in the merged image. Black boxes indicate regions that were imaged in these experiments; the area within the white box is shown at higher magnification in (C) (scale bars: $100\,\mu$m in (B); $10\,\mu$m in (C)). (From Boehm, T., and Zufall, F. (2006). *Trends Neurosci.*, **29**, 100–107).

Gene-mapping studies have been performed to identify chromosomal regions (QTLs) that harbor genes for aggressive behavior. Such studies used recombinant inbred strains of mice that displayed variation in aggressive behavior and identified several QTLs. Although the genes responsible for this phenotypic variation within these QTL regions could not be readily identified, the large number of QTLs that were identified underscores the complex genetic architecture of aggressive behavior.

What are the social triggers that elicit aggressive behavior in mice? Both internal chemical signals, such as androgens and estrogens, and external chemical cues interact to elicit aggressive behavior. Maternal aggression has a hormonal component, but is also dependent on pheromonal cues from conspecifics. Male mice will defend their home cages against intruders, and individual recognition is also dependent on pheromonal cues for these social interactions. Pheromone recognition via the **vomeronasal**

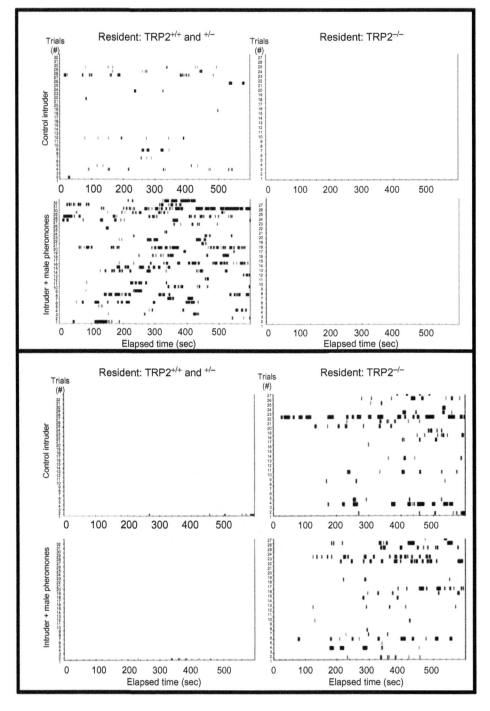

FIGURE 12.18 Abnormal behaviors in Trp2 knockout mice. Left panels show intact controls, right panels, homozygous mutants. The bottom panels show results when the intruder is painted with male pheromone. (A) Homologous recombinant mice that lack the TRP2 channel do not display male–male aggression in a resident/intruder assay. Each black square represents an attack of the resident mouse against the intruder. (B) Resident male TRP2 knock-out mice display sexual behavior toward male intruders. Each black square represents a mounting attempt. (From Stowers, L., Holy, T., Meister, M., Dulac, C., and Koentges, G. (2002). *Science*, **295**, 1493–1500.)

organ plays an important role in mediating aggressive behavior.

Signal transduction in vomeronasal neurons is mediated via the TRP2 channel, a member of the family of **TRP** (transient receptor potential) **channels** which is expressed exclusively in the vomeronasal organ. Catherine Dulac and colleagues at Harvard University generated mice in which the gene that encodes the TRP2 channel had been deleted. Urine-derived pheromones did not elicit electrophysiological responses in the vomeronasal organs of these mice (although other investigators have reported residual activity). Male mice in which vomeronasal input had thus been abolished showed no aggression toward male intruders. Most interestingly, they initiated courtship and sexual advances equally toward male and female partners, as if they had lost their ability for gender discrimination. Thus, in the absence of a functional vomeronasal organ, mating behavior in mice seems to be a default behavior indiscriminant of gender. The vomeronasal organ apparently determines aggressive behavior toward male conspecifics, and is permissive for mating behaviors with regard to conspecific females, while inhibiting mating with males (Figure 12.18).

To begin to identify the pheromone receptors that recognize cues that trigger aggressive behavior, Peter Mombaerts and his colleagues at the Rockefeller University used an innovative genetic approach by generating mice that carried a large deletion of an entire cluster of 16 genes that encode putative pheromone receptors of the V1R type. As there were no intervening genes within this cluster, these mice were homozygous viable, appeared normal, and showed few behavioral abnormalities. However, they showed reduced maternal aggression, suggesting that pheromonal cues that interact with one or more of the deleted receptors provide essential cues for eliciting maternal aggressive behavior (Figure 12.19).

Subsequently, Lisa Stowers and her colleagues at the Scripps Research Institute in La Jolla purified two major urinary proteins from male mouse urine, and showed that each of these proteins was responsible for the aggressive behavior observed against a castrated male painted with male urine. These male–male aggressive responses, however, were mediated via **V2R receptors**, rather than V1R receptors, in the vomeronasal organ (see also Chapter 13).

Aggression in *Drosophila*

Although one may not expect fruit flies to be aggressive animals, aggressive behaviors in *Drosophila* have been described as early as 1975 by Dow and von Schilcher. Flies will display aggression in defense of territory, scarce food resources, or females. Aggressive encounters involve charging, tussling, chasing, kicking, head butts, wing threats, and in extreme cases, boxing with the forelegs (Figure 12.20). Studies by Ari Hoffmann and colleagues

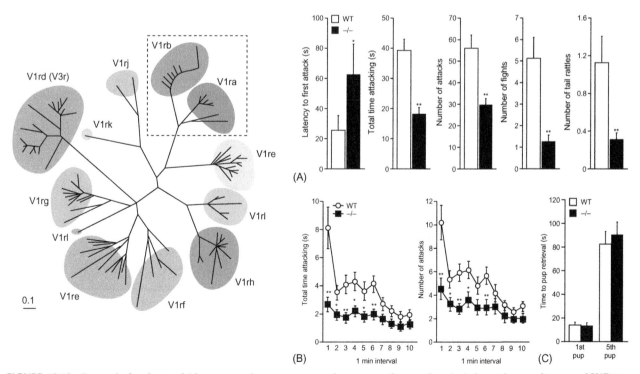

FIGURE 12.19 Removal of a cluster of 16 vomeronasal receptor genes reduces maternal aggression. A phylogenetic tree of groups of **V1R-type vomeronasal receptors** is shown on the left. The genes encoding receptor classes that were eliminated by chromosome engineering are indicated in the dotted box. Panels on the right show that maternal aggression against intruders in terms of latency to attack and number of attacks is greatly reduced in homologous recombinant mice, whereas other maternal behaviors, exemplified by the time it takes to retrieve pups that were removed from the nest, are not affected. (Modified from Del Punta, K. et al. (2002). *Nature*, **419**, 70–74.)

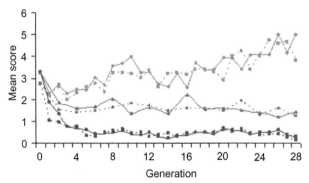

FIGURE 12.21   Artificial selection for increased and reduced aggressive behavior in *Drosophila melanogaster* from a heterogeneous base population. Duplicate lines selected for high aggression ("bullies') are shown in top solid and dashed lines, while lines selected for low aggression are shown in bottom solid and dashed lines ("wimps"). The center lines show unselected control lines to assess genetic drift. The mean score is the mean number of aggressive encounters during the assay period. (From Edwards et al. (2006). *PLoS Genetics*, **2**, e154.)

FIGURE 12.20   Fighting flies.

showed that there is substantial variation for levels of aggression among different Australian populations of flies.

Although several mutants were identified that show increased levels of aggression, including mutations in genes associated with the sex determination pathway (*fruitless* and *dissatisfaction*) and the dopamine biosynthetic pathway, which is also associated with pigmentation (*ebony*), *Drosophila* has only recently become a useful model for large-scale studies on the genetic architecture of aggression. A major impediment has been the lack of a rapid and reliable assay that could measure aggression quantitatively. Video recordings of aggressive encounters were cumbersome, and of limited use for large-scale studies. Genetic background variation can affect aggressive behavior; males and females differ in the extent and manifestation of aggressive behavior and, when aggression is assessed when males fight over females, mating behavior becomes a confounding factor.

The laboratory of Trudy Mackay at North Carolina State University developed a quantitative high-throughput behavioral assay for aggression, simply by measuring the total number of aggressive encounters of previously food-deprived males during a two-minute period after they are placed in a test arena that contained a droplet of food. This assay has opened the door for QTL mapping and large screens of *P*-element insertion lines. Expression microarray experiments of artificially-selected flies for high aggression ("bullies") and low aggression ("wimps") showed differential transcript abundance in a large portion of the genome (Figure 12.21). It is perhaps not surprising that the genetic factors that determine aggressive behavior and underlie variation in aggression in natural populations

are complex, and that understanding the genetic architecture of aggression requires us to look beyond the few neurotransmitter components that have been the focus of most studies to date.

Aggressive behavior in people

It is surprising how little we have learned about the genetic factors that determine aggressive behavior, despite the profound social impact of aggression in human populations. Pathological levels of aggression result in socially-disruptive, violent behaviors. Increased aggression is well documented for alcoholics, patients suffering from neurodegenerative disorders (such as **Alzheimer's disease**), and patients with psychiatric disorders. Like all behaviors, aggression is a complex trait, and its manifestation is determined both by genetic factors and by the interplay between the genome and the environment (e.g. exposure to violence during childhood). Furthermore, suicide can be regarded as a self-directed form of aggression. The social and economic costs that result from violent behavior and efforts to control it are enormous, and it is therefore self-evident that understanding the genetic and neurobiological basis of aggression is important for human welfare.

In parallel with studies on aggressive behavior in mice, serotonin has featured as the focal point for studies on human aggressive behavior. Caspi and colleagues in London performed a classic study. These investigators asked why some children who were maltreated during childhood develop antisocial behavior, whereas others do not. They recruited a large sample of maltreated boys and found that a polymorphism in the gene encoding **monoamine oxidase A**, which metabolizes bioamines, was associated with the development of antisocial behavior in these children. This polymorphism consists of a **variable number tandem**

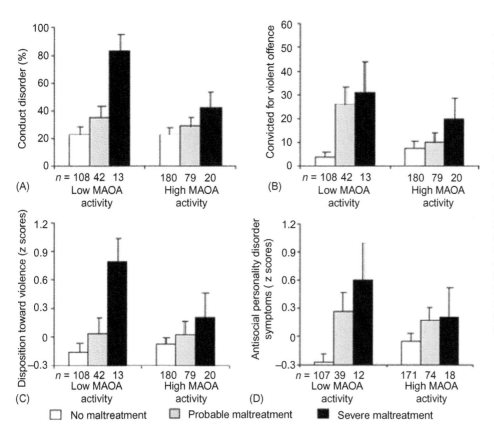

FIGURE 12.22 Association between childhood maltreatment and subsequent antisocial behavior as a function of activity of monoamine oxidase A (MAOA). MAOA activity is defined as expression level of MAOA transcript. Panel A shows the percentage of males meeting the criteria for conduct disorder at age 10–18. Panel B shows the percentage of males convicted of a violent crime by age 26. Panel C shows scores on a psychiatric scale that measure disposition toward violence at age 26. Panel D shows scores on the Antisocial Personality Disorder symptom scale at age 26. Note that maltreatment during childhood predisposes towards aggressive behavior only when the MAOA transcript is expressed at low abundance, due to a VNTR polymorphism at the promoter. (From Caspi, A. et al. (2002). *Science*, **297**, 851–854.)

repeat (VNTR) at the promoter of the monoamine oxidase A gene, and is associated with violent behavior only in individuals who were abused as children. Maltreated children with high levels of this enzyme were less likely to develop antisocial problems than maltreated children with low levels of monoamine oxidase A expression (Figure 12.22).

Caspi and his colleagues also showed that a polymorphism in the promoter region of the serotonin transporter gene is associated with depression, including suicidal tendencies. This polymorphism is characterized as a short allele and a long allele of the serotonin transporter gene. Caspi assessed depression in individuals with different forms of the serotonin receptor gene, and asked them to fill out a questionnaire that tabulated stressful life experiences. He discovered that people with one or two copies of the short allele exhibited more depressive symptoms as a result of stressful life experiences than individuals homozygous for the long allele. Thus, an individual's response to environmental stress is modulated by his or her genotype with respect to alleles of the serotonin transporter gene. Few other investigators have been able to replicate Caspi's study. This may be due to differences in statistical power among the studies, and differences in ethnicity, age, and sex of the study subjects. Nevertheless, these studies provide excellent examples for illustrating gene-by-environment interactions in human behavioral genetics.

Despite some important advances, it is clear that our understanding of the genetics of human aggression is rudimentary at best. Powerful genetic model organisms will be invaluable in future studies, and progress in understanding human aggression will depend on comparative genomic approaches. When we can explain in genetic terms why some of us grow up as dominant bullies and others are content to live their lives as quiet subordinates, we will truly have gained a deeper understanding of human nature itself.

DIVISION OF LABOR: THE GENETICS OF SOCIAL STRUCTURE

There is no doubt that social insects comprise the most successful species on our planet. Hölldobler and Wilson have estimated that up to one third of the terrestrial animal biomass is made up of ants and termites. Social insects provide excellent models for studying the genetics of social organization. The most extensively-studied social insect that has become the "poster child" for **sociogenomics** is the honey bee, *Apis mellifera*. Bees are advantageous insect models because of their large size, which enables neuroanatomical, behavioral, and electrophysiological studies; their haploid–diploid mode of reproduction, which is favorable for genetic studies; and their economic importance as pollinators and honey producers.

A typical honey bee colony contains workers, **drones**, and a single queen. Queens mate in flight with multiple drones, and only the queen produces eggs. Honey bee eggs hatch whether or not they are fertilized by sperm from drones stored in the queen's **spermatheca**. Fertilized eggs give rise to female workers, which are diploid and contain two sets of 16 chromosomes. Unfertilized eggs give rise to drones, which are haploid and contain only a single set of 16 chromosomes. Because drones are haploid, recombination occurs only in the queen. Consequently, all sperm cells produced by a single drone are genetically-identical. Thus, workers that arise from sperm derived from the same drone are sisters that have three quarters of their genes in common by descent.

Sex determination in bees is complex and determined by a locus known as ***complementary sex determiner*** (*csd*). This locus is highly-polymorphic, and an initial study identified 19 different alleles. Fertilized eggs that are heterozygous at the *csd* locus develop into females (workers), whereas **hemizygosity** (i.e. a single gene copy) in unfertilized eggs results in males (drones). Fertilized eggs that are homozygous at the *csd* locus develop into sterile males, but are removed and eaten by workers in the hive, and therefore, have zero fitness. This provides a strong evolutionary driving force for maintaining heterozygosity at the *csd* gene. Similar alleles of the *csd* gene occur in males and females. Thus, there are no male- or female-specific alleles. Rather, polypeptides derived from different alleles may form heteromeric complexes that give rise to female development, while absence of such complexes results in the default pathway, the development of a male. In fact, blocking expression of a *csd* allele with RNAi gives rise to females with male gonads. This type of haplo–diploid reproduction is also found in wasps and ants, and appears to be a conserved feature of most social insects.

Many social interactions within the hive are dependent on pheromones. The queen rules the hive by means of a **pheromone blend** that also acts as a mating attractant for drones. The queen's pheromones also suppress development of ovaries in workers, thus ensuring that the queen is the only reproductive female in the hive. This pheromone blend, known as **queen mandibular pheromone** (QMP), elicits a **retinue response** among the workers, who gather around the queen to tend and nurse her (Figure 12.23). QMP contains 9-oxodec-(E)-2-enoic acid, both enantiomers of 9-hydroxydec-(E)-2-enoic acid, methyl *para*-hydroxybenzoate, and 2-(4-hydroxy-3-methoxyphenyl) ethanol, and is only effective when all five components are present.

Young workers perform their social duties, which include nursing the brood inside the hive. As they age, their tasks change, until at around six weeks of age they make the transition to foragers, who leave the hive to collect nectar. Foragers calculate a map of a food source with respect to the hive by using polarized light that provides information about the position of the sun and adjusting this information for the time of day according to their internal circadian clock. Upon returning to the hive they then communicate the location of the food source to their fellow

FIGURE 12.23   The retinue response with workers surrounding the queen.

bees via an intricate waggle dance, studied extensively by the pioneering Nobel laureate Karl von Frisch (described in Chapter 1). The timing of the transition of bees from nurses to foragers can be experimentally-manipulated by creating conditions in which the hive contains only nurses or only foragers. This results in the precocious maturation of foragers or the reversion of foragers to nurses, respectively, demonstrating that the genetic programs that govern behavioral transitions are plastic, and can be altered by changes in the social environment. Thus, chemical signals and social organization determine the activation of genetic programs that lead to divisions of labor which are optimal for the integrity of the society.

What are the effects of pheromonal signals or social conditions on transcriptional regulation in the honey bee brain which results in shifts in social behaviors? This question was addressed in pioneering studies by Gene Robinson and his co-workers at the University of Illinois, who spearheaded the honey bee genome project that culminated in the complete sequencing of the honey bee genome. Even prior to completion of the honey bee genome sequence, the Robinson's laboratory was able to construct **EST expression microarrays** (see Chapter 10) that represented about 40% of the bee genome. Using these arrays, they could probe transcriptional changes that precede or coincide with changes in behavior. Transcriptional profiles were compared between 5–9-day-old nurses and 28–32-day-old foragers. In addition, effects due to different ages could be taken into account by establishing **single cohort** colonies, in which precocious foragers develop that are the same age as nurses. These experiments revealed genes that are uniquely expressed in, and are diagnostic of, each behavioral stage (Figure 12.24). These

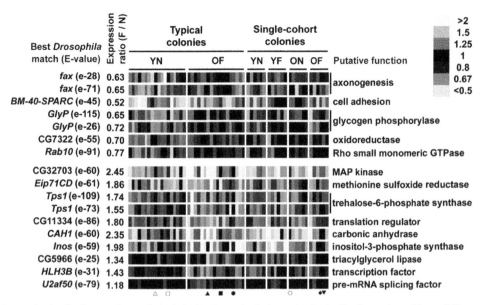

Best *Drosophila* match (E-value)	Expression ratio (F / N)	Typical colonies			Single-cohort colonies				Putative function
		YN		OF	YN	YF	ON	OF	
fax (e-28)	0.63								axonogenesis
fax (e-71)	0.65								
BM-40-SPARC (e-45)	0.52								cell adhesion
GlyP (e-115)	0.65								glycogen phosphorylase
GlyP (e-26)	0.72								
CG7322 (e-55)	0.70								oxidoreductase
Rab10 (e-91)	0.77								Rho small monomeric GTPase
CG32703 (e-60)	2.45								MAP kinase
Eip71CD (e-61)	1.86								methionine sulfoxide reductase
Tps1 (e-109)	1.74								trehalose-6-phosphate synthase
Tps1 (e-73)	1.55								
CG11334 (e-86)	1.80								translation regulator
CAH1 (e-60)	2.35								carbonic anhydrase
Inos (e-59)	1.98								inositol-3-phosphate synthase
CG5966 (e-25)	1.34								triacylglycerol lipase
HLH3B (e-31)	1.43								transcription factor
U2af50 (e-79)	1.18								pre-mRNA splicing factor

Legend: >2, 1.5, 1.25, 1, 0.8, 0.67, <0.5

FIGURE 12.24   Expression levels of genes diagnostic for behavioral stage in the honey bee brain. The figure shows 17 out of 50 expression microarray ESTs with high sequence similarity with functionally-annotated *Drosophila* orthologs that are predictive of behavioral stage. Expression levels were compared in 60 individual honey bee brains from young nurses (YN), old nurses (ON), young foragers (YF), or old foragers (OF) from typical or single cohort colonies. Though diagnostic, changes in transcript levels that accompany these behavioral changes are relatively small. (From Whitfield, C. W. et al. (2003). *Science*, **302**, 296–299.)

genes include transcription factors as expected, but most would not have been predicted to play a role in behavior *a priori*, e.g. carbonic anhydrase, trehalose-6-phosphate synthase. These results illustrate how genes with **pleiotropic effects** are recruited into the genetic architectures that orchestrate behavioral phenotypes. A transcriptional regulator, orthologous to the *Drosophila* transcription factor Krüppel, has been identified as a candidate regulator that may contribute to the behavioral switch that occurs when flies progress from nursing to foraging.

The honey bee system has already taught us some general lessons. Changes in social behavior are not attributable simply to a single gene or a few dedicated genes, but rather are accompanied by widespread changes in the expression of coordinated ensembles of genes that encompass a diverse spectrum of gene products. Individual genes are not necessarily specifically dedicated to a particular behavior, but instead contribute to genetic networks that enable the manifestation of distinct behavioral phenotypes. A complete understanding of the sequence of molecular events and concomitant changes in transcriptional profiles that are causal to behavioral maturation in the honey bee system is a major frontier for future research. General principles that emerge from this elegant model system are likely to find applications in many other species, including humans.

AGGREGATION BEHAVIOR

One of the best examples of how **genotype-by-environment interactions** can modify social behavior comes from

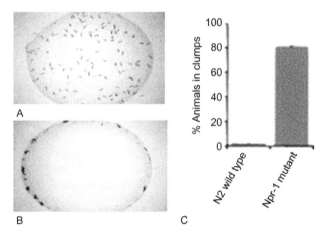

FIGURE 12.25   Solitary (A) and social (B) feeding in *C. elegans*. Solitary feeders spread out over a bacterial lawn, whereas social feeders aggregate in clumps at the borders of the dish where the bacteria are densest. Panel C shows that a mutation in the *npr-1* gene can convert a solitary feeding strain (the N2 strain) into a social feeder. (From de Bono, M., and Bargmann, C. L. (1998). *Cell*, **94**, 679–689.)

feeding behavior of the nematode *Caenorhabditis elegans*. As mentioned earlier (Chapter 9), *Drosophila* larvae can be classified either as "sitters" or "rovers" in terms of their feeding behavior, and these different feeding styles are attributable, at least in part, to a polymorphism in a cyclic GMP-dependent protein kinase, encoded by the *foraging* gene. *C. elegans* displays a similar natural variation in feeding behavior. When nematodes isolated from the wild are placed on a lawn of bacteria, their natural food source, some worms will aggregate into tight social groups, whereas others continue to feed individually (Figure 12.25).

Box 12.4   Behavior without a brain?

The slime mold *Dictyostelium discoideum* exists as single-celled amoebae, which aggregate to form a multicellular organism when nutrients become limiting (Figure A). To mediate this aggregation they release cyclic AMP as an intercellular communication signal. Some investigators have argued that this aggregation behavior should be considered "social behavior," even though the organism lacks a nervous system. Others have argued that this is simply a cell adhesion phenomenon that belongs in the realm of cell biology, and should not be considered a behavior. The term social behavior has also been applied to bacteria that aggregate under conditions of limiting resources. The controversy of whether organisms without a nervous system that respond to changes in their environment should be considered "social" or "behaving" is likely to be ongoing, as opinions on this issue remain polarized.

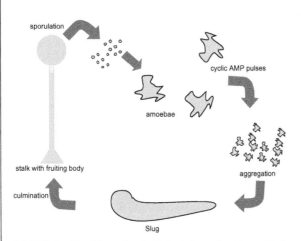

FIGURE A   The life cycle of *Dictyostelium discoideum*. Under adverse environmental conditions, pulses of cyclic AMP are released that trigger aggregation and the formation of a multicellular slug, which is transformed in a fruiting body that releases spores for the next generation of amoebae. Following aggregation, cells undergo specialization to become either spores in the fruiting body or stalk cells. Some investigators argue that this lifecycle should be considered a form of social behavior.

Is there a genetic basis for the difference between solitary and social feeding behavior, and are these behaviors plastic, that is, can social feeders be transformed into solitary feeders or *vice versa* through genetic manipulations? The answer to both questions is yes.

Elegant studies by the laboratories of Mario de Bono at Cambridge University and Cornelia Bargmann at the Rockefeller University have used genetic methods and mutational analysis to dissect the neural circuits that determine feeding behavior in *C. elegans*. They identified a single nucleotide polymorphism in the *npr-1* gene that is associated with solitary or social feeding. The N2 strain of *C. elegans* is

a solitary feeder, but mutations in the *npr-1* gene transform it into a social feeder. This gene encodes a G-protein-coupled neuropeptide receptor that is orthologous to the mammalian receptor for **neuropeptide Y**, which, interestingly, is a potent stimulator of hunger and feeding in people (Chapter 3), and has also been implicated in anxiety and response to stress. In social feeders of *C. elegans*, the NPR-1 receptor has a phenylalanine at position 215 in the fifth transmembrane domain. Substitution of this phenylalanine with valine converts social feeders into solitary feeders. Regulation of feeding behavior via the NPR-1 receptor is achieved through **FMRF-amide**-related peptides. Binding of such ligands to the NPR-1 receptor inhibits social feeding, and results in solitary foraging.

De Bono and his colleagues showed that the NPR-1 receptor is expressed in distinct neurons that are in contact with the body fluid, and that activation of the NPR-1 receptor in these neurons results in inhibition of a signaling pathway that involves a cyclic GMP-gated ion channel, encoded by the *tax2* and *tax 4* genes. Mutations in these genes mimic the effects of NPR-1 receptor activation, in that they disrupt social feeding. What signal generates the cyclic GMP that normally activates the cyclic GMP-gated channel that leads to social feeding? An additional player was identified, a soluble guanylate cyclase, encoded by the *gcy-35* gene. This guanylate cyclase contains a heme group that binds molecular oxygen and functions as an oxygen sensor. Nematodes prefer to live under conditions of 5–12% oxygen. Under hyperoxic conditions, the GCY-35 guanylate cyclase is activated and triggers social feeding behavior (Figure 12.26).

Noxious chemical stimuli also induce social feeding. Ablation of nociceptive neurons (the ASH and ADL chemosensory neurons) transforms social feeders into solitary feeders. In this case TRP channels, encoded by the *ocr-2* and *osm-9* genes, are implicated in nociceptive signaling that influences feeding behavior.

What is the evolutionary significance of solitary and social feeding behaviors? When food resources are scarce, social feeding with minimal energy expenditure for foraging may be advantageous, although it increases competition for food among individuals. When food is plentiful, it may be advantageous for individuals to "go their own way," and maximize their food intake without competitors. Social feeding may bring individuals together for mating. In addition, the risk of predation may be a factor; a large group size could either deter predators or make individuals more conspicuous to predation. In the case of *C. elegans*, adverse environmental conditions promote social feeding, whereas favorable environments encourage solitary feeding. The precise selective advantages that arise from these distinct behaviors are not clear, but the very fact that both behaviors persist in the population and are interchangeable suggests that this behavioral flexibility has significant survival value.

Box 12.5   Social induction of sex change

A small Caribbean reef fish, the bluehead wrasse (*Thalassoma bifasciatum*), presents a fascinating example of the interplay between the social environment, hormones, and behavior. Large males have a distinct coloration, and show aggressive behavior defending territories on the reef where they mate with females. When a male is removed from a group of females, the largest female will undergo a remarkable sex change and physically turn into a male (Figure A). Functional testes that produce fertile sperm develop, along with a change to male-specific coloration and expression of male-specific territorial behavior. These changes are accompanied by changes in the expression of vasotocin (a homolog of vasopressin in fish and other nonmammalian vertebrates) in the brain, and are influenced by steroid hormones. However, the change in behavior is independent from the development of the gonads. This raises some intriguing questions. What are the social cues that promote this profound metamorphosis, and what are the genetic programs that become activated to support the accompanying physical and behavioral

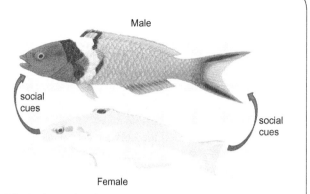

FIGURE A   Socially-induced sex-change in the bluehead wrasse.

changes? Identification of the stimuli and early transcriptional regulators they induce would give us some insights in how the social environment can regulate the transcriptome to switch reproductive physiology and behavior so profoundly.

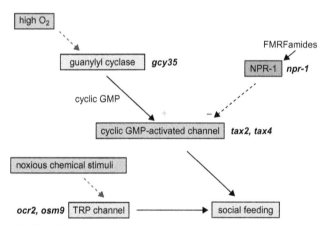

FIGURE 12.26   Diagram of some of the genes and signaling pathways implicated in the control of social feeding in *C. elegans*.

SUMMARY

Social interactions result from competition between individuals for territory, food or mating opportunities, or from cooperation between individuals which leads to the establishment of social organizations. Aggression can result in the formation of stable social hierarchies. Cooperative societies provide individuals the opportunity to optimize their fitness, by promoting the propagation of alleles from closely-related relatives. Courtship rituals have developed to signal fitness to prospective mating partners, as exemplified by the development of birdsong, and the courtship ritual of fruit flies. Social structures can be dynamic and adapt rapidly to changes in the physical or social environment. Furthermore, the role of an individual within a society can change during its development, as in the honey bee. Such behavioral plasticity can be modulated by the social

environment, and is accompanied by widespread changes in the expression of coordinated ensembles of genes that encompass a diverse spectrum of gene products. Polymorphisms that arise in key genes within such a network during the course of evolution can shift the genetic architecture. Such shifts cause alterations in the genes–brain–behavior axis that result in profound effects on neural circuitry and the expression of social behavior. An example is the polymorphism in the promoter region of the vasopressin 1a receptor gene that is associated with monogamous versus polygamous behaviors in voles. Social interactions provide striking examples of genotype-by-environment interactions, exemplified by socially-induced sex change in the bluehead wrasse, and the shift between solitary feeding and social feeding in nematodes induced by adverse environmental conditions. Understanding how environmental stimuli elicit changes in transcriptional regulation that result in altered behavioral phenotypes is an active area of current and future research.

STUDY QUESTIONS

1. Explain why altruistic behavior can have fitness benefits.
2. Explain the term "kin selection."
3. What are the purposes of elaborate courtship rituals that precede mating?
4. Describe the similarities between the development of birdsong and the acquisition of human language.
5. Explain the term "genomic action potential" in relation to birdsong. Could you envision this term being applied to other behaviors? If so, give an example.
6. Moths have evolved volatile pheromones that bring mating partners together from long distances, whereas

flies and certain other insects, such as cockroaches, communicate via cuticular pheromones. Discuss the advantages of cuticular versus volatile pheromones in the ecological context of these insects.

7. Describe the fitness advantages of accessory gland proteins for both male and female *Drosophila*.

8. What is the evidence that regulation of vasopressin expression controls social behavior in voles? What questions does this assertion raise? Is it likely that the vasopressin gene is the only gene responsible for monogamous versus polygamous behavior? What experiments could you design to identify transcriptional differences that may be causal to or associated with propensity towards monogamy or polygamy?

9. Discuss the relationship between aggressive behavior, mating behavior, and fitness.

10. What are the major stumbling blocks for identifying genes that mediate aggression in model organisms and in humans?

11. Describe the haploid–diploid mating system of the honey bee.

12. Give at least five reasons why honey bees are an advantageous model for studying the genetics of social organization.

13. Compare naturally-occurring variation in feeding behavior and associated genetic polymorphisms between nematodes and fruit flies.

14. Describe examples of genotype-by-environment interactions in human behavior.

15. Imagine a genetically-hardwired behavior that is essential for survival and has no phenotypic variation, and another behavior which is plastic and strongly affected by environmental interactions. If both phenotypes can be measured quantitatively, what unique problems would each scenario present in terms of trying to identify the underlying genes that shape each behavior?

RECOMMENDED READING

Bray, S., and Amrein, H. (2003). A putative *Drosophila* pheromone receptor expressed in male-specific taste neurons is required for efficient courtship. *Neuron*, **39**, 1019–1029.

de Bono, M., and Bargmann, C. I. (1998). Natural variation in a neuropeptide Y receptor homolog modifies social behavior and food response in *C. elegans. Cell*, **94**, 679–689.

Caspi, A. et al. (2002). Role of genotype in the cycle of violence in maltreated children. *Science*, **297**, 851–854.

Caspi, A. et al. (2003). Influence of life stress on depression: moderation by a polymorphism in the 5-HTT gene. *Science*, **301**, 386–389.

Hall, J. C. (1994). The mating of a fly. *Science*, **264**, 1702–1714.

Hammock, E. A., and Young, L. J. (2005). Microsatellite instability generates diversity in brain and sociobehavioral traits. *Science*, **308**, 1630–1634.

Jarvis, E. D. (2004). Learned birdsong and the neurobiology of human language. *Ann. N. Y. Acad. Sci.*, **1016**, 749–777.

Kravitz, E. A. (2000). Serotonin and aggression: insights gained from a lobster model system and speculations on the role of amine neurons in a complex behavior. *J. Comp. Physiol.*, **186**, 221–238.

Lim, M. M., Hammock, E. A., and Young, L. J. (2004). The role of vasopressin in the genetic and neural regulation of monogamy. *J. Neuroendocrinol.*, **16**, 325–332.

Markow, T. A., and O'Grady, P. M. (2005). Evolutionary genetics of reproductive behavior in *Drosophila*: connecting the dots. *Annu. Rev. Genet.*, **39**, 263–291.

Mello, C. V., Vicario, D. S., and Clayton, D. F. (1992). Song presentation induces gene expression in the songbird forebrain. *Proc. Natl. Acad. Sci. USA*, **89**, 6818–6822.

Menzel, R., Leboulle, G., and Eisenhardt, D. (2006). Small brains, bright minds. *Cell*, **124**, 237–239.

Robinson, G. E., Grozinger, C. M., and Whitfield, C. W. (2005). Sociogenomics: social life in molecular terms. *Nat. Rev. Genet.*, **6**, 257–270.

Vrontou, E., Nilsen, S. P., Demir, E., Kravitz, E. A., and Dickson, B. J. (2006). *Fruitless* regulates aggression and dominance in *Drosophila*. *Nat. Neurosci.*, **9**, 1469–1471.

Wilson, E. O., and Hölldobler, B. (2005). Eusociality: origin and consequences. *Proc. Natl. Acad. Sci. USA*, **102**, 13367–13371.

Wolf, J. B., Brodie, E. D., III, Cheverud, J. M., Moore, A. J., and Wade, M. J. (1998). Evolutionary consequences of indirect genetic effects. *Trends Ecol. Evol.*, **13**, 64–69.

Wolfner, M. F. (2002). The gifts that keep on giving: physiological functions and evolutionary dynamics of male seminal proteins in *Drosophila. Heredity*, **88**, 85–93.

Chapter 13

Genetics of Olfaction and Taste

OVERVIEW

Chemosensation is essential for food localization and evaluation, avoidance of predators and noxious substances, kin recognition, and for identification of mating partners. The ability to recognize and respond to chemical stimuli is, therefore, critical for the survival and reproduction of most organisms. Sophisticated molecular genetic techniques, and the use of transgenic animals, were instrumental in the discovery of chemoreceptors and in elucidating the functional organization of chemosensory systems. The mouse genome encodes a multigene family of about 1000 odorant receptors, which belong to the superfamily of G-protein-coupled receptors. Olfactory receptor neurons express only a single receptor gene from this repertoire. The axons of olfactory receptor neurons that express the same receptor converge on the same glomeruli in the olfactory bulb, and the chemical structures of odorants are encoded in patterns of activated glomeruli; the organization of insect olfactory systems shows similar glomerular convergence, but with fewer odorant receptors. The mouse vomeronasal organ has a two-layered organization with structurally-distinct pheromone receptors expressed in the apical and basal layers. Their projections to the accessory olfactory bulb are segregated, and form a more diffuse glomerular convergence pattern than that observed in the main olfactory system. In contrast to the olfactory system, taste cells that are receptive to bitter tastants express multiple receptors per cell. These cells can detect aversive stimuli without a need to accurately discriminate among them. Sweet taste receptors are fewer in number, and expressed in different taste cells. Taste recognition in *Drosophila* is also mediated by a multigene family of gustatory receptors that have different projections based on taste modality. The olfactory system of the nematode *Caenorhabditis elegans* is designed to discriminate attractant from repellent odorants, and resembles the mammalian bitter taste system in that cells that mediate attractant and repellent responses express distinct, but multiple, odorant receptors. In this chapter we will describe the genetic approaches that have characterized the molecular, genetic, and neural underpinnings of chemosensation. A more detailed description of the genetic architecture of odor-guided behavior can be found in Chapter 10.

THE STUDY OF OLFACTION

In his 1983 book *Late Night Thoughts on Listening to Mahler's Ninth Symphony*, science philosopher Lewis Thomas wrote: "I should think we might fairly gauge the future of biological science, centuries ahead, by the time it will take to reach a comprehensive understanding of odor. It may not seem a profound enough problem to dominate all the life sciences, but it contains, piece-by-piece, all the mysteries." Thomas could not have predicted at that time that only eight years later Linda Buck and Richard Axel would discover odorant receptors, and that within two decades they would share the 2004 Nobel Prize for their discovery. In only a few years the study of olfaction had been transformed from a seemingly intractable esoteric research field into an area of investigation that would take center stage in sensory neurobiology. This rapid progress was made possible largely due to advances in genetic technologies and to the availability of whole-genome sequences.

THE DISCOVERY OF ODORANT RECEPTORS

Olfactory sensory neurons are **bipolar neurons** that are housed in a **pseudostratified neuroepithelium** in the nose, where they send dendrites with ciliated dendritic knobs to the epithelial surface and axons to the olfactory bulb in the brain (Figures 13.1 and 13.2). As described in Chapter 3, 100–1000 axons from olfactory sensory neurons converge on a single output neuron, forming spherical structures of neuropil, named **glomeruli**. Early electrophysiological and biochemical experiments used the olfactory epithelium of the frog as a study model, since its olfactory neuroepithelium is a readily-accessible flat pigmented sheet

R. Anholt and T. Mackay: Principles of Behavioral Genetics
ISBN: 978-0-12-372575-2

Box 13.1   Worms and the smell of popcorn

The first odorant receptor for which a ligand was discovered was the ODR-10 receptor in the nematode *Caenorhabditis elegans*. In a forward genetic screen, Piali Sengupta and Cornelia Bargmann found that *odr-10* mutants have a 100-fold reduced sensitivity to diacetyl, a volatile compound with the characteristic odor of buttered popcorn.

C. elegans has only 302 neurons, with a total of 32 chemosensory neurons. The nematode genome, however, encodes more than 1000 odorant receptors, a surprisingly large number given the simplicity of its olfactory system. In contrast to the mammalian and insect olfactory systems, chemosensory neurons in nematodes each express a large ensemble of odorant receptors.

The ODR-10 receptor is normally expressed in AWA neurons, and exposure to diacetyl elicits an attractant response. Ectopic expression of an *odr-10* transgene in AWC neurons results in avoidance responses to diacetyl. Thus, the nematode chemosensory system appears to be hardwired, in that activation of AWA neurons via any of the chemoreceptors expressed in these cells directs attractant behavior, whereas activation of AWC neurons triggers repulsion, independent of which one of the chemoreceptors expressed in these cells is activated. In this sense, the olfactory system of *C. elegans* resembles the mammalian bitter taste system. Worms only need to make one decision: to go toward a beneficial environment characterized by an attractant odor, or to avoid a noxious environment. Sophisticated chemosensory discrimination is here not essential.

In addition to volatiles, worms also detect nonvolatile compounds via bilaterally-located taste neurons, known as the ASE left (ASEL) and ASE right (ASER) neurons, each of which expresses a distinct set of chemoreceptors. Oliver Hobert and his colleagues at Columbia University showed that the decision of whether a precursor cell develops into an ASEL or ASER neuron depends on a regulatory network of transcription factors that interact with two microRNAs to stabilize either the ASEL or ASER cell fate.

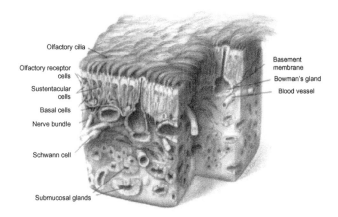

FIGURE 13.1 Diagram of the vertebrate olfactory neuroepithelium. Olfactory receptor cells are surrounded by **sustentacular cells**, which resemble glia and carry **microvilli** at their apical ends. Unmyelinated axons of olfactory receptor cells gather as fascicles that are ensheathed by a single Schwann cell. These fascicles will form the olfactory nerve that projects to the olfactory bulb. Cuboidal cells on top of the basement membrane represent neurogenic stem cells. Note that the epithelium is covered by a layer of cilia that protrude from dendritic tips of olfactory receptor cells to provide a vastly enlarged surface area of chemosensory membrane. Secretions from sustentacular cells, **Bowman's glands**, and numerous submucosal glands in the **lamina propria** contribute to mucus, which is the medium in which interactions between odorants and the chemosensory membrane take place. (From Anholt, R. R. H. (1987). *Trends Biochem. Sci.*, **12**, 58–62.)

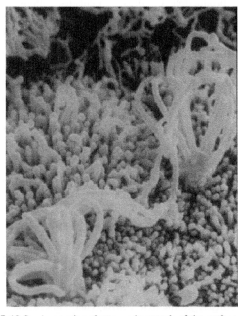

FIGURE 13.2 A scanning electron micrograph of the surface of olfactory neuroepithelium shows the **dendritic knobs** with **olfactory cilia**. Protrusions from the epithelial surface in between the dendritic knobs are microvilli of sustentacular cells. (From Morrison, E. E., and Costanzo, R. M. (1992). *Microsc. Res. Tech.*, **23**, 49–61.)

that is well separated from respiratory epithelium. In the mid-1980s, Doron Lancet at the Weizmann Institute in Israel, and Robert Anholt at the Johns Hopkins University in Baltimore, prepared membrane preparations from dendritic cilia of frog olfactory sensory neurons, and showed that administration of odorants resulted in the production of cyclic AMP. Randall Reed and his colleagues at the

Johns Hopkins University found that olfactory cilia contain a unique adenylyl cyclase subtype and a unique G-protein, closely related to Gs. Using sophisticated **patch-clamp** technology on single olfactory cilia, Geoffrey Gold and his co-workers from the Monell Chemical Senses Center in Philadelphia demonstrated that cyclic AMP directly opens a cation channel. These experiments showed that olfactory transduction is mediated via a cyclic AMP signal transduction pathway, a finding that was further consolidated by elegant electrophysiological experiments on isolated olfactory sensory neurons by Stuart Firestein and Gordon Shepherd at Yale University (see also Chapter 2). The only critical link missing were the actual odorant receptors.

Since G-protein-coupled cyclic AMP transduction pathways are generally controlled by G-protein-coupled receptors (GPCRs), Buck and Axel hypothesized that odorant receptors would be members of this superfamily. Because the olfactory system has to recognize many odorants, all of which appeared to activate adenylyl cyclase, they proposed that odorant receptors would comprise a large family of GPCRs. They hunted for odorant receptors

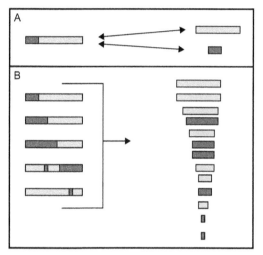

FIGURE 13.3 PCR amplification of odorant receptor encoding cDNAs. Panel A shows that restriction digestion of a homogeneous cDNA gives rise to restriction fragments, of which the sizes when summed equal the size of the original amplicon. Panel B illustrates how a heterogeneous population of same size amplicons contains restriction sites that give rise to different size fragments, the sums of which will be far greater than the size of the originally-observed amplicon.

using PCR amplification with degenerate primers against conserved regions of GPCRs. One challenge they faced was that even if they obtained an **amplicon** with a novel sequence, how would they decide this indeed represented an odorant receptor? Their ingenious PCR strategy was based on the notion that if divergent sequences could be amplified between primers against conserved stretches, the amplification product, which would appear as a single PCR band on the gel, would harbor diverse sequences. Thus, restriction digestion of such amplicons was predicted to give rise to many different fragments, the molecular sizes of which would add up to a size far greater than the original amplicons (Figure 13.3). Furthermore, the corresponding transcripts should be expressed in olfactory sensory neurons. Their hypothesis was correct, and their approach was successful.

Their experiments led to the identification of fragments of putative odorant receptors that were expressed in olfactory sensory neurons. Analyses of the sequences of odorant receptors revealed the predicted seven transmembrane domains of GPCRs, and showed that sequence diversity was greatest in these transmembrane domains, which presumably form the binding pocket for odorants (Figure 13.4). They postulated that receptors with very similar sequences would bind related odorants, whereas receptors that were highly diverse might bind structurally-distinct odorants.

Buck and Axel's discovery enabled them to address a critical question that had been subject to considerable speculation: how many odorant receptors are there? Some investigators had proposed that only a small number of receptors were necessary to recognize many odorants, similar to the three photopigments of the eye that enable perception of a wide color spectrum; others had suggested that odorant receptors would be exquisitely sensitive to specific

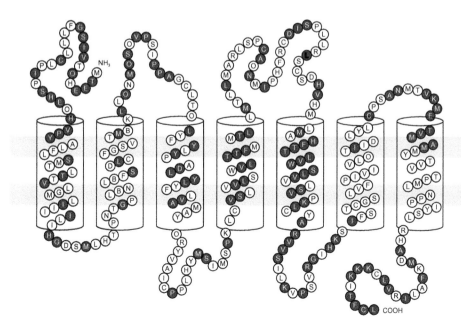

FIGURE 13.4 Diagram of the I15 rat odorant receptor showing its seven transmembrane domains. Positions at which 60% or more of 10 initially-identified receptor cDNAs share the same amino acid residue as I15 are shown as white balls. More variable residues are shown as black balls. Note the high degree of variability in transmembrane domains III, IV, and V, thought to contribute to the odorant binding site. (From Buck, L., and Axel, R. (1991). *Cell*, **65**, 175–187.)

odorants, implying a vast number of tens of thousands of receptors. Buck and Axel were able to estimate that the rat genome contains at least 100 odorant receptor genes, and subsequent studies showed that the mouse genome encodes more than 1000 odorant receptors. Furthermore, when they performed *in situ* **hybridization** to sections through the nose, they observed that the number of cells labeled with a mixture of all receptor probes was the sum of the numbers of cells labeled with each probe individually, and receptors were expressed within one of four distinct dorsal–ventral zones (Figure 13.5). They concluded that each neuron expresses only one receptor, a notion further corroborated by later experiments, and confirmed again that the odorant receptor family in the mouse would comprise as many as 1000 genes by comparing the number of labeled cells and the total number of about 100 million olfactory sensory neurons. Thus, they discovered that there were more odorant receptors than all other GPCRs combined, and that odorant receptor genes represent a substantial (~2.5%) fraction of the genome. Furthermore, olfactory sensory neurons exhibit allelic inactivation of odorant receptor expression; i.e. only one allele of the expressed odorant receptor gene is active, preventing ambiguity of odorant receptor specificities that might arise from heterozygous alleles containing different polymorphisms that affect ligand specificity. This strict one-neuron-one-receptor rule ensures that the brain can deduce which receptor was activated by determining which neuron has been activated.

When the sequence of the mouse genome became available, 1296 olfactory receptor genes were identified, of which 20% were **pseudogenes**, i.e. they accumulated

a mutation that rendered them dysfunctional. Olfactory receptor genes are organized in clusters throughout the mouse genome, except chromosome 12 and the Y chromosome. Phylogenetic analysis classified these odorant receptors into 228 families. In the human genome about two thirds of functional odorant receptor genes have been lost and persist as pseudogenes. However, each of the 228 mouse categories is represented by at least one functional human ortholog, suggesting that humans can perceive the same odor spectrum as mice, but with perhaps less subtle discrimination and sensitivity (Figure 13.6). It should be noted that the designation of pseudogene is based only on one individual sequence from the public database, and

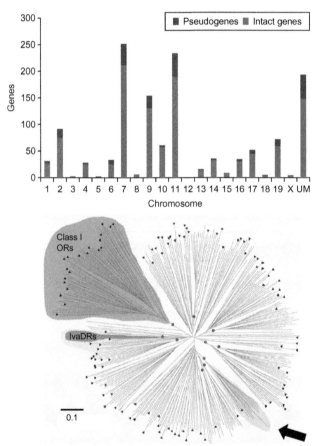

FIGURE 13.6 The mouse and human odorant receptor repertoires. On the top, the number of odorant receptor genes on each mouse chromosome is represented by bars. Gray segments indicate the proportion of pseudogenes. Below is an **unrooted phylogenetic tree** of the consensus protein sequences of mouse and human odorant receptor gene families. Human odorant receptor families are indicated by filled symbols. A few groups with high (>90%) **bootstrap** value are labeled by dots in the center. The Class I odorant receptor families, shaded in dark gray, form a group with more than 90% bootstrap value. A group with high bootstrap value, and including mostly mouse odorant receptor families, is indicated by the arrow. *Iva1* odorant receptors have been implicated in **anosmia** (smell blindness) to isovaleric acid, and are indicated with open diamonds and shading; families close to the *Iva1* families are highlighted. (Adapted from Zhang, X., and Firestein, S. (2002). *Nature Neurosci.*, **5**, 124–133.)

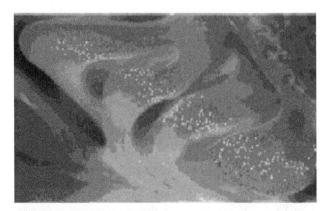

FIGURE 13.5 Diagram of **turbinates** from the rat olfactory epithelium, which illustrates three of four distinct zones within which expression of specific odorant receptors is confined. Expression patterns from *in situ* hybridization for odorant receptors from three different subfamilies (17, F3, and J7) are visualized and define three non-overlapping horizontal zones. The significance of segregation of odorant receptor expression in these zones is not clear. It has been proposed that restricting expression of subgroups of odorant receptors within these dorsal–ventral boundaries may facilitate the wiring of the olfactory system during development, when a vast number of olfactory neurons expressing distinct odorant receptors must be sorted to find their appropriate glomerular projection sites. (From Vassar, R. et al. (1993). *Cell*, **74**, 309–318.)

that functional alleles of pseudogenes may segregate in the population. Functional redundancy among odorant receptor genes would allow segregation of pseudogenes in a population, and the complement of pseudogenes and functional alleles may show individual variation. It has been proposed that pseudogenes that contain a stop codon could conceivably undergo a mutation that would restore them to functional receptors, thereby providing a reservoir of "evolvable" functional receptors.

Following the discovery of odorant receptors in the mouse, the laboratories of John Carlson at Yale University and of Leslie Vosshall and Richard Axel at Columbia University identified odorant receptors in *Drosophila*. These receptors were highly-divergent, seven transmembrane domain GPCRs that showed no evolutionary relationship to the mammalian odorant receptors. They are expressed in a morphological class of sensilla known as **basiconic sensilla**. Each of the approximately 1200 olfactory neurons of basiconic sensilla in the *Drosophila* antenna expresses a unique odorant receptor from a repertoire of 62 odorant receptor (Or) genes, and most neurons also express a common odorant receptor, Or83b ("Or" designates odorant receptor; 83, the cytological band location where the gene is located in the *Drosophila* genome (the right arm of the third chromosome); and, "b" the second *Or* gene found at this location after *Or83a*). The common Or83b receptor forms a heterodimer with the neuron-specific odorant receptor, and this dimerization is essential for transport and insertion of the specific odorant receptors in the chemosensory dendritic membrane. Furthermore, it is thought that this dimeric complex opens an ion channel on binding of an odorant to its cognate receptor, which gives rise to the generator current for activation of the neuron. Like the mammalian olfactory system, neurons that express the same receptor project bilaterally to one of about 43 individually-identifiable symmetrically-located glomeruli in the **antennal lobes**, where olfactory information is encoded in a spatial and temporal pattern of glomerular activation (Figure 13.7). In addition to the *Or* genes, Vosshall and collaborators recently identified a second family of *Drosophila* odorant receptors. This family consists of about 61 predicted genes, which are expressed in a different class of sensilla, the coeloconic sensilla. In contrast to the Or gene family, these receptors resemble glutamate-activated ion channels, and have been named ionotropic receptors, encoded by IR genes. These odorant receptors are generally not co-expressed with Or83b, and appear to respond to certain distinct classes of odorants, including amines. The independent evolution of distinct chemosensory gene families bears testimony to the richly diverse chemosensory perceptions experienced by insects, such as *Drosophila*. Following the discovery of mouse and *Drosophila* odorant receptors, odorant receptors have been found in numerous vertebrates and in insects like mosquitoes, honey bees, and moths.

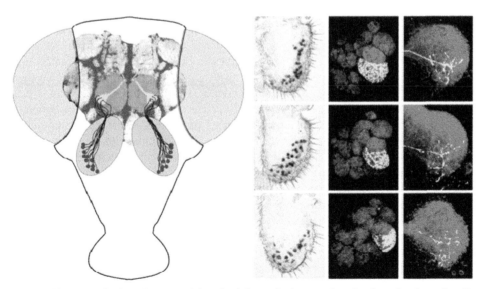

FIGURE 13.7 Mapping olfactory projections in *Drosophila*. The left panel shows a frontal schematic view of a *Drosophila* head that depicts the olfactory circuitry. Olfactory sensory neurons that express the *Or47b* odorant receptor are represented as dots on the antenna. These neurons extend axons (black) toward the brain and synapse in the VA1l/m glomerulus in the AL. Cell bodies of two dorsal projection neurons that synapse with *Or47b*-expressing neurons in the antennal lobe are also shown. The dendrites of these cells innervate the VA1l/m glomerulus and their axons extend dorsally to synapse both in the **mushroom body calyx** and the **lateral horn**. The right panel shows expression of *Or47b* by *in situ* hybridization in a restricted lateral distal region of the third antennal segment. Projections of *Or47b* expressing neurons can be visualized by expression of green fluorescent protein under the *Or47b* promoter, showing axonal convergence upon the VA1l/m glomerulus. Projection neurons send dendrites into the *Or47b* (VA1l/m) glomerulus, and project axons to the ventral region of the lateral horn of the protocerebrum. Three flies are shown to demonstrate the strict stereotypy of olfactory projections in this system. (From Keller, A., and Vosshall, L. B. (2003). *Curr. Opin. Neurobiol.*, **13**, 103–110.)

Box 13.2   Odorant-binding proteins

Odorant binding proteins (Obps) are soluble proteins, which are secreted by nasal glands and have been postulated to facilitate partitioning and transport of hydrophobic odorants in the aqueous mucus that covers the olfactory neuroepithelium. Protecting odorants from degradation or a role in the removal of odorants has also been postulated for these proteins. The mammalian nose contains only a few Obps, with very broad ligand-binding specificities.

Obps are more diverse and prevalent in insect olfactory systems. Here, Obps are secreted by support cells in the sensilla into the perilymph that surrounds the olfactory neurons. Their sequences are divergent, unrelated to mammalian Obps, and characterized by six conserved cysteine residues that are thought to be important for maintaining their three-dimensional structures. The *Drosophila* genome encodes a family of about 50 Obps, which give rise to proteins of 18–20 kD. There is good evidence that some insect Obps are important for transporting odorants in the perilymph to their receptors. As mentioned earlier, flies in which the *lush* gene, which encodes an odorant-binding protein, had been deleted were attracted to high concentrations of ethanol and short-chain alcohols that were repellent to wild-type flies. These flies also were no longer able to respond to the aggregation pheromone *cis*-vaccenyl acetate. Functional and structural studies on the pheromone-binding protein of the silk moth, *Bombyx mori*, revealed a mechanism by which this protein binds and releases the

pheromone bombykol when the pH changes at the surface of the chemosensory dendritic membrane.

Some Obps show sexually-dimorphic expression. For example, in *Drosophila*, Obp99a and Obp99b show profound sex-biased expression in females and males, respectively, suggesting that these genes have evolved through gene duplication of an ancestral gene with subsequent sex-specific functional diversification. The designation Obp99a indicates that this Obp is located at cytological band position 99 on the third chromosome, where it is the first Obp at this location. Altered regulation of expression of some Obps has been observed following mating, during exposure to starvation stress, during the development of alcohol tolerance after exposure to alcohol, and as a correlated response to artificial selection for divergent levels of copulation latency and aggression. These observations suggest that this multigene family of proteins serves important functions that broadly affect the organism's physiology and behavior.

Correlations of olfactory behavioral responses to the odorant benzaldehyde with polymorphisms in *Obp* gene sequences in a natural population of flies showed associations between polymorphisms in several *Obp* genes with phenotypic variation. This suggests that – like odorant receptors – Obps recognize odorants in a combinatorial manner. Thus, Obps may enhance the sensitivity, and modify the response spectrum of odorant-recognition by odorant receptors.

TRANSGENIC APPROACHES TO DETERMINE ODORANT RECEPTOR RESPONSE PROFILES

The discovery of about 1000 odorant receptors in mouse posed a difficult challenge. How could one determine which of the thousands of known odorants would interact with each of those 1000 odorant receptors? Initial experiments aimed at matching odorants to receptors attempted to express odorant receptors in cultured cells. These experiments were largely unsuccessful. It became clear that when odorant receptors were expressed in heterologous cells, the receptor proteins became trapped in the endoplasmic reticulum, and were not inserted into the plasma membrane. Although this problem could, in some cases, be solved by creating a chimeric odorant receptor construct that contained a stretch of N-terminal amino acids of rhodopsin, heterologous expression remained a laborious and unsatisfactory approach. Clearly, olfactory sensory neurons contain specific proteins that regulate transport to, and insertion of, odorant receptors into the chemosensory membrane.

A number of ingenious strategies were devised to determine ligand specificities of several odorant receptors. One strategy was based on the notion that a given mouse olfactory neuron expresses only a single odorant receptor. Thus, olfactory tissue could be dissociated and responses

to odorants measured on isolated olfactory neurons. The responsive cell could then be aspirated in a pipette, and its RNA could be extracted and reverse-transcribed into cDNA. PCR reactions with primers against conserved sequences of odorant receptors could then identify which odorant receptor was expressed in a cell with a known response profile to a battery of odorants. These experiments provided a direct demonstration that odorant recognition is combinatorial, i.e. a given odorant can interact with multiple receptors with different affinities, and a given receptor can recognize a range of odorants (Figure 13.8). This principle of **combinatorial recognition** is perhaps not surprising, given the large surplus of odorants compared to the available number of receptors.

The most extensively characterized vertebrate odorant receptor in terms of its molecular response properties is the rat I7 odorant receptor. This receptor was overexpressed in olfactory neurons of the rat *in situ* by infecting the olfactory neuroepithelium with an adenovirus that carried a construct encoding this receptor. Odorants were applied to the olfactory neuroepithelium, and summed electrophysiological responses ("electro-olfactograms") were recorded. These experiments showed that the I7 receptor responded optimally to octanal. Structure–activity studies showed that this receptor is highly specific for an aldehyde group, but would tolerate structural variations at the hydrocarbon tail

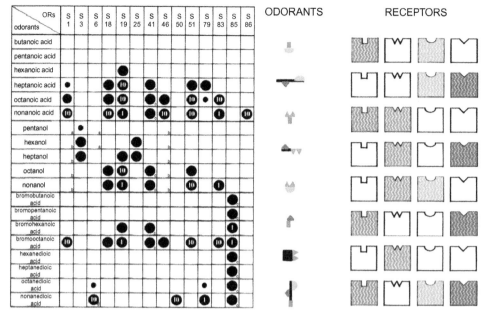

FIGURE 13.8   Combinatorial activation of odorant receptors. A particular odorant activates a subpopulation of neurons, thereby encoding information about its identity into a pattern of neural activity. The extent of activation is illustrated by the size of the circle in the left panel of the figure, and numbers in the circles refer to odorant concentrations. (From Malnic, B. et al. (1999). *Cell*, **96**, 713–723.)

of the molecule (Figure 13.9, p. 216). The mouse ortholog of the rat I7 receptor prefers heptanal over octanal, and this shift in ligand preference results from a single valine-to-isoleucine substitution in the fifth transmembrane domain.

John Carlson and his colleagues at Yale University performed elegant extracellular recordings from sensilla in the **maxillary palps** and **third antennal segments** of *Drosophila melanogaster* to characterize the response properties of olfactory sensory neurons. They found that sensilla can be classified in specific types that contain invariant combinations of neurons with defined **molecular receptive fields**. Each maxillary palp of *Drosophila* contains 120 olfactory neurons that are organized as pairs in 60 basiconic sensilla. When Carlson and colleagues measured **odorant response profiles** of olfactory sensory neurons in the maxillary palp, they noticed that they fall into six functional classes which follow a strict pairing rule such that two neurons with particular response profiles always occur together. Thus, sensilla in the maxillary palps can be classified into three groups, designated pb1, pb2, and pb3 (pb indicates "palp basiconic"). The two neurons in each sensillum can be distinguished based on their spike amplitudes. A similar organization is found in the antenna, where two olfactory sensory neurons are housed together within the basiconic sensilla, with one sensillar type containing four olfactory sensory neurons. As in the maxillary palp, neurons with defined odorant response profiles in the antenna are paired in these sensilla according to invariant pairing rules. Carlson used a diagnostic panel of 12 odorants to discriminate among 16 classes of olfactory sensory neurons housed in seven different types of sensilla in the

antenna, designated ab1-7 (ab indicates "antenna basiconic"). The distribution of the different sensilla varies across the antenna (Figure 13.10, p. 217). In addition, the response kinetics to different odorants varies among olfactory sensory neurons, and both excitatory and inhibitory responses to odorants can be observed.

In an ingenious experimental design Carlson and his colleagues made use of a deletion mutant that lacks the *Or22a* and *Or22b* genes. In these flies, neurons that would normally express these odorant receptors were silent. This enabled them to generate transgenic odorant receptors that could be expressed under the control of the *Or22a* promoter in these otherwise silent mutant neurons. This provided an elegant *in vivo* assay system for a comprehensive characterization of odorant receptor specificities (Figure 13.11, p. 217). Carlson and his colleagues could express both adult and larval odorant receptors in this "empty" neuron, and measure their electrophysiological responses to many odorants.

This successful transgenic strategy led to a comprehensive characterization of the molecular receptive fields and response characteristics of 24 odorant receptors of *Drosophila* with a panel of more than 100 odorants, making the *Drosophila* olfactory system the best-characterized olfactory system in terms of the molecular response profiles and kinetic properties of nearly half of its odorant receptors (Figure 13.12, p. 218). Moreover, one could even express odorant receptors from other insect species in the *Drosophila Or22a/b* deletion mutant. For example, only female mosquitoes of the malaria vector, *Anopheles gambiae*, feed on human blood. When Carlson and colleagues expressed a female specific odorant receptor from this

Box 13.3   Elegant complexity of the *Drosophila* larval olfactory system

The *Drosophila* larva presents perhaps the simplest system for odorant discrimination, and is elegant in its design which maximizes functional ability with anatomical economy. Olfactory sensory neurons are housed in bilaterally-symmetrical dorsal organs in the head. Each dorsal organ is composed of a central "dome" and six sensilla. The dome is perforated, and contains dendritic arbors of only 21 olfactory sensory neurons. These neurons express different odorant receptors, some of which are expressed only in larvae, and all express the common Or83b receptor. Expression of GFP reporter genes under odorant receptor promoters have identified receptors expressed in each neuron, and genetic ablation

studies by driving expression of diphtheria toxin in such neurons have characterized the larval olfactory system functionally (Figure A). One striking observation is that there are more receptors expressed in larvae than there are neuronal sub-types, indicating co-expression of odorant receptors in some cases, which serve as a means by which the larva can extend the molecular receptive range of an olfactory sensory neuron. Behavioral studies, in which olfactory sensory neurons were systematically ablated revealed, even in this simple olfactory system considerable redundancy, to the extent that a single intact olfactory sensory neuron could still mediate chemotaxis to a variety of odorants.

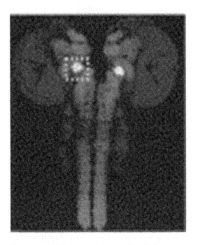 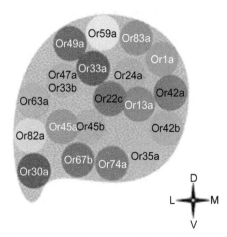

FIGURE A   The larval antennal lobe can be visualized through whole-mount immunofluorescence with an antibody against GFP, when GFP is expressed under the *Or83b* promoter via the *GAL4-UAS* system, which highlights the terminals of olfactory sensory neurons. The antennal lobe is indicated by the dashed square. The entire glomerular map in the larval olfactory system, shown on the right, can be determined using GFP expression under the control of different *Or* promoters. (Adapted from Fishilevich, E. et al. (2005). *Curr. Biol.*, **15**, 2086–2096.)

mosquito, the AgOr1 receptor, in the *Drosophila* silent neuron, they found that this receptor could be activated by 4-methylphenol, a component of human sweat (Figure 13.13, p. 219).

What is the relationship between molecular response profiles of odorant receptors and behavioral responses to odorants? This relationship turns out not to be simple. The *Or22a* deletion mutant displays intact behavioral responses to the majority of odorants that comprise the electrophysiological response profile of the Or22a receptor. This can be explained by the functional redundancy that is built into a system that relies on combinatorial odorant recognition.

In addition to the family of odorant receptors, the *Drosophila* genome also encodes a distinct family of 64 gustatory receptors (Grs), which are expressed in many regions of the fly, including the **proboscis**, the antenna, and the tarsi. Behavioral experiments in which attractant and repellent responses are measured toward compounds that humans experience as sweet or bitter, together with transgenic approaches in which projections

of gustatory neurons can be visualized by green fluorescent protein (GFP) under the control of *Gr* promoters, has generated a gustatory projection map (Figure 13.14, p. 219). The laboratory of Kristin Scott at the University of California at Berkeley showed that distinct cells express sweet taste receptors, e.g. the Gr5a receptor which senses trehalose (the primary blood sugar in insects), and bitter taste receptors with cells of each category co-expressing multiple receptors (Figure 13.15, p. 220). Sweet taste cells and bitter taste cells project to distinct mutually-exclusive projection areas in the **subesophageal ganglion**. Moreover, axons from taste cells expressed in different body parts project to topographically distinct regions of the subesophageal ganglion. Because of the vast number of gustatory receptors in flies, and the notion that it is not clear exactly where to draw the line between olfaction and gustation when considering insect chemosensation, we must be cautious not to underestimate the sophistication of these relatively "simple" chemosensory systems.

Box 13.4 Sensing carbon dioxide

Carbon dioxide (CO_2) is an important chemical signal for many insects with diverse ecological significance. When CO_2 levels increase in the hive, honey bees show a characteristic fanning response to promote airflow. The hawk moth, *Manduca sexta*, uses CO_2 to evaluate the quality of the flower from which it collects nectar. Newly-opened flowers produce more CO_2 and better quality nectar, and are preferred. Blood-feeding insects, such as ticks and mosquitoes, are directed to their hosts by the CO_2 they emit. For the fruit fly, *Drosophila melanogaster*, CO_2 when sensed by specialized gustatory neurons in the proboscis can indicate the presence of fermenting food (such as yeast), but when sensed by olfactory neurons in the antenna serves as an alarm signal, and elicits an avoidance response. Flies can sense minute elevations of CO_2 in vials in which other flies have recently been subjected to stress.

Chemosensory information about CO_2 in *Drosophila*, when sensed by gustatory neurons, is relayed to a specialized region of the suboesophageal ganglion and an attractant response is elicited. However, volatile CO_2 is sensed by olfactory neurons in the antenna, which project to a single glomerulus in the antennal lobe, designated glomerulus V. This glomerulus receives input from a select population of antennal chemosensory neurons (ab1C neurons) that express the gustatory receptors Gr21a and Gr63a. Co-expression of *Gr21a* and

Gr63a transgenes in unrelated olfactory sensory neurons conveys CO_2 sensitivity, whereas deletion of the *Gr63a* gene by homologous recombination obliterates electrophysiological and behavioral responsiveness to CO_2. The neurons that express *Gr21a* and *Gr63a* do not express the ubiquitous Or83b odorant receptor, which serves as a coreceptor for other odorant receptors. Since both Gr21a and Gr63a are necessary and sufficient for sensing CO_2, and are expressed in the same neurons, it has been postulated that these receptors dimerize to form a functional complex. It is not clear whether CO_2 itself is the ligand for these receptors, or whether these receptors are activated by bicarbonate or a change in pH, which would accompany increased levels of CO_2.

Blood-feeding mosquitoes have orthologs of both Gr21a and Gr63a which are expressed in neurons in their maxillary palps, which serve as CO_2 sensors. Understanding the mechanisms by which blood-feeding insect disease vectors, such as the malaria mosquito *Anopheles gambiae*, locate their hosts may lead to the development of disease-prevention strategies. Olfactory receptors and gustatory receptors represent related branches of a chemoreceptor superfamily. Recognition of carbon dioxide by gustatory receptors in the antenna, and by receptors in the proboscis, illustrates that the boundaries between olfaction and taste in insects cannot always be clearly defined.

TRANSGENIC APPROACHES TO MAP PROJECTION PATTERNS OF OLFACTORY SENSORY NEURONS

Early studies showed that some mRNA that encodes olfactory receptors is transported to the axonal termini in the **olfactory bulb** and, because of axonal convergence in glomeruli of the olfactory bulb, could generate a strong enough signal to be detected by *in situ* hybridization. These studies gave the first evidence that those axons from neurons that express the same receptors project to the same glomeruli, in each case one medial and one lateral glomerulus. Subsequent studies with transgenic knock-in mice have provided deeper insights into the organization of the chemosensory projection from the periphery to the olfactory bulb. Peter Mombaerts and colleagues at the Rockefeller University made knock-in mice in which the gene that encodes the P2 odorant receptor had been replaced by a bicistronic construct, in which the coding region of the P2 odorant receptor is separated from the coding sequence of a lacZ or GFP reporter sequence by an internal ribosome entry site (IRES). This ensures that the P2 odorant receptor is always co-expressed in the same cells with the reporter gene (Figure 13.16, p. 220). The reporter constructs also contain a sequence that corresponds to the tau protein, which is a microtubule-associated protein. Consequently, the tau-lacZ or tau-GFP reporter proteins

are associated with microtubules and transported along the axon, staining the entire projection of those cells that co-express the reporter with the P2 odorant receptor. These elegant experiments demonstrated, once again, that all cells that expressed the P2 receptor converged on two glomeruli, one medial and one lateral (Figure 13.17, p. 220). When the P2 coding region was replaced by tau-GFP or lacZ alone, this functional organization was lost, demonstrating that the receptor itself has an instructive role in establishing correct neuronal connectivity.

Next, Mombaerts and colleagues conducted experiments in which open reading frames of odorant receptors were swapped. They might have predicted that axons would now project to a location that would precisely correspond to the native glomerulus to which axons from cells that usually express the swapped receptor would normally project. This was, however, not the case. The transgenic receptor steered the axons towards the general direction of its native glomerulus, but the placement was not precise (Figure 13.18, p. 221). These experiments led to the conclusion that additional factors other than the odorant receptor are needed to generate precise projections. Such factors are likely to include differentially-expressed components of the **extracellular matrix** and their interactions with axon guidance molecules. For example, knock-out mice which were deficient in the extracellular matrix components **ephrin A** and **ephrin B** showed distortions in the

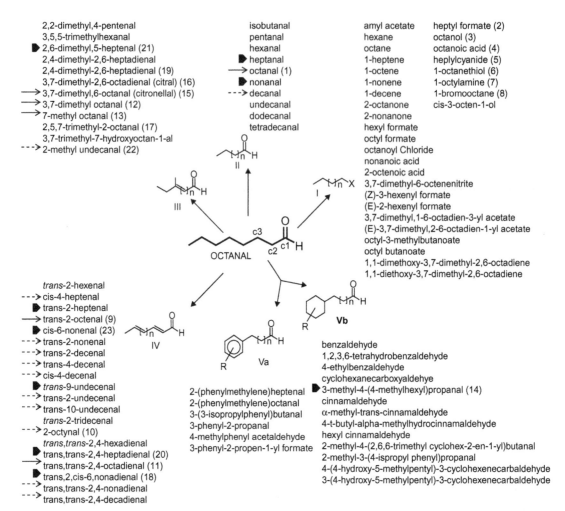

2,2-dimethyl,4-pentenal
3,5,5-trimethylhexanal
▶ 2,6-dimethyl,5-heptenal (21)
2,4-dimethyl-2,6-heptadienal
2,4-dimethyl-2,6-heptadienal (19)
3,7-dimethyl-2,6-octadienal (citral) (16)
→ 3,7-dimethyl,6-octanal (citronellal) (15)
→ 3,7-dimethyl octanal (12)
→ 7-methyl octanal (13)
2,5,7-trimethyl-2-octanal (17)
3,7-trimethyl-7-hydroxyoctan-1-al
---▶ 2-methyl undecanal (22)

isobutanal
pentanal
hexanal
▶ heptanal
→ octanal (1)
▶ nonanal
---▶ decanal
undecanal
dodecanal
tetradecanal

amyl acetate
hexane
octane
1-heptene
1-octene
1-nonene
1-decene
2-octanone
2-nonanone
hexyl formate
octyl formate
octanoyl Chloride
nonanoic acid
2-octenoic acid
3,7-dimethyl-6-octenenitrite
(Z)-3-hexenyl formate
(E)-2-hexenyl formate
3,7-dimethyl,1-6-octadien-3-yl acetate
(E)-3,7-dimethyl,2-6-octadien-1-yl acetate
octyl-3-methylbutanoate
octyl butanoate
1,1-dimethoxy-3,7-dimethyl-2,6-octadiene
1,1-diethoxy-3,7-dimethyl-2,6-octadiene

heptyl formate (2)
octanol (3)
octanoic acid (4)
heplylcyanide (5)
1-octanethiol (6)
1-octylamine (7)
1-bromooctane (8)
cis-3-octen-1-ol

trans-2-hexenal
---▶ cis-4-heptenal
▶ trans-2-heptenal
→ trans-2-octenal (9)
▶ cis-6-nonenal (23)
---▶ trans-2-nonenal
---▶ trans-2-decenal
---▶ trans-4-decenal
---▶ cis-4-decenal
▶ trans-9-undecenal
---▶ trans-2-undecenal
---▶ trans-10-undecenal
trans-2-tridecenal
---▶ 2-octynal (10)
trans,trans-2,4-hexadienal
▶ trans,trans-2,4-heptadienal (20)
→ trans,trans-2,4-octadienal (11)
▶ trans,2,cis-6,nonadienal (18)
---▶ trans,trans-2,4-nonadienal
---▶ trans,trans-2,4-decadienal

2-(phenylmethylene)heptenal
2-(phenylmethylene)octanal
3-(3-isopropylphenyl)butanal
3-phenyl-2-propanal
4-methylphenyl acetaldehyde
3-phenyl-2-propen-1-yl formate

benzaldehyde
1,2,3,6-tetrahydrobenzaldehyde
4-ethylbenzaldehyde
cyclohexanecarboxyaldehye
▶ 3-methyl-4-(4-methylhexyl)propanal (14)
cinnamaldehyde
α-methyl-trans-cinnamaldehyde
4-t-butyl-alpha-methylhydrocinnamaldehyde
hexyl cinnamaldehyde
2-methyl-4-(2,6,6-trimethyl cyclohex-2-en-1-yl)butanal
2-methyl-3-(4-isopropyl phenyl)propanal
4-(4-hydroxy-5-methylpentyl)-3-cyclohexenecarbaldehyde
3-(4-hydroxy-5-methylpentyl)-3-cyclohexenecarbaldehyde

FIGURE 13.9 Determining the molecular receptive range of the rat I7 receptor overexpressed via an adenovirus in rat olfactory epithelium. Using octyl aldehyde as a template, potential agonists were selected according to the number of carbons, degree of saturation, branching, and substitutions. Compounds are classified according to substitutions of the aldehyde group (I), *n*-aldehydes (II), *n*- and branched singly-unsaturated aldehydes (III), singly- and doubly-unsaturated *n*-aldehydes (IV), and cyclic and aromatic aldehydes (Va and Vb). Compounds that elicited a significant electrophysiological response when applied to the transfected olfactory epithelium are indicated from the highest value (solid arrows) to the lowest (dashed arrows), with arrow heads indicating intermediate activation. (From Araneda, R. C. et al. (2000). *Nat. Neurosci.*, **3**, 1248–1255.)

chemotopic map, suggesting that ephrin gradients may play a role in determining connectivity. In addition, different levels of expression of the membrane receptor neuropilin 1 in olfactory sensory neurons and interactions with its axon guidance ligand semaphorin 3A appear to be instrumental in axon sorting as axons from olfactory sensory neurons make their way to the olfactory bulb.

How stereotypic is the formation of the chemosensory map? This question could be answered by double-labeling experiments, in which closely-related receptors that project to neighboring glomeruli are visualized through the use of fluorescent bicistronic constructs that emit light at different wavelengths. These experiments showed that there is, in fact, individual local variation in the precise projection patterns between axons from neurons that express similar receptors (presumably recognizing similar odorants) within a globally universal chemosensory map.

THE USE OF TRANSGENIC REPORTER GENES TO VISUALIZE ODOR CODING

Transgenic technologies have been developed both in flies and mice to introduce probes, for example **cameleon** or **GCaMP**, which can monitor neural activity in real-time, while animals are exposed to odorants. Cameleon is a synthetic protein that consists of an enhanced cyan fluorescent protein (ECFP) and an enhanced yellow fluorescent protein (EYFP), separated by a calmodulin sequence and the calmodulin target peptide M13. On binding of calcium to calmodulin, interaction with the calmodulin-binding peptide will bring the two GFP moieties close together. This results in a shift in the ratio of EYFP to ECFP emission, as a result of fluorescence resonance energy transfer (FRET) from ECFP to EYFP (Figure 13.19, p. 221). Introduction of a transgenic construct in *Drosophila* that

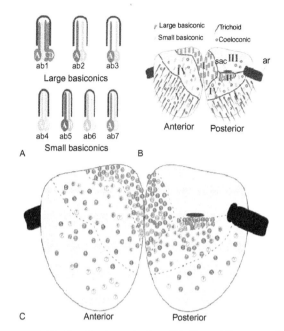

A Large basiconics

Small basiconics

B

Large basiconic /Trichoid
Small basiconic •Coeloconic

Anterior Posterior

C Anterior Posterior

FIGURE 13.10 Functional organization of the *Drosophila* antenna. Panel (A) shows the 16 functional classes of olfactory sensory neurons and their stereotyped organization within seven functional types of sensilla. Panel (B) shows the distribution of morphological categories of sensilla across the antenna. The arista (ar) is a mechanosensory structure that extends from the antenna. Panel (C) shows the distribution of functional types of sensilla. Each circle represents a recording from one sensillum; question marks represent recordings that could not be classified. The dotted lines are those shown in (B), representing boundaries between morphological categories of sensilla. Dorsal is at top; medial is along the center of the diagram. (From de Bruyne, M. et al. (2001). *Neuron*, **30**, 537–552.)

expresses cameleon in neurons will now allow imaging of neural activity in behaving animals (Figure 13.20, p. 221).

A further advance was made with the development of **synapto-pHluorin**, which is a fusion of a pH-sensitive green fluorescent protein variant with the mouse synaptic vesicle-associated protein VAMP-2. In transgenic animals synapto-pHluorin becomes associated with synaptic vesicles in presynaptic terminals. The fluorescent domain is exposed to the acidic lumen of the vesicles, and is twenty times less fluorescent than at neutral pH. The pH inside the vesicles changes to neutral when they fuse with the presynaptic membrane to release neurotransmitter, and this will generate a fluorescent signal that can be monitored in live animals (Figure 13.21, p. 222). Both transgenic flies and transgenic mice that express synapto-pHluorin have been constructed and used to monitor synaptic activity during odor-conditioned learning in the mushroom bodies and olfactory signaling in the olfactory bulb, respectively. In principle, this method can be used to monitor a wide variety of neural pathways that are activated in animals displaying different behaviors in response to environmental stimuli.

The ability to genetically-tag neurons that express a defined odorant receptor with a fluorescent reporter enabled the identification of ligands for such receptors, as ligand-activated calcium signals could be directly associated with a fluorescent glomerulus. Such studies which utilized optical imaging approaches on transgenic animals showed that the medial and lateral glomeruli that receive

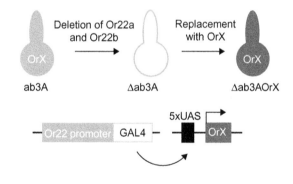

Deletion of Or22a and Or22b → Replacement with OrX

OrX
ab3A → Δab3A → Δab3AOrX

Or22 promoter GAL4 — 5xUAS OrX

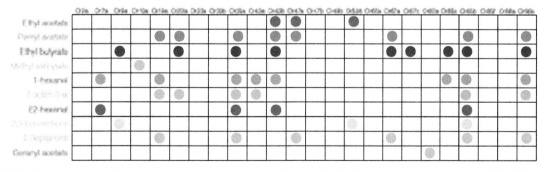

FIGURE 13.11 An *in vivo* transgenic assay system for determining molecular receptive fields of odorant receptors in *Drosophila*. Olfactory sensory neurons in the ab3A sensilla normally express the closely-linked Or22a and Or22b receptors. Deletion of these *Or* genes results in a silent ab3A sensilla. Using an *Or22a* promoter to drive GAL4 allows expression of any transgenic *Or* sequence cloned behind UAS sequences to be expressed in the ab3A sensilla. Responses recorded from such ectopically-expressed odorant receptors faithfully reproduced their native odorant response profiles. The lower panel shows odorant response specificities of odorant receptors expressed in the ab3A sensilla of the *Or22a/Or22b* deletion mutant. (Modified from Hallem, E. A., and Carlson, J. R. (2004). *Trends Genet.*, **20**, 453–459.)

information from the same odorant receptor are connected via an interneuronal circuitry in the bulb, suggesting that temporal oscillations or coincidence circuits may aid in odorant discrimination (Figure 13.22, p. 222).

What is the mechanism by which neurons select the odorant receptor they express? Does the selection of a particular odorant receptor involve reorganization within the genome, as is the case for immunoglobulins, where physical

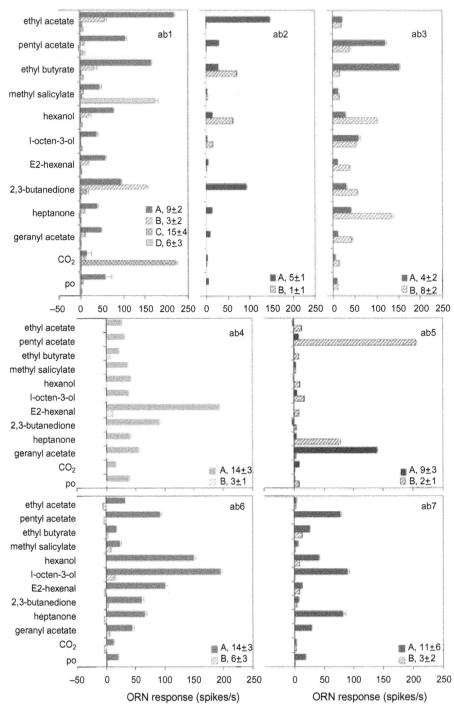

FIGURE 13.12 Response profiles of olfactory sensory neurons for the three types of large basiconic sensilla (ab1, 2, and 3) and the four types of small basiconic sensilla (ab4, 5, 6, and 7) in the *Drosophila* antenna, using a set of 11 diagnostic stimuli and the solvent control, paraffin oil (po). The indicated response is measured as the increase (or decrease) in spike(s) compared to the spontaneous firing frequency. Spontaneous frequencies of the neurons in each sensillum type are indicated in the lower right corner. (From de Bruyne, M. et al. (2001). *Neuron*, **30**, 537–552.)

alterations in the genome ensure that a given lymphocyte expresses only one specific antibody? In a *tour de force* experiment, mice were cloned from postmitotic olfactory neurons. The functional organization of their olfactory system was indistinguishable from that of normal mice, with olfactory neurons each expressing a unique different odorant receptor rather than all expressing the

same receptor; thus, the choice of odorant receptor does not involve irreversible alterations in genome structure. Studies using homologous recombinant mice have shown that expression of the odorant receptor protein provides negative feedback that inhibits expression of additional odorant receptors. In *Drosophila*, such negative feedback has not been observed. Instead, odorant receptor choice in flies appears to depend on combinations of transcription factors acting on distinct promoter elements (Figure 13.23, p. 223).

It had been hypothesized that in mice an activation complex formed in a promoter region, designated the H region, stochastically chooses an odorant receptor gene promoter site by random collision, activating one particular odorant receptor gene (Figure 13.23). Once the activated gene is expressed, the functional odorant receptor molecules would then transmit a currently unknown inhibitory signal to block the further activation of additional odorant receptor genes. However, knock-out mice that lack the H region showed normal expression patterns of odorant receptor genes; only a few odorant receptor genes in a nearby cluster were affected, along with minor effects on four odorant receptor genes further removed from the homologous recombinant site. These results suggest that the H region exerts at best a limited *cis*-regulatory effect, and that other currently unknown mechanisms must determine odorant receptor gene choice.

In a further twist on the "monoclonal nose," the laboratory of Richard Axel generated transgenic mice in which greater than 95% of the olfactory sensory neurons express a single odorant receptor, the mouse M71 receptor, which recognizes the odorant acetophenone. Such mice can still discriminate odorants using endogenously expressed odorant receptors, albeit less effectively, but they cannot distinguish acetophenone, despite the fact that all glomeruli in the olfactory bulb are activated by this odorant. Thus, differential activation of glomeruli is critical for odorant recognition.

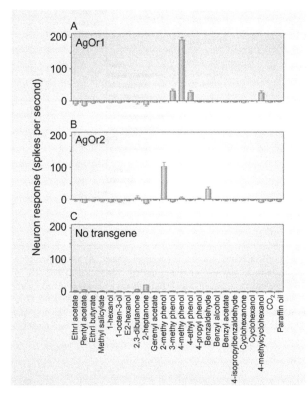

FIGURE 13.13   Expression of odorant receptors from the malaria-transmitting mosquito, *Anopheles gambiae*, in the silent neuron of the *Or22a* deletion mutant. Expression of the AgOr1 receptor elicits electrophysiological responses to 4-methylphenol, an odorant found in human sweat. (From Hallem, E. A. et al. (2004). *Nature*, **427**, 212–213.)

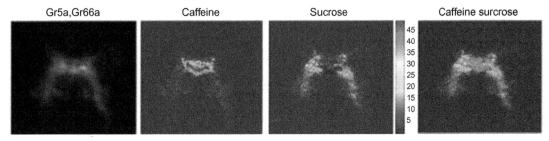

FIGURE 13.14   Images of a live fly preparation used for monitoring taste-induced, G-CaMP fluorescent changes. Removal of antennae and surrounding cuticle allows visual access to the fly brain. G-CaMP expression in both sugar-sensitive Gr5a neurons and bitter-sensitive Gr66a neurons reveals that the projections of these neurons to the suboesophageal ganglion are spatially segregated. The first image shows initial G-CaMP fluorescence, second is intensity increase after 100 mM caffeine, third is intensity increase after 1 M sucrose stimulation, fourth is overlay of caffeine-induced and sucrose-induced fluorescent change. Neurons in the proboscis that express the Gr5a receptor send axons to the anterior suboesophageal ganglion and do not cross the midline. Gr66a neurons send axons that terminate in the medial suboesophageal ganglion in a ringed web. (Modified from Marella, S. et al. (2006). *Neuron*, **49**, 285–295.)

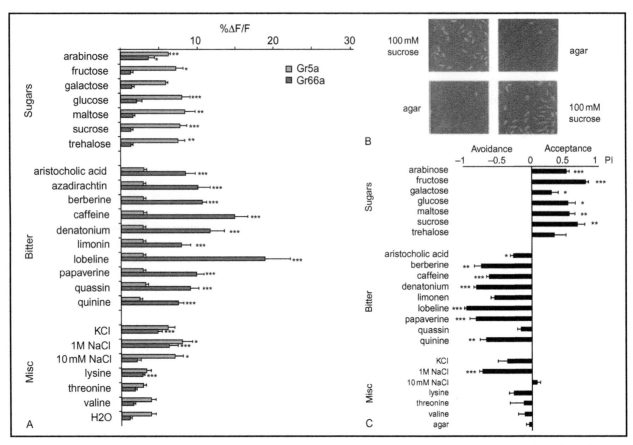

FIGURE 13.15 Responses of flies to substances that activate Gr5a ("sweet") and Gr66a ("bitter") expressing neurons. Panel (A) shows fluorescent changes for Gr5a projections and Gr66a projections shown in Figure 13.14, in response to 23 "sweet-" or "bitter-tasting" compounds. Panels (B) and (C) show the corresponding behavioral responses. Panel (B) shows the adult taste preference assay, in which flies, given a choice between agarose versus agarose plus 100 mM sucrose, flock to the sucrose quadrants. Panel (C) shows preference indices of wild-type flies ranging from +1 (100% attraction) to −1 (100% avoidance) against 23 compounds. Stars denote significant differences (**Student's t test**, ***p < 0.005, **p < 0.01, *p < 0.05). **Tastants** that activate the Gr5a projection elicit attractant responses, whereas tastants that activate the Gr66a projection show avoidance responses. (Modified from Marella, S. et al. (2006). *Neuron*, **49**, 285–295.)

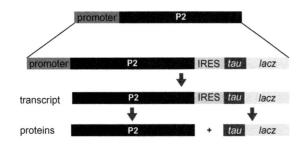

FIGURE 13.16 A knock-in construct in homologous recombinant mice gives rise to a bicistronic construct where a tau-*lacZ* construct is linked to transcription of the P2 open reading frame via an internal ribosome entry site (IRES).

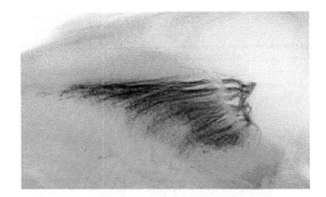

FIGURE 13.17 P2-IRES-tau-lacZ mice, stained with X-Gal. Whole mount view of the wall of the nasal cavity and the medial aspect of the bulb of a 4-month-old mouse. P2-IRES-tau-lacZ is a bicistronic construct, which means that two proteins are translated from the same mRNA. The internal ribosome entry site (IRES) ensures that each cell which expresses the P2 receptor also expresses the tau-lacZ marker. Note the convergence of the axons of olfactory sensory neurons as they project from the olfactory neuroepithelium to the olfactory bulb. (From Mombaerts, P. et al. (1996) *Cell*, **87**, 675–686.)

Identification of Pheromone Receptors and Functional Organization of the Accessory Olfactory (Vomeronasal) System

In addition to the main olfactory system, rodents and other mammals (except higher primates) rely on the **accessory olfactory system** for chemosensory input from their environment. As indicated in Chapter 12, the accessory olfactory system mediates gender discrimination and kin recognition, and contributes to social behavior (Figure 13.24, p. 223). The portal to the accessory olfactory system is the **vomeronasal organ**, an elongated cigar-shaped blind sac that is located in the vomer bone above the palate and

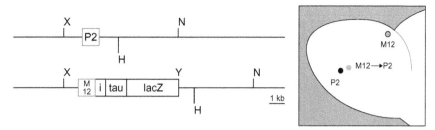

FIGURE 13.18   When the coding region of the mouse M12 odorant receptor is replaced by the P2 receptor, olfactory neurons expressing this transgene, together with *lacZ*, project to a location near, but not identical to, that of the original P2 glomerulus. (Adapted from Mombaerts, P. et al. (1996). *Cell*, **87**, 675–686.)

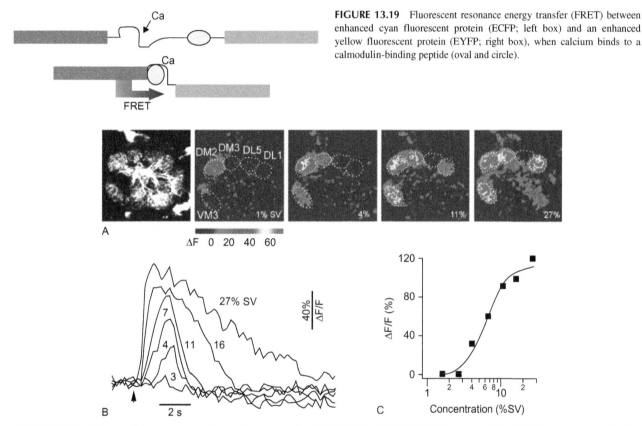

FIGURE 13.19   Fluorescent resonance energy transfer (FRET) between enhanced cyan fluorescent protein (ECFP; left box) and an enhanced yellow fluorescent protein (EYFP; right box), when calcium binds to a calmodulin-binding peptide (oval and circle).

FIGURE 13.20   Patterns of glomerular activation in flies bearing the *GH146-Gal4* and *UAS-GCaMP* transgenes. The *GH146* promoter strongly drives expression of GAL4 in projection neurons in the *Drosophila* antennal lobe. Panel A shows temporal activation of glomeruli on exposure to isoamyl acetate. The concentration dependence and time course of the glomerular response to isoamyl acetate in a single glomerulus, the VM3 glomerulus, is shown in panels B and C, respectively. (From Wang, J. W. et al. (2003). *Cell*, **112**, 271–282.)

communicates with the external environment via a duct that opens above the upper lip, or in some species in the nasal cavity. The duct is lined with vascular tissue that acts as a pump to draw liquid into the vomeronasal organ. The organ is lined on one side with neural tissue, which consists of supporting cells and vomeronasal chemosensory neurons. The latter are distinct from olfactory sensory neurons in that they do not have a ciliated dendritic knob, but instead extend microvilli into the lumen. Their axons project to a region in the brain that is dorsocaudal to the main olfactory bulb, referred to as the **accessory olfactory bulb**.

Like the main olfactory bulb, the accessory olfactory bulb is a laminated structure which has glomeruli, but the glomeruli are here not as clearly defined as in the main olfactory system. Output neurons from the accessory olfactory bulb do not project to the olfactory cortex, but rather form a direct connection with limbic brain regions, the amygdala and the hypothalamus, where they can elicit behavioral and neuroendocrine responses. Thus, due to the lack of a cortical connection, vomeronasal sensations are thought to be "subconscious," and to drive "instinctive" behaviors in response to chemical signals, described in Chapter 12.

A seminal discovery was made by Mimi Halpern and her colleagues in 1995, when she found that the vomeronasal organ consists of two segregated subdivisions, which

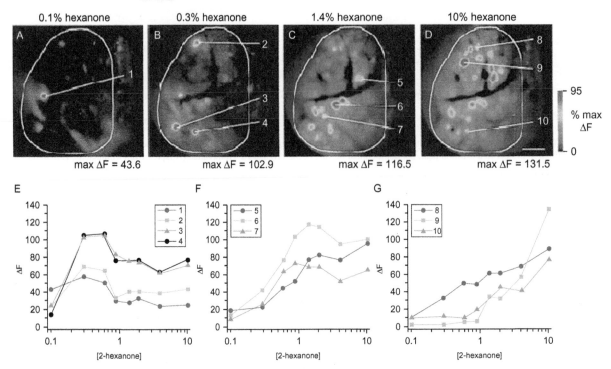

max ΔF = 43.6   max ΔF = 102.9   max ΔF = 116.5   max ΔF = 131.5

FIGURE 13.21   Real-time fluorescent images in mice that express synapto-pHluorin in olfactory sensory neurons in response to different concentrations of 2-hexanone. Expression of synapto-pHfluorin is driven by the promoter of **olfactory marker protein** (OMP), a protein expressed in the olfactory pathway specifically in olfactory sensory neurons. Fluorescent signals reveal spatial and temporal patterns of glomerular activation that become more pronounced as the odorant concentration is increased (A–D). Dose-responses for activation of each of the responding glomeruli can be measured separately (E–G). (From Bozza, T. et al. (2004). *Neuron*, **42**, 9–21.)

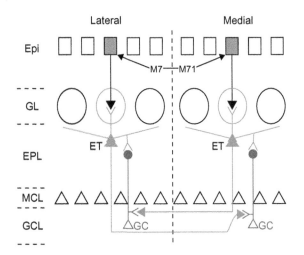

FIGURE 13.22   The coding region of the mouse M71 receptor was replaced by the rat I7 receptor, which has a known odorant-specificity profile, to direct the expression of this receptor to the dorsal surface of the olfactory bulb, where glomerular activity can be easily visualized by optical imaging. The lateral and medial glomeruli to which neurons expressing the transgenic receptor project were found to be connected via a reciprocal neuronal circuitry are illustrated in the diagram (Epi, epithelium; GL, glomerular layer; EPL, **external plexiform layer**; MCL, **mitral cell layer**; GCL, **granule cell layer**; ET, external tufted cell; GC, granule cell). (From Belluscio, L. et al. (2002). *Nature*, **419**, 296–300.)

could be classified on the basis of the G-proteins they express. Neurons that express the $G_{\alpha i2}$ subunit were located in the apical layer of the vomeronasal neuroepithelium and project axons to the rostral region of the accessory olfactory bulb, whereas neurons that express the $G_{\alpha o}$ subunit are located in the basal layer of the vomeronasal neuroepithelium, and project to the caudal region of the accessory olfactory bulb. This discovery was not only important because it suggested that the two subdivisions of the accessory olfactory system perform different tasks, but it also set the stage for the subsequent discovery of vomeronasal chemoreceptors. The first family of about 300 putative **pheromone receptors** in the vomeronasal organ was identified simultaneously by three laboratories and was based on the expectation that vomeronasal neurons that express $G_{\alpha i2}$ would express distinct pheromone receptors, and that such pheromone receptors would be members of the family of G-protein-coupled receptors. Single-cell PCR and differential hybridization among $G_{\alpha i2}$ expressing cells resulted in the identification of a new family of chemoreceptors, known as the **V1R receptors**. Like the odorant receptors, V1R receptors have seven transmembrane domains, but that is where the similarity stops. They are phylogenetically entirely distinct from odorant receptors.

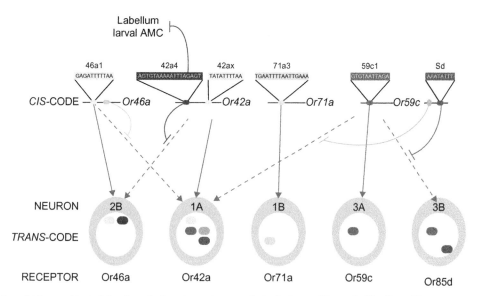

FIGURE 13.23   A model for combinatorial coding of odorant receptor gene choice in the maxillary palp. The figure illustrates conserved gene-specific regulatory elements in the *Or46a*, *Or42a*, *Or71a*, and *Or59c* genes and the olfactory receptor neurons in the maxillary palp, in which they are expressed. Both positive and negative regulatory elements are shown. Solid arrows connect regulatory elements to the olfactory receptor neurons in which the corresponding genes are expressed. These arrows do not imply that these elements alone are capable of directing proper expression. The curved lines show elements that repress gene expression and the dashed arrows show ectopic expression in the absence of repression. AMC designates the antenno–maxillary complex, which contains the larval dorsal organs. (From Ray, A., van der Goes van Naters, W., and Carlson, J. (2008). *PLoS Biol.*, **6**, e125.)

FIGURE 13.24   A mare displaying a **Flehmen response**, characterized by curling of the upper lip to enable volatile pheromones to be drawn into the vomeronasal organ (also known as **Jacobson's organ**). Stallions will exhibit the Flehmen response when smelling the urine of a mare in heat.

They are expressed only in neurons in the apical zone of the vomeronasal organ, and regulation of their expression obeys the same one-neuron-one-receptor rule as in the main olfactory epithelium.

Subsequent to the discovery of the V1R receptors, a second family of about 50 putative pheromone receptors was discovered and named **V2R receptors**. They are expressed in the basal layer of the vomeronasal neuroepithelium along with $G_{\alpha o}$. Again, V2R receptors are structurally distinct from V1R and odorant receptors. They are seven transmembrane domain receptors characterized by a large extracellular N-terminal domain, which resembles **metabotropic glutamate receptors** and **calcium sensing receptors**. It became clear that independent families of chemoreceptors have evolved to fulfill distinct chemosensory functions (Figure 13.25).

When transgenic mice were engineered which express GFP in vomeronasal neurons together with vomeronasal receptors, neural projections to the accessory olfactory bulb visualized a more diffuse pattern than the distinct convergence evident in the main olfactory bulb. Consequently, glomeruli in the accessory olfactory bulb are smaller and less well-defined than their counterparts in the main olfactory bulb. The projection pattern in the accessory olfactory bulb suggest that decoding of chemosensory information, which occurs in cortex in the olfactory system, happens earlier in the vomeronasal projection, perhaps already at the level of the first synapse in the accessory olfactory bulb (Figure 13.26).

Whereas the main olfactory system and the vomeronasal system were once considered functionally entirely separate chemosensory systems, it is now clear that common odorants can activate vomeronasal neurons, and that vomeronasal stimuli can activate neurons in the main olfactory epithelium. A subset of olfactory sensory neurons, in fact, responds to components of urine and projects to the ventral olfactory bulb. These neurons are unique, in that they express a member of the TRP-channel family, TRPM5, which also plays a role in taste transduction, as described in the next section.

TASTE AND GUSTATORY RECEPTORS

When we comment on the excellent taste of a fine wine, we are in reality referring to an integrated chemosensory sensation that consists of gustatory, olfactory, and trigeminal stimulation, in which the sweetness or sourness of the wine is assessed by gustatory mechanisms, the "punch" of the alcohol is a trigeminal sensation and the complex "bouquet" arises from the activation of olfactory sensory neurons. There are five gustatory modalities, sweet, sour, salty, bitter, and umami. These have evolved, presumably, to assess the nutritional value of food by sensing the sugar content (sweet), its pH (sour), its salt content for osmoregulation (salty), and detection of poisonous compounds that should not be ingested (bitter). People of Asian origin are especially sensitive to the meaty taste of monosodium glutamate, a taste modality designated as **umami**, which, like the sweet taste modality, may have evolved to assess nutritional value through the detection of amino acids.

Studies on genetically-engineered mice, pioneered by the laboratory of Robert Margolskee at the Mount Sinai Hospital in New York and Charles Zuker and his group at the University of California at San Diego, have been instrumental in identifying how taste cells expressing distinct gustatory receptors trigger gustatory behavioral responses. The first gustatory receptors that were identified were

---

Box 13.5   Organization of the mammalian gustatory system

Taste perception is mediated by taste cells, which are organized as groups of 40–60 cells, known as **taste buds**. Taste cells are elongated cells, isolated from one another via **tight junctions** that separate the apical membrane from the basolateral membrane. The apical membrane forms microvilli to provide an expanded surface for the detection of tastants. From the base of the taste cell, neurotransmitter can be released onto nerve fibers that synapse onto the taste cell. The top of the taste bud communicates with the external world via an opening, the **taste pore**. Taste buds are found throughout the oral cavity, but in the tongue are contained within **papillae**, peglike projections of the lingual epithelium. There are four types of papillae: filiform, fungiform, foliate, and circumvallate. Filiform papillae do not contain taste buds, but contribute roughness to the lingual surface to help the mechanics of food processing. Fungiform papillae are found all over the tongue, especially in the anterior region, and contain one or a few taste buds. They mediate perception of sweet, salty, and sour taste. Foliate papillae are restricted to the edges of the posterior regions of the tongue. Circumvallate papillae are lined up near the base of the tongue (Figure A). Whereas the rat tongue contains only a single circumvallate papilla, there are about 7–12 circumvallate papillae in a chevron-like arrangement on the human tongue. About half of the 2000 taste buds of the human tongue are in the circumvallate papillae, and are specialized for bitter taste perception.

Taste cells on the anterior two thirds of the tongue are innervated by the **chorda tympani** branch of the facial nerve (cranial nerve VII). Taste cells on the posterior part of the tongue (including the cicumvallate papillae) are innervated by the glossopharyngeal nerve (cranial nerve IX). Like olfactory neurons, taste cells undergo continuous functional replacement, and new taste cells are generated from basal cells in the taste bud. Thus, new synapses are continuously being formed in the taste bud. Gustatory information is relayed initially to the **nucleus of the solitary tract** in the brain stem. From there, taste information is relayed to nuclei in the pons and thalamus before it is relayed to the cortex. Gustatory information is also sent from the pons to the hypothalamus, where it can modulate feeding behavior (ingestion or rejection reflexes).

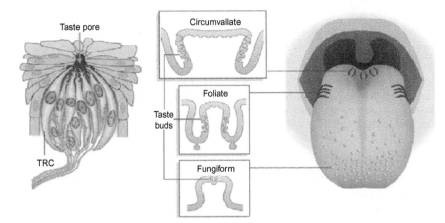

FIGURE A   Organization of taste cells on the human tongue. Elongated taste receptor cells (TRC) are organized in taste buds that communicate with the external environment via a taste pore, and taste buds are located in papillae on the human tongue. (From Chandrashekar, J. et al. (2006). *Nature*, **444**, 288–294.)

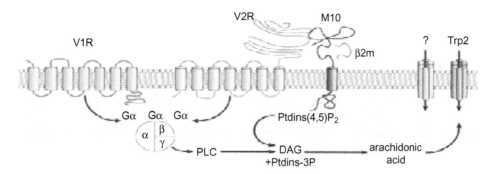

FIGURE 13.25   A model for signal transduction in the vomeronasal organ. Binding of pheromone ligands to G-protein-coupled receptors (V1R and V2R) results in stimulation of a heterotrimeric G-protein, which in turn triggers a signaling cascade through activation of phospholipase C (PLC) that generates inositol 1,4,5-triphosphate (PtIns3P) and arachidonic acid as second messengers. The latter controls opening of the Trp2 cation channel. An additional un-characterized channel may also play a role. PtdIns(4,5)P2, phosphatidyl inositol bisphosphate; β2 m, β2-microglobulin. Note that this diagram does not intend to imply that both V1R and V2R receptors are expressed in the same cells, which is not the case. (Modified from Dulac, C., and Torello, A. T. (2003). *Nat. Rev. Neurosci.*, **4**, 551–562.)

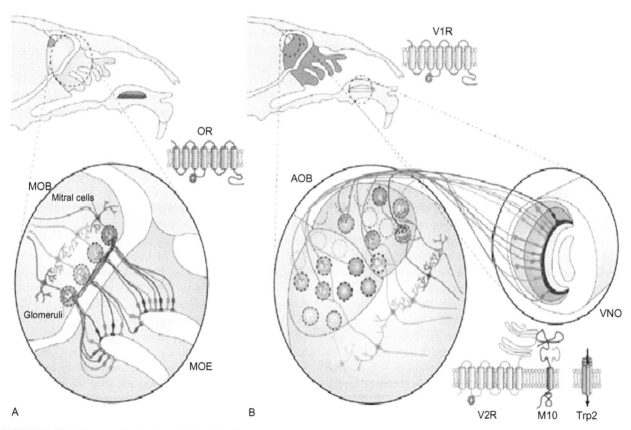

FIGURE 13.26   Diagram of projections in the main olfactory system (A) and the accessory olfactory system (B). Axons from olfactory neurons that express odorant receptors in the main olfactory epithelium (MOE) converge on distinct glomeruli in the main olfactory bulb (MOB). The vomeronasal organ (VNO) is divided into two subdivisions, which express a family of V1R receptors and V2R receptors, respectively. V2R receptors are associated with the nonclassical major histocompatibility complex class 1b protein, M10 (the classical **major histocompatibility complex** (MHC) encodes surface antigens associated with self-recognition and cell–cell interactions in the immune system). V1R receptors appear to recognize small ligands, whereas urinary peptides are recognized by the V2R-M10 complexes (as described in Chapter 12). Projections from the VNO to the AOB are diffuse, forming glomeruli that are less well-defined and receive input from neurons with different pheromone receptor specificities. (From Dulac, C., and Torello, A. T. (2003). *Nat. Rev. Neurosci.*, **4**, 551–562.)

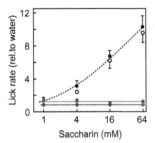
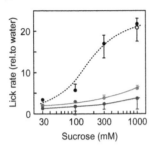

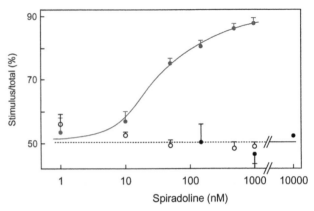

FIGURE 13.27 Taste preferences of control mice (open circles, dashed lines), T1R1 knock-out mice (gray circles, dashed), T1R2 knock-out mice (upper solid line), and T1R3 knock-out mice (lower solid line) were measured relative to water, by measuring the number of licks from a drinking source. T1R1 knock-out mice are equivalent to controls, whereas T1R2 and T1R3 knock-out mice showed complete loss in preference for artificial sweeteners and D-amino acids, but retained residual responses to high concentration of natural sugars, which were eliminated in double T1R2/T1R3 knock-outs. (Modified from Zhao, G. Q. et al. (2003). *Cell*, **115**, 255–266.)

FIGURE 13.28 Expression of RASSL in T1R2-cells generates animals that exhibit specific behavioral attraction to spiradoline (solid line). Note that no responses are seen in control mice (dotted line). Values are means ± SEM (n = 7). (Modified from Zhao, G. Q. et al. (2003). *Cell*, **115**, 255–266.)

receptors for sweet taste and umami. These became known as the **T1R receptors**, and they are distantly related to the V2R pheromone receptors, metabotropic glutamate receptors, and extracellular calcium-sensing receptors. It quickly became evident that an ensemble of only three T1R receptors is sufficient for the recognition of all natural and artificial sweet compounds, and for recognition of the amino acid glutamate that conveys the sensation of umami. Surprisingly, T1R receptors are functional as dimers. Association of T1R1 and T1R3 constitutes the receptor for umami, whereas the complex between T1R2 and T1R3 forms the sweet taste receptor. Mice that lack the T1R1 or T1R3 receptors cannot taste umami as assessed by **two-bottle preference tests**, in which mice can choose to drink from a bottle that contains a tastant, and a bottle that contains only water (Figure 13.27). Early QTL mapping studies had identified a chromosomal region that contained a locus that determines variation among mice strains in their sensitivity to the sweetener saccharin, designated as the *sac* locus. The gene responsible for this phenotypic variation at the *sac* locus was found to encode the T1R3 receptor.

In addition to the sweet and umami taste receptors, a second group of G-protein-coupled receptors was discovered that mediate bitter taste perception. These receptors represent a family that is distinct from the T1R receptors and highly divergent. They are designated T2Rs. There are 25 T2R receptors in humans, and 35 in the mouse. Whereas the T1R1 and T1R2 receptors are expressed in non-overlapping cell populations, each of which co-expresses T1R3, bitter taste receptors are co-expressed in the same cells. This explains why many structurally-distinct aversive compounds all generate a similar bitter taste sensation. As the objective of the bitter taste modality of the gustatory system is to avoid ingestion of any potentially harmful compound, there has been no evolutionary pressure to discriminate among bitter tastants. Expression of a diverse array of bitter taste receptors in cells that are hardwired to

generate an avoidance response when activated represents the simplest, most economical, solution.

An elegant demonstration that sweet and bitter responses are indeed hardwired and mediated via distinct neural pathways came from a study in which a modified **κ-opioid receptor** RASSL (Receptor Activated Solely by a Synthetic Ligand) was expressed in taste cells of genetically-engineered mice. RASSL binds to the opioid agonist spiradoline, a compound which mice never encounter naturally. When RASSL was expressed under a T1R promoter in sweet taste cells, mice preferred to drink water supplemented with spiradoline. In contrast, when RASSL was expressed in bitter taste cells, mice would avoid spiradoline (Figure 13.28).

The clear demarcation between sweet and bitter responses was also demonstrated by selective ablation of each cell type by generating transgenic mice in which expression of diphtheria toxin was targeted to either sweet taste cells or bitter taste cells using constructs with T1R or T2R promoters. Ablation of sweet taste cells obliterated sensitivity to sweet taste, but left other taste modalities intact. Ablation of bitter taste cells obliterated avoidance responses to bitter compounds, and similarly left responses to other taste modalities unaffected.

Expression studies of taste receptors in cell culture have identified ligands for some bitter taste receptors. For example, the T2R5 receptor responds to cycloheximide *in vitro*, and T2R5 knock-out mice no longer avoid drinking water that contains cycloheximide (Figure 13.29). However, many T2R receptors are still **orphan receptors**, that is, receptors for which a ligand has not yet been identified. In some cases, natural polymorphisms can be helpful in matching receptors to bitter tastants. This was illustrated most vividly by a gene-mapping study for the bitter compound **phenylthiocarbamide** (PTC). This compound occurs naturally in certain vegetables, such as artichokes, and is experienced as intensely bitter by about 70% of people,

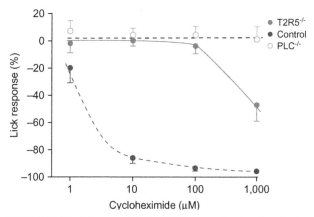

FIGURE 13.29 Mice in which the T2R5 receptor has been knocked-out by homologous recombination are insensitive to the aversive bitter tasting compound cycloheximide compared to control mice. Mice that lack the essential transduction enzyme phospholipase C are also unable to taste cycloheximide (dotted line). Responses of the $T2R5^{-/-}$ mice to other bitter tastants are, however, unaltered as are their responses to other taste modalities. (Modified from Mueller, K. L. et al. (2005). *Nature*, **434**, 225–229.)

whereas others cannot taste PTC at all. It has been postulated that taste sensitivity to PTC has evolved in primates to enable them to avoid ingestion of certain poisonous plants. The PTC taste phenotype segregates as a nearly **Mendelian trait**, and was mapped to a region on chromosome 7q. This chromosomal region contains a *T2r* receptor gene, and association analyses in which taster and non-taster phenotypes were associated with distinct haplotypes identified this gene, which encodes the T2R38 receptor, as the source of the observed phenotypic variation for PTC taste ability (Figure 13.30). An ortholog of this gene occurs in chimpanzees. Behavioral tests in which chimpanzees were given a choice between normal apples or apples laced with PTC showed that a taster/non-taster dimorphism is also present in chimpanzees. However, the polymorphisms that are associated with taster status in humans and chimpanzees are different. Such distinct evolutionary trajectories that lead to similar variation in bitter taste phenotypes is perhaps not surprising, as chemoreceptors are thought to evolve rapidly to adapt to

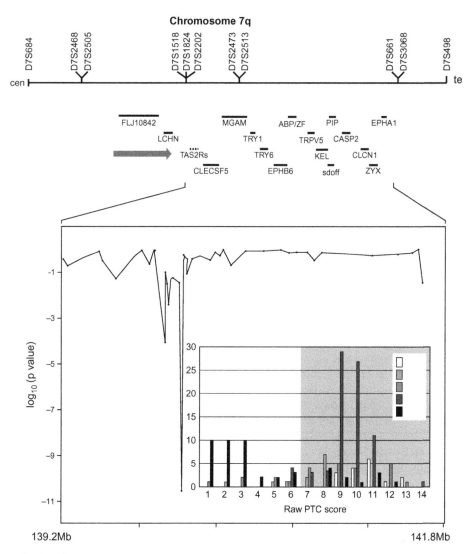

FIGURE 13.30 Localization of the locus for PTC sensitivity to a region on human chromosome 7q, which contains a cluster of bitter taste receptors (arrow). The inset shows that haplotypes in the T2R38 gene under the large QTL peak correspond to taster (shaded area) or non-taster (white background) status of a Utah population. (Modified from Kim, U. K. et al. (2003). *Science*, **299**, 1221–1225; see also Chapter 8.)

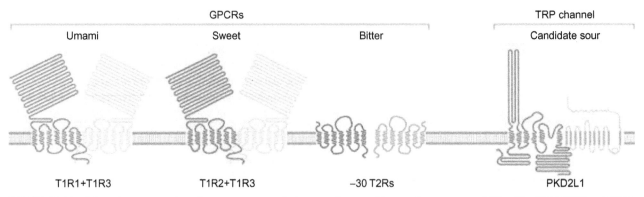

FIGURE 13.31 Schematic representation of taste receptors mediating umami, sweet, bitter, and sour taste modalities. The T1R1-T1R3 and T1R2-T1R3 dimers are illustrated, although responses to high concentrations of sugars are also detected by T1R3 alone. The gray T2R receptor is hypothetical, illustrating the possibility that T2Rs may also function as heteromeric complexes. Similarly, the receptor next to PKD2L1 depicts a PKD1-family member as a candidate partner. (From Chandrashekar, J. et al. (2006). *Nature*, **444**, 288–294.)

changes in the chemical environment. Final proof that the T2R38 receptor is indeed the receptor for PTC came from transgenic mouse studies. Mice are normally nonresponsive to PTC. However, when the human T2R38 receptor that conveys PTC taster status was expressed in mouse bitter taste cells, they acquired an aversion to PTC. It is likely that PTC taster status associated with the T2R38 receptor is merely an indicator of many associations between polymorphisms in taste receptors and phenotypic variation for gustatory perception, but such phenotypes have not been as clearly delineated or identified, and may in many cases be masked by functional redundancy among members of the T2R receptor family.

Gustatory transduction for sweet, bitter, and umami taste follows a similar pathway, in which binding of the ligand to the receptor results in the activation of a G-protein, which is either a taste cell-specific member of the G-protein family known as **gustducin** or a more widely-expressed G-protein, G_{i2}. Activation of either G-protein results in stimulation of phospholipase Cβ2, and the breakdown of phosphatidylinositol biphosphate with release of the second messengers inositol 1,4,5-triphosphate and diacylglycerol, and mobilization of calcium. The final step in this signal transduction cascade is the activation of an ion channel, which is a member of the TRP channel family, designated TRPM5. Mice in which the gene for TRPM5 or phospholipase Cβ2 had been knocked-out were no longer able to respond to sweet, bitter, and umami tastants, but retained the ability to perceive salty and sour stimuli, which are mediated by different mechanisms (Figure 13.31).

The mechanism that regulates salt taste perception is the least well-understood, but appears to involve a sodium channel that can be inhibited by the diuretic **amiloride**. Sour taste is mediated by other members of the TRP channel family that are sensitive to protons, and are designated

Box 13.6   Some like it hot . . .

The trigeminal nerve is dedicated to the detection of **somatosensory** and **nociceptive** information from the face. In addition to its classical role as the "dentist's nerve," the trigeminal nerve contributes to chemosensation by detecting texture, astringency, and nociceptive chemical stimuli, for which humans can develop an uncanny preference. The pungency of hot chili peppers, mustard, curry, and horseradish is a taste quality that arises from trigeminal stimulation. Cell bodies of trigeminal nerve fibers are located in the **trigeminal nucleus** in the brain stem. The trigeminal nerve divides into three main branches, the ophthalmic branch, the maxillary branch, and the mandibular branch. The ophthalmic branch innervates the cornea of the eye. The other branches provide extensive innervation to the nasal and oral cavities, where trigeminal input contributes to the quality of olfactory and gustatory sensations (Figure A).

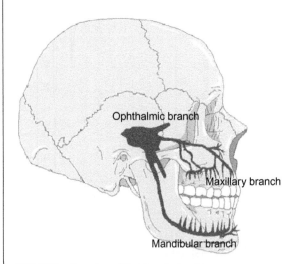

FIGURE A Branches of the trigeminal nerve.

PKD1L3 and PKD2L1. An interesting and fascinating twist to the discovery of these sour taste channels was the fortuitous discovery that the PKD2L1 channel is also expressed in neurons that line the central canal of the spinal cord, and may be physiologically important as chemosensors that monitor **intraventricular pH**. Thus, similar mechanisms may be employed to sense both internal pH and the pH of external nutrients.

For many years, researchers in the taste field found themselves divided into two camps, those that believed that taste modalities were separate and hardwired to taste perceptions via segregated neural mechanisms (the **labeled line theory**), and those that believed that individual taste cells could respond to multiple types of tastants and that the relative activity across the population of taste cells gives rise to the gustatory perception (the **cross-fiber pattern theory**). Although this debate still lingers, and the controversy is by no means resolved, evidence derived in the last decade from studies on transgenic mice increasingly favors the labeled line hypothesis. The main question that needs to be addressed in the future is how taste modalities are represented and processed in the brain, and how they are integrated with olfactory and trigeminal input to provide the complex chemosensory sensation which we associate with coffee, an Indian curry, or a fine wine.

SUMMARY

The importance of chemoreception for survival and reproduction is underscored by the large proportion of the genome that is dedicated to chemical recognition. In most mammals, different unrelated families of gustatory, vomeronasal, and olfactory receptor genes have evolved that are dedicated to recognition of chemical signals that mediate food evaluation, close-range social interactions, and long-range assessment of the chemical environment, respectively. In both the mammalian main olfactory system and the accessory olfactory system, each neuron expresses only one chemoreceptor gene enabling the brain to decode which substance is recognized by determining which neurons – and by inference which receptors – are activated. The mouse olfactory receptor repertoire contains approximately 1000 functional genes and a negative feedback mechanism exists that ensures that only one allele of a single odorant receptor gene is expressed in each olfactory sensory neuron. Odorant recognition by the brain proceeds through combinatorial activation of odorant receptors that recognize different features of odorant molecules, and activation of subpopulations of olfactory neurons results in distinct spatial and temporal patterns of glomerular activation in the olfactory bulb that encode the identity and concentration of the odor.

The vomeronasal organ is compartmentalized into superficial and basal cell layers which express V1R and V2R pheromone receptors, respectively, coincident with the expression of the α subunits of G_{i2} with V1R and G_o with V2R receptors. Vomeronasal neuronal projections to the accessory olfactory bulb are more diffuse than the characteristic axonal convergence of the main olfactory system. V1R receptors appear to recognize small organic compounds, whereas V2R receptors bind peptides. The precise functional differentiation and interactions among the two vomeronasal compartments, and the interaction between the vomeronasal organ and the main olfactory system still remains to be precisely defined.

The best-characterized olfactory system is represented by the antenna and maxillary palps of *Drosophila melanogaster*, which is numerically simpler than the mammalian olfactory system. In this system, transgenic approaches have enabled electrophysiological characterization of response profiles of a large proportion of the odorant receptor repertoire and mapping of projections of neurons expressing defined odorant receptors to individual glomeruli in the antennal lobes. A different family of gustatory receptors has also been identified in flies, and distinct projections of neurons expressing sweet (attractant) and bitter (aversive) compounds have been mapped to topographically-defined non-overlapping projection areas of the suboesophageal ganglion.

Like all chemoreceptors, mammalian gustatory receptors for sweet, bitter, and umami taste are G-protein-coupled receptors. The T1R receptors form dimers; the T1R1-T1R3 complex is associated with recognition of glutamate (umami), whereas the T1R2-T1R3 complex mediates responses to sweet tastants. A separate family of bitter taste receptors, the T2R receptors, is expressed in a distinct population of taste cells. Such cells express multiple T2R receptors per cell and, therefore, generate an aversive response to any stimulus that interacts with its cognate receptor, generating a bitter taste experience without concern for the precise chemical structure of the tastant. The evidence from experiments on homologous recombinant mice supports a labeled line theory for gustatory perception, where each taste modality transfers information to the brain independently. The ultimate conscious perception of a chemical stimulus results from a combination of olfactory, gustatory, and trigeminal input.

STUDY QUESTIONS

1. What were the hypotheses and the experimental approach used by Buck and Axel to discover odorant receptors?

2. Why is it generally assumed that the binding pockets for odorants are formed by the transmembrane domains of odorant receptors? Is there experimental evidence for this notion?

3. What insights regarding the evolution of odorant receptor genes can be obtained by comparing the repertoire of functional odorant receptors in mice and humans?

4. What are the perceptual and behavioral consequences for a system that consists of sensory neurons that each express only a single chemoreceptor versus one that consists of sensory neurons that each express multiple chemoreceptors?

5. In the mammalian olfactory system, olfactory sensory neurons that express the same odorant receptor are not clustered, but distributed in expression zones along the anterior–posterior axis of the olfactory neuroepithelium, while their axons converge on the same glomeruli in the olfactory bulb. How does this organization impact the sensitivity of odorant recognition?

6. Describe the transgenic method that was employed to study molecular response profiles of odorant receptors in *Drosophila*.

7. Describe the genetic transfection approach that was used in rats to elucidate the molecular response profile of an odorant receptor. What have we learned from this study about the breadth of ligand-specificity of odorant receptors? If the odorant receptor were a V1R receptor instead, would you expect a similar, narrower, or broader molecular response profile?

8. What is the strategy that was used to identify the V1R family of putative pheromone receptors?

9. Describe the structural differences between V1R and V2R receptors.

10. Describe how transgenic mice that express reporter genes in olfactory or vomeronasal sensory neurons have contributed to elucidating the neural projections of both systems. To what extent do olfactory and vomeronasal projections differ, and what could be the implications of these differences for information processing?

11. What is the advantage of using transgenic calcium-sensitive reporters to visualize glomerular activation in the olfactory bulb, as compared to conventional application of calcium-sensitive fluorescent dyes? What are the limitations of real-time fluorescent imaging studies in the olfactory system?

12. What conclusion regarding the olfactory system could be drawn from studies on mice that were cloned from a single olfactory neuron?

13. Describe the genetic approaches that were used to identify the human taste receptor for phenylthiocarbamide.

14. Discuss the labeled line versus the cross-fiber pattern theories for taste recognition, and indicate how results from studies with knock-out mice affect these opposing theories.

15. What is the proposed function of odorant-binding proteins, and how could their presence or absence affect odorant sensitivity?

16. What is known about the recognition of carbon dioxide in *Drosophila*? How could understanding olfaction in the malaria mosquito *Anopheles gambiae* be exploited for the development of new disease control strategies?

RECOMMENDED READING

Bargmann, C. I. (2006). Comparative chemosensation from receptors to ecology. *Nature*, **444**, 295–301.

Buck, L., and Axel, R. (1991). A novel multigene family may encode odorant receptors: a molecular basis for odor recognition. *Cell*, **65**, 175–187.

Chandrashekar, J., Hoon, M. A., Ryba, N. J., and Zuker, C. S. (2006). The receptors and cells for mammalian taste. *Nature*, **444**, 288–294.

Dulac, C., and Wagner, S. (2006). Genetic analysis of brain circuits underlying pheromone signaling. *Annu. Rev. Genet.*, **40**, 449–467.

Fleischmann, A. et al. (2008). Mice with a "monoclonal nose:" perturbations in an olfactory map impair odor discrimination. *Neuron*, **60**, 1068–1081.

Hallem, E. A., and Carlson, J. R. (2006). Coding of odors by a receptor repertoire. *Cell*, **125**, 143–160.

Kwon, J. Y., Dahanukar, A., Weiss, L. A., and Carlson, J. R. (2007). The molecular basis of CO_2 reception in Drosophila. *Proc. Natl. Acad. Sci. USA*, **104**, 3574–3578.

Malnic, B., Hirono, J., Sato, T., and Buck, L. B. (1999). Combinatorial receptor codes for odors. *Cell*, **96**, 713–723.

Mombaerts, P., Wang, F., Dulac, C., Chao, S. K., Nemes, A., Mendelsohn, M., Edmondson, J., and Axel, R. (1996). Visualizing an olfactory sensory map. *Cell*, **87**, 675–686.

Ray, A., van der Goes van Naters, W., and Carlson, J. (2008). A regulatory code for neuron-specific odor receptor expression. *PLoS Biol.*, **6**, e125.

Serizawa, S., Miyamichi, K., and Sakano, H. (2004). One neuron-one receptor rule in the mouse olfactory system. *Trends Genet.*, **20**, 648–653.

Vosshall, L. B., and Stocker, R. F. (2007). Molecular architecture of smell and taste in *Drosophila*. *Annu. Rev. Neurosci.*, **30**, 503–533.

Wang, Z., Singhvi, A., Kong, P., and Scott, K. (2004). Taste representations in the *Drosophila* brain. *Cell*, **117**, 981–991.

Zhang, X., and Firestein, S. (2002). The olfactory receptor gene superfamily of the mouse. *Nat. Neurosci.*, **5**, 124–133.

Zhang, Y. et al. (2003). Coding of sweet, bitter, and umami tastes: different receptor cells sharing similar signaling pathways. *Cell*, **112**, 293–301.

Chapter 14

Learning and Memory

OVERVIEW

The ability to associate beneficial or adverse sensory stimuli with behavioral experiences is important for survival. Storing and retrieving information about previous events enables an animal to elicit proactive behaviors. Forming associations with advantageous or deleterious experiences, such as remembering the location of food and water, anticipating strategies of predators based on prior experience, and learning and respecting social relationships are important attributes to an animal's survival skills, and thus contribute to an individual's fitness. Consolidation of information can be classified into short-term and long-term memories. The resulting memories, in turn, can be categorized as declarative or procedural. Different types of memories are processed by different neural circuits, and circuits that mediate memory recall are distinct from those that mediate memory storage. Evidence suggests that memories are plastic, and that they can be modified during recall. Genetic studies in model organisms have provided insights into some of the cellular and neural pathways that regulate cognitive processes in the nervous system, such as the formation and retrieval of memories. Investigators have also begun to examine the genetic networks that underlie these cognitive processes. Studies on humans have shown that learning ability, as reflected in IQ scores, is highly heritable. As our population ages and our environment becomes more complex, it becomes increasingly important to understand how the interplay between genotype and environment affects learning ability, and to identify alleles that are associated with developmental learning disabilities, mental retardation, or memory loss during neurodegeneration.

FORMING MEMORIES

Studies carried out on patient H.M., who was unable to form new memories after bilateral removal of his medial temporal lobes (described in Chapter 3), demonstrated vividly the distinction between the acquisition of short-term memories via the hippocampus and their consolidation in long-term memories. Storage and retrieval of long-term memories no longer requires the hippocampal formation.

Short-term and long-term memories serve different functions. For example, when we park our car in front of the supermarket we only need to remember its precise location for the duration of the shopping trip. However, its location is no longer relevant once we leave the parking lot and, therefore, is not committed to long-term memory. However, if we have an assigned daily parking spot in the university parking garage, we will remember its location even years after we have moved to a different institution. Only events that make a major impact on our lives, or that recur repeatedly, will be stored in long-term memory.

When different learning paradigms for the formation of long-term memories are compared in a variety of model organisms, one invariably finds that repeated training sessions separated by time intervals (**spaced learning**) are far more effective for the consolidation of long-term memory than a single continuous intensive training session (**massed learning**) (Figure 14.1). Thus, an effective strategy for students would be to study course material repeatedly at intermittent

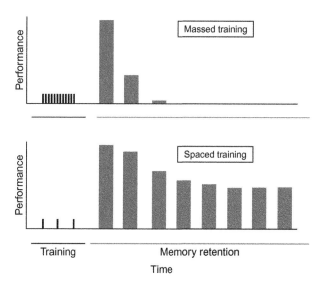

FIGURE 14.1 Graphical representation of the effectiveness of memory formation during massed learning (top) and spaced learning (bottom).

R. Anholt and T. Mackay: *Principles of Behavioral Genetics*
ISBN: 978-0-12-372575-2

times, rather than indulging in a single cramming session the night before an examination. The spaced learning approach will result in storage of the study material in long-term memory, whereas the massed learning approach may result in memory retention during the short-term, enough to pass the examination the next day. Soon, however, this information will be forgotten.

Long-term memory can be classified as **declarative memory** and **procedural memory**. Declarative memory refers to the storage of factual information (e.g. Paris is the capital of France), whereas procedural memory refers to the execution of skills that cannot be readily verbalized (e.g. learning to play a musical instrument, or learning to swim). Declarative memory can further be classified as **semantic memory** or **episodic memory**. Semantic memory refers to conceptual knowledge that is independent of specific facts (e.g. "vegetables are good for you"), whereas episodic memory refers to recollections of specific events, times, and places (e.g. an enjoyable fishing trip). Once formed, episodic memories are not static, but can be modified and reconsolidated on recall.

Studies on the cellular and neural mechanisms that mediate learning and memory have made extensive use of behavioral assays that are based on **associative learning** or **operant conditioning**. These experimental designs hark back to the classical experiment of Pavlov, who observed his dogs salivating in the absence of food simply by an auditory signal, which they had previously been exposed to in association with imminent arrival of food (see also Chapter 1). **Conditioned fear** responses are another example of such an associative learning paradigm. In this paradigm, a mouse or rat is subjected to an electric foot shock, paired with a sound or odor. After several trials the animal will show a characteristic freezing posture in anticipation of the electric shock when only the conditioning stimulus (sound or smell) is applied (Figure 14.2). The first studies that provided insights into the molecular mechanisms that mediate associative learning came from studies by Eric Kandel and his colleagues at Columbia University, who used an unlikely experimental animal, the sea slug *Aplysia californica*.

Learning in *Aplysia*

It was Stephen Kuffler who, in the 1950s, realized the potential of marine invertebrates for electrophysiological studies. Their neurons have unmyelinated axons that are large, to increase propagation speed in the absence of salutatory conduction (see also Chapter 2); because of their larger size, these axons are easier to impale with electrodes. Furthermore, marine invertebrates have simpler nervous systems. In many species, such as the sea slug *Aplysia californica*, they form defined neural circuits in which individual neurons can readily be identified. The influence of Kuffler's work, and the advantages offered by *Aplysia* for neurophysiological experimentation, led Eric Kandel on his Nobel Prize-winning quest to use this organism for his groundbreaking studies on learning (Figure 14.3).

FIGURE 14.3 Pioneers of neuroscience, Stephen Kuffler (left) and Eric Kandel (right).

How and What Does a Sea Slug Learn?

The simplest reflexes that arguably could be considered forms of learned behaviors are **habituation** and **sensitization**, both of which can be readily-demonstrated in *Aplysia*. Habituation is characterized by a decline in the response to a repeatedly-delivered stimulus. *Aplysia* will withdraw its gill in response to a gentle touch to its siphon. This **gill withdrawal reflex** decreases in magnitude and duration on repeated stimulation. Persistence of habituation of the gill withdrawal reflex depends on the number of training sessions. When few repeated stimuli are given, habituation is observed for only a short time. However, when training

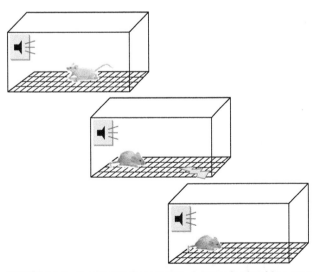

FIGURE 14.2 Conditioned fear learning. A mouse is placed in a cage with an electrified grid that can deliver a foot shock. An auditory signal (or an olfactory or visual cue) is delivered simultaneously with the foot shock. After several trials the mouse will adopt a freezing posture, characteristic of fear, as soon as the conditioning stimulus (in this diagram a sound) is delivered, even in the absence of the electric shock.

Box 14.1   Smart worms

With its small nervous system of 302 individually-identifiable neurons, and its sequenced genome, the nematode *Caenorhabditis elegans* presents an attractive model system for biological studies. Nematodes respond to chemical, thermal, and tactile environmental stimuli. But can they learn? Indeed, they display fundamental elements of learning and memory: sensitization and habituation to environmental stimuli. When a mechanical stimulus is applied to a Petri dish in which worms are swimming, they reverse direction and swim backwards. This **tap-withdrawal response** can be modified by repeated stimulation to elicit sensitization and habituation, not unlike the gill-withdrawal reflex of *Aplysia*. Moreover, different phases of memory, corresponding to short-term and long-term memory, can be elicited depending on the training regimen, in this case the intervals between mechanical taps. Like in other organisms, spaced training is more effective for inducing long-term habituation than massed training. Under optimal conditions, long-term memory in worms can extend for more than 24 hours, and its formation is dependent on protein synthesis.

Can worms form associative memories? The answer is yes. Nematodes feed on bacteria. When nematodes are exposed for a period of four hours to pathogenic bacteria, which would ultimately kill them, they will avoid these bacteria in a subsequent choice test between pathogenic and non-pathogenic bacteria. This response is not unlike our own behavior. We will avoid a restaurant which in the past has served us food that made us sick. The nematodes appear to respond to the chemical signature of the pathogenic bacteria, and retain an association between that signature and the physiological ill-effect they experienced. In the case of the bacterium *Serratia marcescens*, the aversive compound is the cyclic lipodepsipentapeptide, serrawettin W2. Worms who carry a mutation in the tyrosine hydroxylase gene fail to form these associations, due to failure to produce serotonin. Exposure to serrawettin increases serotonin in chemosensory neurons (designated ADF neurons). These findings implicate serotonin as a neurotransmitter in mediating the conditioned aversion response.

What are the molecular entities and neural circuits that mediate the integration of information required for associative learning in *C. elegans*? While this question is still unanswered, it is possible that learning in worms, as in rodents, may be dependent on activation of glutamate receptors. This hypothesis is based on the finding that genes which code for glutamate receptors (*glr-1*, homologous to the AMPA receptor; *nmr-1*, homologous to the NMDA-type receptor; and *avr-15*, a glutamate-gated chloride channel) are expressed in neurons that are essential for the tap-withdrawal response.

It appears that sensitization and habituation, the basic components of selective attention (i.e. the ability to respond to relevant stimuli in a complex environment) were established early in the animal kingdom as universal features of living organisms from which more complex learning paradigms could evolve.

---

sessions are repeated multiple times, habituation can be stable for weeks, a phenomenon that has been considered akin to the formation of long-term memory.

Sensitization in *Aplysia* can be demonstrated by giving a mild electric shock shortly before touching its siphon. The electric shock sensitizes the animal, so that the magnitude and duration of the gill withdrawal reflex will increase when its siphon is subsequently touched. Sensitization, like habituation, can persist for a long time, depending on the frequency of the training regimen (Figure 14.4, overleaf).

The third and most complicated form of learning that can be demonstrated in *Aplysia* is classical conditioning. Here, the animal learns to associate stimulation of its siphon with a shock to its tail. The conditioning stimulus is a light touch to the siphon, enough to produce a slight withdrawal of the siphon and the gill. This stimulus can now be paired with an unconditioned stimulus, a strong electric shock to the tail. The animal learns to associate the slight touch to its siphon with the noxious sensation of the electric shock, and consequently will show an enhanced withdrawal reflex on tactile stimulation of the siphon alone. This learned behavior is acquired rapidly (within 15 trials) and, once acquired, persists for several days (Figure 14.5).

MEMORY FORMATION: CELLULAR MECHANISMS

Short-term memories are formed as training proceeds. Consolidation of long-term memories takes place subsequent to the training session. Both forms of memory are accompanied by modification of synaptic function, but in contrast to short-term memory, the formation of long-term memory requires protein synthesis. This depends on gene expression, which ultimately leads to long-lasting structural changes in neuronal connections.

What are the cellular mechanisms that mediate short-term and long-term memory formation during habituation, sensitization, and classical conditioning? Several decades of elegant experimentation by Kandel and his associates have implicated voltage-sensitive calcium channels and cyclic AMP-mediated regulation of gene expression as some of the pathways that mediate learning in *Aplysia*. Most importantly, the cellular pathways identified in learning in *Aplysia* appear to be conserved in other organisms, including *Drosophila* and mice, and by inference, probably also humans. This provides a striking example of how studies on a simple model organism can provide insights into the mechanisms that underlie complex cognitive functions in higher vertebrates.

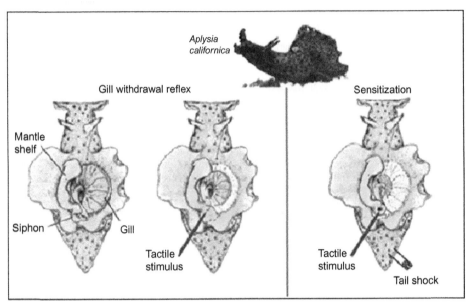

FIGURE 14.4 The gill withdrawal reflex (left) and sensitization of the gill withdrawal reflex (right) of *Aplysia californica*. The diagrams show a dorsal view. Touching the siphon with a probe causes the siphon to contract, and the gill to withdraw. The gill withdrawal response is intensified when a noxious stimulus is applied elsewhere to the body, such as an electric shock to the tail. (Modified from Kandel, E. R. (2001). *Science*, **294**, 1030–1038.)

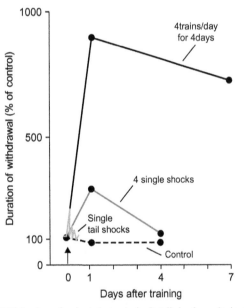

FIGURE 14.5 Learning in *Aplysia*. A single tail shock results in a small transient gill withdrawal reflex (light gray trace). The duration of the gill withdrawal is enhanced by administering four single shocks (massed training). However, when four successive shocks are administered at daily intervals (spaced training), enhancement of the gill withdrawal reflex persists even after a week, indicating the formation of long-term memory. (Modified from Kandel, E. R. (2001). *Science*, **294**, 1030–1038.)

Kandel and co-workers showed that habituation involves inactivation of presynaptic voltage-gated calcium channels. Remember that influx of calcium through these channels is necessary to trigger neurotransmitter release

(Chapter 2). Thus, limiting calcium influx at a presynaptic sensory neuron will decrease the release of neurotransmitter onto postsynaptic motor neurons that mediate the gill withdrawal reflex. The precise mechanisms that regulate the activity of these presynaptic voltage-sensitive calcium channels are not understood. They are thought to involve different mechanisms for modulation of channel activity during short-term habituation and long-term habituation. How changes in gene expression during long-term habituation lead to modification of presynaptic calcium permeability remains unknown.

Short-term sensitization is also mediated via modulation of ion channel activity. Here, facilitating interneurons release **serotonin** onto axons of sensory neurons via **axo-axonic synapses**. Serotonin triggers the formation of cyclic AMP, which activates a cyclic AMP-dependent protein kinase. This enzyme phosphorylates K⁺-channels, prolonging the duration of action potentials by reducing the rate of repolarization (see Chapter 2). As a result of the prolonged membrane depolarization, more calcium can enter through presynaptic voltage-gated calcium channels. This promotes the release of more neurotransmitter at the synaptic junction between the sensory neuron and the motor neuron. Here, in contrast to habituation, increase of neurotransmitter release facilitates the gill withdrawal reflex. Note that this short-term modulation of synaptic transmission does not require protein synthesis.

Long-term sensitization is mediated via the same serotonin-dependent activation of cyclic AMP. When activation of the cyclic AMP-dependent protein kinase is sustained over a prolonged time, its catalytic subunit is translocated to the nucleus of the cell. Once there, it phosphorylates a transcription factor that binds to enhancer sequences which contain **cyclic AMP response elements** (CRE). The transcription factor is known as **CRE binding protein** (CREB), and it recognizes the DNA consensus sequence TGACGTCA. Phosphorylation of CREB results in activation of transcription. Repression of transcription by an inhibitory CREB isoform has also been documented (Figure 14.6).

The Targets of CRE Binding Protein

What are the target genes activated by CREB and how do their gene products contribute to the strengthening of synapses? The complexity of this question is self-evident, if we recall that changes in gene expression perturb complex co-regulated genetic networks (Chapter 10). CREB activation in *Aplysia* results in the activation of a CCAAT enhancer-binding protein, which is homologous to the mammalian **immediate early gene** *c-fos*, another transcriptional activator. Thus, activation of CREB triggers a cascade of events that alters the transcriptional profile of the cell. It should be noted that CREB has not only been implicated in learning, but also in a range of other biological processes, from tumor formation to depression.

Activation of CREB was one of the first mechanisms identified as an intermediate in gene regulation during the formation of long-term memory, but it is not the only

pathway. Thomas Carew and his colleagues at the University of California at Irvine discovered that long-term memory formation in *Aplysia* is also mediated by a **receptor tyrosine kinase** pathway. Tyrosine kinase triggers the activation of **mitogen-activated protein kinase** (MAPK), which triggers an enzymatic cascade that results in regulation of gene expression. Classically, MAPK, also known as **extracellular signal-regulated kinase** (ERK), initiates a kinase cascade in response to stimulation of cells by growth factors. Phosphorylation of the inhibitory isoform of CREB by MAPK has been postulated to suppress CREB inhibition, and to contribute to CREB activation of gene expression.

The mechanisms that mediate long-term sensitization in *Aplysia* are essentially the same as those that mediate classical conditioning. The only difference is the critical dependence on the time interval between the conditioning stimulus and the unconditioned stimulus. In this case, the conditioning stimulus causes an increase in presynaptic calcium release, which facilitates activation by the unconditioned stimulus via the serotonin-mediated formation of cyclic AMP during the time at which intracellular presynaptic calcium remains elevated. This mechanism of classical conditioning is somewhat reminiscent of the opening of **NMDA-type glutamate receptor** channels, described in Chapter 3, which is contingent on stimulation by the presynaptic neurotransmitter glutamate simultaneously with depolarization of the postsynaptic neuron.

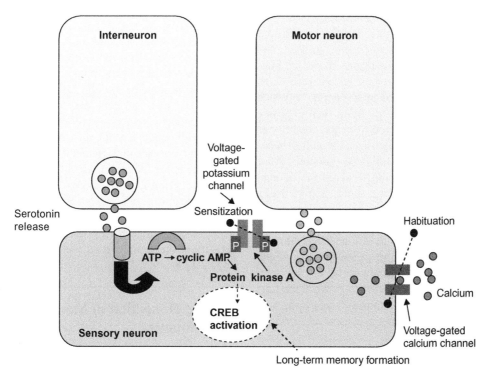

FIGURE 14.6 Diagram of the processes that underlie habituation and sensitization in *Aplysia*. The cylinder represents the serotonin receptor, and the half-moon adenylyl cyclase. Habituation results from blockage of the presynaptic voltage-gated calcium channel. Short-term sensitization results from inactivation of potassium channels through cyclic AMP-dependent protein kinase (protein kinase A). Persistent activation of this enzyme will lead to activation of CREB in the nucleus, which mediates long-term sensitization (memory formation).

What Did We Learn From a Sea Slug?

Looking back at the last 30 years, it is remarkable how much we learned about memory formation from the simple sea slug. *Aplysia* displays short-term memory, which is mediated primarily by modulation of ion channels that impact neurotransmitter release, and long-term memory, which results from activation of defined kinase pathways that regulate specific transcription factors. These transcription factors alter gene expression to give rise to long-term synaptic modifications. These general principles appear to be evolutionarily-conserved, and apply to a broad range of both invertebrate and vertebrate organisms.

HARNESSING THE POWER OF GENETICS: LEARNING IN *DROSOPHILA*

The neurobiological and molecular approaches that were applied to *Aplysia* have sparked complementary genetic approaches in *Drosophila*. One paradigm to study learning in flies is based on male courtship behavior. Males are paired with previously-mated females and, after being rejected repeatedly, they are less eager to court virgin females. The memory of previous rejections lasts for several days. Although this learning paradigm may have ecological significance, since males may learn to distinguish unsuitable previously-mated females from available virgin females, the assay is restricted, as it can only be applied to one sex and requires observations of single flies.

Alternative learning paradigms in *Drosophila* are similar to conditioned behaviors elicited in *Aplysia*, and utilize a conditioned olfactory avoidance response. Flies are placed in an electrically-wired tube, and receive an electric shock as they are exposed to an odorant to which they would ordinarily be attracted. After such exposure they are presented with a choice between the conditioned odorant (for example, a commonly used odorant is 4-methylcyclohexanol), and an unconditioned odorant. After a period of training, the flies learn to avoid the odorant that they associate with the electric shock (Figure 14.7). The percentage of flies that learn to avoid the conditioning odorant is quantified as an **avoidance index**, which in wild-type flies can be as high as 0.9, and in mutants is often lower than 0.3. Alternative to a punishing paradigm, associative learning can also be demonstrated by reinforcement through a sugar reward. In either case, when the time course over which memory is retained is assessed, both short-term and long-term memory phases are evident. Furthermore, similar to *Aplysia*, spaced training paradigms are more effective than massed training in eliciting long-term memory. Massed training that consists typically of 10 sessions administered one immediately after the other produces a short-term memory that can be retained for up to three days, but is not sensitive to the protein synthesis

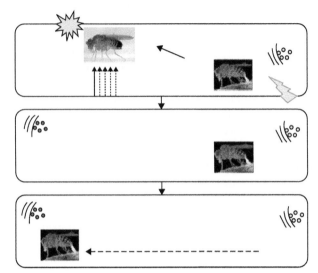

FIGURE 14.7 Schematic representation of the olfactory learning paradigm in *Drosophila*. The conditioning odorant stimulus is represented by light circles on the right and the unconditioned odorant by dark circles on the left. In the top panel flies are exposed to an electric shock paired with the conditioning odorant. After a period of training they are then given a choice between the conditioned and unconditioned stimulus.

inhibitor **cycloheximide**. In contrast, spaced training, which can be performed by administering 10 sessions separated by 15 minute rest intervals, results in long-term memory, which lasts for more than a week and is dependent on protein synthesis.

The mushroom bodies represent the major central integrative centers for learning and memory in *Drosophila*. Ablation of the mushroom bodies during development eliminates learning, even though animals are still able to sense, and respond to, stimuli. Reversible silencing of neurons in the alpha and beta lobes of the mushroom bodies by expressing a temperature-sensitive dominant-negative form of *shibire* prevents memory retrieval, but has no effect on memory acquisition (remember from Chapter 9 that *shibire* encodes dynamin, which is required for synaptic vesicle recycling). This suggests that neuronal circuitry outside the mushroom bodies contributes to the acquisition of memories, whereas the storage and retrieval of associations once stored in long-term memory depends on the integrity of the neural circuit within the mushroom bodies (Figure 14.8).

Genetic Dissection of Memory with Mutants

Genetic approaches in *Drosophila* have identified mutants that are defective in learning and memory. Such mutants have enabled a more precise dissection of memory phases into short-term, middle-term, anesthesia-resistant, and long-term memory. The first learning mutant in *Drosophila*

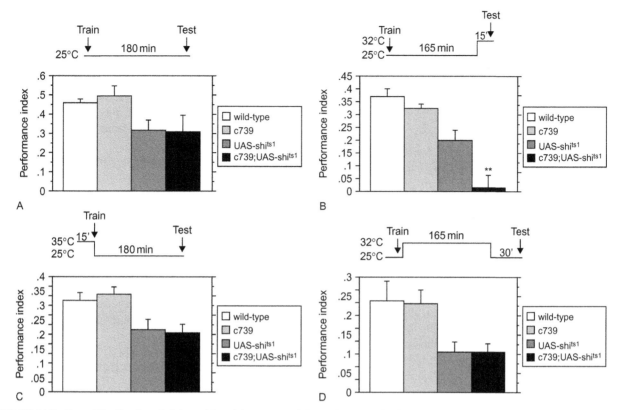

FIGURE 14.8 Reversible silencing of alpha and beta lobe neurons of the mushroom bodies in *Drosophila* by cell-specific expression of temperature-sensitive *shibire*. The c739 driver line expresses GAL4 in alpha and beta lobe neurons of the mushroom bodies to drive expression of temperature sensitive *shibire* under the UAS promoter. Flies are not able to retrieve memories when tested at the non-permissive temperature. Training and testing protocols are schematically indicated above each panel. (From McGuire, S. E. et al. (2001). *Science*, **293**, 1330–1333.)

was discovered in Seymour Benzer's laboratory in 1974, and was named "*dunce (dnc)*." The *dunce* gene is complex. It spans 150 kb and encodes multiple alternative transcripts of a cyclic AMP phosphodiesterase, an enzyme which is responsible for the inactivation of cyclic AMP via conversion into 5'AMP (Figure 14.9). Mutants are defective in normal short-term memory formation, and can form memories only briefly. The *dunce* gene is expressed in the mushroom bodies, which have been implicated as the integrative centers in associative learning in the fly brain.

Further evidence for participation of the cyclic AMP pathway in learning in *Drosophila* came from the identification of a second mutant with a defect in a calmodulin-dependent adenylyl cyclase, named "*rutabaga (rut)*." Here, a single base change, which results in an amino acid substitution from glycine to arginine, eliminates adenylyl cyclase activity. Phenotypically, *rut* mutants are similar to *dnc* mutants, in that they can only acquire rapidly-decaying memories. This defect can be rescued by driving expression of a wild-type *rut* transgene in the mushroom bodies, especially in the γ lobes. These transgenic studies were particularly elegant, because expression of the transgene could be controlled not only spatially, but also temporally (i.e. it could be turned on or off in the same individuals)

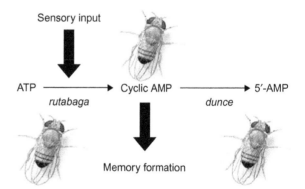

FIGURE 14.9 The first *Drosophila* mutants that provided insights into the mechanisms of memory formation were *rutabaga* and *dunce*, which regulate levels of cyclic AMP in the mushroom bodies.

by using a temperature-sensitive GAL80 repressor that suppresses activation of a GAL4-linked driver via the TARGET system, or a progesterone-sensitive spatially-restricted GAL4 driver via the Gene-Switch method (these spatiotemporally controllable expression systems have been described in Chapter 9).

Note the parallel between learning in *Drosophila* and *Aplysia* in the critical involvement of cyclic AMP.

Accordingly, long-term memory formation in flies is dependent on activation of cyclic AMP-dependent protein kinase and the *Drosophila* homolog of CREB, again recapitulating the cellular pathway identified in *Aplysia*.

Middle-Term and Anesthesia-Resistant Memory

The temporal sequence of memory formation turned out to be more complicated than initially expected when two *Drosophila* mutants were found with deficient memory phenotypes that affected neither short-term nor long-term memory. These mutants demonstrated the existence of two intermediate forms of memory: **middle-term memory** and **anesthesia resistant memory** (that is, memory that persists after a period of cold-shock induced anesthesia) (Figure 14.10). The first mutant, named "*amnesiac* (*amn*)," shows near-normal memory retention immediately after a single training session, and again about seven hours later. Interestingly, however, between these time points, memory retention in *amnesiac* mutants is lower than normal, indicating that the *amn* mutation disrupts memory at a stage intermediate between initial learning and memory consolidation. The *amnesiac* gene encodes a peptide that resembles a **pituitary adenylyl cyclase activating peptide**, known as **PACAP** (activation of adenylyl cyclase by this gene product, however, has not been demonstrated). The cells that express *amnesiac* are intriguing. The formation of short-term and long-term memories appears to involve the mushroom bodies. However, *amnesiac* is expressed in only a few neurons outside the mushroom bodies, the **dorsal paired medial** (DPM) **neurons**. These are large neurons that send projections to the mushroom body lobes, and are thought to be cholinergic. Transgenic expression of *amnesiac* in DPM neurons rescues the mutant memory phenotype, indicating that these neurons are indeed associated with the formation of middle-term memory.

The second mutant, named "*radish*," specifically abolishes anesthesia-resistant memory, while long-term memory remains intact. The *radish* gene encodes a **phospholipase A2**, and has a complex expression pattern. Interestingly, *radish* is not expressed in Kenyon cells of the mushroom body, which are thought to integrate the coincident detection of the conditioned and unconditioned stimulus during associative learning. Thus, prior to the consolidation of long-term memories, intermediate forms of memory are mediated by neural circuits that are associated with, but external to, the mushroom bodies.

The Role of the Mushroom Bodies

It is clear from the *amnesiac* and *radish* mutants that processing of information into memory requires not only the mushroom bodies, but also neural circuitry that is extraneous to the mushroom bodies. Within the mushroom bodies Kenyon cells are instrumental in integrating information during associative learning (see also Chapter 3).

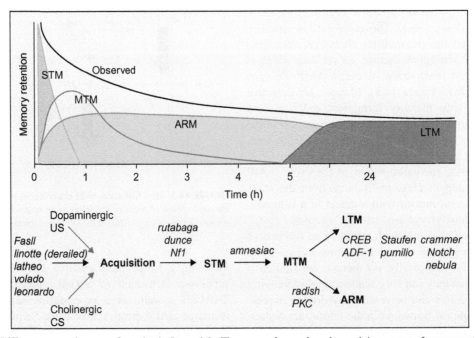

FIGURE 14.10 Different stages of memory formation in *Drosophila*. The apparently seamless observed time course of memory retention reflects four discrete underlying stages of memory formation, short-term memory (STM), middle-term memory (MTM), anesthesia-resistant memory (ARM), and long-term memory (LTM). These stages of memory formation can be distinguished through the use of mutations in different genes that selectively affect these processes, as illustrated in the diagram below the graph. (Modified from Margulies, C. et al. (2005). *Curr. Biol.*, **15**, R700–R713.)

These cells function as coincidence detectors, as they receive input from several different sensory modalities, including cholinergic inputs from the antennal lobes, and dopaminergic and octopaminergic inputs that form synaptic connections both in the **calyx** and the **mushroom body lobes**. The dopaminergic neurons appear to convey the electric shock sensation, whereas cholinergic projections from the antennae mediate olfactory sensation, and octopaminergic neurons mediate sensory perception of a sugar reward during association through positive reinforcement. Studies with temperature-sensitive mutants of DOPA decarboxylase, which is essential for the synthesis of dopamine and serotonin in flies (Chapter 3), have illustrated that associative learning is, in fact, dependent on dopaminergic neurotransmission (Figure 14.11).

Molecular evidence for **synaptic remodeling** during long-term memory formation in the mushroom bodies came from studies on a trypsin-like serine protease, known as **Tequila**. Tequila is an ortholog of human neurotrypsin. Mutations in neurotrypsin are associated with autosomal recessive **nonsyndromic mental retardation**, i.e. mental retardation with normal brain development and no clinical features. Studies on conditioned learning in *Drosophila* showed that *Tequila* mRNA is transiently up-regulated in the mushroom bodies during the formation of long-term memory, and that suppression of *Tequila* expression by reversible mushroom body-specific expression of *Tequila* RNAi using the GeneSwitch system (Chapter 9) prevents the formation of long-term memory. The effects of Tequila were only observed during spaced training, but not during massed training (Figure 14.12). These experiments suggest that Tequila provides a proteolytic function essential for synaptic remodeling during the interval at which short-term memories are converted into long-term memory storage.

IDENTIFYING GENETIC NETWORKS FOR LEARNING AND MEMORY

Several independent mutant screens have identified a diverse array of genes that affect learning and memory. These genes generally fall into two categories: genes that prevent short-term memory formation (i.e. "learning genes"), and genes that affect the retention of information in long-term memory ("memory genes"). The former include *latheo*, *linotte*, *14-3-3 (leonardo)*, *scabrous (volado)*, *fasII*, and *DC0* (which encodes cyclic AMP-dependent protein kinase). The latter include *dCREB2*, *Adf1 (nalyot)*, *Notch*, *crammer*, and *nebula*. While many of these genes are intrinsically interesting, it is not immediately obvious how they function together in a mechanistic pathway that mediates the storage of learned information. Most parts of the puzzle are still missing.

The central questions surrounding mechanisms of learning in flies are: which genes are associated with elevation of cyclic AMP; and which genes are downstream targets for activation by CREB? Tim Tully and Josh Dubnau at the Cold Spring Harbor Laboratories addressed these questions by performing a large scale mutagenesis screen for mutants defective in one-day memory after spaced training, and then using expression microarray analysis to identify transcripts that showed altered regulation during long-term memory formation in wild-type flies. They identified new genes, some of which encoded transcripts

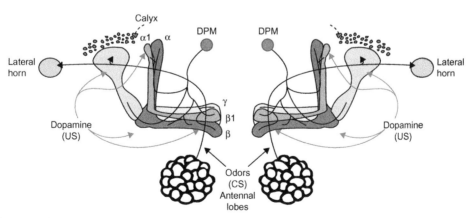

FIGURE 14.11 Schematic diagram of the neural circuitry involved in olfactory associative memory in *Drosophila*. Associations between the conditioned stimulus (CS, odors) and the unconditioned stimulus (US, electric shock) are formed in the mushroom bodies. Olfactory information is processed in the antennal lobes and output neurons from the antennal lobe project to the lateral horn and mushroom bodies via several different projection neuron tracts. The unconditioned stimulus arrives at the mushroom bodies via dopaminergic inputs to the calyx and lobes. Neuromodulation by dorsal-paired medial neurons (DPM) is required after training for memory consolidation. These are the neurons which express *amnesiac*. The calyx of the mushroom body contains the dendritic field, whereas axons extend into the lobes. There are three types of Kenyon cells: α, β neurons, whose axonal projections comprise the α and β lobes; α′ β′ neurons, whose projections enter the α′ β′ lobes; and γ neurons whose projections form the γ lobes. (Adapted from Margulies, C. et al. (2005) *Curr. Biol.*, **15**, R700–R713.)

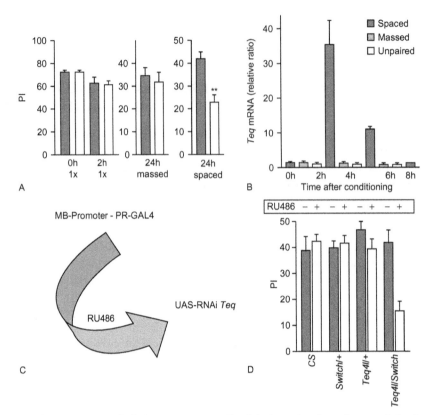

FIGURE 14.12 Involvement of the protease tequila in long-term memory. Panel A shows that the performance index for olfactory conditioning in *Drosophila* is impaired in *Tequila* mutants (open bars) compared to controls (solid bars) after 24 hours of spaced training, which results in long-term memory, but not after massed training, which only generates short-term memory. In wild-type flies there is a transient peak of *Tequila* mRNA expression during spaced training, which coincides with the formation of long-term memory (Panel B). Panel C shows how an RNAi, which silences the *Tequila* gene, can be driven under the control of a mushroom body (MB)-specific promoter that drives expression of a fusion construct of the progesterone receptor (PR) and GAL4 (the GeneSwitch binary expression system, see Chapter 9). In the presence of the progesterone receptor ligand RU486 the expression of this RNAi will be turned on in the mushroom bodies. This leads to inhibition of long-term memory formation during olfactory conditioning (bar graphs in Panel C: CS designates Canton S, the control *Drosophila* strain). (Adapted from Diderot, G. et al. (2006). *Science*, **313**, 851–853.)

of unknown function. These genes were named in honor of Pavlov after his experimental dogs (e.g. *krasavietz* and *milord*). Studies by Tully and his colleagues showed that **NMDA receptor** activation may lead to activation of the cyclic AMP pathway, an intriguing parallel to the involvement of NMDA receptors in long-term potentiation in the hippocampus of the vertebrate brain (see Chapter 3). This suggests that **glutamatergic neurotransmission** may play a role in learning in *Drosophila*.

Both mutagenesis studies and expression analysis implicated a translational repressor, encoded by *pumilio*, as a potential downstream target for CREB. Expression of *pumilio* is up-regulated after spaced training, and *P*-element insertions in this gene interfere with the acquisition of memory. The *pumilio* gene is a developmental gene which regulates the spatial distribution of mRNA. It has been associated with the regulation of another regulator of mRNA translation, *staufen*. Temperature-sensitive *staufen* mutants confirmed that this gene product is necessary for the formation of long-term memory. Thus, long-term memory formation depends on a pathway that regulates mRNA

transport and translation. It is important to note, however, that these observations only scratch the very tip of a large iceberg, and that multiple integrated cellular pathways, which remain yet to be defined, are likely to mediate the cellular mechanisms that underlie the modulation of neural pathways for the formation and storage of memory.

What Have We Learned From Flies?

Learning and memory were once considered intractable cognitive processes of inexplorable complexity. Simple model systems, like *Aplysia* and *Drosophila*, have demonstrated that the neural circuits that mediate learning and memory, and the genetic networks that regulate gene expression in these neural circuits, can be dissected. Moreover, broadly-applicable general principles have emerged from these systems. Memory acquisition occurs in temporally-distinct phases that are mediated by distinct cellular mechanisms that can be separated experimentally. The formation of associative memories involves not only integrative brain centers, such as the mushroom bodies, but also neural pathways that communicate with, but are external to, such structures. Memory retrieval proceeds along different neural pathways than memory formation. Finally, cyclic AMP-dependent processes and CREB-regulated transcription play an important role in the formation of long-term memories. The formation of such memories is likely accompanied by widespread changes in transcription, not unlike that observed in the honey bee brain during changes in behavioral status (Chapters 10 and 12).

Mammalian Models for Learning and Memory

Rats and mice are used extensively as vertebrate models for learning and memory. Rats are generally easier to train, and are eminently suitable for behavioral studies of learning. Mice, however, are more amenable for genetic approaches, due to the well-established gene knock-out technologies (Chapter 9) and the ever-increasing number of publicly available genetic resources. Studies using conditional knock-out mice have, by-and-large, recapitulated the cellular mechanisms that mediate learning and memory identified in invertebrate models. Maze learning has been a popular behavioral paradigm to assess learning and memory in rodent models. For example, in the **Morris water maze**, mice learn to use external cues to locate a hidden platform in an opaque pool of water (Figure 14.13).

The hippocampus has long been considered the gateway for the acquisition of memories, and long-term potentiation (LTP) in the hippocampal circuitry (Chapter 3) is considered to be the physiological manifestation of this process. As a consequence, many studies on learning and memory have focused on NMDA-type glutamate receptors as the molecular entities that integrate information during learning. Activation of NMDA receptors in hippocampal

CA1 neurons is especially important for the acquisition of memories. As described in Chapter 3, NMDA receptors act as coincidence detectors during LTP that integrate presynaptic and postsynaptic activation. This coincidence detection was first postulated by Donald Hebb in 1949 to be essential for the strengthening of synaptic connections (Figure 14.14). The consequence of NMDA-type

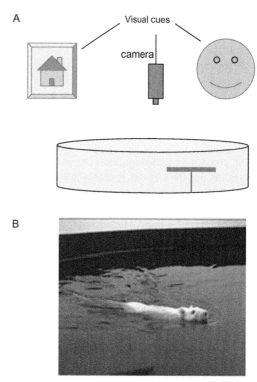

FIGURE 14.13   The Morris water maze (Panel A) with a submerged hidden platform. Panel B shows a rat performing the water maze task. An alternative to the Morris water maze is the radial arms maze, which also assesses spatial learning and provides a food reward when the animal chooses the correct arm of the maze.

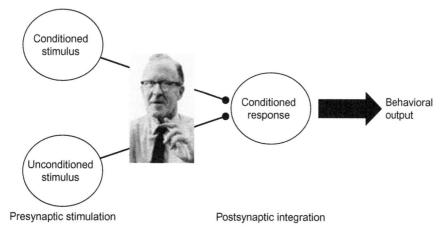

FIGURE 14.14   The notion that learning depends on coincidence detection via postsynaptic integration of the neural response to the conditioned and the unconditioned stimulus was first proposed by Donald Hebb (pictured). Synapses of the type pictured in the diagram have become known as **Hebbian synapses** (see also Chapter 2).

receptor activation by glutamate during depolarization of the postsynaptic cell is an influx of calcium into the postsynaptic neuron. This results in activation of calcium-calmodulin-dependent protein kinase II (CaMKII), which leads to strengthening of the synaptic connection through, among other mechanisms, phosphorylation of postsynaptic densities. CaMKII can undergo **autophosphorylation**, thereby perpetuating its own activation. In addition to CaMKII, other kinases, such as cyclic AMP-dependent protein kinase A, and phosphatases, such as calcineurin, play a role in memory acquisition in the hippocampus. Genetic disruptions of the balance of kinase and phosphatase activities by hippocampus-specific knock-out or overexpression of kinases and calcineurin show that an excess of phosphatases can lead to reduced LTP and learned behaviors, while a reduction of phosphatase activity enhances LTP and learning. Mutations that reduce or eliminate CamKII activity can be generated with conditional and spatially targeted transgenic constructs to manipulate gene expression levels. These systems use the Cre-lox recombination system combined with a **tamoxifen-sensitive promoter** (Chapter 9), or a promoter that can be activated by a tetracycline agonist, such as doxycycline, administered in the food or drinking water, or injected. Administration of tamoxifen or doxycycline activates the expression of Cre, and if Cre is linked to a hippocampus-specific promoter, this will result in the excision of the floxed target gene in a spatially- and temporally-controlled manner. Such genetic manipulations impair performance in the Morris water maze, as overexpression of calcineurin also does.

Phosphorylation of a different form of calcium-calmodulin-dependent protein kinase, Ca^{2+}/calmodulin-dependent protein kinase IV (CaMKIV), triggers the initial phosphorylation of CREB. However, phosphorylation by the mitogen-activated protein kinase (MAPK) is required for persistent activation of CREB. After MAPK is activated in the cytoplasm, it phosphorylates a kinase, known as **pp90 ribosomal protein S6 kinase** (Rsk). Activated MAPK and Rsk then form a complex that translocates to the nucleus, where Rsk phosphorylates CREB (Figure 14.15).

Mice in which the predominant isoforms of CREB had been deleted developed normally, and had normal short-term memory. However, they were severely impaired in associative and spatial long-term memory. In conditioned fear tests they were unable to retain the association between a conditioning stimulus and a foot shock, and in the Morris water maze they were unable to remember the position of the hidden platform. Mice in which the CaMKIV gene has been deleted are also unable to store conditioned fear memories.

A vast number of transgenic mouse strains with learning and memory deficits have been identified, and the general rules governing mechanisms of learning and memory gleaned from invertebrate models appear to hold up as central components of mammalian learning and memory

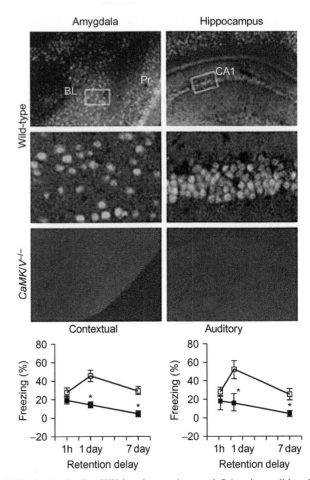

FIGURE 14.15 CamKIV knock-out mice are deficient in conditioned fear learning. The top panel shows coronal brain sections of CaMKIV-labeled neurons in the amygdala and hippocampus. The basolateral amygdala is instrumental in promoting fear-conditioned learning (see also Box 14.2 Figure A). High-magnification images in the center row show staining in the basolateral amygdala (BL) and the pyramidal cell layer of the CA1 region of the hippocampus. No immunostaining is detected in the $CaMKIV^{-/-}$ mice (bottom row). Retention of contextual or auditory fear at 1 hour, 1 day, and 7 days after training shows that $CaMKIV^{-/-}$ mice (closed symbols, n = 7) do not retain memories of fear conditioning, compared to wild-types (open symbols, n = 8; * $P < 0.05$). (Modified from Wei, F. et al. (2002). *Nature Neurosci.*, **5**, 573–579.)

(Figure 14.16). However, in addition to NMDA receptor-, CaMKII/CaMKIV-, and MAPK-mediated-activation of CREB in the hippocampus, homologous recombination in mice has pointed to the involvement of several other neurotransmitter pathways in learning and memory, including muscarinic and nicotinic cholinergic, as well as serotonergic and dopaminergic mechanisms. It is clear that the formation and retrieval of memories is orchestrated by complex neural circuits, and understanding the relationship between these circuits and the genetic networks that enable them to function is of central interest in behavioral genetics. The complexity of the genetic underpinnings that serve learning and memory, and the importance of brain regions

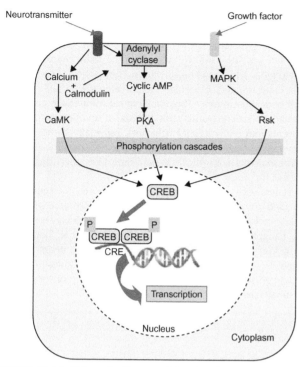

FIGURE 14.16 Schematic diagram of the phosphorylation reactions that regulate activation of CREB. This diagram is simplified, and shows only some of the primary transduction pathways that regulate phosphorylation of CREB during the formation of long-term memories. Regulation of transcription of immediate early genes (such as *fos*) by CREB results in transcriptional responses of downstream target genes. The pathways shown in the diagram are conserved between simple model organisms, such as flies, and vertebrates (PKA, protein kinase A; CRE, cyclic AMP-response element).

other than the hippocampus are especially evident from studies on human learning disabilities and neurodegenerative diseases, which implicate a diverse array of candidate genes that are not directly associated with the hippocampal NMDA-CaMKII/CaMKIV-CREB pathway of transcriptional regulation. In the next sections, we will highlight two well-described human cognitive disorders, **dyslexia** as an example of a learning disability, and compromised cognitive function during **Alzheimer's disease** as an example of neurodegenerative loss of memory.

LEARNING DISABILITIES

The Genetics of Dyslexia

Speech evolved as an intrinsic behavioral trait during human evolution, but writing and reading are cultural innovations that appeared in human history only a few thousand years ago. Acquiring reading skills to understand written language is, therefore, purely a learned behavior. Genetic studies on dyslexia have been informative in identifying alleles that impact the ability to learn written language (Figure 14.17).

Individuals with dyslexia find it difficult to acquire and consolidate written information. **Phonological deficiencies**, that is, the inability to process speech sounds and recognition of words and their meaning, are also common. However, in contrast to learning disabilities that result from genetic defects that cause mental retardation, such as Down syndrome, people with dyslexia show normal performance with regard to other cognitive functions, and have normal or higher than normal intelligence (some often-disputed claims assert that famous historical personalities, such as Albert Einstein and Winston Churchill, were dyslexic).

Estimates of the incidence of dyslexia vary widely, depending on different tests used. They range between 5% and 17% in the United States, but the National Institutes of Health has estimated that as many as 20% of school children may be affected. Thus, this learning disability is extremely common, and consequently it is relatively easy to recruit subjects for genetic studies. The **heritability** of dyslexia has been estimated at about 40% to 70%; therefore, genetic variation contributes prominently to this learning disability.

The prevalence of dyslexia and its substantial heritability provide a favorable scenario for human genetic studies. Such studies have used either comparisons between **monozygotic twins** (which are genetically-identical) and **dizygotic twins** (which share 50% of alleles in common), linkage studies based on transmission of the disorder in families, and association studies in populations. The non-random association between an allele and a phenotype in family studies is known as **transmission disequilibrium**. Combinations of such studies have resulted in the identification of several chromosomal regions that contain candidate genes which harbor polymorphisms that may

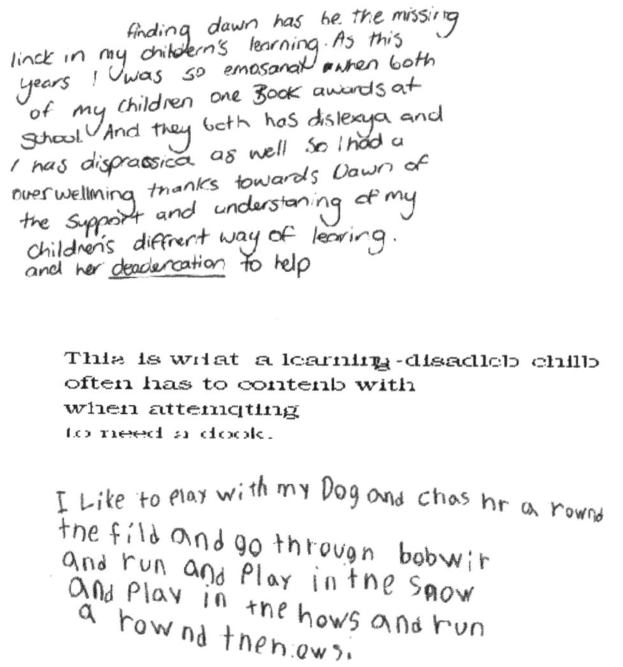

[Handwritten text:]
finding dawn has be. the missing linck in my chitolern's learning. As this years I was so emosanal when both of my children one Book awards at shool. And they both has dislexya and I has disprassica as well. So I had a overwellming thanks towards Dawn of the support and understaning of my childrens diffrent way of learing. and her deaderication to help

[Text with distorted/degraded typeface:]
This is what a learning-disabled child often has to contend with when attempting to read a book.

[Handwritten text:]
I Like to play with my Dog and chas hr a rownd the fild and go throvgh bobwir and run and Play in the Snow and Play in the hows and run a rownd the hows.

FIGURE 14.17 Examples of dyslexia.

Box 14.2   Fragile X syndrome

Fragile X syndrome is the second most common genetic mental retardation disorder after Down syndrome. It results from an expansion of a trinucleotide repeat (CGG) as the *FMR1* (fragile X mental retardation 1) gene on the long arm of the X-chromosome (Xq27.3). The protein encoded by the *FMR1* gene (FMRP) is an mRNA-binding protein, which mediates translocation of mRNA from the nucleus to the cytoplasm. FMRP is expressed in many tissues, but especially in testes and brain, where it may be important for synaptic plasticity that occurs during learning.

The *FMR1* gene normally contains between 6 and 55 CGG repeats. In the case of fragile X syndrome, more than 200 CGG repeats may be present. This trinucleotide expansion undergoes methylation, which results in silencing of the *FMR1* gene, and in an altered "fragile" appearance of the X-chromosome under the microscope (hence the name **fragile X syndrome**).

Because males carry only one X-chromosome, they are more susceptible to developing symptoms of fragile X syndrome, and their symptoms are generally more severe than those of affected females. Boys with the syndrome show characteristic anatomical and behavioral features. They often have unusually large testicles, elongated facial features with large ears, and poor muscle tone. These anatomical features are accompanied by severe learning disabilities. In addition, fragile X patients may show asocial behaviors, especially shyness and avoidance of eye contact, as well as nervous speech patterns. Some patients with fragile X syndrome meet the diagnostic criteria for autism spectrum disorder (ASD).

Males with fragile X syndrome will not transmit the affected chromosome to their sons, as their sons receive only the parental Y-chromosome from their father. However, all of the daughters will inherit the fragile X-chromosome. Since the expanded *FMR1* gene is dominant with variable penetrance, females who are heterozygous for the expanded *FMR1* gene can show some symptoms of the disorder, or they can be unaffected.

Box 14.3   Genes and intelligence

Few issues in behavioral genetics have fueled as much controversy in the **nature versus nurture** debate in the genetics of intelligence. A fundamental problem is that an unbiased concept of what constitutes intelligence is not easy to define and to measure quantitatively. A major breakthrough in the study of intelligence came in 1939, when David Wechsler developed his now widely-used intelligence test.

Born in 1896 to a Jewish family in Romania, Wechsler immigrated as a child with his parents to the United States, and would become one of the most famous psychologists of the twentieth century. During World War I, while studying under Charles Spearman and Karl Pearson, Wechsler developed psychological tests to screen soldiers for enlistment. In 1932, he was appointed as Chief Psychologist at Bellevue Hospital, where he developed the Wechsler–Bellevue Intelligence Test.

Wechsler's major conceptual breakthrough was his realization that the popularly-held concept of intelligence as a single characteristic was flawed, and that there are different aspects to intelligence, which can be assessed using different criteria. Wechsler defined intelligence as "the global capacity to act purposefully, to think rationally, and to deal effectively with his or her environment." He divided the concept of intelligence into two main areas, verbal and performance (non-verbal) areas, each of which was further subdivided into seven different categories.

The first Wechsler Adult Intelligence Scale (WAIS) was developed in 1939, it was expanded in 1949 with a Wechsler test for children, and again in 1967 with the Wechsler Preschool and Primary Scale of Intelligence (WPPSI) tests.

In 1981, WAIS was standardized with a sample of 1880 US subjects divided into nine age groups between 16 and 74 years of age. The Wechsler test measures an IQ (intelligence quotient) value which is adjusted so that the median of the population corresponds to a value of 100 with a standard deviation of 15. In a normally-distributed population the IQ scores of approximately 68% of adults range between 85 and 115. One should keep in mind that judging intelligence based on performance in an IQ test alone can be misleading, as cultural differences can influence performance in the test, and talents, such as aptitude for musical, artistic, or athletic skills, are not evaluated as components of standard IQ tests.

WAIS scores are highly reliable, and have provided a quantifiable phenotype for genetic studies, although it is still disputed to what extent IQ scores on the WAIS test reflect "intelligence," depending on one's concept of intelligence. Family studies have shown that the heritability of performance on standardized intelligence tests can be as high as 0.86. However, despite the large genetic contribution to variation in intelligence, efforts to identify "intelligence genes" have been largely unsuccessful. Whole genome scans in 725 individuals from 329 Australian families, and 225 individuals from 100 Dutch families, identified only a single QTL region on chromosome 2q and a suggestive QTL region on chromosome 6p. These regions are large and contain many genes (including candidate genes implicated in autism and dyslexia). It appears from studies to date that the diverse manifestations of phenotypes, which we collectively refer to as "intelligence," are regulated by many genes of small effect with likely few, if any, genes of large effect.

contribute to susceptibility for dyslexia. The most notorious among these candidate genes are *DYX1C1*, *KIAA00319*, *DCDC2*, and *ROBO1*. They were identified in independent genetic studies, but are functionally-related. Their gene products mediate cell migration, axon guidance, and cell adhesion, functions that impact the formation of the cortex in early development as neurons migrate from a proliferative zone near the ventricle (the **subventricular zone**) along the processes of **radial glia** toward the surface of the nascent brain. Here they will form part of the modular cerebral cortex. Indeed, subtle cortical anomalies and brain malformations have been reported in postmortem brains of people with dyslexia.

The first dyslexia-related candidate gene, *DYX1C1* (which stands for dyslexia-susceptibility-1, candidate-1), was identified as the result of a rare chromosomal **translocation** in a Finnish family, which involved chromosomes 2 and 15. The breakpoint of the translocation on chromosome 15 disrupted the *DYX1C1* gene, which encodes a gene product with protein–protein interaction motifs, known as tetratricopeptide repeat domains. The father of this family and three of his four children carried this chromosomal translocation, and all showed characteristic symptoms of dyslexia. When polymorphisms in the *DYX1C1* gene were characterized in a larger Finnish population, two rare SNPs were found to occur with higher frequency in patients with dyslexia than in unaffected controls. However, studies on *DYX1C1* in other populations did not replicate this effect. This might be due to genetic background effects. The Finnish population derives from a small **founder population** that colonized Finland, and as a result is genetically more homogeneous than most other populations (see also the discussion on bottlenecks and genetic drift in Chapter 16). Thus, the strongest evidence for association between the *DYX1C1* gene and dyslexia rests only on the four affected individuals from the original family study (Figure 14.18).

The search for dyslexia-associated genes continued, and another gene on chromosome 6, *KIAA0319*, was found to be associated with dyslexia in large numbers of families in two different populations, one a British population from Berkshire, and the other an American population from Colorado. This gene encodes a surface protein implicated in cell adhesion. The story became more complicated when a gene located adjacent to *KIAA0319* was also implicated in dyslexia. This gene, *DCDC2*, had been associated previously with brain disorders that result from deficits in neuronal migration. Alleles of this gene differ in the number of short tandem repeat sequences in an intron. Significant associations were observed between the copy number of these short tandem repeats and the incidence of dyslexia. *DCDC2* was further confirmed as a candidate gene for dyslexia in **linkage studies** of German families. It appears that regulation of gene expression, rather than changes in the structure of the encoded protein, may be associated with susceptibility to dyslexia for *KIAA0319*, *DCDC2*, or both (Figure 14.19).

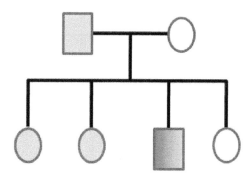

FIGURE 14.18 The pedigree of the Finnish family carrying the chromosome 15-2 translocation. Circles represent females and squares represent males. Diagnosis of dyslexia in the son is tentative, as indicated by the shading, since the child had additional severe learning disabilities unrelated to dyslexia [□].

The discovery of a fourth candidate gene that predisposes to dyslexia again came from two fortuitous circumstances. One was an unusually large Finnish family that spanned four generations with dyslexic and unaffected family members, and the second was another rare chromosomal translocation, this time between chromosome 3 and chromosome 8. The Finnish family consisted of 74 members, 27 of which were dyslexic. Analysis of transmission of this phenotype was unexpectedly simple, indicating a single dominant gene linked to a region on chromosome 3. The translocation pinned the phenotype down to a disruption in *ROBO1*, an ortholog of the *Drosophila* gene *Roundabout* (*Robo*) that mediates axon extensions across the midline of the fly's nervous system during development. It appears that a reduction in expression of the human *Robo* ortholog, *ROBO1*, during development of the nervous system predisposes to dyslexia, even in heterozygotes that contain only one risk allele. As with the other dyslexia-risk alleles, the *ROBO1* story, although highly suggestive, is not iron-clad. The child that carried the chromosomal translocation had a sister who did not carry this translocation, but had severe dyslexia. Furthermore, linked genes in the region around the *ROBO1* locus cannot be entirely excluded as harboring risk alleles.

Despite the caveats surrounding these four candidate genes, the fact that all of them relate to cell migration, cell adhesion, and axon growth, which can explain some of the cortical malformations seen in some cases of dyslexia, establishes them as interesting candidate genes (Figure 14.20). As is the case with many leading-edge discoveries, the possible involvement of these genes in dyslexia raises more questions than it provides answers. Precisely which sequence variants are associated with dyslexia, and how predictive are these alleles for the manifestation of this learning disorder? Why should genes that are expressed widely throughout the developing brain selectively affect a particular cognitive function mediated by distinct cortical regions that are associated with speech and audition? Do these genes only function during development, or do they also have functions in specific

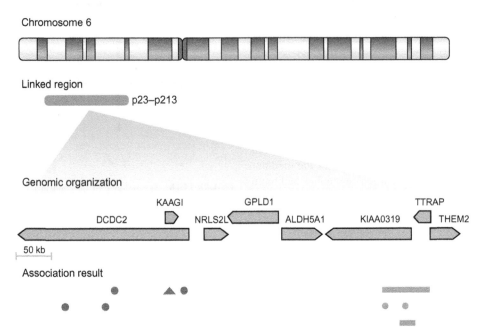

FIGURE 14.19 Genomic organization of the chromosome-6 region associated with reading disability. Dots, triangles and bars indicate the positions of SNPs, insertion–deletion polymorphisms, and genomic regions associated with learning disability, respectively. (From Paracchini, S. et al. (2007). *Annu. Rev. Genomics Hum. Genet.*, **8**, 57–79.)

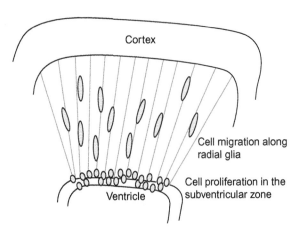

FIGURE 14.20 During formation of the cortex, neurons are born in a proliferative zone near the ventricle (the **subventricular zone**) from where they migrate along processes of **radial glia** toward the nascent cortex. Once there, they will be assembled as modular arrays that form the neural circuitry of the six-layered cortex. Compromised neuronal migration during development might be one factor that gives rise to dyslexia.

cortical regions in the adult brain? What other genes interact with these four candidate genes to enable writing and reading ability? One thing, however, is clear. Our ability to communicate through writing did not require the evolution of new genes, but was made possible through the recruitment of existing genetic networks of **pleiotropic** genes. These genes are necessary for the formation of neural circuits that happen to be suitable for the performance of this cognitive task.

NEURODEGENERATION AND MEMORY IMPAIRMENT: ALZHEIMER'S DISEASE

In 1906, a German psychiatrist, Dr Alois Alzheimer, described a patient with progressive **senile dementia**, which would later become known as Alzheimer's disease, and is now recognized as the most prevalent form of senile dementia. The incidence of Alzheimer's disease is likely to increase sharply as the population ages, and represents a major healthcare challenge for the twenty-first century.

Diagnosis of Alzheimer's Disease

Alzheimer's disease affects men and women equally, and is a progressive neurodegenerative disease that is usually diagnosed in people over the age of 65, although familial early-onset forms of Alzheimer's disease can be diagnosed in people in their 40s or 50s. Early symptoms of the disease are forgetfulness and progressive memory impairment. Behavioral changes may also occur, such as uncharacteristic outbursts of violence or unresponsiveness. In the later stages of the disease, patients become progressively weaker and more debilitated, and require constant care.

Diagnosis of possible Alzheimer's disease is based primarily on neuropsychiatric and physiological criteria. Final diagnosis can only be done postmortem, by examining the brain of the deceased for characteristic neurodegenerative lesions, known as **neuritic plaques** and **neurofibrillary tangles**. These lesions are most apparent in cortical

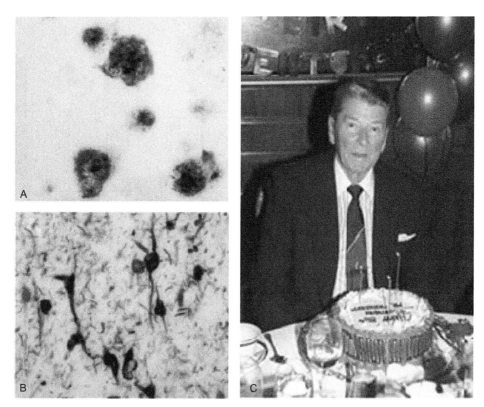

FIGURE 14.21   Characteristic neuritic plaques (A) and neurofibrillary tangles (B) in the postmortem brain of a patient with Alzheimer's disease. One of the most well-known personalities afflicted with Alzheimer's disease was President Ronald Reagan (C), here pictured at his 84th birthday after he was diagnosed with Alzheimer's disease. He died on June 4, 2004, at the age of 93. His death sparked awareness of the potential for stem cell research to develop cures for neurodegenerative diseases.

regions of the brain, where they especially affect cholinergic neurons. These lesions often start in the **entorhinal cortex**, and spread through the hippocampus and temporal and frontal lobes (Figure 14.21).

Neuritic Plaques and Tangles: Animal Models

The neuritic plaques that are characteristic of Alzheimer's disease contain β-**amyloid** protein, a 42-amino acid proteolytic fragment of a larger **amyloid precursor protein** (APP). Two transmembrane proteolytic enzymes, presenilin 1 and presenilin 2, contribute to the proteolysis of APP and the formation of the 42-amino acid β-amyloid fragment. The tangles are formed by twisted fragments of the tau protein. This is a microtubule-associated protein, which is regulated by phosphorylation. Hyperphosphorylation of tau results in the formation of neurofibrillary tangles.

The mechanisms that lead to the formation of brain lesions in Alzheimer's disease are difficult, if not impossible, to study in humans. Therefore, genetic animal models have been established to investigate the mechanisms of neurotoxicity of β-amyloid and tau, and the interrelationships between these proteins. For such models to be useful, they must recapitulate the behavioral and neurodegenerative features of the disease. Mouse models of Alzheimer's disease

have been generated by overexpressing human APP or tau in neurons. These mice develop symptoms that are characteristic of Alzheimer's disease, including the formation of neuritic plaques and neurodegeneration, as well as progressive deterioration in performance in the Morris water maze (Figure 14.22, overleaf). Reduction of the expression of the endogenous tau protein provides protection against the effects of overexpressed APP. This suggests an interaction in the etiology of the disease between the neurofibrillary tangles formed by tau and the β-amyloid plaques.

Surprisingly, *Drosophila melanogaster* is a good model for what is considered a quintessentially human neurodegenerative disorder. Flies have a gene that encodes an APP-like protein that is expressed in the nervous system, and a single presenilin gene. Expression of the human 42-amino acid β-amyloid fragment in the *Drosophila* brain gives rise to symptoms that are remarkably similar to Alzheimer's disease. The brain of the fly accumulates amyloid deposits, which results in neurodegeneration. At the same time, flies show an age-dependent decline in learning and memory. They perform progressively worse in the conditioned avoidance learning paradigm, in which exposure to an odorant is paired with an electric shock. The precise function of APP, or its *Drosophila* ortholog, is not known. In the absence of a functional gene, flies are viable, fertile,

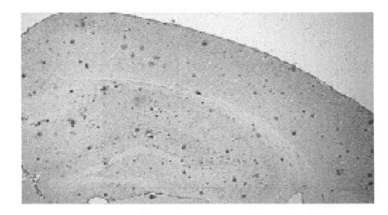
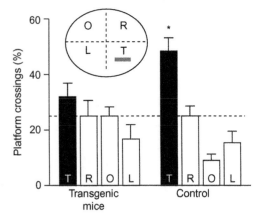

FIGURE 14.22   Neuritic plaques in the brain of a mouse that expresses human APP. The bar graphs on the right show the performance of mice that express a mutant form of human APP in the Morris water maze. The four quadrants of the maze are labeled T, containing the target platform; O, oppo-site; R, right; and L, left. Transgenic mice are impaired in their ability to learn and remember the location of the hidden platform compared to controls. (Modified from Hyde, L. A. et al. (2005). *Behav. Brain Res.*, **160**, 344–355.)

and morphologically-normal. Overexpression studies in *Drosophila* suggest a role for APP during development of the nervous system, or stabilizing and maintaining axonal connections.

Studies on the effects of APP on neurodegeneration in *Drosophila* were paralleled by studies on the tau protein. When wild-type and mutant forms of the human tau pro-tein were overexpressed in the *Drosophila* brain, cholin-ergic neurons underwent progressive degeneration, which was more pronounced with mutant than with wild-type tau protein. However, the accumulation of the tau protein did not lead to the formation of neurofibrillary tangles seen in Alzheimer's disease, suggesting that the formation of such tangles is not necessary for neurotoxicity (Figure 14.23). Hyperphosphorylation of tau in the transgenic flies exac-erbated the observed neurodegeneration. Thus, *Drosophila* provides a remarkable *in vivo* model for studies on the mechanisms that link overexpression of APP or tau to neu-ronal cell death.

Human genetics of Alzheimer's disease

Alzheimer's disease is not a single well-defined disease, but rather a spectrum of disorders with similar diagnostic features. It is a complex disease, with both genetic and envi-ronmental contributions. Familial early-onset Alzheimer's disease (before the age of 60) is rare, and is often caused by dominant genes that run in families. Nearly 200 different mutations in the presenilin-1 or presenilin-2 genes (on chro-mosomes 14 and 1, respectively) have been documented, in over 500 families. In addition, more than 20 different mutations in the APP gene on chromosome 21 have been implicated in early-onset Alzheimer's disease. It is of inter-est that the brains of people with Down syndrome, which results from trisomy of chromosome 21, also contain neu-ritic plaques similar to those found in Alzheimer's disease.

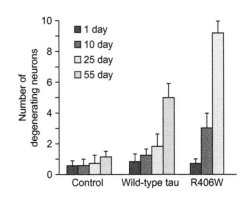

A

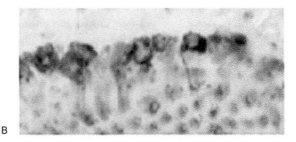

B

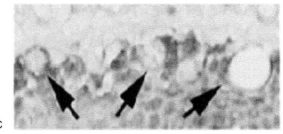

C

FIGURE 14.23   Tau-induced neurodegeneration in transgenic flies. Panel A quantifies the number of degenerating neurons as a function of age in the brains of control flies and transgenic flies that express the human tau protein (wild-type tau) or a mutant form of tau (R406W) that exacerbates the neurodegenerative effect. Panels B and C show cholinergic neurons in the optic lamina of 1-day-old and 30-day-old tau-expressing transgenic flies, respectively. The arrows point at degen-erating neurons with large vacuoles. (Adapted from Wittmann, C. W. et al. (2001). *Science*, **293**, 711–714.)

Box 14.4   Memory reconsolidation

There is substantial evidence that memories can be modified during recall. This process of memory modification has been termed **reconsolidation**, and can serve to enhance or update memories. Re-exposure of a subject to a conditioned cue or context can restore partially-extinct memories.

Fear conditioning paradigms have been especially useful in the study of memory reconsolidation. Like the initial formation of a memory, its reconsolidation during recall requires protein synthesis. When protein synthesis inhibitors are injected during memory retrieval in the amygdala, a region of the brain closely associated with fear responses, reconsolidation is disrupted.

Reconsolidation of memory involves transcription factors and kinases that are similar to those that contribute to the initial formation of long-term memories, including CREB and cyclic AMP-dependent protein kinase (protein kinase A) (Figure A).

Understanding memory reconsolidation can have important clinical applications, since disruption of this process might attenuate harmful memories, as is the case in posttraumatic stress disorder. Memory reconsolidation may also play a role in addiction, by strengthening maladaptive memories during retrieval, thereby encouraging long-lasting relapses of adverse behaviors.

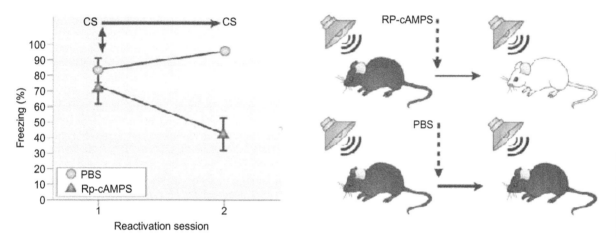

FIGURE A Memory reconsolidation is dependent on protein kinase A (PKA). Rats were trained to display an auditory fear response by pairing an electric shock with a tone. At 24 hours after training they were placed in a novel context before the conditioned stimulus (CS) was presented to them (reactivation session 1). Immediately after this reactivation session, the PKA inhibitor Rp-cyclic AMPS, or phosphate buffered saline (PBS) as a control, was administered through cannulae into the basolateral amygdala. Another 24 hours later, the rats were placed in the same reactivation context and presented with the tone. The intensity of the color of the rat in the diagram indicates the intensity of the fear response, which is represented in the graph as percentage of time that the rat spent in its fearful freezing posture. Inhibition of PKA after retrieval disrupted the reconsolidation of the auditory fear memory. (Modified from Tronson, N. C. et al. (2006). *Nature Neurosci.*, 9, 167–169.)

Familial early-onset Alzheimer's disease represents only about 5% of all Alzheimer's disease cases. The vast majority of Alzheimer's disease patients does not have an obvious familial risk, and develop the disease at an advanced age. Margaret Pericak-Vance and Alan Roses at Duke University embarked on a series of **linkage analyses** and **association studies** that identified an allele of the gene that encodes **Apolipoprotein E** (ApoE) on chromosome 19 as a major susceptibility factor for late-onset Alzheimer's disease. ApoE is the main protein component of **chilomicra**, lipid particles that transport cholesterol and triglycerides and that facilitate their uptake in the liver and peripheral tissues. The studies by Pericak-Vance and Roses showed that the *ApoE- 4* allele constitutes a major risk factor for sporadic late-onset Alzheimer's disease. It should be noted, however, that at least one third of patients with Alzheimer's disease do not carry the *ApoE- 4* allele, and some people who are homozygous for this allele never

develop the disease. Thus, the *ApoE- 4* allele is only one of many risk factors that confer susceptibility for Alzheimer's disease, and the mechanism of its involvement in the pathogenesis of Alzheimer's disease is unknown.

More than five million Americans are afflicted with Alzheimer's disease. This number is expected to increase three- to four-fold by the middle of the century, as the population ages. The economic and social burden of Alzheimer's disease is huge. In the decades ahead, identifying genetic and environmental risk factors to improve early diagnosis of Alzheimer's disease and developing more effective interventions, are major challenges for healthcare of the elderly.

SUMMARY

Using operant conditioning paradigms, studies on model organisms, such as *Aplysia* and *Drosophila*, have provided

Box 14.5 The genetic conundrum of autism

Autism is one of the most common mental retardation disorders, with heritability greater than 90%. It is a neurodevelopmental disorder, which is usually diagnosed before the age of three and is characterized by impairments in social interaction and communication, restricted interests and repetitive behaviors. In fact, diagnosis of autism is based entirely on these characteristic abnormal behaviors. Such behaviors can be stereotypic (i.e. not serving any direct purpose, such as body-rocking and hand-flapping), compulsive (e.g. insisting that objects are always positioned in a certain orientation; preoccupation with a television program), or ritualistic, in which there is insistence on invariant procedures (e.g. dressing the same way or eating the same food every day). Self-injury is also common.

Despite the high heritability of autism, **linkage studies** and **association analyses** have been largely inconclusive, although some studies implicate components of postsynaptic densities as risk factors for autism. There is a general agreement that the genetic basis for autism encompasses many genes that appear to be sensitive to gene–gene and gene–environment interactions. In addition to single nucleotide polymorphisms, variation in the number of copies of certain genes (**copy number variants** (CVNs)) has also been suggested as a factor that may predispose to autism. Interestingly, copy number variants throughout the genome have also been implicated in schizophrenia. Autism is not localized to any specific brain area, but affects widespread regions of the brain. These observations suggest that the genetic risk for developing autism is expressed during early development of the central nervous system.

The last decades have seen a sharp increase in the reported incidence of autism. This is likely due to better and earlier diagnoses, improved availability of medical and social services for autistic patients, and increased public awareness. The existence of autism as a defined disorder was only documented in 1938, by the Viennese pediatrician Hans Asperger, and was not well-accepted as a separate syndrome for decades thereafter. Today, however, autism is recognized as a common neurodevelopmental disorder. It is the most heritable of all psychiatric conditions.

In addition to classical autism, a number of similar disorders, often with milder syndromes, have been identified as **autism spectrum disorders** (ASD). The best studied ASDs are **Asperger syndrome** and **Rett syndrome**. The former resembles autism, and is characterized by impaired social interactions, restricted interests, and repetitive activities. Rett syndrome is an X-linked disease that almost exclusively affects girls. The onset of autism-like behavior occurs between 6 and 18 months of age. The rate of head growth slows down at this time, resulting in an undersized head (microcephaly), and patients often display breathing irregularities, such as breath-holding or hyperventilating. In contrast to autism, genetic studies to identify the culprit for Rett syndrome have been highly-successful, and implicated in most cases a mutation in the *MECP2* (*methyl-CpG-binding protein-2*) gene, which encodes a transcriptional repressor that binds to methylated DNA.

significant insights into the cellular mechanisms that mediate learning and memory. In such studies, massed training results in the formation of short-term memories and is primarily mediated by modulation of ion channels that impact neurotransmitter release. Spaced training results in the conversion of short-term memories into long-term memories through a process that requires protein synthesis. Genetic studies in *Drosophila* have identified mutants that enabled the dissection of temporally-distinct phases of memory acquisition. Here, the formation of associative memories depends on the mushroom bodies, as well as structures that communicate with, but are external to, the mushroom bodies. The formation of long-term memories depends on phosphorylation reactions through cyclic AMP-dependent protein kinase and MAP kinases that result in activation of transcription factors, such as CREB. Activation of CREB regulates gene expression, and gives rise to long-term synaptic modifications. These cellular pathways are evolutionarily-conserved.

In vertebrates, the hippocampus plays a critical role in the acquisition of memories. Here, long-term potentiation, mediated via NMDA-type glutamate receptors that act as molecular coincidence detectors, is the physiological substrate, which leads to synaptic modifications that underlie the formation of memories. Studies with homologous

recombinant mice have identified calcium-calmodulin-dependent protein kinases, as well as CREB and other transcriptional regulators in long-term memory consolidation. Once stored, memories can be modified during recall in a process known as memory reconsolidation.

Learning disabilities and memory loss during senile dementia are major social problems. Human genetic studies have made substantial advances in identifying risk genes for several learning and memory disorders, including dyslexia, in which candidate genes that mediate neural cell migration and adhesion during development have been implicated: Fragile X syndrome, attributed to a trinucleotide expansion in the *FMR1* gene; Rett syndrome, in which a mutation in the *MECP2* gene has been implicated; and Alzheimer's disease, in which mutations in the amyloid precursor, the tau protein, and presenilin genes have been associated with early-onset familial Alzheimer's disease, and in which an allele of *ApoE* has been implicated in late-onset Alzheimer's disease. Despite these substantial advances, it is clear that the genetic networks which mediate learning and memory are only partially-understood, and that the genes thus far implicated in human learning and memory represent only a small fraction of the genome that participates in these cognitive processes.

STUDY QUESTIONS

1. What is the difference between massed learning and spaced learning? Which procedure is more effective in generating long-term memory?
2. What are the advantages and limitations of studies on learning and memory in *Aplysia*?
3. Describe how mutations have been used to identify the different phases of memory consolidation in *Drosophila*.
4. Which cellular pathways that regulate gene expression during memory formation have been identified in all three models of *Aplysia*, *Drosophila*, and mice?
5. What is the role of phosphorylation and dephosphorylation in acquisition of memory in the hippocampus?
6. Describe the conditioned fear learning paradigm.
7. Discuss neurogenetic methods that can be used to investigate the role of the mushroom bodies in learning in *Drosophila*.
8. What is a Hebbian synapse? Describe the role of synaptic coincidence-detection in mediating long-term potentiation.
9. Describe two behavioral paradigms that are commonly used to assay spatial memory in mice and rats.
10. What is dyslexia? How strong is the experimental support for the *DYX1C1* gene as a candidate dyslexia gene?
11. What appears to be a common functional theme among dyslexia genes which have thus far been identified?
12. What are common problems encountered in human genetic studies aimed at identifying alleles that contribute to learning disorders?
13. What is the genetic basis for Fragile X syndrome?
14. If heritability for performance on the Wechsler IQ test is high, does that mean *per se* that there are genes of major effect that contribute to intelligence? Explain.
15. Which is the best-established gene with risk alleles for late-onset Alzheimer's disease?
16. What are the criteria for animal models for Alzheimer's disease, and what genetic strategies can be used to generate such models? What behavioral assays would be appropriate to monitor neurodegeneration?

RECOMMENDED READING

Carew, T. J. (2000). *Behavioral Neurobiology: The Cellular Organization of Natural Behavior*. Sinauer Associates, Inc, Sunderland, MA.

Corder, E. H., Saunders, A. M., Strittmatter, W. J., Schmechel, D. E., Gaskell, P. C., Small, G. W., Roses, A. D., Haines, J. L., and Pericak-Vance, M. A. (1993). Gene dose of apolipoprotein E type 4 allele and the risk of alzheimer's disease in late-onset families. *Science*, **261**, 921–923.

Diderot, G., Molinari, F., Tchénio, P., Comas, D., Milhiet, E., Munnich, A., Colleaux, L., and Preat, T. (2006). Tequila, a neurotrypsin ortholog, regulates long-term memory formation in *Drosophila*. *Science*, **313**, 851–853.

Giles, A. C., Rose, J. K., and Rankin, C. H. (2006). Investigations of learning and memory in *Caenorhabditis elegans*. *Int. Rev. Neurobiol.*, **69**, 37–71.

Glessner, J. T. et al. (2009). Autism genome-wide copy number variation reveals ubiquitin and neuronal genes. *Nature*, **459**, 569–573.

Kandel, E. R. (2001). The molecular biology of memory storage: a dialogue between genes and synapses. *Science*, **294**, 1030–1038.

Kandel, E. R. (2006). *In Search of Memory: The Emergence of a New Science of Mind*. W. W. Norton & Company, New York, NY.

Mace, N. L., and Rabins, P. V. (2001). *The 36-Hour Day: A Family Guide to Caring for Persons with Alzheimer's Disease, Related Dementing Illnesses, and Memory Loss in Later Life*. Warner Books, Clayton, AUS.

Margulies, C., Tully, T., and Dubnau, J. (2005). Deconstructing memory in *Drosophila*. *Curr. Biol.*, **6**, R700–R713.

Martinez, J. L., and Kestner, R. P. (2007). *Neurobiology of Learning and Memory*. Elsevier, New York, NY.

Paracchini, S., Scerri, T., and Monaco, A. P. (2007). The genetic lexicon of dyslexia. *Annu. Rev. Genomics Hum. Genet.*, **8**, 57–79.

Quinn, W. G., Harris, W. A., and Benzer, S. (1974). Conditioned behavior in *Drosophila melanogaster*. *Proc. Natl. Acad. Sci. USA*, **71**, 708–712.

Tonegawa, S., Nakazawa, K., and Wilson, M. A. (2003). Genetic neuroscience of mammalian learning and memory. *Philos. Trans. R. Soc. Lond. B Biol. Sci.*, **358**, 787–795.

Tronson, N. C., and Taylor, J. R. (2007). Molecular mechanisms of memory reconsolidation. *Nat. Rev. Neurosci.*, **8**, 262–275.

Wittmann, C. W., Wszolek, M. F., Shulman, J. M., Salvaterra, P. M., Lewis, J., Hutton, M., and Feany, M. B. (2001). Tauopathy in *Drosophila*: neurodegeneration without neurofibrillary tangles. *Science*, **293**, 711–714.

Chapter 15

Genetics of Addiction

OVERVIEW

The vertebrate brain has evolved neural reward circuits that form pleasurable associations with activities which are beneficial to survival, such as eating and sex. Such rewarding sensations that are associated with previous experiences reinforce these activities. Drugs that cause addiction impinge on neurotransmitter mechanisms in the reward pathway. Alcohol, nicotine, stimulants (such as cocaine or amphetamine), and narcotics (such as morphine and heroin), stimulate the reward circuit, and the resulting sensation of euphoria reinforces their continued use. However, excessive intake of such substances has detrimental physiological consequences that cause severe withdrawal symptoms when their use is discontinued. It is the combination of euphoria (psychological dependence) and the avoidance of such withdrawal symptoms (physiological dependence) that leads to and sustains addiction. The magnitude of addiction resulting from substance abuse worldwide is staggering, and carries a huge socioeconomic cost. Addictive behaviors are strongly influenced by the environment, but also have a significant genetic risk. Along with research on people, studies on model organisms have yielded useful insights in the physiological responses to addictive substances, and the genes that are associated with genetic variation in such responses. In this chapter we will explore the neural mechanisms and the genetic factors that predispose to addictive behaviors. We will limit our discussion to the most common addictive substances, alcohol, nicotine, cocaine, and opioids.

HALLMARKS OF ADDICTION

Addiction can be defined as a recurring compulsion by an individual to engage persistently in some specific activity, despite adverse consequences to his or her health, mental state, or social life. Addictive behaviors include the intake of alcohol, nicotine, or drugs, but these are not the only forms of addiction. Gambling, compulsive eating, and a variety of other **compulsive obsessive behaviors** with potentially harmful outcomes can be considered addictions.

Hallmarks of addiction are **dependence** and **withdrawal**. Dependence can be either physical, psychological, or both. Physical dependence is evident when sudden discontinuation of the use of an addictive substance results in physiological withdrawal symptoms, which are characterized by intense physical discomfort, and which can be severe. For example, alcohol withdrawal symptoms or withdrawal symptoms from sedatives, such as barbiturates, can lead to seizures and even death. Withdrawal symptoms are not restricted to physical dependence, but can also accompany psychological dependence, albeit in different manifestations. Here, withdrawal symptoms result in cravings that lead to irritability, insomnia, depression, or eating disorders, such as anorexia. The initial driving force for the development of addiction is the induction of **euphoria**, a sense of pleasure. However, as addiction progresses, the principal factor for maintaining the addiction increasingly becomes the avoidance of withdrawal symptoms. Thus, addiction escalates to the point where addicted individuals will forego food, sleep, and sex for continued access to the addictive substance.

What causes addiction? Social, physiological, and genetic factors all play a role. Genetic factors have been demonstrated convincingly by twin studies, in which **monozygotic twins** and **dizygotic twins** are compared, and twins reared in the same environment can be compared to twins reared in different environments, and by studies in which adopted children are compared to siblings raised by their biological parents. Such studies show a clear genetic contribution to addictive behavior, but also demonstrate the importance of environmental factors. For example, growing up in a family of heavy drinkers increases the probability of alcohol addiction. There is substantial phenotypic variation in susceptibility to addiction. Some people may become addicted soon after they start drinking or smoking, whereas others may be social drinkers or smokers for their entire life, and never develop addiction. Individuals who are at risk for addiction to one substance (e.g. alcohol) show correlated susceptibility to other substance addictions (e.g. smoking, cocaine). This may, in large part, be related to the fact that many types of addiction are mediated by shared neural mechanisms.

The major neural pathway associated with addiction is the **mesolimbic** dopaminergic pathway that projects from the **ventral tegmental area** in the midbrain to the **nucleus accumbens**, which is part of the ventral striatum (see also Chapter 3). The nucleus accumbens can be viewed as the "pleasure center" of the brain. When electrodes are implanted in the nucleus accumbens of rats and connected to a lever that activates a stimulating current, the animals will obsessively press the lever. The nucleus accumbens mediates pleasure sensations associated with food and sex, and thus promotes beneficial behaviors. Addiction impinges on this innate reward pathway (Figure 15.1).

Output neurons from the nucleus accumbens are mostly GABAergic and, via a relay in the thalamus, form a projection to the **orbitofrontal cortex.** The orbitofrontal cortex is a region of prefrontal association cortex located above the orbits of the eye. Although this area of the cortex is still poorly-understood, it is thought to integrate sensory input with affective input, such as pleasure sensations from the nucleus accumbens, to make decisions that lead to reinforcement or avoidance of behaviors. Damage occurring to the orbitofrontal cortex, as in the famous case of Phineas

Gage (see Chapter 3), results in patterns of disinhibited behaviors, such as excessive alcohol use and smoking, using foul language, compulsive gambling, and poor social interactions.

Excessive drinking, smoking, use of addictive drugs, compulsive gambling, and other addictive behaviors lead to an increase in dopamine in the nucleus accumbens. This can either be due to stimulating the release of dopamine or, as is the case for cocaine, inhibition of dopamine reuptake at dopaminergic synapses (see Chapter 2). Increased levels of dopamine result in down-regulation of dopamine receptors, and it is this loss of responsiveness that gives rise to **anhedonia,** that is, a reduction in pleasure sensation that requires addicts to use increasingly higher doses of the addictive substance to elicit the same sensation of euphoria. Altered synaptic sensitivity, as a result of modulation of neurotransmitter receptor expression, also accounts for withdrawal symptoms and physical dependence. In the case of alcoholism, development of **tolerance** to alcohol (that is, the ability to drink increasingly more alcohol

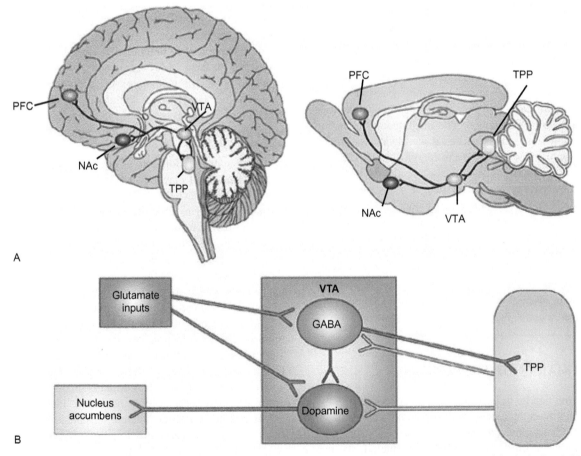

FIGURE 15.1 Neural pathways that mediate reward and reinforcement in the human (panel (A), left) and rat (panel (A), right) brain. Panel (B) shows neurotransmitter inputs in the ventral tegmental area (VTA). The ventral tegmental area communicates with the tegmental pedunculopontine nucleus (TPP) in the brainstem, and sends a dopaminergic projection to the nucleus accumbens (NAc). The nucleus accumbens, in turn, projects to the prefrontal cortex (PFC). GABAergic neurons inhibit the dopaminergic output from the nucleus accumbens. The balance of GABAergic and dopaminergic activity is also regulated by glutamatergic neurons that project to the VTA. (From Laviolette, S. R., and van der Kooy, D. (2004). *Nat. Rev. Neurosci.,* **5,** 55–65.)

before adverse effects are experienced) is a preamble for the development of addiction. Most Western societies have enacted legislation to regulate or prevent the use of addictive drugs. However, due to a history of cultural acceptance, the most commonly-abused substances, alcohol and nicotine, remain largely unregulated. In terms of the genetics of addictive behavior, alcoholism is one of the best studied examples.

ALCOHOLISM

Alcoholism presents widespread social and health problems throughout the industrialized world. The National Institute on Alcohol Abuse and Alcoholism has estimated that approximately 17.6 million people in the United States suffer from alcoholism. Alcohol abuse is responsible for a vast number of traffic accidents, resulting in serious injury or death. Alcoholism is accompanied by neurological, gastrointestinal, and cardiovascular disorders, and is the culprit for many socioeconomic problems, including increased

aggressive behavior, with adverse impacts on parenting and marital stability.

As mentioned above, alcohol sensitivity, the development of tolerance to alcohol, and susceptibility to addiction vary in the population. Environmental factors, such as stress and social experience, contribute to individual variation in sensitivity to chronic alcohol consumption, but genetic factors have also been implicated. Genetic studies have demonstrated that alcohol sensitivity is determined by multiple genes. Thus, vulnerability to alcohol use shows all the hallmarks of a quantitative trait, as variation in the propensity to consume alcohol is attributable to multiple interacting genes with individually small effects, whose expression depends on the environment.

Although studies aimed at identifying risk alleles for the development of alcoholism have made some progress, relatively little remains known about the genetic architecture that predisposes to addiction to alcohol (or any other addictive substance). The reasons are: (1) difficulty in obtaining unambiguous, quantitative phenotypic values; (2)

Box 15.1   Pathological gambling

The American Psychiatric Association defines pathological gambling as an impulse control disorder that is a chronic and progressive mental illness. Addiction to gambling shows symptoms that are similar to other forms of addiction, namely preoccupation with thoughts about gambling; tolerance, in that larger or more frequent wagers are required to provide the same "rush;" and withdrawal, which is revealed as restlessness or irritability when compulsive gamblers attempt to stop gambling. Pathological gamblers tend to risk money on whatever game is available. They prefer fast games, which end quickly and provide a constant temptation to play again (e.g. the roulette or slot machines), over games which require a long period of waiting to experience the results (e.g. the state lottery). Pathological gamblers, like other addicts, will risk ruining their lives and their relationships in order to satisfy their compulsion.

Neurobiological studies have implicated the same motivation and reward pathways in gambling as in alcohol and

drug addiction. When subjects received a monetary reward in a gambling-like experiment, fMRI studies revealed activation in brain regions such as the prefrontal cortex that closely resembled those activated when cocaine is administered to cocaine addicts. It is, therefore, not surprising that predisposition to pathological gambling is often accompanied by risk for alcohol dependence and/or drug addiction. Dopaminergic and serotonergic pathways, as well as opioid receptors, have been implicated in pathological gambling, and several studies have pointed at dopamine receptor gene alleles (DRD1, DRD2, and DRD4) as genetic risk factors for pathological gambling. As addictions appear to depend on similar neurobiological and genetic mechanisms, understanding the genetic and environmental factors that confer risk for alcohol and drug addiction is likely at the same time to provide insights in the pathophysiology of impulse disorders, such as pathological gambling.

Box 15.2   Alcohol withdrawal

Excessive alcohol intake can lead to a variety of adverse health effects, including cirrhosis of the liver, heart disease, pancreatitis, epilepsy, dementia, increased chance of cancer, and sexual dysfunction. It is common for alcoholics to continue drinking even after serious health problems become manifest. Abrupt cessation of drinking, however, is not without risk for alcohol-dependent patients, because alcohol withdrawal symptoms can be fatal. This is due to the fact that heavy alcohol consumption reduces the production of the inhibitory neurotransmitter GABA. Thus, abrupt cessation of

drinking can lead to overstimulation of synapses in the brain, and this can result in hallucinations, seizures and convulsions, and possible heart failure. These withdrawal symptoms are known as **delirium tremens**, and they can be prevented by a carefully-controlled medically-supervised detoxification and rehabilitation program. Because of the modulation by alcohol of GABAergic neurotransmission, several human genetic studies have focused on identification of polymorphisms that might be associated with alcoholism in the GABA receptor.

confounding effects of **genotype-by-environment interactions**; and (3) lack of power of detection due to insufficient sample sizes and/or **population admixture** (see Chapters 8 and 16). In the next sections, we will discuss these problems, and examine what strategies can be employed to alleviate them in studies on human populations. We will also examine how studies in animal models, where these issues can be circumvented, can complement human genetics studies.

Phenotyping

The term "alcoholism" was first used by the physician Magnus Huss in 1849 to describe the persistence of drinking despite adverse health effects. Although this term has become popular, "alcohol dependence" is a more accurate description of alcohol addiction. Alcohol intake spans a gamut of phenotypes, from abstinence to normal alcohol use without ill effects, regular drinking, heavy drinking, and addiction. Alcoholism is a chronic disease characterized by impaired control over drinking, preoccupation with alcohol, and use of alcohol despite adverse consequences. Whereas it is easy to discriminate established alcoholics from abstainers, precisely quantifying levels of alcohol intake or extent of dependence is not straightforward.

Assessments of alcohol dependence rely on self-reports (Figure 15.2). The simplest questionnaire is the "CAGE" questionnaire, which asks the following four questions: (1) have you ever felt you needed to **C**ut down on your drinking; (2) have people **A**nnoyed you by criticizing your drinking; (3) have you ever felt **G**uilty about drinking; and (4) have you ever felt you needed a drink first thing in the morning (**E**ye-opener) to steady your nerves or to get rid of a hangover? If the answer to two or more of these questions is "yes," further follow-up is warranted.

Whereas more sophisticated psychiatric tools have been developed, such questionnaires can be confounded by other **comorbid** disorders, such as depression, anxiety, neuroticism, or schizophrenia, which often accompany alcoholism. Other confounding factors in predicting alcohol dependence as a consequence of alcohol use are environmental variation and genotype-by-environment interactions. The amount and frequency of alcohol intake needed to develop alcoholism varies greatly among people, and depends on social factors, physical and emotional health, and genetic risk. Consequently, precise quantification of alcohol intake and its effects, as is possible in model organisms, is difficult to accomplish in human genetic studies and such studies have, therefore, often relied primarily on a qualitative categorization of subjects into alcoholics or controls based on self-reported questionnaires.

Twin Studies

Monozygotic twins are genetically-identical, whereas **dizygotic twins** share at least 50% of their genomes.

Short Michigan Alcoholism Screening Test-Geriatric version (SMAST-G)

In the past year:

1) When talking with others, do you ever underestimate how much you actually drink?
2) After a few drinks, have you sometimes not eaten or been able to skip a meal because you do not feel hungry?
3) Does having a few drinks help decrease your shakiness or tremors?
4) Does alcohol sometimes make it hard for you to remember parts of the day or night?
5) Do you usually take a drink to relax or calm your nerves?
6) Do you drink to take your mind off your problems?
7) Have you ever increased your drinking after experiencing a loss in your life?
8) Has a doctor or a nurse ever said that they were worried about your drinking?
9) Have you ever made rules to manage your drinking?
10) When you feel lonely, does having a drink help?

FIGURE 15.2 Example of a simple self-report questionnaire to assess possible alcohol addiction. Positive answers to three or more questions are indicative of a recent or current alcohol use problem.

Therefore, studies on twins provide an opportunity to assess the genetic and environmental contributions to phenotypic variation, including susceptibility to addiction. Thus, twin studies can readily generate heritability estimates. In the case of well-established alcohol dependence, such estimates range between 50–60%, indicating both a substantial genetic risk and a similar environmental contribution to variation in alcohol dependence. Heritability estimates tend to be slightly higher for men than women, and they may vary depending on the procedure used for phenotyping and on the study population. In addition to heritability estimates, twin studies, as well as family studies, can be used for linkage mapping by comparing which regions of the genome that are contributed by either parent to their offspring is consistently-associated with increased risk for development for alcohol dependence. Whereas it is clear from these and other studies that risk for alcohol dependence is contributed by the cumulative effects of many genes of small effect, a major contribution to predisposition for alcoholism is conferred by alleles of alcohol dehydrogenase.

Alcohol dehydrogenase converts ethanol into acetaldehyde, which is then further metabolized by aldehyde dehydrogenase. Aldehydes, rather than alcohol itself, are the culprit of the alcohol-induced malaise commonly known as a "hangover." Different alleles of alcohol dehydrogenase differ in enzymatic effectiveness, and are correlated with the extent of tolerance to alcohol consumption (Figure 15.3). High tolerance provides increased risk for addiction. Asian populations have a generally lower risk for developing alcohol dependence than European populations, due to differences in frequencies of alleles for alcohol dehydrogenase and aldehyde dehydrogenase, designated ADH2(2) and ALDH2(2), respectively. Alcohol consumption by individuals with the ALDH2(2) allele (the essentially non-functional allele in many Asians) causes an immediate and very strong dysphoria, with flushing, tachycardia, and nausea. Inhibitors of aldehyde dehydrogenase, such as disulfiram (popularly known as "Antabuse") are used clinically to treat alcoholism. Such drugs will induce sickness only when alcohol is consumed, and this negative reinforcement is aimed at preventing drinking. However, such treatment is often ineffective due to patient non-compliance.

Alcohol and drug addiction are perhaps the best-studied traits in terms of the influence of environmental effects on developing substance dependence. One approach has been adoption studies in which susceptibility for alcohol dependence in children raised by adoptive parents is compared to their biological parents. A critical stage in the development of alcohol dependence is the time of onset of drinking. This depends to a large extent on cultural and social factors. For example, in countries where the use of alcohol is illegal, alcoholism is virtually non-existent. Peer pressure, e.g. emulating the behavior of an older sibling, may lead to initiation of drinking and, depending on genetic predisposition, escalate in various stages toward alcohol dependence.

Linkage studies on twins and nuclear families have mostly led to the identification of large chromosomal regions (QTLs) that may harbor risk alleles for alcoholism, including the region on chromosome 4 that contains a tandem array of seven alcohol dehydrogenase genes, but identifying the actual genes responsible for alcoholism within such regions has been challenging. Since it is addiction to alcohol, rather than innocuous social drinking, that gives rise to socioeconomic and medical problems, geneticists have focused their attention primarily on identifying genes that encode critical neurotransmitter components that may participate in neural pathways known to be associated with addiction. Such genes have been targeted as candidate genes for association studies in the hope of identifying polymorphisms in these genes that are associated with alcohol addiction in human populations (see also Chapter 8).

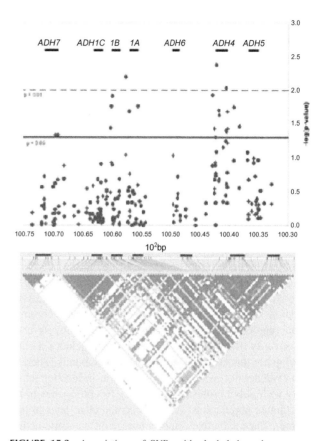

FIGURE 15.3 Associations of SNPs with alcohol dependence across the seven human ADH genes clustered on chromosome 4. Levels of statistical significance are indicated by the horizontal solid and dashed lines. Note the significant associations especially in the ADH4 gene. Diamonds represent results with the DSM-IV (Diagnostic and Statistical Manual of Mental Disorders by the (American Psychiatric Association) definition of alcohol dependence, and circles represent results using COGA criteria. The genes are transcribed from left to right; the centromer is towards the right. The bottom of the figure shows pairwise comparisons between all SNPs in the ADH genes to illustrate the extent of linkage disequilibrium across the cluster, which reflects its history of recombination. The positions of the genes are indicated at the top. SNPs in which particular combinations of alleles are commonly inherited together have high linkage disequilibrium and are indicated as dark boxes, while SNPs that occur together essentially at random show less linkage disequilibrium, and are indicated as white boxes. SNPs within genes tend to show greater linkage disequilibrium with each other than with SNPs in neighboring genes. One region of moderately-high linkage disequilibrium spans most of the genes except ADH7, where we can see a "hot spot" of recombination. Frequent recombination at this location has occurred so linkage disequilibrium has been disrupted and SNPs upstream of the hot spot are randomly associated with SNPs downstream. (Modified from Edenberg, H. J. et al. (2006). Hum. Mol.Genet., 15, 1539–1549; see also Chapter 8.)

LINKAGE AND ASSOCIATION STUDIES

The principles that underlie linkage and association studies have been described in Chapter 8. To perform whole-genome linkage analysis and association studies in human populations, several large data sets have been collected.

The most prominent of these is a large family-based data set of alcohol phenotypes and DNA samples from a project known as the Collaborative Study on the Genetics of Alcoholism (COGA). Other studies on alcoholism have recruited individuals from ethnically homogeneous populations with high prevalence of alcoholism, such as native American Indians from US Indian reservations, and people from the Finnish population. Human gene-mapping and association studies have focused on two classes of genes: genes for enzymes associated with alcohol metabolism; and genes that code for neurotransmitter receptors, transporters, and biosynthetic enzymes. The first class of candidate genes includes alleles of alcohol dehydrogenase and aldehyde dehydrogenase, which revealed not only in twin studies, but also in studies on human populations, significant associations of polymorphic markers with alcohol sensitivity. The second class of genes includes components of neurotransmitter pathways relevant to the neural reward circuit. One polymorphism of the dopamine 2 receptor, known as the Taq1A allele, has been identified in some studies as a risk factor for alcoholism; involvement of the dopamine 2 receptor based on this evidence is, however, controversial, as the actual polymorphism appears to be in the neighboring gene (ANKK2). Several other neurotransmitter-receptor genes have also been implicated, most notably the GABA receptor. QTL studies in mice have identified a chromosomal region that is associated both with alcohol preference and severity of withdrawal symptoms that includes a cluster of genes that encode $GABA_A$ receptor subunits. There is also evidence for association of variation in alcohol sensitivity with polymorphisms in $GABA_A$ receptor genes in rat and human populations. As will become clear later in this chapter, since addictive drugs converge on the same neural reward circuit, candidate genes associated with alcohol-related phenotypes are also candidate genes for other forms of addiction, including addiction to nicotine, cocaine, and opioids.

ALCOHOL SENSITIVITY IN MODEL ORGANISMS

Problems encountered in studying the genetic basis of alcoholism in people can be circumvented in model organisms, where alcohol-related phenotypes can be measured precisely, and where the genetic background and environment can be controlled. Such studies have identified several genes involved with response to alcohol exposure that are conserved from nematodes to flies, mice, and humans. Worms, Caenorhabditis elegans, can be grown rapidly in large numbers, stored frozen and readily-manipulated genetically. Drosophila melanogaster presents an advantageous model, because alcohol sensitivity and tolerance can be quantified in large numbers of individuals in controlled genetic backgrounds and under standardized environmental conditions. Mice and rats are harder to grow in large numbers, and genetic manipulations often rely on "one-gene-at-a-time" knock-out procedures, or on QTL mapping studies that yield large genomic regions in which it is often difficult to pinpoint the relevant candidate genes. However, the effects of alcohol in rodents mimic the effects of alcohol on people more closely, including withdrawal and addiction. In addition, the neural reward circuit in rats and mice is similar to that in the human brain. All three model systems have yielded insights into the genetic underpinnings of the response to alcohol exposure that are broadly applicable. These studies reveal complex genetic architectures for alcohol-related traits, aspects of which are likely to be conserved in people.

Responses to Alcohol Exposure in the Nematode, Caenorhabditis elegans

The nematode C. elegans provides a genetic model that is amenable to large unbiased screens for the identification of genes that mediate its response to ethanol. When worms are exposed to ethanol they show a decrease in locomotion. In the continued presence of ethanol, however, locomotor behavior is gradually restored, which reflects adaptation. This acute sensitivity to ethanol exposure, and the subsequent development of tolerance, is reminiscent of similar reactions to ethanol exposure in flies, mice, and humans.

The locomotor impairment observed as an acute response to alcohol exposure appears to be mediated, at least in part, by a potassium channel, known as the BK potassium channel coded by the slo-1 gene. Worms that carry loss-of-function mutations in this gene are resistant to the effects of ethanol, whereas animals that overexpress the BK potassium channel show defective locomotion that is reminiscent of ethanol-intoxicated animals (Figure 15.4). Orthologs of the slo-1 gene have also been implicated in the acute response to alcohol exposure in flies and mice. Studies on natural variation in acute alcohol sensitivity in wild strains of C. elegans have implicated the **NPR-1 pathway**. This is perhaps not surprising since this pathway generally appears to mediate behavioral responses to environmental conditions including nutrients and the ambient level of oxygen (see also Chapter 12). NPR-1 is an ortholog of the mammalian neuromodulator neuropeptide Y (see Chapter 3). The NPR-1 pathway regulates the development of tolerance to ethanol, and its general function may be to enable the nematode to move away from a noxious chemical stimulus, such as ethanol, to prevent immobilization and, ultimately, death. Despite the excellent genetic resources available for studies on nematodes, relatively few genes have been identified to date that impact the response to alcohol. In contrast, considerable progress has been made in another powerful genetic model system, Drosophila melanogaster.

The Genetics of Acute Alcohol Sensitivity and Development of Tolerance in Drosophila

Ethanol, which is a by-product of fermented fruit, is a normal component of the environment of fruit flies. However, when exposed to concentrations of ethanol that are higher than those usually encountered, such as concentrated ethanol vapors, flies show behavioral responses that are

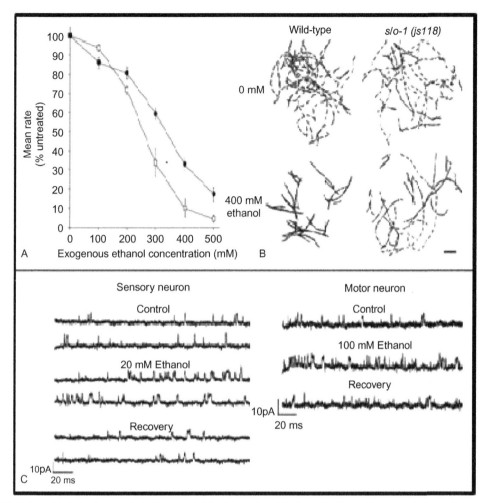

FIGURE 15.4   Alcohol sensitivity in the nematode *Caenorhabditis elegans*. Panel A shows that increasing concentrations of ethanol inhibit speed of movement (closed symbols) and frequency of egg laying (open symbols) of the wild-type (N2) strain of *C. elegans*. Panel B shows locomotion tracks of wild-type and *slo-1(js118)* animals in the presence and absence of ethanol. Fast movement is evident from a track of distinct animal outlines, whereas a solid line shows slow locomotion. Untreated animals (top left) move at a slightly higher speed than untreated *slo-1(js118)* mutants (top right). Ethanol-treated *slo-1(js118)* mutants (lower right) have a greater average speed than ethanol-treated wild-types (lower left) (scale bar: 2 mm). Panel C shows a patch-clamp recording (see Chapter 2) of single SLO-1 channels in excised patches from a sensory neuron and a motor neuron. Ethanol activates channel opening, and this activation is reversible when ethanol is washed out. (Modified from Davies, A. G. et al. (2003). *Cell*, **115**, 655–666.)

reminiscent of human intoxication. First, they become hyperactive, but soon thereafter they lose mobility and postural control, and enter a state of sedation. Several assays have been developed to quantify the effects of ethanol on *Drosophila*. The most popular assay uses a device designed in 1986 by Ken Weber, then a doctoral student at Harvard University, known as the **inebriometer** (Figure 15.5, overleaf). This device consists of a 4-foot-long vertical glass tube with slanted mesh partitions. Flies are introduced into the top of the tube and exposed to concentrated vapors of ethanol by bubbling compressed air through a flask with alcohol. As the flies become intoxicated they can adhere to the mesh partitions, but ultimately they become sedated and fall through the column. They are collected from the bottom of the tube at one minute intervals. A mean elution

time can be calculated from their elution profile, and this turns out to be a reproducible measurement of their sensitivity to alcohol. Flies recover soon after they elute from the column. When they are re-exposed within several hours after the first exposure, their mean elution time shifts, and the difference in mean elution times between the initial and second exposures is a measure of the induction of tolerance to alcohol. The shift in mean elution time peaks at around two hours following the initial exposure, and dissipates over a 24–36 hour period. Induction of tolerance is dependent on age and genetic background, and is more evident in flies that are sensitive to ethanol than flies that are relatively more resistant to acute exposure (Figure 15.6).

In contrast to the complex questionnaires that investigators must rely on when attempting to measure

FIGURE 15.5 An inebriometer. A tank bubbles air through a flask that contains ethanol to generate vapor in the top of the glass column. Flies can adhere to the mesh partitions, and as they become intoxicated will fall through the funnel at the bottom of the column.

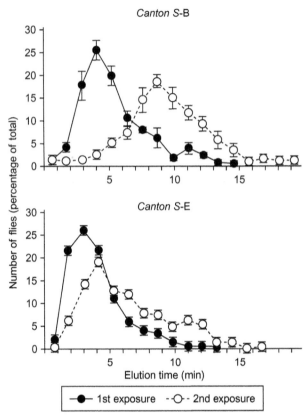

FIGURE 15.6 Development of tolerance in *Drosophila* on exposure to ethanol. The figure shows an elution profile of flies from an inebriometer after an initial exposure to ethanol and a second exposure two hours later. Note the shift in the elution profile. The magnitude of this shift is dependent on the genetic background, and is more pronounced in the *Canton S* B than in the *Canton S* E flies. (From Morozova, T. V. et al. (2006). *Genome Biol.*, **7**, R95).

alcohol-related phenotypes in people, the inebriometer assay in *Drosophila* provides an objective and straightforward method for quantifying sensitivity to acute alcohol exposure and the development of tolerance.

The Candidate Gene Approach

The development of the inebriometer enabled pioneering studies by Ulrike Heberlein and her colleagues at the Gallo Institute of the University of California in San Francisco. These investigators used *Drosophila* mutants to identify several cellular pathways associated with responses to alcohol. The first pathway implicated in alcohol sensitivity involves cyclic AMP signaling, and overlaps with the cellular mechanisms implicated in learning and memory (see also Chapter 14). One of the first mutants identified by the Heberlein laboratory, named *cheapdate*, showed increased sensitivity to alcohol and is an allele of *amnesiac*, which encodes a neuropeptide implicated in olfactory memory and is produced by two dorsal paired medial cells, from where it is released onto the mushroom bodies (see Chapters 3 and 14). The amnesiac neuropeptide is thought

to activate the cyclic AMP-signaling pathway. Indeed, flies with decreased cyclic AMP-dependent protein kinase activity are less sensitive to alcohol exposure. The *rutabaga* gene, which encodes a calcium-calmodulin-dependent adenylyl cyclase (Chapter 14), has also been implicated in alcohol sensitivity.

Neurons of the central complex (Chapter 3) that mediate locomotion play a role in the induction of tolerance. Catecholaminergic neurotransmitter pathways have also been implicated in sensitivity and tolerance to alcohol in *Drosophila*. Dopamine has been associated with acute sensitivity, and octopamine, the invertebrate homolog of noradrenaline (Chapter 3), with induction of tolerance. Tolerance is reduced in flies with mutations in the gene that encodes tyramine β-hydroxylase, a critical enzyme for the biosynthesis of octopamine.

Exposure to high concentrations of ethanol can be considered an environmental stress condition, not unlike other forms of environmental stress such as heat stress or cold stress. Indeed, when flies were subjected to heat stress prior to exposure to ethanol, the heat stress response

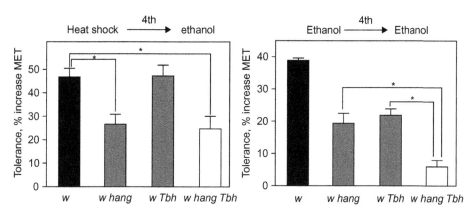

FIGURE 15.7   Crosstalk between stressors in alcohol tolerance induction in *Drosophila*. Exposure of flies to heat shock results in relative tolerance to a subsequent exposure to ethanol (left panel). Control flies (*w*) and flies lacking octopamine (*w Tbh*) show normal cross-tolerance, whereas hangover mutants (*w hang*) and the double mutant (*w Tbh hang*) show reduced cross-tolerance. Asterisk, *P* <0.0001. The right panel shows a similar experiment in which ethanol tolerance is assessed after an initial exposure to ethanol itself instead of heat shock. The induction of tolerance is essentially eliminated in the *Tbh hang* double mutants. Asterisk, *P* <0.001. *w* indicates a *white* mutant background as a visible marker. (Modified from Scholz, H. et al. (2005). *Nature*, **436**, 845–847.)

elicited the induction of tolerance to the subsequent alcohol exposure. Thus, both heat stress and ethanol-induced stress activate a similar cellular pathway. Heberlein and colleagues identified a transcription factor, named *hangover* (*hang*), as a component of this pathway, and showed that *hangover* mutants developed less tolerance to alcohol. Double mutants of *hangover* and the tyramine β-hydroxylase gene showed complete elimination of tolerance (Figure 15.7).

The Genomics Approach

In addition to the single mutant candidate gene approach, *Drosophila melanogaster* has also been used as a model to survey the genome-wide transcriptional response to alcohol exposure. Tatiana Morozova and colleagues at North Carolina State University in Raleigh used expression microarrays to show that acute exposure to ethanol results in the immediate down-regulation of genes affecting olfaction, and rapid up-regulation of expression of **biotransformation enzymes**. In addition to these rapid transcriptional changes, however, a second slower adjustment in gene expression occurred while the flies developed tolerance. This involved transcriptional regulators, proteases, and metabolic enzymes. The latter included enzymes associated with fatty acid biosynthesis and regulation of the fate of pyruvate, a central metabolite at the crossroads of carbohydrate and fatty acid metabolism, and a precursor for the entry of **acetyl-CoA** into the energy-generating combustion machinery of the **Krebs cycle**. It appears that, during the development of tolerance, a metabolic switch is activated that promotes the conversion of malic acid into pyruvate to drive the synthesis of fatty acids (Figure 15.8). The conversion of malate into pyruvate by malic enzyme generates NADPH, an essential co-factor for fatty acid biosynthesis. Malic

enzyme was also implicated with ethanol resistance when whole-genome expression levels were analyzed in lines artificially selected for alcohol sensitivity (Figure 15.9). The altered regulation of intermediary metabolism that may accompany the development of tolerance in flies is of interest for human alcoholism, as it is reminiscent of metabolic changes that occur in heavy drinkers who are prone to develop **fatty liver syndrome**, a reversible condition characterized by the accumulation of fat in the liver.

Because of evolutionary conservation of function, it is likely that human orthologs of genes affecting alcohol sensitivity in *Drosophila*, such as malic enzyme, may contribute to alcohol-associated phenotypes in humans. Because, in flies, the genetic background and environment can be tightly-controlled, and large numbers of individuals can readily be reared, one can perform genome-wide analysis in flies with high statistical power. Based on the outcome of such studies, one can then select human orthologs of *Drosophila* genes implicated in alcohol sensitivity or tolerance as candidate genes for association studies in human populations. Such a two-step approach provides an attractive alternative to **genome-wide association studies**, since it avoids the high multiple-testing penalty inherent in such studies.

ALCOHOL-RELATED PHENOTYPES IN RODENT MODELS

Mice and rats have been the most extensively-studied model systems for studies on the genetics of alcoholism, as their alcohol-related behaviors most closely resemble those of people. Different strains of mice, such as the C57BL/6 J and DBA/J2 strains, differ in their preference for alcohol intake. These mouse strains served as progenitor strains for the generation of 26 **recombinant inbred** (RI) **lines**, which contain a mosaic of chromosomal segments from

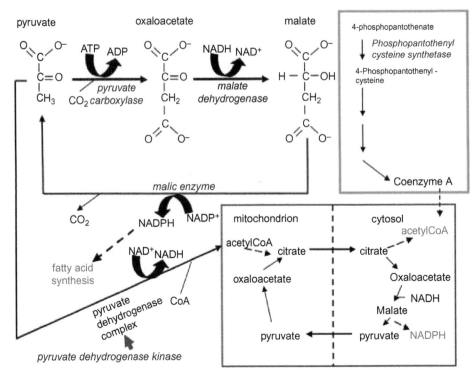

FIGURE 15.8 Diagram of the network of pathways that regulate the metabolism of pyruvate. Phosphopantothenyl cysteine synthetase, malic enzyme and pyruvate dehydrogenase kinase undergo altered expression levels during the development of tolerance to ethanol. Acetyl-CoA and NADPH can be generated in the cytosol for the biosynthesis of fatty acids as a consequence of the transport of citrate and pyruvate across the mitochondrial membrane. Malic enzyme is critical for the production of NADPH. (From Morozova, T. V. et al. (2006). *Genome Biol.*, **7**, R95.)

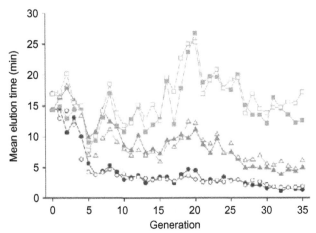

FIGURE 15.9 Artificial selection for alcohol sensitivity from a natural population of *Drosophila melanogaster*. Duplicate selection lines were generated for low sensitivity (upper traces, square symbols) and high sensitivity (lower traces, round symbols) to alcohol exposure as assessed by the mean elution time from an inebriometer. The response to selection is rapid. The center traces (triangles) denote unselected controls, which show increased sensitivity to alcohol due to inbreeding depression. (From Morozova, T. V. et al. (2007). *Genome Biol.*, **8**, R231.)

the parental strains, and which have been systematically-genotyped for more than 1500 genetic markers across the 20 mouse chromosomes. These BXD RI strains have been used extensively for QTL mapping studies in mice. The most common measurement for alcohol intake in mice or rats is the **two-bottle preference test**, in which animals have a choice between drinking from a bottle that contains water and one that contains water with a certain concentration of alcohol (Figure 15.10). Intoxication in mice can be assessed by loss of the righting reflex, i.e. difficulty in recovering an upright posture after the animal is turned on its back. Mice can also develop physiological dependence on alcohol, as evidenced by withdrawal symptoms. Such withdrawal symptoms are manifest as seizures when animals are picked up by the tail, and their severity can be scored and quantified.

Genetic studies on alcohol-related phenotypes fall into two categories. Many studies have used transgenic approaches to look at the effects of both knock-out and knock-in candidate genes on alcohol consumption. The second type of approach has focused on identifying QTLs that harbor genes that underlie variation in alcohol-related phenotypes between divergent strains (Figure 15.11, p. 264). Since alcohol is known to affect motor coordination through its effects on GABA receptors, many studies have targeted GABA receptors to gain insights into the neurotoxic mechanisms of alcohol intake by creating knock-outs of different GABA receptor subunits. In addition, dopamine receptor knock-outs and knock-out mice for several other neurotransmitter receptors, including serotonin receptors,

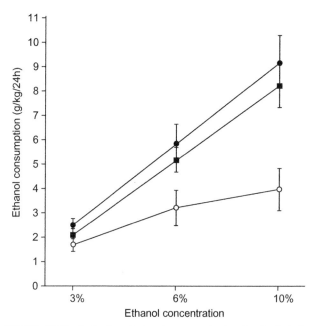

FIGURE 15.10   Reduction of ethanol consumption in mice that lack the dopamine D$_2$-receptor. Mice that lack the dopamine D2-receptor (open circles) consume less ethanol averaged over a four-day period than their heterozygous (filled squares) or homozygous wild-type (filled circles) controls in a two-bottle preference test. The positions of the bottles were switched after the second day. (From Phillips, T. J. et al. (1998). *Nature Neurosci.*, **1**, 610–615.)

opioid receptors, and corticotrophin releasing factor receptors have been generated to study the contributions of these neurotransmitter pathways to alcohol-associated phenotypes. Although these studies have clearly pointed at GABA receptors and dopamine receptor subtypes (the D1 and D2 dopamine receptors) in alcohol-related symptoms, many of these studies have yielded contradictory results and progress has by-and-large been disappointing. This is due to a number of factors: (1) the effects of knock-outs are sensitive to genetic background, and are often also sensitive to environmental conditions; (2) functional redundancy of alternative pathways may obscure the full spectrum of effects of the knocked-out candidate gene; (3) since the effects of alcohol on the animal's physiology and its nervous system are diverse, it is difficult to interpret studies that focus on only one piece of a complex puzzle; the "one-candidate-gene-at-a-time" approach, which focuses repeatedly on a limited number of the same candidate genes, cannot easily uncover unknown mechanisms or reveal complex interconnections.

In principle, QTL mapping studies would overcome many of these limitations. However, a vast number of QTL studies for different alcohol-related phenotypes have implicated large regions of the genome, and narrowing such regions down to the level of single quantitative trait genes (QTG), let alone single quantitative trait nucleotides (QTN), has been challenging. One QTL study, however, has met

this challenge. Kari Buck and her coworkers at the Oregon Health and Science University identified QTLs for alcohol withdrawal on mouse chromosomes 1, 4, 11, and 19, and were able to identify the *Mpdz* gene on chromosome 4 as a QTG with large effect (Figure 15.12, p. 265). This gene encodes a multi-PDZ domain encoding protein (MPDZ/ MUPP1). PDZ domains are conserved motifs that consist of 80–90 amino acids, named according to the first letters of three proteins in which they were initially identified (postsynaptic density protein PSD95, Drosophila disc large tumor suppressor DlgA, and zonula occludens-1 protein Zo-1). PDZ domains are found in a large number of proteins, and mediate protein–protein interactions. They have been implicated in synaptic organization, where PDZ domains can anchor neurotransmitter receptors such as GABA receptors or serotonin receptors, and transduction enzymes. The gene identified by Buck and colleagues is expressed in a neural circuitry associated with withdrawal symptoms that involves the basal ganglia and the limbic system. This study represents an important advance, in that it has led to the identification of a central synaptic component in a neural circuit associated with a well-defined alcohol-related phenotype, namely withdrawal. This study presents an example of how quantitative genetic analyses can lead to insights in neural mechanisms for complex behaviors.

SMOKING

The second most common addiction to alcohol is smoking. Tobacco leaves are harvested from plants of the genus *Nicotiana*, and cured to produce the dry leaves that are used for smoking. The use of tobacco originated in North and South America among native Americans for ritualistic purposes, and dates back as far as 2000 bc. Tobacco was discovered by Christopher Columbus and his crew, who introduced smoking to Spain, from where the habit spread through the rest of Europe and the world. After vending machines were invented, which enabled the mass production and distribution of cigarettes in the late-nineteenth century smoking became widespread and was considered fashionable.

One of the first people to condemn smoking as a detrimental habit was King James I of England, who in 1604 wrote a treatise "A Counterblaste to Tobacco" in which he argued that smoking is "A custome lothsome to the eye, hatefull to the Nose, harmfull to the braine, dangerous to the Lungs, and in the blacke stinking fume thereof, nearest resembling the horrible Stigian smoke of the pit that is bottomelesse." With the rise in smoking in the early-nineteenth century, the incidence of lung cancer, previously a rare disease, increased to near epidemic proportions. In 1912, an American physician, Dr Isaac Adler, was the first to suggest a direct link between smoking and lung cancer, but it was not until decades later that the connection between smoking and lung cancer,

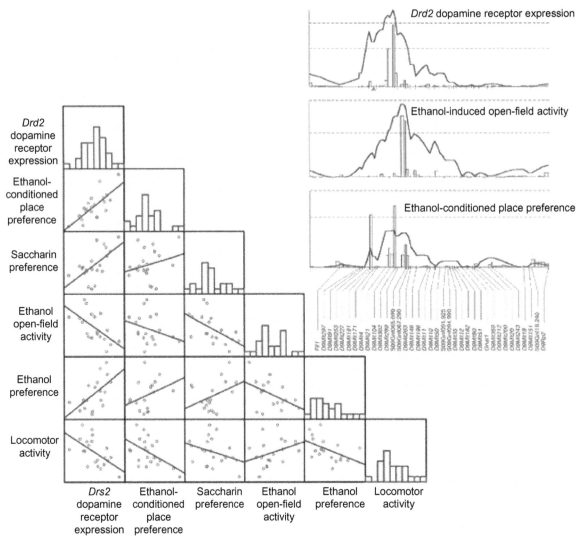

FIGURE 15.11   Genetic correlation of expression of the dopamine receptor *D2* (*Drd2*) gene, based on microarray data with several behavioral phenotypes documented for BXD recombinant inbred strains of mice. The correlated traits are ethanol-conditioned place preference, saccharin preference, ethanol-induced open field activity, ethanol preference and locomotor activity. The inset shows that the same QTL region on chromosome 9 is associated with variation in *Drd2* expression, ethanol-induced open field activity and ethanol-induced conditioned place preference. The allelic effect is in the opposite direction for open field activity, a trait that is negatively correlated with *Drd2* expression. (From Chesler, E. J. et al. (2005). *Nat. Genet.*, **37**, 233–242.)

along with emphysema and heart disease, became generally recognized (Figure 15.13). An important turning point came in 1964, with the release of a report on Smoking and Health from the United States Surgeon General, which resulted in a change in public attitude towards smoking, the banning of certain advertisements, and the requirement for warning labels on tobacco products. Today, smoking tobacco remains the leading cause of preventable death and, according to the World Health Organization, is responsible for the deaths of about five million people per year worldwide.

Neural Mechanisms of Nicotine Addiction

The addiction to smoking is due to nicotine and its by-products. Nicotine is a highly-addictive psychoactive substance. Whereas most of the nicotine in cigarettes is combusted, enough survives to give rise to addiction. Nicotine crosses the blood–brain barrier, and binds to nicotinic acetylcholine receptors in the ventral tegmental area of the brain. Activation of these receptors in turn leads to increased release of dopamine in the nucleus accumbens and activation of the mesolimbic reward pathway, the same neural circuit that mediates positive reinforcement that underlies addiction to alcohol and psychoactive drugs of abuse. Repeated activation of nicotinic acetylcholine receptors in the ventral tegmental area leads to up-regulation of these receptors, which causes withdrawal symptoms on cessation of smoking. Thus, dependent smokers constantly feel the urge to smoke to regulate their blood levels of nicotine in order to avoid withdrawal

A

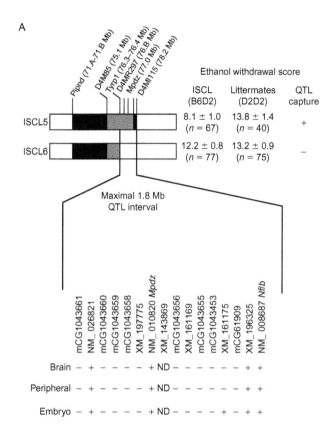

	Ethanol withdrawal score		
	ISCL (B6D2)	Littermates (D2D2)	QTL capture
ISCL5	8.1 ± 1.0 (n = 67)	13.8 ± 1.4 (n = 40)	+
ISCL6	12.2 ± 0.8 (n = 77)	13.2 ± 0.9 (n = 75)	−

Maximal 1.8 Mb
QTL interval

B

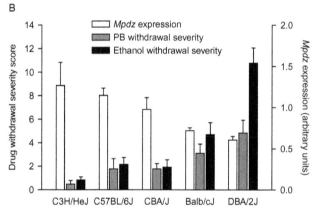

FIGURE 15.12 Identification of the *Mpdz* gene as a quantitative trait gene for alcohol withdrawal. Panel A illustrates fine mapping of the target QTL for seizures induced by alcohol withdrawal (see also Chapter 8). The QTL interval was introgressed in an inbred parental background and interval-specific congenic lines (ISCL) of mice, i.e. mice with recombinations only in this interval, were compared to their respective noncongenic (D2D2) littermates for severity of alcohol withdrawal. ISCL5 congenic mice had significantly less severe alcohol withdrawal than noncongenic littermates, pinpointing the candidate gene for the QTL effect within a narrow interval, whereas ISCL6 congenic mice did not differ from noncongenic littermates, indicating a location between the markers *Tyrp1* and *D4Mit115*. Expression of genes in this interval was evaluated using a panel of cDNAs for brain, peripheral tissues, and embryo. Panel B shows that expression of *Mpdz* within this region is genetically-correlated, with severities of withdrawal from alcohol and pentobarbital (PB) among standard inbred mouse strains. (From Shirley, R. L. et al. (2004). *Nat. Neurosci.*, **7**, 699–700.)

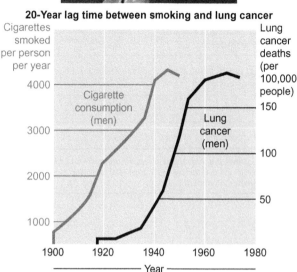

FIGURE 15.13   Correlation between smoking and the incidence of lung cancer.

symptoms and, consequently, breaking a smoking addiction is extremely difficult.

Genetic Risk for Smoking and Nicotine Addiction

Like alcoholism, smoking is a complex behavioral trait with both genetic and environmental components. Similar to alcohol use, environmental circumstances can be critical determinants in initiating smoking. In societies where smoking is prohibited or culturally not accepted, the risk for nicotine addiction will be much smaller than in societies where smoking is common. Peer pressure, a family with a history of smoking, or psychological factors, such as depression or neuroticism, can lead to smoking (Table 15.1). In permissive environments, once smoking is initiated genetic factors contribute to the risk of addiction, with heritability estimates that range widely from 11–78% among different studies. This wide range of heritability estimates is due to differences in populations and phenotypes examined (such as regular tobacco use, occasional smoking, age of initiation, nicotine dependence).

Because of the known importance of nicotine metabolism, and the effects of nicotine on the reward pathways in the brain in eliciting addiction, most genetic studies have

TABLE 15.1 Comorbidity of alcohol addiction with psychiatric disorders and other forms of drug dependence

Disorder	Odds
Anxiety disorders	2.6 ×
Mood disorders (especially major depression)	4.1 ×
Personality disorders	4.0 ×
Antisocial personality disorders	7.1 ×
Drug dependence	36.9 ×
Nicotine dependence	6.4 ×

The table shows the odds (fold) of co-occurrence in an addicted person versus general incidence. Source: National Institute of Alcoholism and Alcohol Abuse.

focused on candidate genes associated with neurotransmitter pathways and nicotine metabolism. Candidate genes associated with neurotransmission recapitulate the "usual suspects" implicated in alcoholism, including *DRD2*, which encodes the dopamine D2 receptor, the serotonin transporter, GABA receptor, the μ-opioid receptor, encoded by the *OPRM1* gene (discussed further below), and, of course, the nicotinic acetylcholine receptor. A vast number of studies that found associations between polymorphisms in these genes and smoking phenotypes have been reported, but they could not be reproduced. This is due to common problems with human genetic studies, including sample sizes that do not provide sufficient statistical power, **ethnic stratification** (see also Chapters 8 and 16), and environmental effects that differ between populations. In addition, the recruitment and phenotyping of study subjects can be highly diverse. For example, subjects from the general population can be phenotyped by a questionnaire sent out and returned by mail, or they can be recruited via a clinic. To make matters worse, definitions of smoker phenotypes are not always clear. For example, in terms of genetic risk how does one classify an occasional infrequent smoker from a person who quit smoking a few years ago, or a person with nicotine dependence who still smokes from a person with nicotine dependence who quit smoking and has not yet relapsed, but could do so? Nonetheless, the preponderance of studies implicates sequence variations in the dopamine pathway with liability for nicotine addiction.

About 80% of the nicotine that is ingested during smoking is inactivated by a specialized **cytochrome P450** enzyme, CYP2A6, which converts nicotine into cotinine, and further into *trans*-3-hydroxycotinine, which can be excreted in the urine. CYP2A6 also acts on other compounds in tobacco smoke, which result in the production of carcinogens. Thus, reduced levels of CYP2A6 are expected to be associated with decreased risk of smoking-related cancer, and, conversely, increased CYP2A6 activity is expected to increase risk of smoking-associated cancer. This indeed appears to be the case. The *CYP2A6* gene, located on chromosome 19, is highly polymorphic. A large number of studies have tried to associate polymorphisms in the *CYP2A6* gene with smoking status, but these studies have been inconclusive. Like many complex behavioral traits, genetic risk for smoking appears to be due to a large number of genes with small effects that are difficult to identify with sufficient power in genetic studies on human populations.

DRUG ADDICTION

Although alcohol abuse and nicotine addiction together place a large socioeconomic burden on society, alcohol consumption and smoking have a long history of cultural acceptance and, therefore, are legal in most countries. In contrast, the use of recreational drugs such as cannabis (also known as hashish or marijuana), cocaine, and opioids, e.g. heroine, is criminalized and stigmatized in many Western societies. For these reasons it is difficult to recruit subjects for human genetic studies aimed at identifying polymorphisms in candidate risk alleles for addiction to illegal recreational drugs, other than to solicit participation in such studies from people who have undergone or are undergoing rehabilitation. Furthermore, as in the case of alcoholism and nicotine addiction, drug addiction can be confounded by psychiatric disorders.

Recreational drugs such as opioids, cocaine, and lysergic acid diethylamide (LSD) interact with well-defined neurotransmitter receptors, which mediate their responses. For example, the analgesic effects of opioids are mediated via opioid receptors that normally bind enkephalins or endorphins; cocaine is produced by the leaves of the coca plant as a protection against host insects, since it is an antagonist of octopamine, an insect neurotransmitter – it inhibits the reuptake of dopamine, noradrenaline, and serotonin in the nervous system, which accounts for the cocaine "buzz" LSD shows structural similarity to serotonin, and elicits its hallucinogenic effects by binding to serotonin receptors. The euphoria experienced by the use of such drugs is mediated by the same mesolimbic dopaminergic reward pathway involved in alcohol and nicotine addiction. It is therefore not surprising that susceptibility to addiction is usually not restricted to a single substance, but applies to a wide spectrum of addictive compounds. Thus, alcoholism, smoking, and addiction to recreational drugs are often **comorbid**. Extensive studies on the mechanisms of addiction to cocaine and opioids have been done in rodent model systems, and some association studies have also been carried out with human subjects. We will briefly look at the genetics that underlie susceptibility to the most widely-used illegal recreational drugs, cocaine and opioids.

Cocaine

Coca leaves were first brought to Europe by the Spanish, who had learned about their stimulant and anesthetic properties from South American natives who chewed coca leaves for ceremonial and medicinal purposes. The active ingredient of coca was isolated in 1855, and this alkaloid, cocaine, was at first viewed favorably for its medical applications and stimulating effects (Figure 15.14). A "pinch of coca leaves" was an ingredient in John Styth Pemberton's original recipe for Coca-Cola for about 20 years, until it was deleted from the formulation in 1906 (Figure 15.15). Sigmund Freud lauded the use of cocaine, and mistakenly considered the drug non-addictive. He described cocaine as causing:

". . . exhilaration and lasting euphoria, which in no way differs from the normal euphoria of the healthy person . . . You perceive an increase of self-control and possess more vitality and capacity for work . . . In other words, you are simply normal, and it is soon hard to believe you are under the influence of any drug . . . Long intensive physical work is performed without any fatigue . . . This result is enjoyed without any of the unpleasant after-effects that follow exhilaration brought about by alcohol . . . Absolutely no craving for the further use of cocaine appears after the first, or even after repeated taking of the drug . . . "

Contrary to Freud's assessment, the addictive properties of cocaine became all too clear. Cocaine use can lead to severe physical and psychological dependence and a wide spectrum of adverse health consequences, including heart attack and stroke. Cocaine today is recognized as the second most addictive controlled substance after heroin, and it is the second most popular recreational drug (after marijuana) in the United States.

The addictive effects of cocaine arise from its inhibitory action on dopamine reuptake receptors in the dopaminergic projection from the ventral tegmental area of the brain to the nucleus accumbens. Blocking of dopamine reuptake causes accumulation of dopamine in the synaptic cleft and, thus, prolonged activation of the reward signaling pathway. Rats can be trained to self-administer cocaine. When dopamine antagonists are introduced via a canula into their nucleus accumbens, reinforcement for self-administration is abolished, and the animals no longer show the addictive behavior, providing direct evidence for the central role of dopamine in mediating cocaine addiction. Prolonged use of cocaine results in down-regulation of postsynaptic dopamine receptors. This gives rise to **tolerance**, in other words, the same dose of cocaine is no longer sufficient to achieve the expected sense of euphoria, and higher dosages are necessary to accomplish the rewarding sensation that reinforces the drug-seeking behavior.

Human genetics studies have focused on identifying associations between various cocaine responses and polymorphisms in candidate genes associated with

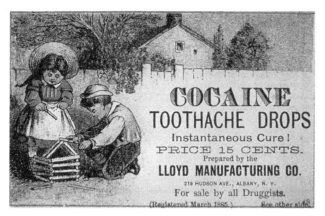

FIGURE 15.14 A label for over-the-counter cocaine toothache drops illustrates the early generally-accepted medicinal use of cocaine.

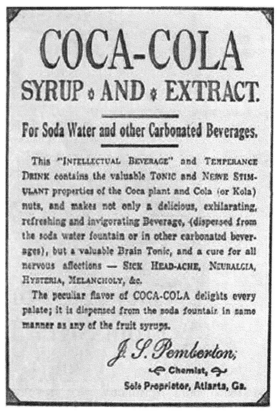

FIGURE 15.15 An early advertisement for Coca-Cola formulated in 1885 by J. S. Pemberton.

dopaminergic neurotransmission, including genes for dopamine β-hydroxylase (DBH), which converts dopamine into norepinephrine, the dopamine transporter (DAT), and the dopamine type 2 receptor (DRD2). Although several polymorphisms have been identified in such studies, animal models have been more informative in gaining insights into cocaine addiction. Rats and mice will readily self-administer cocaine by pressing a lever and cocaine intake results in hyperactivity. A popular behavioral paradigm to measure cocaine-induced reward is the **conditioned place preference** test, in which mice are given a choice between a compartment in which they have been exposed previously to cocaine, and another compartment not associated with the effects of cocaine (Figure 15.16). As in humans, genetic studies on the effects of dopamine on mediating reward sensations in cocaine-addicted mice have focused on dopamine receptors and the dopamine transporter that mediates the reuptake of dopamine into the presynaptic

cell, which is the pharmacological target for cocaine and amphetamine, another psychotropic drug. Although various effects of the five subtypes of dopamine receptor knock-outs on cocaine exposure have been documented extensively, the wide range of developmental and physiological effects (e.g. locomotion) has made interpretation of the observed phenotypes often difficult. Surprisingly, however, mice in which the dopamine transporter has been deleted are hyperactive, but still show conditioned place preference when exposed to cocaine, indicating that compensatory mechanisms can mediate the drug's reward effects. Indeed, the place preference response was completely eliminated in double knock-out mice in which both the dopamine transporter and serotonin transporter had been deleted. Further pharmacological studies showed that both noradrenergic and serotonergic mechanisms can come into play to mediate the hedonic effects of dopamine in the absence of a functional dopamine reuptake mechanism. This redundancy provides robustness for the reward circuit, which bears testimony to the importance of reward reinforcement under normal physiological conditions.

Opioids

The opium poppy plant, *Papaverum somniferum*, was cultivated in lower Mesopotamia as long ago as 3400 bc, and opium harvested from the unripe poppy seed has been used for medicinal and ritualistic purposes for many centuries, with widespread use from Europe to China (Figure 15.17). A major advance in the use of opium as an **analgesic** came in the nineteenth century, when chemical analysis revealed that the analgesic properties of opium could be attributed to two ingredients, codeine and morphine. The availability of pure morphine was important, as it enabled physicians to precisely control the dose of analgesic they administered

FIGURE 15.16 A place preference test in which the animal has experienced drug effects in either the white or the black compartment, and subsequently is allowed to chose in which compartment it wishes to spend time.

FIGURE 15.17 Botanic temptations. Many addictive substances are naturally-occurring compounds. Illustrated are tobacco plants (A), coca plants (B), and the poppy flower (C) with the unripe pod from which opium is extracted (inset). It is reasonable to speculate that the neuroactive compounds produced by these plants evolved as defenses against herbivores.

to their patients. Today, morphine, codeine, and their derivatives are still the most commonly-used pain killers.

In 1874 an English chemist, C.R. Alder Wright, boiled morphine with acetic anhydride and produced a more potent, acetylated form of morphine, diacetylmorphine, commonly known as heroin (Figure 15.18). The German pharmaceutical company Bayer began marketing heroin in 1898, among others, as a cough syrup (Figure 15.19). As awareness of the addictive effects of opioids grew, their use became increasingly regulated and in 1914 morphine, heroin, and cocaine were declared controlled substances in the United States.

Morphine and heroin are highly-addictive drugs. Heroin is more potent than morphine, since the two acetyl groups which distinguish it from morphine allow it to cross the **blood–brain barrier** and act more rapidly. Once in the brain, the acetyl groups are removed, and heroin is converted to morphine. In the brain, morphine binds to opioid receptors. It is primarily binding of morphine to a subtype of opioid receptors, the μ-opioid receptor, which causes the decrease in pain perception and the sense of **euphoria**. The latter arises from modulation via the μ-opioid receptor of GABAergic inhibition on dopamine release in the meso-limbic dopaminergic projection. In the 1970s, the peptides β-endorphin, Leu-enkephalin, and Met-enkephalin were identified as endogenous ligands for the μ-**opioid receptor** for modulation of **nociception** (see also Chapter 3). It was also discovered that repeated exposure to heroin leads to down-regulation of μ-opioid receptors. This reduction in the number of μ-opioid receptors leads to tolerance and physiological dependence, as fewer receptors require increasingly higher doses of the drug to elicit the same sensation of euphoria.

Both physical and psychological withdrawal symptoms are severe. Morphine and heroin addicts have the highest relapse rates among all drug users. Nevertheless, morphine withdrawal symptoms are less life-threatening than alcohol withdrawal symptoms, and death usually results not from opioid withdrawal itself, but from the combined use of alcohol or other drugs during withdrawal.

The central role of the μ-opioid receptor in mediating analgesic, addictive, and withdrawal responses to morphine was demonstrated in homologous recombinant mice in which the gene encoding the μ-opioid receptor had been deleted. The mice appeared healthy and behaved normally, but they were deficient in morphine-induced analgesia, which is measured by the time and stimulus threshold it takes for the animal to flick its tail or jump up from a hotplate when exposed to a painful temperature (up to 52–54°C). The tail flick or hotplate responses of wild-type mice were attenuated by morphine, but morphine had no effect on the knock-out mice. Similarly, the knock-out animals did not show a conditioned place preference when given a choice between a chamber in which they had been previously exposed to morphine versus an environment without morphine administration. Wild-type mice, on the other hand, developed a morphine-associated place preference. Similarly, wild-type mice treated repeatedly with increasing doses of morphine showed severe withdrawal symptoms, while μ-opioid receptor knock-out mice did not show withdrawal symptoms after repeated injection with morphine. These results showed that the μ-opioid receptor is necessary and sufficient for mediating the effects of morphine (Figure 15.20, overleaf). QTL mapping studies in recombinant inbred strains of mice generated from DBA/2J and C57BL/6J parental lines, which differ widely in a range of behaviors, identified a QTL for morphine preference drinking and morphine analgesia on chromosome 10. This region contained the *Oprm* gene, which encodes the μ-opioid receptor, clearly an excellent candidate gene for the observed variation in morphine-associated phenotypes.

These mouse studies were followed by a large number of human association studies, which targeted the μ-opioid receptor as a candidate gene. A common single nucleotide polymorphism (SNP) in the first exon of the coding region of the μ-opioid receptor gene (*OPRM1*), which alters an alanine into a glycine (A118G), has been implicated in opiate addiction, both in a Chinese population and in a Swedish population. Other studies, however, have not replicated this association, either because of lack of statistical power, or because environmental effects and genetic backgrounds may modulate the effect of this polymorphism in

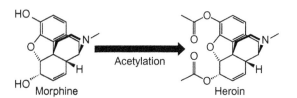

FIGURE 15.18 Conversion of morphine to heroin.

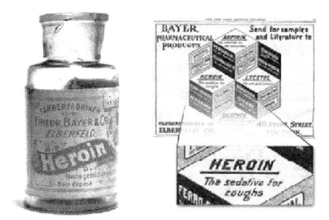

FIGURE 15.19 Bayer cough syrup with heroin, marketed by the company one year before they introduced aspirin.

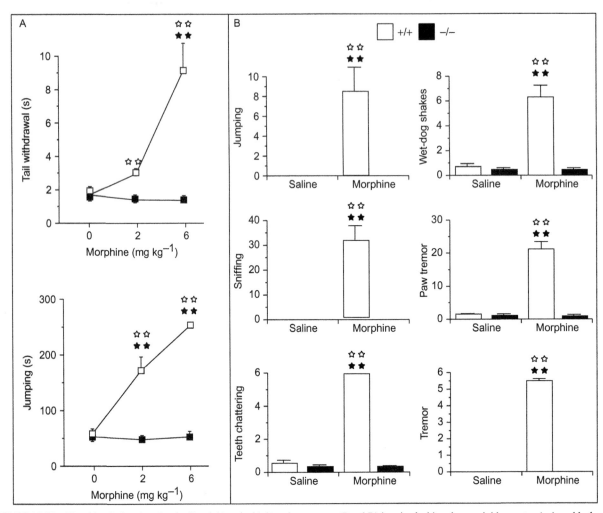

FIGURE 15.20 Morphine-induced analgesia (Panel A) and withdrawal symptoms (Panel B) in mice lacking the μ-opioid receptor (−/ − ; black symbols) and their control wild-type littermates (+/ + ; open symbols). The *y*-axes in Panel A show response latencies to tail withdrawal when the tail is immersed in 52°C water, and to the jumping response when mice are placed on a hotplate. Mutant mice do not show morphine-induced analgesia. Panel B shows measurements of various withdrawal symptoms after mice are treated with the morphine antagonist naloxone. Saline-treated animals serve as baseline controls for morphine-treated animals. Mutant mice do not show any of the morphine-induced withdrawal symptoms that are evident in wild-types. The stars show levels of statistical significance at $P < 0.01$. (Modified from Matthes, H. W. D. et al. (1996). *Nature*, 383, 819–823.)

the *OPRM1* gene on opiate addiction. The fact that allele frequencies of this polymorphism vary between 2% and 50% among populations may also contribute to difficulties in finding associations in small samples of populations where the allele frequency is low. Other candidate genes, for example in the dopaminergic neurotransmitter signaling pathway, have also been proposed as contributing to variation in opiate addiction, but results from association studies on other candidate genes have thus far been inconclusive.

SUMMARY

The neural reward pathway that includes the dopaminergic projection from the ventral tegmental area to the nucleus accumbens has evolved to reinforce beneficial behaviors

that are essential for survival, such as feeding and reproduction. Exogenous substances that interfere with this neural circuit by inappropriately promoting dopamine signaling give rise to pleasurable sensations that reinforce their continued use, despite adverse physiological effects. Modulation of expression of receptors that bind these compounds results in dependence on increasing amounts of the addictive substance to prevent physiological withdrawal symptoms. Since a variety of addictive compounds converge on the same neural reward circuit, multiple drug addiction is predicted and, indeed, common.

The probability of becoming addicted to a particular substance depends strongly on environmental factors, such as availability, cultural and social acceptance, role models among close relatives and friends, and conditions of psychological stress. Once substance use has been initiated, genetic factors contribute to the likelihood of developing addiction, with substantial heritability. Twin studies and adoption studies can help tease apart the environmental and genetic contributions that predispose to addiction.

Precise assessment of phenotypes is often challenging in human populations, and recruitment of addicted individuals and ethnic diversity of the study population further complicate genetic association studies. Animal models facilitate precise phenotyping through standardized behavioral assays, and allow key components of neurotransmitter signaling pathways to be manipulated genetically. Studies on alcohol sensitivity in flies have identified cyclic AMP signaling in the brain, and implicated pathways that respond to environmental stress, as well as regulation of pyruvate and fatty acid metabolism, in the induction of tolerance. Isoforms of the alcohol dehydrogenase gene have been correlated with alcohol use in human populations. Enzymes that regulate the bioconversion of other addictive substances, such as cytochrome P450 2A6, which metabolizes nicotine, also represent important factors in determining addictive or adverse health effects of drugs.

Association studies and linkage studies in human populations and mice, and homologous recombination studies in mice, have focused on a core set of candidate genes, including dopamine receptors and the dopamine transporter, as well as receptors that regulate the dopaminergic mesolimbic pathway, including the μ-opioid receptor, which mediates the addictive effects of opium and heroin, and GABA receptors which are down-regulated by prolonged heavy intake of alcohol. As experimental and statistical tools become ever more sophisticated, genome-wide association studies are likely to become more powerful in extending the analyses of the relationship between genetic variation and predisposition to addiction from a small group of focal genes to genome-wide ensembles of genes.

STUDY QUESTIONS

1. What are the hallmarks of addiction?
2. How does habitual eating of bacon and eggs for breakfast differ from addiction? Or does it?
3. Why would genetic risk for addiction to gambling be accompanied by genetic predilection for substance abuse?
4. How can one disentangle environmental and genetic contributions to alcohol abuse in people?
5. What have we learned about alcohol sensitivity and tolerance from candidate gene approaches and whole-genome transcriptional profiling studies in *Drosophila*?
6. A small set of candidate genes has been the target for association studies on addiction to several different substances. What are these genes and why have they been the focus for these studies?

7. What have we learned about cocaine addiction from dopamine transporter knock-out mice?
8. Imagine that you are a scientist interested in identifying genetic variants associated with compulsive gambling. What experimental and conceptual problems would you have to consider in your experimental design?
9. Is the metabolism of addictive substances an important factor in developing physiological dependence? If so, give examples.
10. What behavioral assays are used to measure analgesia in mice, and how do mice in which the μ-opioid receptor gene has been deleted respond in these assays?
11. Several controlled substances have medical applications. Give two examples.
12. What are the advantages and limitations of genome-wide association studies?

RECOMMENDED READING

Davies, A. G., Pierce-Shimomura, J. T., Kim, H., VanHoven, M. K., Thiele, T. R., Bonci, A., Bargmann, C. I., and McIntire, S. L. (2003). A central role of the BK potassium channel in behavioral responses to ethanol in *C. elegans*. *Cell*, **115**, 655–666.

Edenberg, H. J. et al. (2006). Association of alcohol dehydrogenase genes with alcohol dependence: a comprehensive analysis. *Hum. Mol. Genet.*, **15**, 1539–1549.

Freud, S. (1884). Über Coca. Centralblatt für die ges. *Therapie*, **2**, 289–314.

Goldman, D., Oroszi, G., and Ducci, F. (2005). The genetics of addictions: uncovering the genes. *Nat. Rev. Genet.*, **6**, 521–532.

Kieffer, B. L., and Gavériaux-Ruff, C. (2002). Exploring the opioid system by gene knockout. *Prog. Neurobiol.*, **66**, 285–306.

Koob, G. F., and Le Moal, M. (2005). *Neurobiology of Addiction*. Elsevier, Burlington, MA.

Matthes, H. W. D. et al. (1996). Loss of morphine-induced analgesia, reward effect and withdrawal symptoms in mice lacking the μ-opioid receptor gene. *Nature*, **383**, 819–823.

Morozova, T. V., Anholt, R. R. H., and Mackay, T. F. C. (2006). Transcriptional response to alcohol exposure in *Drosophila melanogaster*. *Genome Biol.*, **7**, R95.

Mulligan, M. K. et al. (2006). Toward understanding the genetics of alcohol drinking through transcriptome meta-analysis. *Proc. Natl. Acad. Sci. USA*, **103**, 6368–6373.

Scholz, H., Ramond, J., Singh, C. M., and Heberlein, U. (2000). Functional ethanol tolerance in *Drosophila*. *Neuron*, **28**, 261–271.

Shirley, R. L., Walter, N. A., Reilly, M. T., Fehr, C., and Buck, K. J. (2004). *Mpdz* is a quantitative trait gene for drug withdrawal seizures. *Nat. Neurosci.*, **7**, 699–700.

Snyder, S. H. (1989). *Brainstorming: The Science and Politics of Opiate Research*. Harvard University Press, Cambridge, MA.

Thorgeirsson, T. E. et al. (2008). A variant associated with nicotine dependence, lung cancer and peripheral arterial disease. *Nature*, **452**, 638–642.

Chapter 16

Evolution of Behavior

"Nothing in evolution makes sense except in the light of population genetics."

Michael Lynch, paraphrasing Theodosius Dobzhansky

OVERVIEW

Behaviors evolve as organisms adapt to their environments. Such adaptations are driven by changes in allele frequencies and other characteristics of the genome. Recombination, chromosomal rearrangements, the effects of deleterious and beneficial mutations, and selective forces acting on such mutations all contribute to genome evolution. Genetic drift causes population differentiation in the absence of selection. In sexually reproducing organisms, recombination continually shuffles genetic polymorphism in the population. Individuals in a population differ in their fitness (their proportionate contribution of offspring to the next generation). When differences in fitness are associated with different genotypes at a locus, selection operates to change allele frequencies at the locus. When subpopulations of a species become isolated, for example due to a geographical barrier, individuals are prevented from interbreeding. Such "prezygotic" isolation may lead to "postzygotic isolation," in which genetic changes occur that prevent interbreeding individuals from the subpopulations from giving rise to fertile offspring ("hybrid sterility"). The populations can now evolve as different species, which can compete for different ecological niches, and evolve different adaptive behaviors. As illustrated in earlier examples in this book, the evolution of new behaviors need not always involve a large number of genes, but can sometimes be driven by single mutations of large effect. As described in Chapter 10, single mutations can exert wide-ranging effects on the expression of co-regulated networks of genes in the genome. Understanding the genetic mechanisms that underlie the evolution of behaviors requires an appreciation of population genetics. This chapter will discuss only the most fundamental principles of population genetics theory and the focus on behavior as a vehicle for evolution. For greater detail on population genetics theory, the reader is referred to specialized textbooks.

POPULATION GENETICS AND EVOLUTION

The population genetic definition of evolution is a change in allele frequency over time. The two major evolutionary forces causing changes in allele frequency are natural selection and random **genetic drift**. In addition, mutation continually adds new variation to a population. Since evolution depends on the existence of genetic variation in a population, the question of how this variation is maintained, and the relative importance of the major forces of evolution are inter-related.

Natural Selection

Individuals may differ in both viability and fertility, and so contribute different numbers of offspring to the next generation. The **fitness** of an individual is defined as its proportional contribution of offspring to the next generation. If differences in individual fitness are on average associated with an allele, then selection acts on the allele. To understand how selection alters allele frequencies, we define the coefficient of selection, s, as the proportionate reduction of the gametic contribution of a selected genotype compared with the most favored genotype. Thus the relative fitness of the favored genotype is usually 1, and of the selected genotype is $(1 - s)$. The value of s gives an idea of the intensity of selection with respect to the locus under consideration. If $s = 1$, then the fitness of the selected genotype is $1 - 1 = 0$; the genotype is lethal and the individuals carrying it do not survive (or are sterile). As s approaches 0, the fitness of the selected genotype approaches 1; when $s = 0$ the gene has no effect on the fitness of the individuals carrying it, and is selectively neutral.

Consider the case of a deleterious recessive genotype (Box 16.1). The proportional contribution of each genotype to the next generation is computed by multiplying the genotype frequency before selection, which is the Hardy–Weinberg equilibrium frequency (see Chapter 5), by the relative fitness. Note that the total contribution of all genotypes is no longer 1, but is reduced by $-sq^2$, caused by selective deaths of A_2A_2 homozygotes. The reduction in total population fitness due to selection is called the **genetic load**, or cost of natural selection (see below). The frequency of the A_2 allele in the next generation is no longer q, but is less than q. The exact expression describing the change in allele frequency with selection (Δq) depends on the degree of dominance of the selected allele with respect to fitness. Box 16.2 shows changes of allele frequency after one generation of selection for different models of degree of dominance with respect to fitness.

The amount by which selection changes allele frequency depends on both the initial allele frequency and the selection coefficient for all models. Selection is a potent force for evolutionary change when allele frequencies are intermediate, but not when allele frequencies are extreme (Figure 16.1). This is particularly obvious for the case of selection against a recessive homozygote. Selection will be able to drive the selected allele to relatively low frequencies, but as the allele becomes rare most copies will be in heterozygotes and will thus escape selection.

The long-term effects of selection depend on the relative fitnesses of the three genotypes. **Directional selection** occurs when selection favors either the A_1 or A_2 allele. Directional selection reduces genetic variation, and the population will evolve to become homozygous for the favored genotype (i.e. all individuals will be A_1A_1 or A_2A_2). Directional selection also increases population fitness, because in each generation there are fewer of the maladapted genotypes, and more of the best adapted genotypes. Directional selection is viewed as positive selection if a new mutation is favored; and negative selection if a new mutation is deleterious (Figure 16.2).

Box 16.1 Change of allele frequency with natural selection against a recessive allele

Genotype	A_1A_1	A_1A_2	A_2A_2	Total
Initial frequency	p^2	$2pq$	q^2	1
Relative fitness	1	1	$1-s$	
Gametic contribution	p^2	$2pq$	$q^2(1-s)$	$1-sq^2$

The frequency of the A_2 allele after one generation of selection is the frequency of the A_2A_2 homozygotes plus half the frequency of the A_1A_2 heterozygotes, divided by the total, or $q_1 = [q^2(1-s) + pq]/(1-sq^2)$. The change in frequency of the A_2 allele after one generation of selection is $\Delta q = q_1 - q = -sq^2(1-q)/(1-sq^2)$.

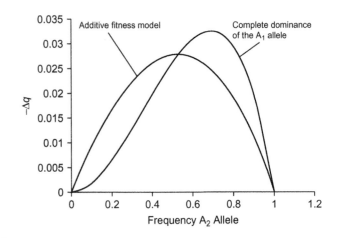

FIGURE 16.1 Change of allele frequency with selection as a function of initial allele frequency, for $s = 0.2$. The figure shows curves both for the additive fitness model and for the case of complete dominance of the A_1 allele.

Box 16.2 Change of allele frequency after one generation of selection for different selection models

Model	Genotypes A_1A_1	A_1A_2	A_2A_2	Change of Allele Frequency, Δq
Additive	1	$1-s/2$	$1-s$	$-\dfrac{0.5sq(1-q)}{1-sq}$
Recessive	1	1	$1-s$	$-\dfrac{sq^2(1-q)}{1-sq^2}$
Over-dominance	$1-s_1$	1	$1-s_2$	$\dfrac{pq(s_1p - s_2q)}{1-s_1p^2 - s_2q^2}$

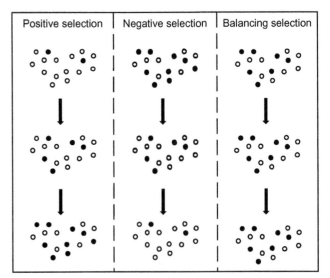

FIGURE 16.2 Schematic representation of a change in allele frequency when an allele (represented by the dark circle) experiences selection over evolutionary time.

Selection can also result in the maintenance of genetic variation, when there is selection against both homozygotes, and the heterozygote is the fittest genotype. This is called **heterozygote advantage**, or **overdominance**. The equilibrium allele frequency is easily derived from the expression for Δq with overdominance (Box 16.2). At equilibrium, $\Delta q = 0$. This occurs when the numerator of the expression for $\Delta q = 0$, or when $p = 0$, $q = 0$, and $(s_1 p - s_2 q) = 0$. Solving the expression $s_1 \hat{p} = s_2 \hat{q}$ (where the "hat" denotes equilibrium allele frequencies) gives $\hat{q} = s_1/(s_1 + s_2)$. The equilibrium allele frequency with overdominance is determined by the relative magnitudes of the selection coefficients against the homozygous genotypes, and not their absolute values. It is a stable equilibrium, in the sense that the gene frequency will return to its equilibrium value after any perturbation. Overdominance is an example of **balancing selection**, because genetic variation is maintained at equilibrium (Figure 16.2). It is easy to show that population fitness is the maximum it can be under this kind of selection at equilibrium gene frequencies, but there is a cost to maintaining two alleles in the population because there are $-s_1 p^2 - s_2 q^2$ genetic deaths each generation.

One example of heterozygote advantage is the maintenance of a point mutation at position 6 in the β-globin chain of hemoglobin (a glutamic acid to valine substitution) that gives rise to sickle cell anemia in African populations. One might imagine that this allele would be eliminated from the population over time, as it clearly has deleterious effects, since individuals homozygous for this allele are afflicted with sickle cell anemia. However, typical red blood cells of individuals that are heterozygous for the mutant allele have only a slightly altered shape, and do not give rise to disease symptoms. They are, however, resistant to infection by the malaria parasite *Plasmodium falsiparum* that is transmitted by the mosquito *Anopheles gambiae*. Thus, in the heterozygous condition, the sickle cell allele provides a distinct survival advantage. Because of this heterozygous advantage, the mutant and wild type alleles are both maintained at intermediate frequencies in the population.

Selection can also act against heterozygotes. Such heterozygote inferiority, or underdominance, may be true for some chromosomal polymorphisms, such as translocations, for which heterozygotes are sterile. The model is similar to the one for heterozygote superiority, except here we let s_1 and s_2 be negative. There are two stable equilibria at $\hat{q} = 0$ and $\hat{q} = 1$, as before, and an unstable equilibrium at $\hat{q} = s_1/(s_1 + s_2)$. This equilibrium is unstable, because any perturbation will drive the population away from the equilibrium values and towards $\hat{q} = 0$ or 1. Here, fitness is minimal at the unstable equilibrium, leading to the counterintuitive situation in which selection causes a population to be maladapted to its environment.

Random Genetic Drift

Allele frequencies will change not only due to selection, but also as a consequence of genetic drift. Random genetic drift is the change of allele frequencies that result from chance sampling of alleles in a finite population. All real populations are of finite size, N. Thus, allele frequencies can change at random from one generation to the next due to sampling of gametes. We expect the allele frequency to remain constant from one generation to the next in an infinitely large population, but in finite populations with $2N$ gametes sampling error will introduce a variance in allele frequencies. This is easily visualized in terms of a coin toss experiment, where we expect on average 50% heads and 50% tails. If we toss the coin only twice, a result of two heads or two tails is very likely. If we toss the coin 1000 times, the variance in the number of heads will be much smaller. Thus the sampling variance is greater, the smaller the number of tosses, or in the population case, the smaller the number of gametes.

The consequences of genetic drift can be seen if we imagine that a hypothetical infinitely large population with allele frequencies p_0 and q_0 at a single locus is divided into many small subpopulations with N individuals each. In every subsequent generation, only N individuals breed in each subpopulation, and there is no migration between subpopulations. This is equivalent to the coin toss experiment, except we are sampling alleles. Thus, in the first generation, the subpopulations will not all have the same allele frequency as their progenitor population. Rather, the allele frequencies will vary among the subpopulations according to the binomial sampling formula: $\sigma^2_{\Delta q} = p_0 q_0/2N$. As in the coin toss example, the variance in allele frequencies after one generation of sampling is inversely proportional to the size of the subpopulations (Figure 16.3).

If this process is repeated over time, the sampling process is repeated anew in each subpopulation. However, the starting point each generation is the current allele frequency, not the allele frequency in the progenitor population. Thus, over time, the variance in allele frequency among the subpopulations increases. After t generations of random drift, the variance in allele frequency is expected to be $\sigma^2_{qt} = p_0 q_0 [1 - (1 - 1/(2N))^t]$. In real populations, the population size that gives rise to the variance in allele frequencies after one or t generations of random drift is not the census population size, N, but the effective population size, N_e. N_e is the number of breeding individuals that give the same variance of allele frequencies as an ideal population of size N, and accounts for differences between numbers of males and females in the population, variation in population size from generation to generation, and variation in the number of progeny contributed by different individuals.

Thus, the consequence of one generation of genetic drift is that allele frequencies in each subpopulation will diverge from that of the progenitor population, and the magnitude of the difference depends both on the allele frequencies on the progenitor population and the size of the subpopulation. This is called a **population bottleneck**, or a **founder effect**. Continued small population size exacerbates the initial sampling event. The variance in allele frequencies continues to increase, or drift apart, in the different subpopulations. The ultimate consequence of genetic drift is that all populations become homozygous for one or the other allele in the progenitor population. The fraction of subpopulations that are homozygous for the A_1 allele is p_0, and for the A_2 allele is q_0. Thus, considered over all subpopulations, genetic diversity is preserved, but within each population, genetic diversity is depleted.

Because the end-point of genetic drift is homozygosity, the consequences of drift for a behavioral trait are the same as inbreeding. We have seen in Chapter 5 that finite populations lead to inbreeding, because the smaller the population size, the less remote is the common ancestor of any individual of the population. Thus, genetic drift in small populations can lead to inbreeding depression for behavioral traits. In the human population the incidence of genetic diseases can sometimes be attributed to genetic drift. For example, the Afrikaner population of Dutch settlers in South Africa descended from a few colonists. This population still has an unusually high frequency of Huntington's disease. Genetic drift and inbreeding are not always deleterious, since they can also lead to the purging of deleterious recessive alleles from a population. For example, cheetahs have experienced a population bottleneck and, as a consequence, have a high degree of homozygosity. Nonetheless, cheetahs are the fastest hunting cats in the world (although their birth rate is low). Northern elephant seals were hunted intensively in the 1890s, to the extent that as few as 20 individuals survived at the end of the nineteenth century. The population has since rebounded to over 30 000 individuals, but they still show markedly-reduced genetic variation as a consequence of the bottleneck (Figure 16.4).

MODELS OF EVOLUTION

We observe genetic variation in behavior both within and between populations. How can this be explained? Most

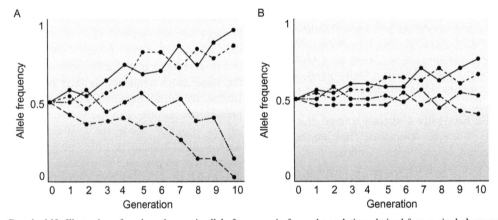

FIGURE 16.3 Genetic drift. Illustration of random changes in allele frequency in four subpopulations derived from a single large population at generation 0, in which the frequency of the A_2 allele was 0.5, for subpopulations with a small effective population size (A), and with a larger effective population size (B). Note that in both panels the variance of allele frequency increases over time, and the average allele frequency over all subpopulations remains 0.5.

FIGURE 16.4 Bottleneck species with reduced genetic heterozygosity, the cheetah (left) and northern elephant seals (right).

models postulate that this pattern of genetic variation can be accounted for by a balance of evolutionary forces. Two broad categories of evolutionary models have been proposed, those involving a balance between drift and forces tending to introduce variation (mutation and migration), and those involving a balance of selective forces, or selection and mutation.

Neutral Mutation: Random Drift Balance

Until the late 1960s it had been assumed that mutations are either deleterious or beneficial, and that natural selection would favor the elimination of deleterious alleles and the fixation of beneficial ones. Kimura proposed that the vast majority of mutations are selectively neutral, and do not affect the fitness of the individual. An example of such mutations are nucleotide substitutions at redundant third codon positions that do not change the amino acid composition of the encoded protein, or mutations in non-coding regions of the genome that do not affect regulation of gene expression. Since such silent mutations do not affect the biology of the organism, they are not expected to undergo natural selection. The notion that mutations could occur that would evade natural selection was provocative to the neo-Darwinists, even though in Kimura's own words "the theory does not deny the role of natural selection in determining the course of adaptive evolution."

The **neutral theory** predicts that new alleles carrying silent mutations would accumulate in the genome as they become fixed through **genetic drift** (see also Chapter 5), because they are not subject to the forces of natural selection. Thus, the neutral theory predicts that genetic drift would play a major role in genome evolution. The neutral theory makes two key predictions. First, the rate of neutral evolution is equal to the neutral mutation rate. To understand this, consider a new mutation in a finite population. When it arises, there is only one copy in the population, and the allele frequency is $1/(2N_e)$. The probability that this allele will be fixed in the population by random drift is equal to its initial allele frequency, or $1/(2N_e)$. The rate of evolution, k, is equal to the number of new mutations per generation, multiplied by the probability of fixation. The number of new mutations is $2N_e u$, where $2N_e$ is the effective number of gametes in the population, and u is the neutral mutation rate. Thus $k = 2N_e u \times 1/(2N_e) = u$. On average, it takes about $4N_e$ generations for a neutral mutation to be fixed by random drift. This is important, because it means that neutral evolution proceeds in a clock-like manner, depending only on the neutral mutation rate. Therefore, the number of neutral substitutions between species can be used to estimate their time of divergence from a common ancestor. The assumption that the mutation rate remains constant over evolutionary time is not necessarily

correct, however, and estimates of species divergence based on mutation rate are often corroborated or preceded by other lines of evidence from the fossil record (e.g. carbon dating).

The second prediction of the neutral theory is that the fraction of heterozygotes (H) at any locus (or averaged over all loci) is a function of the neutral mutation rate (u) and the effective population size (N_e): $H = 4N_e u/(4N_e u + 1)$. Thus, the neutral theory predicts that most molecular variation we observe in natural populations is not evolutionarily relevant, but consists of neutral alleles in the process of being fixed or lost by genetic drift.

The theory of neutral evolution is especially useful because it provides a null hypothesis for evidence for natural selection. One can compare the number of observed differences between specific DNA sequences of two species with the number of differences that would be predicted according to neutral expectations, given the independently estimated divergence time. If the observed number of differences deviates from the predicted number, the null hypothesis has failed and the sequence examined may have experienced some form of selection. A variety of tests for deviation from neutral expectations have been developed to test whether certain genomic sequences have been under selection and, if so, to attempt to reconstruct their evolutionary history. The designs and assumptions underlying some of the most commonly-used tests for deviation from neutrality are discussed briefly below.

The neutral model of evolution has the attractive features that it accounts for variation within and between populations, and there is no "cost" to neutral evolution. The major problem with the neutral model is that it assumes populations are homozygous for all evolutionarily important genes, so evolutionary advance, in the sense of increased adaptation, depends on the occurrence of rare favorable mutations, which will be quickly fixed by natural selection. Under this model, adaptation and speciation are slow processes dependent on the mutation rate.

SELECTION MODELS

Deleterious Mutation: Directional Selection Balance

Directional selection removes unfavorable alleles from a population, but in reality they are constantly being reintroduced by recurrent mutation. Selection can thus never completely eliminate deleterious alleles, but will prevent them from reaching very high frequencies. When both directional selection and recurrent mutation operate in a population, the selected allele will come to an equilibrium frequency at which the changes in frequency each generation due to both processes exactly balance.

The equilibrium frequency of the deleterious allele, $\hat{q}$, is a function of the ratio of the mutation rate to the selection coefficient, but the exact expression depends on the degree of dominance with respect to fitness. If the fitness effect is completely recessive, $\hat{q} \approx \sqrt{u/s}$. A balance between selection removing recessive deleterious alleles and mutation reintroducing them will cause recessive mutant alleles to persist at low levels in the population. Note that if the incidence (frequency of homozygous affected genotypes) and selection coefficients can be estimated in the population, using this expression gives an indirect way of estimating the mutation rate. For an additive fitness model, $\hat{q} \approx 2u/s$. Thus, deleterious mutant alleles that are not completely recessive have much lower equilibrium frequencies than the recessive case, because there is selection against them in heterozygotes. Finally, if the mutation has dominant effect on fitness, $H \approx 2u/s$, where H is the frequency of heterozygotes. A dominant deleterious mutation will be selected against immediately in heterozygotes, and so will almost never be present in homozygotes.

A balance between mutation and selection can explain the occurrence in populations of very rare alleles, but cannot account for alleles at intermediate frequencies. The low frequency of most deleterious mutant alleles in natural populations is in accord with what would be expected from the joint action of mutation and selection.

Balances Involving Selection

We have seen above that selection favoring heterozygotes can lead to the maintenance of genetic variation in a population. There are many other possible models in which genetic variation can be maintained in theory; all involve heterozygote superiority when averaged over all conditions, or marginal overdominance. These models include frequency-dependent selection, density-dependent selection, spatially- and temporally-heterogeneous environments, and epistasis. In these models, an allele is selected for in one condition, and selected against in another. Selective balances can also involve variable selection between young and old individuals, or between males and females. If an allele has opposite effects on the viability and fertility components of fitness it can persist in populations; this form of selective balance is called **antagonistic pleiotropy**.

Models of evolution involving selection assume that different genetic variants are all perceived by natural selection. Natural selection is then a stronger force than genetic drift or migration, and so response to natural selection dominates evolution. Under selectionist models, the genetic variation observed within populations is determined by a balance of selective forces, and genetic variation arises between populations because they have adapted genetically to different environments.

Cost of Genetic Variation

A major problem of models invoking selective balances for maintaining genetic variation is the cost of maintaining deleterious alleles in the population in terms of reduced population fitness. This cost was termed the **genetic load (L)** by Muller, who defined load as the difference between the average fitness of a population ($\bar{w}$) and its maximal possible fitness: $L = 1 - \bar{w}$, where $\bar{w}$ is the sum of the products of genotype frequencies and fitnesses over all genotypes.

For example, the equilibrium frequency of deleterious recessive alleles is $\hat{q} \approx \sqrt{u/s}$. The fitness of the population carrying the deleterious recessive is $\bar{w} = 1 - sq^2$, or $1 - u$. The load, $1 - \bar{w}$, is then simply u. For n deleterious recessives, with the probability of death from each locus independent, $\bar{w} = (1 - u)^n$, and $L = 1 - (1 - u)^n$, which could be quite large.

With overdominance, $\bar{w} = 1 - s_1 s_2/(s_1 + s_2)$ and $L = 1 - \bar{w} = s_1 s_2/(s_1 + s_2)$. This could be very large if selection is strong. For example, in the sickle cell anemia example discussed above, the variant hemoglobin allele causing sickle cell anemia is present at frequencies of approximately 0.2 in human populations living in areas in which malaria is endemic. The selection coefficients against the two homozygous genotypes have been estimated as 1 and 0.15, respectively. The genetic load on populations harboring the variant hemoglobin allele at equilibrium frequencies is 0.13; that is, 13% of the equilibrium population dies either of anemia or of malaria before reproduction. Thus, there is a limit to the number of overdominant loci a population can support before becoming extinct. It is difficult to reconcile the large amounts of genetic variation observed within populations with maintenance by appreciable overdominance at each and, in practice, overdominance is rarely found, although it has been extensively looked for. However, maintenance of large amounts of variation by heterozygote advantage should not necessarily be rejected, because the load calculations for multiple loci assume the loci act independently. For n loci, this is equivalent to comparing the average fitness of the population to the n-fold multiple heterozygote, which for large n will never exist in a real population. If the genetic load incurred by overdominant selection at multiple loci is compared to the most fit genotypes that actually exist in the population, not to the hypothetically most fit genotype, greater amounts of polymorphism can be maintained by heterozygote advantage without prohibitive genetic load. In addition, the models of selective balance involving marginal overdominance are associated with less genetic load than purely overdominant loci.

The attraction of selection models of evolution is that populations will be heterozygous for many evolutionarily important genotypes. Because evolutionary potential exists in every population, evolution may be rapid. A disadvantage of many selection models of evolution is the genetic load associated with selective maintenance of genetic variation.

ASSESSING DEVIATIONS FROM NEUTRALITY

It is clear that any one model of evolution will not apply to all loci. Thus, we would like to know what loci are evolving according to the neutral model, what loci have deleterious alleles segregating from a balance between mutation and selection, what loci are under balancing selection, and what loci are under positive selection. These questions can be addressed by various tests comparing observed DNA sequence variation in samples of alleles with the predicted pattern of variation from the neutral theory.

According to the neutral theory, nucleotide substitutions at degenerate third codon positions of coding regions that do not alter the encoded amino acid would have no effect on the phenotype, and therefore would not be under selection. Such mutations are known as synonymous substitutions (note, however that this assumption may not be entirely correct, as synonymous substitutions conceivably can affect the structure of mRNA and thus its transport from the nucleus, stability, or translation efficiency). Mutations that lead to a change in the encoded amino acid of a protein (non-synonymous substitutions), on the other hand, may give rise to a deleterious or beneficial phenotypic effect. Several tests for deviation from neutrality are based on comparing the ratio of non-synonymous substitutions (K_a) to synonymous substitutions (K_s) of a gene. If $K_a/K_s = 1$, then the null hypothesis is confirmed, i.e. there is no deviation from neutrality. However, if the ratio of non-synonymous versus synonymous substitutions is statistically different from one, selection is inferred.

When a gene is under directional selection, the frequency in the population either of a favorable allele increases (**positive selection**), or of a deleterious allele decreases (**negative or purifying selection**). When a new beneficial mutation occurs in the genome, it will be in **linkage disequilibrium** with adjacent polymorphic markers in the genome, since it takes time for recombination to break down **linkage disequilibrium** around the site of the new mutation (see also Chapter 8). Genomic regions with polymorphic markers that show an excess of linkage disequilibrium compared to other regions in the genome may be "hitchhiking" with a recently-arisen beneficial mutation under positive selection (Figure 16.5). Here, positive selection results in an excess of high-frequency derived alleles compared to neutral expectations around the selected allele which has swept to high frequency. Fay and Wu devised a statistical analysis that generates a test statistic, designated H, to detect such a recent **selective sweep**.

In contrast to directional selection, **balancing selection** results in the maintenance of polymorphisms (or multiple alleles) within a population. Increased levels of genetic variation between alleles or haplotypes in a species can be interpreted as a sign of balancing selection.

One approach for determining if and what kind of selection has influenced the evolution of a gene is Tajima's method. It relies on comparing two different estimates of the **population level mutation rate** θ ("theta"), where $\theta = 4N_e u$. One approach for estimating θ is to measure the average pairwise nucleotide differences between two sequences, a parameter known as "π." Another approximation for θ is to measure the proportion of segregating polymorphic sites, known as "S." In a fixed-size population in **Hardy–Weinberg equilibrium**, the values of π

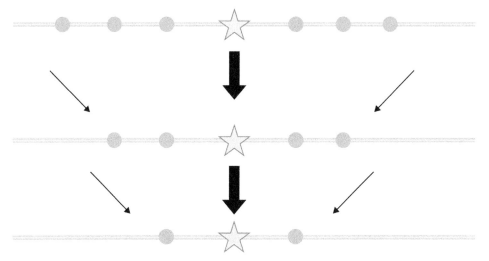

FIGURE 16.5 "Hitchhiking" will be observed when a new mutation (star) has arisen recently and will be in linkage disequilibrium initially to all alleles on the chromosome (circles). Recombination events (arrows) will lead to gradual decay of linkage disequilibrium over time, and the time course of this decay is inversely-related to the distance in cm around the mutant allele under study, as the probability for recombination decreases as the region around the focal mutation becomes narrower.

and S will be correlated for neutrally-evolving sites. The difference between the values of θ estimated from π and S will therefore be zero, if the sequence under study is evolving according to neutral expectations. However, if the estimates of θ differ significantly between π and S, the sequence has experienced selection. Fumio Tajima developed a statistical measure for significant deviation of the difference between θ_π and θ_S from zero (the designation θ_w is also used instead of θ_S). This statistic is known as Tajima's D. If D is not significantly different from 0, the null hypothesis, i.e. neutrality, is confirmed. The 95% confidence interval for the distribution of Tajima's D is $-2 < D < 2$. If Tajima's D is positive (larger than 2), balancing selection is inferred ($\theta_\pi > \theta_S$), whereas values of Tajima's D below -2 ($\theta_S > \theta_\pi$) would indicate a **selective sweep**. Fu and Li extended Tajima's analysis, and developed the test statistics D^* and F^*, by taking into account polymorphisms observed only once in the population (singletons). Tajima's D and Fu and Li's D^* and F^*, can be applied not only to an entire coding region, but also to overlapping fragments of such a region to detect highly-localized deviations from neutrality. Such an analysis is known as **sliding window analysis**.

Hudson, Kreitman, and Aguadé developed an analysis, the **HKA test**, which assesses whether a genomic region harbors statistically significantly more polymorphisms than a neutrally-evolving region. Whereas the HKA test can imply deviation from neutrality, it cannot make inferences about the nature of selection acting on the region. McDonald and Kreitman extended this analysis, and developed what is today perhaps the most popular method for detecting signatures of selection within a protein coding sequence, the McDonald–Kreitman test. This test was first applied to the extensively-studied alcohol dehydrogenase gene of *Drosophila melanogaster*, and compares the number of synonymous substitutions to the number of non-synonymous substitutions within a species and between species (e.g. *D. melanogaster* and *D. simulans*). If a protein coding region develops neutrally, the ratio of synonymous to non-synonymous substitutions within a species (recent evolution) should not be different from that between species (distant evolution). Significant deviations from this neutral expectation can indicate directional selection, if the ratio of synonymous to non-synonymous substitutions within a species is less than that between species, or balancing selection or positive selection if the ratio of synonymous to non-synonymous substitutions within a species is greater than that among species. Note that it may not be possible to distinguish balancing selection from positive selection based on the McDonald–Kreitman test alone. It should also be noted that recombination can affect the outcome of tests for deviation from neutrality, such as the HKA-test. The **population recombination rate** is $4N_ec$, where N_e is the effective population size and c is the recombination rate per generation. Whenever possible,

historical recombination should be taken into account when attempting to reconstruct evolutionary history.

BEHAVIOR AS A VEHICLE FOR EVOLUTION

The process of evolution depends on the ability of organisms to transfer their genes to the next generation. In this sense, behaviors provide the stage on which natural selection acts. Selection of mating partners, mating, taking care of offspring, and survival until reproduction has been completed are the ultimate mechanisms that lead to changes in allele frequencies in a population. Behaviors are determinants of fitness, and behavioral adaptations for survival and competition for mating opportunities mediate the "survival of the fittest." Some behaviors are stereotypical, and under normal conditions show little phenotypic variation. Examples are nest building behavior by some birds, and web weaving by spiders. Other behaviors show phenotypic plasticity, which is the ability to change a behavior in response to changes in either the physical or social environment. An example of such behaviors is the change in aggressive behavior of a subordinate individual, when a dominant individual is removed from the population. Variation of **phenotypic plasticity** as a function of genetic background is a hallmark of **genotype-by-environment interaction** (GEI). A consequence of GEI is that the most phenotypically-plastic individuals can gain an advantage when environmental conditions change. Thus, GEI provides a platform for natural selection, and for behavioral evolution.

Perhaps one of the best examples of the central role behaviors play in evolution is **sexual selection**. Here, morphological features or courtship behaviors are assessed as proxies of reproductive fitness by prospective mating partners, and in the process of sexual competition reach extravagant dimensions (see Chapter 8, Figure 8.3). The magnificent peacock plumage, the extravagantly-wide separation between the eyes of stalk-eyed flies, the elaborate mating rituals and mating structures built by bowerbirds, and complex birdsongs of songbirds all are examples of sexual selection. In some cases, extravagant displays may incur risks, such as increased exposure to predators, or can be co-opted as a protective survival trait, for example the peacock display can be used to intimidate potential enemies.

A reproductive strategy that provides an alternative to sexual selection and competition for mating partners is **sperm competition**. In this case, sperm from two or more males compete for the fertilization of an ovum. All else being equal, the more sperm a male produces and the more females he inseminates, the greater will be the chance that his alleles will be spread through the population. There is, however, a trade-off between investing energy in the expensive production of spermatozoa, or investing energy in other activities, for example excluding other males by defending a territory.

Sperm competition is prevalent among promiscuous animals in societies where males mate with many females, but do not contribute to care for the offspring. Primates provide excellent examples. Chimpanzees mate promiscuously, but because each female mates with many males, individual males are not *de facto* assured of the survival of their offspring and, hence, do not participate in parental care. In contrast, gorillas are polygynous, that is a male gorilla will mate with many females, but these females belong to his social group, his harem, to which he is faithful and which he protects. It is of interest that the promiscuous behavior of male chimpanzees compared to the polygynous lifestyle of the gorillas is reflected in the size of their testes, which are larger in chimpanzees to sustain high volumes of sperm production than in gorillas, even though the body size of a male gorilla is four times that of a chimpanzee (Figure 16.6).

Genotype-by-sex-interaction can cause a situation in which a genotype that is beneficial in one sex is suboptimal in the opposite sex. An example of such **sexual antagonism** is provided by the wild population of red deer on the Isle of Rum in Scotland, which has been studied extensively for several decades by Tim Clutton-Brock and his colleagues. They found that male red deer with high fitness generated, on average, daughters with reduced fitness, and that selection favored males with lower breeding values whose genotypes are more optimal for female fitness. Sexually-antagonistic selection is a form of balancing selection that can contribute to maintaining genetic variation in a population.

Whereas the genetic architectures that support behaviors can be complex, there are some instances in which

changes in one or few genes can have dramatic effects. As described in Chapter 12, the difference in the promoter region of the vasopressin receptor gene between monogamous and polygamous voles alters its expression pattern, and consequently social behaviors. Similarly, the dazzling eyespots on butterfly wings, a complex morphological trait, develop by co-opting a relatively small number of developmental and pigmentation genes. Because it is impossible to discuss the evolution of the entire rich spectrum of behaviors in the animal kingdom in a single chapter, we will focus the remainder of this chapter on behaviors that exemplify aspects of adaptive behavioral evolution, including insect–host plant interactions, co-evolution of sexual communication systems, burrowing behavior in wild mice, and the adaptive radiation of cichlids and sticklebacks.

INSECT–HOST PLANT INTERACTIONS: AN EXAMPLE OF EVOLUTIONARY ADAPTATION

Feeding strategies of insects and their choices of **oviposition** sites provide a classic example of adaptive evolution. Insect herbivores can be classified as generalists and specialists. Generalists will feed and lay eggs on a wide range of host plants, whereas specialists exploit only a few, or even a single, host plant species. The advantage of being a generalist is the wide choice of feeding and oviposition opportunities. What could be the evolutionary advantage of host plant specialization? The answer is simple. If an insect can utilize a host plant that is toxic to other species, it will gain the exclusive noncompetitive use of this resource. Furthermore, a unique relationship may evolve between the plant and the insect, in which the plant can provide protection to the insect. For example, the moth *Heliothis subflexa* feeds exclusively on plants of the genus *Physalis*, which includes the tomatillo. Caterpillars from eggs deposited on this host plant are sheltered from predation inside a lantern structure of the plant during larval development. A closely-related species of moths, *Heliothis virescens*, is a generalist that can feed on at least 14 families of host plants, and is considered a major agricultural pest (Figure 16.7).

These different species of heliothine moths can be intercrossed. There is no recombination in female *Lepidoptera*, and crosses between females of *H. virescens* and males of *H. subflexa* yield viable offspring. This provides an opportunity to identify QTL regions associated with host selection, an area of great agricultural interest. Since the genome of *Heliothis* is highly-fragmented into 31 chromosomes, a **linkage group** can be considered a chromosome, and corresponds to approximately 3% of the genome, already a higher level of resolution than found in many other QTL analyses.

Although comparatively little is known about the ecology of *Drosophila* in the wild, the availability of whole-genome

FIGURE 16.6 This 25-year-old chimpanzee has huge testes, the most prominent feature of mature males, an evolutionary product of its promiscuous mating system that results in sperm competition.

FIGURE 16.7 Two closely-related species of moths, the specialist *Heliothis subflexa* (A), and the generalist *H. virescens* (B). The host plant for *H. subflexa* of the genus *Physalis* provides a protective lantern around its fruit in which *H. subflexa* larvae can develop (C).

sequence information for 12 *Drosophila* species has helped to provide insights into the molecular genetic mechanisms that lead to host plant specialization. *Drosophila sechellia* is a *Drosophila* species found on an island of the Seychelles, where it has undergone a unique host plant specialization on the fruit of *Morinda citrifolia*, also known as Tahitian Noni. The ripe fruit of *M. citrifolia* produces hexanoic and octanoic acid, which are toxic and repellent to *D. melanogaster* and *D. simulans*. In contrast, *D. sechellia* feeds exclusively on *M. citrifolia*, and has developed an attraction for the fruit's odorants. Homologous recombinant *D. melanogaster* flies that lack genes that encode the odorant-binding proteins OBP57d and OBP57e no longer show the characteristic olfactory avoidance response for hexanoic acid and octanoic acid. Matsuo and colleagues at Tokyo Metropolitan University made interspecies hybrids between *D. melanogaster* deficiency-mutants that lack the *Obp57d* and *Obp57e* genes. When *D. simulans* was crossed to the *D. melanogaster* **deficiency stock**, the hybrids generated an avoidance response to hexanoic acid and octanoic acid. However, offspring from an interspecies cross between *D. sechellia* and the *D. melanogaster* deficiency line showed attraction and oviposition behavior in the presence of these odorants characteristic of *D. sechellia*. Additional experiments showed that polymorphisms in the odorant-binding protein *57e (Obp57e)* gene determine the behavioral difference between the species. A 4 bp insertion in the upstream regulatory region of the *D. sechellia Obp57e* gene was found to disrupt the expression of the gene, thereby eliminating the negative olfactory perception that elicits avoidance of the *Morinda* fruit. These studies show that mutations in a single gene, encoding an odorant-binding protein, can dramatically alter the relationship between an insect and its food source (Figure 16.8). It should be noted, however, that other as yet unknown genetic differences between *D. sechellia* and its sister species must have evolved to ameliorate the toxic effects of the *M. citrifolia* fruit.

The finding that odorant-binding proteins may be key mediators of food source and oviposition site localization is perhaps not surprising as chemical signals play a dominant role in insect–host plant interactions. Population genetics analyses of *Obp* gene sequences in natural populations in some cases showed the persistence of segregating null alleles in the population, indicating that combinations of odorant-binding proteins likely participate in odorant recognition, and that the lack of one member of such an ensemble can be compensated so that the individual's ability to respond to critical chemical signals is not significantly impaired. Tests for deviations from neutrality showed that different odorant-binding proteins have experienced different evolutionary histories, with some showing signatures of balancing selection or positive selection. Analyses of sequences of genes encoding odorant-binding proteins, odorant receptors, and gustatory receptors among different *Drosophila* species indicate that these chemoreceptor genes generally evolve rapidly, as would be predicted if evolution following speciation is accompanied by adaptation to rapidly-changing chemical environments.

The apple maggot, *Rhagoletis pomonella*, provides an example of recent **sympatric** isolation due to host plant specializations (Figure 16.9). Before the introduction of apples into North America from Europe, *R. pomonella* fed on hawthorns. After the introduction of apple orchards between 1850 and 1880, a race of *Rhagoletis* emerged that

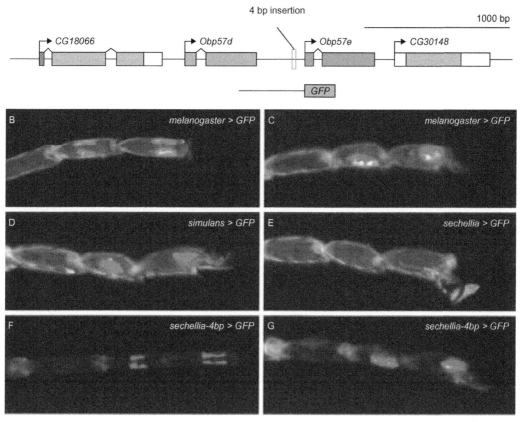

FIGURE 16.8   A 4 bp upstream insertion in the *Obp57e* gene in *D. sechellia* prevents its expression. Panel A shows genomic organization of the *Obp57d/e* region, indicating a 4 bp CCAT insertion in *D. sechellia*. Arrows indicate translation initiation sites. A GFP construct was made that replaces the *Obp57e* coding region driven by upstream promoter sequences of *D. melanogaster*, *D. simulans*, or *D. sechellia*. Expression is seen in a few cells in the tarsi when the reporter gene is driven by promoter sequences of *D. melanogaster* (panel (B) and (C), dorsal and lateral view, respectively) or *D. simulans* (panel (D), dorsal view), but not under the *D. sechellia* promoter. Removal of the CCAT insertion from the *D. sechellia* upstream sequence restores the expression of *Obp75e* ((F) and (G), dorsal and lateral view, respectively). (From Matsuo, T. (2007). *PLoS Biol.*, **5**, e118.)

Box 16.3   Gene duplication and evolution

The Japanese-American evolutionary biologist Susumu Ohno was a main protagonist of the long-held hypothesis that gene duplication plays a major role in evolution. Duplication of a gene can occur as a result of unequal crossing-over during recombination, or as a consequence of the activity of **retrotransposons**. When a gene acquires a duplicate redundant copy, selective pressure on this second copy of the gene is relaxed, because mutations in this gene will not cause deleterious effects. This enables the duplicated gene to accumulate mutations and evolve faster than a functional single-copy gene. This can have three possible consequences: (1) one of the two genes may accumulate mutations that render it dysfunctional, generating a **pseudogene**; (2) the genes can undergo **subfunctionalization**, in which they undergo divergent evolution to contribute to different aspects of a phenotype. An example of subfunctionalization can be seen in the family of odorant receptor genes, where gene duplication events have given rise to odorant receptors with divergent molecular response profiles; (3) the duplicated gene can evolve to acquire a novel function,

which is known as **neofunctionalization**. Recently-duplicated genes are usually located in tandem on the chromosome, and such tandem arrays of homologous duplicated genes can give rise to additional unequal crossing-over events, generating gene clusters (Figure A). Subsequent chromosomal rearrangements over evolutionary time can lead to break up of clusters, and dispersion of related genes through the genome, as has been the case for the rapidly evolving families of chemoreceptors.

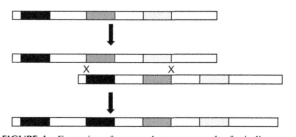

FIGURE A   Expansion of a gene cluster as a result of misalignment of closely-related genes during recombination.

FIGURE 16.9 The body shape and markings of the apple orchard fruit fly *Rhagoletis pomonella* mimic a jumping spider to deter predators of this vulnerable insect.

feeds on apples. The *Rhagoletis* population split into two groups, one that feeds normally exclusively on hawthorns, and one that feeds only on apples. These two populations that occupy the same habitat do not normally interbreed, and may be an indication of incipient **speciation**.

In addition to chemoreceptors, a second multigene family is critical in responding to environmental chemicals. These are the **cytochrome P450** genes, which encode enzymes that oxidize **hydrophobic** compounds, enabling subsequent conjugation by sulfatases, attachment of glucuronic acid residues, or conjugation to the tripeptide glutathione to render them soluble. In addition to detoxification of foreign chemicals, these enzymes also perform metabolic reactions for the synthesis of steroid hormones, such as estrogen and testosterone, and the insect hormone ecdysone. May Berenbaum and her colleagues at the University of Illinois at Urbana-Champaign have studied the role of cytochrome P450 enzymes in the evolution of insect–host plant interactions by assessing how members of the cytochrome P450 family detoxify plant defense chemicals, furanocoumarins, produced by herbaceous plants of the *Umbelliferae* family (Figure 16.10). A variety of herbivorous insects, such as the black swallowtail butterfly *Papilio polyxenes*, have developed the ability to utilize these plants as hosts by producing cytochrome P450s that are able to detoxify furanocoumarins. Berenbaum has provided evidence for a co-evolutionary association between substrate specificities of cytochrome P450 members with host plant furanocoumarins. Like the chemoreceptor genes, cytochrome P450 (*Cyp*) genes are rapidly evolving.

It is of interest to note that many insect species have large *Cyp* gene families, often more than 80 members, but that the honey bee genome contains remarkably few *Cyp* genes. There is an interesting evolutionary explanation for this unusual observation. Honey bees are social insects, and their social organization can be viewed as a "superorganism." The only reproducing individual is the queen, who remains in the hive sheltered from the external chemical environment. Workers who are in contact with the external environment are not reproductively active and, therefore, are dispensable from an evolutionary perspective. Thus, there has been no selection pressure on honey bee colonies to evolve a large family of detoxification enzymes. This lack of detoxification ability may be the reason for a recent crisis in the agriculturally- and commercially-important beekeeping industry, where foragers were not returning to the hive, leading to **colony collapse**. It is thought that pesticides or environmental chemicals might be to blame, by killing foragers or interfering with their navigation abilities.

Adaptations of co-evolving genes that enable interactions between an insect and its host plant often evolve through adaptations and counteradaptations. This evolutionary pattern is sometimes referred to as an **evolutionary arms race**, and is common for co-evolution between predators and their prey. An example are molluscs that have developed hard shells to avoid being eaten, and crabs that have evolved powerful claws that are able to break those shells. An evolutionary arms race can also occur between a parasite and its host, where the development of the parasite will kill the host, and the immune system of the host adapts to kill the parasite. An example of this type of evolutionary arms race is aphids and their parasitic wasps, where aphids that have adapted to live on different host plants show different susceptibility to parasitoids.

CO-EVOLUTION OF SEXUAL COMMUNICATION SYSTEMS

Localization of appropriate mating partners is essential for procreation. As speciation occurs, divergent sexual communication systems evolve. This poses an evolutionary conundrum. Because the signal-emitter and signal-recipient must be finely-tuned to each other for optimal recognition, sexual communication signals are generally under strong stabilizing selection. For example, if males of a moth species are finely-tuned to the pheromone blend emitted by the female, any change in the signal produced by the female, or in the recognition of this signal by the male, will lower the chance of successful mating. Thus, changes in pheromone blends can contribute to **prezygotic isolation** between populations, but for speciation to occur it is essential that the change in pheromone blend is accompanied by a change in response to the new pheromone composition. This requires overcoming the force of stabilizing selection acting on the maintenance of this sexual communication

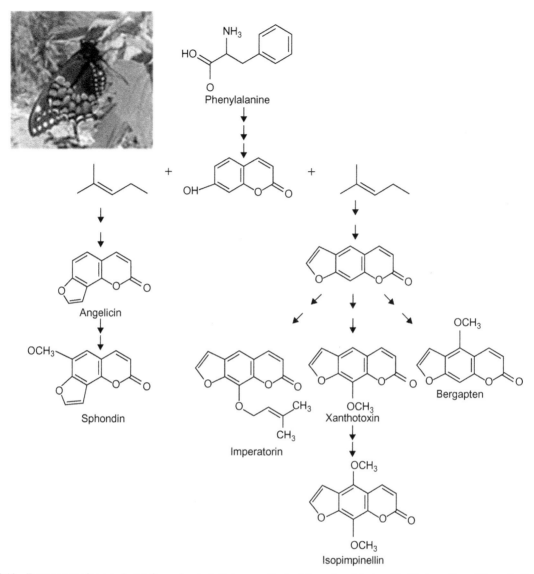

FIGURE 16.10 Furanocoumarins are plant defense chemicals that are synthesized from phenylalanine. The black swallowtail butterfly *Papilio polyxenes*, shown in the figure, can feed on plants that produce furanocoumarins, because it has evolved cytochrome P450 isoforms that can detoxify these plant chemicals.

system. This could potentially be accomplished through a shift of responding males to components of a pheromone blend that were previously not critical for mate attraction, which could exert a directional selection force to drive the evolution of such a chemical communication system.

Again, the closely-related moth species *Heliothis virescens* and *Heliothis subflexa* provided an opportunity to test hypotheses for co-evolution of sexual communication systems. *H. virescens* and *H. subflexa* occupy the same habitats. They are said to be **sympatric** as compared to **allopatric**, i.e. occupying different geographical regions, but hybridization does not occur in nature. When *H. subflexa* females are crossed with *H. virescens* males in the laboratory, male offspring are sterile, and females have low fecundity. Thus, hybridization between the species in

nature would come at a high fitness cost, as the offspring from such matings would have essentially zero fitness. It is, therefore critical that prospective sexual partners of these sympatric species accurately discriminate the correct mating partner. This requires exquisite tuning to the pheromone blend. The pheromone blends of both species contain the same major component, (Z)-11-hexadecenal, but differ in the composition of minor components. Discrimination of some of these minor components is essential for appropriate mate recognition.

A major difference between these pheromone blends is the presence of three acetate esters released by *H. subflexa* females, which are absent in *H. virescens*. Astrid Groot, Fred Gould, and Coby Schal at North Carolina State University introgressed a QTL region that results in

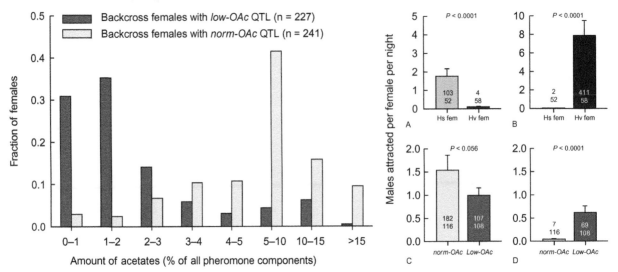

FIGURE 16.11 The left panel shows the compositions of pheromone blends as the percentage of minor components (acetate esters, OAc) of females from a backcross of *H. virescens* and *H. subflexa* hybrids to a genetic background with a QTL that gives rise to a pheromone blend with normal acetate ester content (light gray) or a low acetate ester content (dark gray). The right panel shows the behavioral response of males to females that produce different pheromone blends. Panels A and B show that *H. subflexa* (Hs) and *H. virescens* (Hv) males respond virtually exclusively to their corresponding females. *H. subflexa* males were marginally more attracted to backcross females with normal acetate ester levels than to females with the low acetate ester QTL (Panel C). Although dramatically fewer *H. virescens* males were attracted to both types of backcross females, about 10 times more were caught in traps baited with backcross females containing the low acetate ester QTL than in traps baited with females that produce a normal pheromone blend (Panel D). (From Groot, A. T. et al. (2006). *Proc. Natl. Acad. Sci.* USA, **103**, 5858–5863.)

low production of acetate esters in the genome of *H. sub-flexa*, and assessed to what extent *H. subflexa* females that produce a high acetate ester pheromone cocktail, and those that produced a pheromone blend with a low acetate ester content, would be able to attract *H. subflexa* and *H. virescens* males. Surprisingly, *H. virescens* males were much more attracted to *H. subflexa* females that produce few acetate esters, suggesting that the acetate esters produced by *H. subflexa* females antagonize attraction by *H. virescens* males. Thus *H. virescens* males can exert directional selection on *H. subflexa* females to produce high amounts of acetate esters in their pheromone blend, to prevent mating with closely-related, but inappropriate, males (Figure 16.11). It is the production of antagonistic components in the pheromone blend that facilitates the evolution of this chemical communication system (Figure 16.12).

Sexual communication has been studied extensively in *Drosophila*, where it involves an elaborate courtship ritual, described in Chapter 12. Wing vibration by the male is a prominent feature in courting the female. Wing displays and courtship songs are likely to be under intense selective pressures, and differ between *Drosophila* species. Sean Carroll and his colleagues at the University of Wisconsin noted that some species have a pigmented spot on their wings, likely to attract attention during the courtship display (Figure 16.13).

Carroll compared the genealogical relationships among 77 species of fruit flies, and assessed when and how their wing spots evolved. He found that wing spots evolved

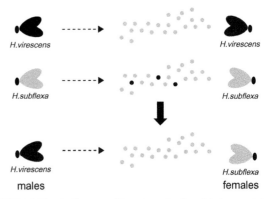

FIGURE 16.12 A diagram of how two closely-related sympatric species, *H. virescens* and *H. subflexa* can undergo prezygotic isolation. Pheromone molecules are represented as circles. Dark small circles represent a minor pheromone component produced by *H. subflexa* females that antagonizes *H. virescens* males. When the minor component is abolished by a mutation in an enzyme necessary for the synthesis of the minor component or by experimental manipulation, as described in the text, the prezygotic isolation barrier is lowered and *H. virescens* males show greater attraction to *H. subflexa* females.

independently at least twice along the evolutionary tree. Moreover, the wing spot was also lost on several occasions. For example, *D. melanogaster* has no wing spots, but its ancestor had spotted wings (Figure 16.14). What would be the mechanism that enables wing spots to be turned on and off so readily during evolution? Carroll discovered that the *yellow* gene is responsible for the wing spot pigmentation pattern, and that mutations in ***cis*-regulatory** regions are responsible for the spotted wing or blank wing phenotypes. Mutations that enable transcription factors to activate the *yellow* gene in the distal region of the wing have arisen independently in *D. biarmipes* and *D. tristis*, and mutations in the regulatory region of the *yellow* gene have caused

FIGURE 16.13 The spots on the wings of fly species like *Drosophila elegans* can accentuate the courtship ritual during mating, as illustrated in the figure (female on the left, male on the right). (Courtesy of B. Prud'homme and S. Carroll.)

loss of the wing spots in *D. gunungcola* and *D. mimetica*. The acquisition and loss of wing spots thus depends on the integrity of a regulatory element that controls expression of a pigmentation gene. Whereas the appearance of the wing spot may provide an advantage for the courtship rituals of those species in which the spots have been acquired and retained, loss of wing spots in other species suggest that the spot itself is not an essential component when already elaborate courtship songs and mating rituals exist.

The effect of a change in the transcription pattern of the *yellow* gene in related species of *Drosophila* is reminiscent of the earlier described change in the promoter region of the *Obp57e* gene of *D. sechellia* and **microsatellite** insertions in the promoter region of the vasopressin receptor gene that controls monogamous versus polygamous social behaviors in voles (Chapter 12). Whereas evolutionary changes in regulation of a single gene of major effect can produce profound differences in behavioral phenotypes, one should remember that those critical genes themselves form part of genetic networks, and that changes in these critical "hubs" are accompanied by extensive alterations in expression of co-regulated genes, thereby remodeling the

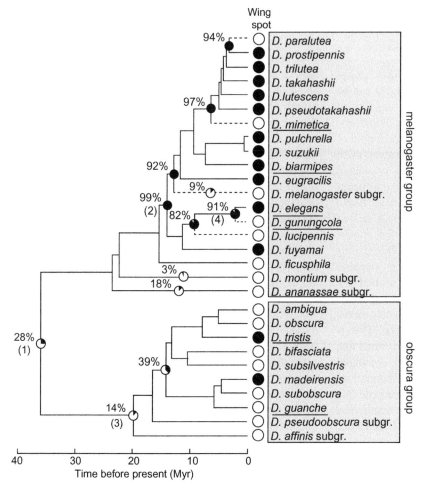

FIGURE 16.14 Phylogenetic tree of the *D. melanogaster* and *D. obscura* groups showing the presence (black circle) or absence (open circle) of wing spots, and the probability that a common ancestor had a wing spot (pie-charts on the nodes of the branches). Analysis of this phylogeny shows that wing spots were gained twice and lost five times during evolution (Myr = Million years ago). (From Prud'homme, B. et al. (2006). *Nature*, **440**, 1050–1053.)

Box 16.4   Stalk-eyed flies and sexual selection

Stalk-eyed flies (*Diopsidae*) provide a classical example of sexual selection. These flies generate long bilateral eyestalks that are sexually dimorphic, with males having much longer eyestalks than females. The length of the eyestalks reflects fitness quality of the males, and females prefer to mate with males with the longest eyestalks (Figure A). Females who mate with males with "good genes" benefit by being able to generate genetically fit offspring.

Gerald Wilkinson at the University of Maryland demonstrated the occurrence of **meiotic drive** in populations of stalk-eyed flies. The term meiotic drive refers to any process which causes some alleles to be overrepresented in the gametes which are formed during meiosis, thereby distorting normal Mendelian chromosome segregation. In some populations of stalk-eyed flies, females carry a meiotic drive gene on their X chromosomes that causes female-biased sex ratios in the off-spring. Males that have genes that suppress this meiotic drive have longer eyestalks. When females mate with these males they will gain a genetic benefit by producing disproportionately

more male offspring in a female-biased population. The phenomenon of meiotic drive is sometimes referred to as an example of **intragenomic conflict**.

FIGURE A   The most widely-studied species of diopsid, *Teleopsis dalmanni*, from Malaysia.

genetic architecture that underlies the behavior (see also Chapter 10).

THE EVOLUTIONARY GENETICS OF BURROWING AND NEST BUILDING

Insights into the evolution of behaviors requires studying natural populations in the field, as few insights can be obtained from captive animals that have been maintained and are often inbred for many generations in the laboratory. Wild mice of the genus *Peromyscus* provide an example of how established resources developed for laboratory mice can be utilized to gain insights into behaviors of related species in the wild. Species of the genus *Peromyscus* occur widely in a variety of habitats through much of North America. The possibility for crossbreeding closely-related species of *Peromyscus*, the availability of the mouse genome sequence, and **synteny** between the *Peromyscus* and the laboratory mouse *Mus musculus* genomes enable resources developed for the latter to be recruited for studies on *Peromyscus*, both in the wild and in captivity. *Peromyscus sp.* have undergone remarkable adaptations, many of which are designed to avoid predation. This is immediately evident from the light coat color of mice that live on sandy beaches near the Gulf of Mexico, compared to the darker pigmentation of mice that live in vegetation-covered soil. A single nucleotide mutation in the gene that encodes the melanocortin-1 receptor has been implicated in the adaptation to a lighter coat color in the beach mice (Figure 16.15).

Wallace Dawson, in the 1980s, pioneered studies on *Peromyscus* and, among other traits, studied their burrowing

behavior. Burrowing is an essential survival behavior. Burrows provide protection from predators, and a reproductive nest where progeny can develop safely. The old field mouse, *Peromyscus polionotus*, inhabits beach dunes and sandhills in the southeastern United States. As their habitat offers little protection against predation, *P. polionotus* constructs an elaborate burrow, which consists of a long entrance tunnel that descends at a slope of about 45° to a nest chamber, typically located 40–80 centimeters below the surface (Figure 16.16). The entrance of the tunnel is hidden by a plug of soil while the animal is hiding in its burrow. In addition to the entrance tunnel, the burrow also contains an escape tunnel that rises at a steep slope to within 2–3 centimeters of the surface. This escape tunnel ensures that the occupant is not trapped in its nest chamber when cornered by an invading enemy. The nest chamber is about 12 centimeters in diameter, and the young are reared within the nest until they are weaned. In contrast to *P. polionotus*, the deermouse, *Peromyscus maniculatus*, inhabits territories covered with vegetation, which provide ample hiding places. These mice construct much simpler and shallower burrows.

To what extent is burrow-building genetically determined? Hopi Hoekstra and her colleagues located burrows of *P. polionotus* in the wild. Filling the burrow with a solidifying foam generated three-dimensional full-size replicas that enabled precise measurements of the lengths and slopes of the entrance and escape tunnels, as well as the depth and size of the nest chamber. This provided quantitative measurements of the behavioral phenotype that could be subjected to genetic analyses. Hoekstra made several critical observations. First, mice that had grown up in a captive laboratory population and had never

FIGURE 16.15 Pigmentation differences among the Santa Rosa Island beach mouse (*Peromyscus polionotus leucocephalus*) (left) and the oldfield mouse (*P. p. subgriseus*) that lives inland in Florida (right). (Courtesy of Dr Hopi Hoekstra.)

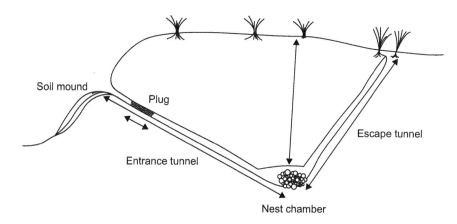

FIGURE 16.16 Diagram of a typical burrow of *Peromyscus polionotus*. (Courtesy of Dr Hopi Hoekstra.)

been exposed to the wild would faithfully construct the same burrows in the laboratory as wild-reared animals built in the sand dunes. This innate ability to construct an architecturally-predefined burrow without prior experience is reminiscent of nest building in birds, or the construction of underground cities of geometrically-precise interconnected corridors and brood chambers by leafcutter ants, and indicates a remarkable genetically-hardwired program that directs a complex behavior. By crossing closely-related species of *Peromyscus*, Hoekstra showed that the construction of complex burrows is dominant over the construction of simple burrows, and is probably controlled by relatively few genes of large effect. Furthermore, the origins of complex burrowing behavior could be traced to branch points in the evolutionary history of the genus *Peromyscus*.

It is truly astounding that the genome specifies a blueprint for the architecture of a complex burrow and contains all the elements necessary to direct its flawless implementation. Whereas until recently understanding the nature of this genetic specification might have been considered an intractable daunting challenge, the availability of the sequenced mouse genome and comparative genomic technologies, together with the possibility of quantifying the burrow parameters precisely, make it possible to

solve this problem in principle, given enough time and resources. Ultimately, we may understand the spatial and temporal dynamics of the genetic networks that direct the nervous system of *P. polionotus* to construct a burrow of great complexity, and be able to identify the critical genetic switches that can control these networks that have evolved to mediate the transition from simple to complex burrowing behavior.

THE ASTOUNDING DIVERSITY OF CICHLIDS AND STICKLEBACKS

Extensive diversification of an ancestral lineage in response to divergent ecological conditions produces differentiated populations that evolve into new species, which are collectively termed an **adaptive radiation. Cichlid fishes** and **sticklebacks** present remarkable examples of rapid evolution, and morphological and behavioral diversification. Studies on these adaptive radiations can address the question of whether the same genes repeatedly undergo selection to contribute to similar phenotypes among different related species (**parallel evolution**), or whether alleles of entirely different genes give rise to similar phenotypes (**convergent evolution**).

Box 16.5 Speciation and synteny

The concept of what constitutes a species has been remarkably controversial. The most accepted definition today was formulated by the naturalist Ernst Mayr, who defined a species as all the individual organisms in a natural population that generally interbreed in the wild, and whose interbreeding produces fertile offspring. Note that this leaves ambiguity about what constitutes a species for organisms that reproduce asexually, or the designation of infertile hybrid offspring, such as mules. According to Mayr's definition, speciation occurs when members of two populations can no longer interbreed to provide fertile offspring.

Speciation can occur when a population splits into two geographically-separate locations. This is known as **allopatric speciation**. Once geographically separate, populations can diverge independently by mutation accumulation or genetic drift, and develop phenotypically-distinct characteristics, including behaviors. Island populations provide good examples for allopatric speciation, with Darwin's finches of the Galapagos Islands (Chapter 1) being perhaps the most well-known example. When speciation occurs in the same habitat it is known as **sympatric speciation**. This can occur when members of a population occupy different ecological niches within the same habitat, for example different insect–host plant specializations.

The conditions in which members of a population are geographically- or behaviorally-isolated and will not interbreed,

although they could still do so and provide viable hybrid offspring, is known as **prezygotic isolation**. When the genomes have diverged to the point where they are no longer compatible for the production of fertile offspring, **postzygotic isolation** has occurred, which consolidates the speciation process.

As early as the 1930s Theodosius Dobzhansky postulated that the process of postzygotic isolation might involve chromosomal rearrangements and reorganization of the genome, a hypothesis that has gained substantial support in more recent studies. When corresponding genomic regions are on different chromosomes, meiosis and recombination can no longer occur, and species are reproductively-isolated.

In modern comparative genomics the term synteny is used to describe the preserved order of genes on chromosomes of related species that have descended from a common ancestor. Thus, closely-related species will have the same amount of genetic material, but housed on a different number of chromosomes (Figure A). Synteny is useful, as it facilitates applying information from a sequenced genome of a laboratory species to syntenic regions of a different species. An example of synteny is the highly-preserved order of genes in the "Hox cluster," which are master segmentation genes that determine the body plan throughout the animal kingdom.

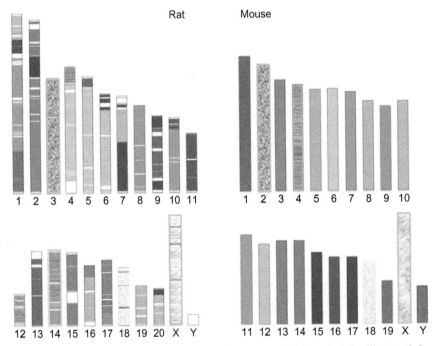

FIGURE A Synteny between the mouse and rat genomes. The mouse chromosomes (right) are shaded to illustrate their corresponding homologous segments mapped onto the rat genome (left). Note the synteny between mouse chromosome 2 and rat chromosome 3, between chromosomes 18, and between the X-chromosomes. Note also how segments of mouse chromosome 5 have been redistributed among rat chromosomes 4, 12, and 14. It is evident that during evolution the order of large sequences of genes has been preserved, but the chromosomal organization has extensively diversified, rendering the two rodent genomes incompatible for recombination.

The Cichlidae are one of the most species-rich families of teleosts, comprising more than 3000 species. They are found in Central and South America, Africa, Madagascar, and India, but the greatest diversity occurs in the great lakes of East Africa: Lake Victoria; Lake Tanganyika; and Lake Malawi (Figure 16.17). These lakes formed after the last ice age, and represent independent evolutionary incubators for the divergence of species of cichlid fishes. Here, cichlids have undergone adaptive specializations that have led to a wide variety of different morphological and behavioral specializations, including a wide array of reproductive and social behaviors, such as maternal and paternal mouth brooding, biparental and uniparental brood care, and nest guarding. Evolution of cichlids has been rapid, and presents a rich resource for investigating the evolutionary mechanisms that drive morphological and behavioral adaptations in different ecological niches. A cichlid genome consortium has been established, and genomic and bioinformatics resources are being developed to explore this potentially-rich evolutionary model.

Hans Hofmann at the University of Texas in Austin, one of the protagonists of this endeavor, has used genetic and genomic approaches to study the evolution of behavior in cichlids, using *Astatotilapia burtoni* as a "cichlid poster child" (Figure 16.18, overleaf). *A. burtoni* is a mouth-brooding fish; larvae are protected by developing in the parent's mouth. *A. burtoni* live in shallow, temporary ponds on the lake margins, a constantly changing habitat. Dominant males are brightly colored, and aggressively defend small territories. A dominant male will solicit and court a female with a courtship ritual that involves the display of its coloration, large tail movements, and quivering of its anal fin in front of the female, ultimately guiding her to the spawning site in his territory. Dominant males, however, comprise only 20–30% of the male population. The remaining males are sexually repressed, and camouflaged

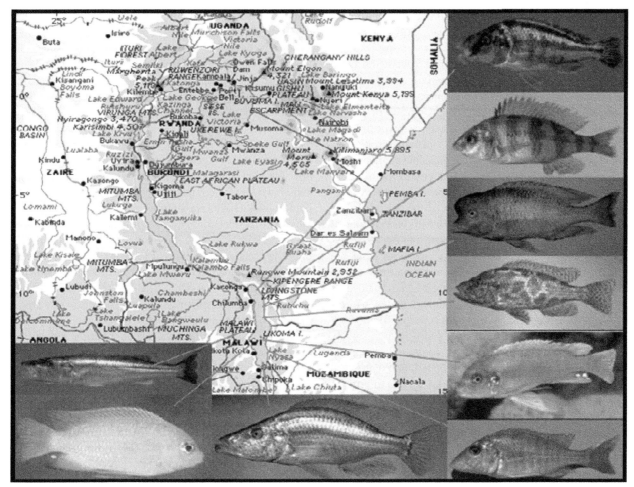

FIGURE 16.17   The great lakes of East Africa are among the oldest in the world. Lake Tanganyika is the oldest, and has been estimated to be 5–7 million years old. Lake Malawi is younger, about 1 million years old, and Lake Victoria likely formed a few hundred thousand years ago, and might have mostly dried out during the last ice age, after which it was re-colonized about 12500 years ago. The radiation of cichlids in these lakes has resulted in an amazing variety of fish morphologies, illustrated here for cichlids from Lake Malawi. (Fish photographs are from the Cichlid Genome Consortium website http://hcgs.unh.edu/cichlid/.)

FIGURE 16.18   A brightly colored male *Astatotilapia burtoni*.

to look like the less-colorful females to avoid provoking aggression and to be allowed to feed deceptively in territories of dominant males. However, because the territories are frequently disrupted by predators or changing environmental conditions, social relations can change rapidly, and transitions between territorial and nonterritorial males are dynamic; in this community social status can change within minutes.

What are the genetic programs that govern the rapid adaptation to a changing social environment? Hofmann and his colleagues used expression microarray analysis to assess differences in gene expression as a result of change in social status. They found that the expression of the neuropeptides arginine vasotocin (AVT) and somatostatin was altered when social status changed, and showed that somatostatin and AVT are involved in the control of behavior in *A. burtoni*. Arginine vasotocin expression levels in different regions of the **preoptic area** of the hypothalamus corresponded with territorial or non-territorial behavior of males. Somatostatin appears to modulate aggressive behavior, and somatostatin neurons in the **preoptic area** are larger in dominant territorial than in subordinate nonterritorial males. Hofmann found that an **agonist** for somatostatin decreased aggressive behavior in dominant males, while a somatostatin **antagonist** increased aggression. These findings suggest that somatostatin may function as a switch, through which a male *A. burtoni* can rapidly reduce aggressive status when he is "demoted" socially and aggressive displays are no longer adaptive. It is of interest to remember again the previously-described involvement of a related neuropeptide (vasopressin) in contributing to the differences in monogamous versus polygamous behaviors in voles (Chapter 12). The involvement of similar neuropeptides in the rapid behavioral adaptations seen in *A. burtoni* suggests that the genes that mediate rapid adaptation to social change within a species are likely to be the same genes that are under selection during the evolution of different social behaviors among related species.

Another example of an adaptive radiation with diverse phenotypic differentiation are sticklebacks. The first studies on stickleback behavior were done on the three-spined stickleback, *Gasterosteus aculeatus*, by the Dutch ethologist and Nobel laureate, Niko Tinbergen. At the end of the last ice age, thousands of postglacial lakes and streams arose in North America, Northern Europe, and Northern Asia. Many of these lakes contain stickleback populations that have evolved independently from a common marine ancestor, and studies of these populations can again address the question whether morphological and behavioral traits evolved along parallel or convergent evolutionary trajectories. The stickleback radiation occurred only 10 000–15 000 years ago, and resulted in a vast number of freshwater forms that were originally classified as over 40 different species, because of their dramatic morphological differences. However, the major barrier for sympatric stickleback morphs to interbreed appears to be mostly due to prezygotic isolation as a result of behavioral differences in courtship and mating. The different morphs can interbreed in the laboratory, and give rise to fertile F_1 offspring. This opens the way for genetic analyses to identify alleles that contribute to phenotypic variation in behavior.

Limnetic sticklebacks live in open water. They have slender bodies with extensive skeletal body armor with long spines in the dorsal and pelvic fins, and bony lateral plates to deter potential fish and bird predators. Lake-dwelling benthic sticklebacks live in shallower water. Their body shape is more rounded, and they have greatly reduced body armor.

Limnetic male sticklebacks acquire a bright red throat and belly in the breeding season. This coloration has a dual function. It sends a dominance signal to other males and also mediates competition for females (Figure 16.19). Female sticklebacks are attracted to redder males, an example of sexual selection. After getting the female's attention, males perform a zig-zag dance as part of the courtship ritual. The female will enter and lay eggs in a nest structure built by the male, and once fertilized the male will protect the eggs and offspring. There is substantial variation among courtship displays and aggressive behaviors among males in different populations.

In contrast to the limnetic males, **benthic** sticklebacks do not show the bright nuptial coloration of their limnetic counterparts, but tend to be drab during courtship. Here, courtship is initiated by females, who swim to the male and press their abdomens against the male's dorsal spines and then move in circles. This dorsal pricking behavior in benthic populations contrasts with the typical and conspicuous zig-zag dance displayed exclusively by limnetic males. The different behaviors of benthic and limnetic sticklebacks may reflect the presence or absence of cannibalism. Limnetic sticklebacks feed almost exclusively on plankton, and are not cannibalistic. In contrast, benthic males feed on invertebrates and form foraging groups that cannibalize young, including eggs in nests. Thus, benthic males tend to avoid drawing attention to their nest and,

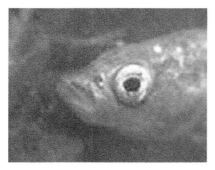

FIGURE 16.19 A male stickleback with its characteristic red throat and belly is attracted to a gravid female, and will fertilize her eggs as he follows her to the nest.

in fact, display diversionary behaviors in response to the approach of potentially-cannibalistic groups, including very visible upright or erratic swim-root, snout tap, shimmer, and digging behaviors.

What are the genes that give rise to this behavioral variation? Are the genes that contribute to variation within a population the same as those that give rise to different behaviors among closely-related species? In recent years, sticklebacks have become a model in which these questions can be addressed.

Catherine Peichel and David Kingsley from Stanford University crossed a benthic female and a limnetic male from Priest Lake in Canada, and a single F_1 male to a second benthic female. They genotyped 92 individuals of the offspring, and found polymorphic microsatellite markers from which they constructed a genomewide **linkage map** for QTL analyses with a **polymorphic marker** about every 4 cM (Figure 16.20, p. 295). Sticklebacks have 21 chromosomes and a genome size of about 0.58–0.70 Gb, which is relatively small for fishes.

QTL analyses have thus far focused on the stickleback's characteristic morphological features. The length of the protective spines in the dorsal and the pelvic fins, and the number of lateral plates, appears to be under the control of a small number of QTL with large effects on different chromosomes. One candidate gene, *Pitx*, has been implicated in loss or reduction of pelvic armor in benthic sticklebacks. Studies on *Pitx1* in stickleback populations again support the notion of parallel evolution in the construction of protective armor.

As genomic resources are being developed to support genetic studies on sticklebacks, identification of genes that contribute to variation in behavior, which is the key to understanding speciation and prezygotic isolation in the stickleback radiation, will not be far behind. The greatest challenge in working with this interesting model system is the long generation time, as sticklebacks only breed once per year and the number of offspring that can be reared, maintained, and phenotyped in the laboratory is limited. Large sample sizes will be needed to gain sufficient statistical power for detecting not only QTL of large effect,

but also QTL of smaller effect that may contribute significantly to phenotypic variation in the population, and may represent important targets for natural selection through genotype-by-environment interactions.

UNDERSTANDING THE EVOLUTION OF BEHAVIOR: HOW MUCH DO WE REALLY KNOW?

So, how do behaviors evolve? Are the genes that regulate adaptation of behaviors in different environments during an individual's lifetime the same genes that are targets for behavioral adaptations over evolutionary time? Are the genes that cause phenotypic variation within a species the same genes that cause variation among species, i.e. are they the targets for speciation? These are but some of the profound unanswered questions.

Do behaviors evolve as a consequence of incremental changes due to many mutations of small effects that spread gradually through the population, or does evolution of behavior occur in leaps and bounds, where few mutations of large effect result in abrupt changes in behavior? This question is reminiscent of the controversy that existed in quantitative genetics, where genetic variation could either be explained by a vast number of genes of small effect (the "infinitesimal model") or few genes of large effect. The generally accepted model today is Alan Robinson's insight that relatively few genes with alleles of large effect account for the majority of genetic variation, but that large numbers of genes with alleles of small effect account for the remaining genetic variation. A similar concept may apply to the evolution of behavior. Thus, single mutations that alter the regulation of expression of one or few focal genes may have profound effects on the manifestation of a behavior (e.g. host selection by *Drosophila sechellia*), while changes in allele frequencies of numerous additional genes contribute to consolidate the behavioral adaptation. This model, however, is speculative and may not be universally applicable.

Box 16.6   Camouflage and mimicry

Many species have evolved colors or color patterns that allow them to blend in with their environment, either to escape predation or to facilitate capturing prey. A classic example of protective camouflage are the stripes on zebras, which make it difficult for their main predators, lions, who are color-blind, to distinguish them among tall grass in their grassland habitat, or to distinguish single individuals among a herd. Tigers, on the other hand, have turned camouflage to their hunting advantage, and their stripes hide them from their prey in their jungle habitat. Camouflage coloration is not always permanent, but can sometimes change according to season. For example, the arctic fox has a white coat in winter and a brown coat in summer (Figure A).

Mimicry is distinct from camouflage. Here, a species imitates another, often closely-related, species to gain a survival advantage. Examples are butterflies that mimic unpalatable species to avoid predation without investing the cost of producing toxins themselves. An example of behavioral **mimicry** is the firefly "femme fatale" of the genus *Photuris*. Fireflies are beetles that produce **bioluminescent** signals by releasing energy from ATP via the **luciferase** reaction to attract mates. The firefly femme fatale mimics the light signal of a different species, and when an unfortunate male attempts to mate with her, kills and eats him (Figure B).

FIGURE A   The arctic fox in summer (left) and in winter (right).

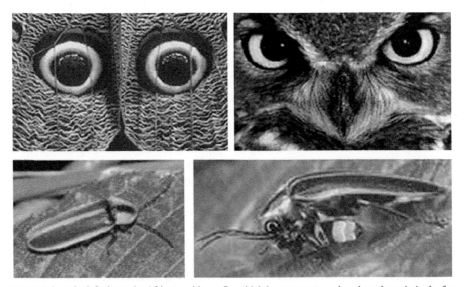

FIGURE B   The upper panel on the left shows the African owl butterfly, which has eye spots on its wings that mimic the face of an owl, shown on the right, in an attempt to intimidate potential predators. The lower panel of the left shows a firefly of the genus *Photinus*. On the right, a *Photinus* male has been lured by a *Photuris* femme fatale female into a kiss of death, instead of the expected embrace of love.

It is important to remember that genes do not act independently, but are integrated as functional networks. In Chapter 10 we described how a single mutation in the genome can give rise to altered regulation of the expression of a suite of transcripts. Thus, a single polymorphism that arises can generate a ripple effect through the transcriptome. If one assumes that a single SNP affects the expression of about 50 other genes, it follows that multiple polymorphic differences between two individuals will result in extensive transcriptome-wide differences in expression levels. This

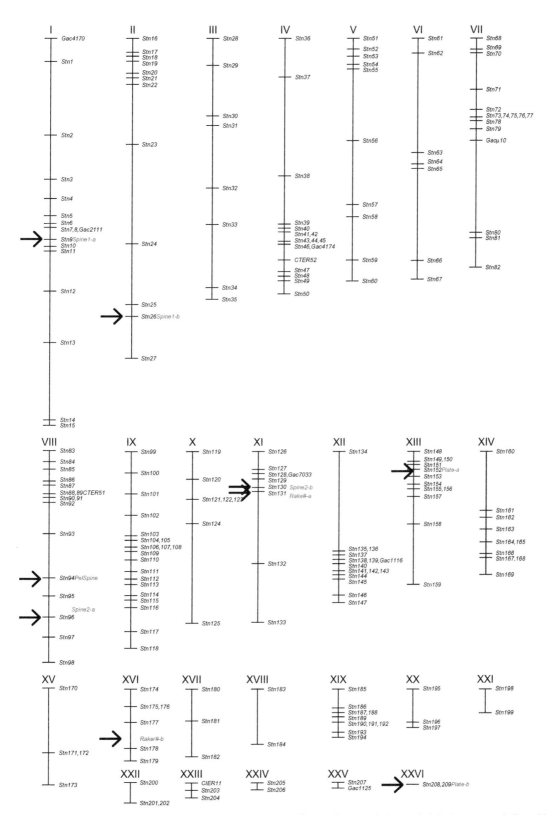

FIGURE 16.20 Linkage map of the stickleback genome. QTL locations that affect feeding morphology and skeletal armor are indicated by arrows. (Modified from Peichel, C. L. et al. (2001). *Nature,* **414,** 901–905.)

is indeed what is observed when allele frequencies diverge during bidirectional artificial selection for a behavioral trait from a base population. Thus, the evolution of behavior is intimately linked to genome evolution and cannot be analyzed at the single gene level, but must be analyzed in terms of selection on and adaptation of complex transcriptional networks.

Although we have learned a lot about the mechanisms of natural selection since the days of Charles Darwin, the complexity of the mechanisms that drive the evolution of life on our planet is still poorly-understood, and we realize that the more we learn, the greater the complexity we are faced with. In this respect Albert Einstein's famous quote is both sobering and reassuring:

"One thing I have learned in a long life: that all our science, measured against reality, is primitive and childlike – and yet it is the most precious thing we have."

SUMMARY

Evolution progresses through the transfer of genes to successive generations, with changes in allele frequencies through both adaptive mechanisms (natural selection) and nonadaptive mechanisms (e.g. genetic drift). Genetic drift is a nonadaptive process by which random variation in sampling of particular alleles in a population causes some alleles to be under- or overrepresented in the next generation, which ultimately leads to homozygosity.

According to Kimura's neutral theory for molecular evolution, new alleles carrying such neutral mutations are expected to accumulate in the genome as they become fixed through genetic drift. The theory of neutral evolution provides a null hypothesis for tests for deviations from neutral expectations that can assess the role natural selection may have played during evolutionary history. Understanding the evolutionary history of a gene requires estimates of the mutation rate, the effective population size, and the population history (e.g. migration).

Behaviors represent the instruments for evolutionary change, as they support an individual's survival and its procreation. Behaviors can exert strong evolutionary pressures. Sexual selection is a process by which fitness of a potential mating partner is evaluated on the basis of anatomical or behavioral displays, driving evolution of extravagant morphological features and intricate courtship behaviors. Sperm competition drives the evolution of polygamous mating systems, where males attempt to inseminate as many females as possible to optimize their chance of transmitting their genes to progeny. Rapidly-evolving gene families that mediate chemosensation and detoxification appear especially important in adaptive evolution, such as interactions between insects and their host plants.

Evolution of chemical communication systems can establish reproductive barriers for sympatric prezygotic isolation. Subsequent chromosomal rearrangements can then lead to postzygotic isolation and the evolution of new species. Adaptive radiations of cichlid fishes or stickleback populations provide attractive models for the study of speciation and evolution of behaviors. The diversity of these fishes evolved through subdivision of ancestral populations in lakes formed during the recent ice age, where each lake represents an independent "natural experiment in evolution."

Whereas evolutionary changes in regulation of a single gene of major effect can produce profound differences in behavioral phenotypes, one should remember that those genes form part of genetic networks. Thus, regulatory changes at a single critical gene can lead to extensive remodeling of the genetic architecture that underlies the behavior.

STUDY QUESTIONS

1. What was the innovation and impact of Kimura's neutral theory of evolution?
2. What is the difference between the HKA test and the McDonald–Kreitman test, and what are the limitations of each?
3. What is a "recent selective sweep," and why does it leave a characteristic signature in the genome?
4. What is the difference between prezygotic and postzygotic isolation?
5. How does sexual selection contribute to the evolution of behavior? Give an example.
6. Why is polygamy advantageous under conditions in which sperm competition plays a role in reproduction?
7. What is the conundrum for co-evolution of sexual communication systems, and how can this be overcome?
8. What are the advantages and disadvantages of generalism versus specialism in insect–host plant interactions?
9. Which gene families appear to be critical in the evolution of insect–host plant adaptations?

10. Explain Hamilton's rule.
11. Discuss the advantages of studying wild populations to gain insights in the evolution of behavior as compared to standard laboratory animals. Give an example.
12. Explain genetic drift and how it impacts allele frequencies. What is a bottleneck?
13. Why does population admixture confound tests for deviation from neutrality?
14. Explain the concepts of convergent evolution and parallel evolution.
15. How did Mayr define a species? How does his definition relate to behavior? Are there limitations to this definition?
16. What would be a signature of possible balancing selection in a population? Can you think of behaviors that might be under balancing selection?

RECOMMENDED READING

Berenbaum, M. R. (2002). Postgenomic chemical ecology: from genetic code to ecological interactions. *J. Chem. Ecol.*, **28**, 873–896.

Carroll, S. B. (2005). Evolution at two levels: on genes and form. *PLoS Biol.*, **3**, e245.

Coyne, J. A., and Orr, H. A. (2004). *Speciation*. Sinauer Associates, Inc, Sunderland, MA.

Falconer, D. S., and Mackay, T. F. C. (1996). *Introduction to Quantitative Genetics*, 4th edition. Prentice Hall, Harlow, UK.

Fay, J. C., and Wu, C. I. (2000). Hitchhiking under positive Darwinian selection. *Genetics*, **155**, 1405–1413.

Foster, S. A., and Baker, J. A. (2004). Evolution in parallel: new insights from a classic system. *Trends Ecol. Evol.*, **19**, 456–459.

Fu, Y. X., and Li, W. H. (1993). Statistical tests of neutrality of mutations. *Genetics*, **133**, 693–709.

Gompel, N., Prud'homme, B., Wittkopp, P. J., Kassner, V. A., and Carroll, S. B. (2005). Chance caught on the wing: cis-regulatory evolution and the origin of pigment patterns in *Drosophila*. *Nature*, **433**, 481–487.

Greenwood, A. K., Wark, A. R., Fernald, R. D., and Hofmann, H. A. (2008). Expression of arginine vasotocin in distinct preoptic regions is associated with dominant and subordinate behaviour in an African cichlid fish. *Proc. Biol. Sci.*, **275**, 2393–2402.

Groot, A. T., Horovitz, J. L., Hamilton, J., Santangelo, R. G., Schal, C., and Gould, F. (2006). Experimental evidence for interspecific directional selection on moth pheromone communication. *Proc. Natl. Acad. Sci. USA*, **103**, 5858–5863.

Hartl, D. L., and Clark, A. G. (2007). *Principles of Population Genetics*, 4th edition. Sinauer Associates, Inc, Sunderland, MA.

Hoekstra, H. E., Hirschmann, R. J., Bundey, R. A., Insel, P. A., and Crossland, J. P. (2006). A single amino acid mutation contributes to adaptive beach mouse color pattern. *Science*, **313**, 101–104.

Hudson, R. R. (1987). Estimating the recombination parameter of a finite population model without selection. *Genet. Res.*, **50**, 245–250.

Hudson, R. R., Kreitman, M., and Aguadé, M. (1987). A test of neutral molecular evolution based on nucleotide data. *Genetics*, **116**, 153–159.

Jiggins, C. D., and Bridle, J. R. (2004). Speciation in the apple maggot fly: a blend of vintages? *Trends Ecol. Evol.*, **19**, 111–114.

Kimura, M. (1983). *The Neutral Theory of Molecular Evolution*. Cambridge University Press, Cambridge, UK.

King, M. C., and Wilson, A. C. (1975). Evolution at two levels in humans and chimpanzees. *Science*, **188**, 107–116.

Li, X., Schuler, M. A., and Berenbaum, M. R. (2007). Molecular mechanisms of metabolic resistance to synthetic and natural xenobiotics. *Annu. Rev. Entomol.*, **52**, 231–253.

Lynch, M. (2007). *The Origins of Genome Architecture*. Sinauer Associates, Inc., Sunderland, MA.

Matsuo, T., Sugaya, S., Yasukawa, J., Aigaki, T., and Fuyama, Y. (2007). Odorant-binding proteins OBP57d and OBP57e affect taste perception and host-plant preference in *Drosophila sechellia*. *PLoS Biol.*, **5**, e118.

McDonald, J. H., and Kreitman, M. (1991). Adaptive protein evolution at the *Adh* locus in *Drosophila*. *Nature*, **351**, 652–654.

Mead, L. S., and Arnold, S. J. (2004). Quantitative genetic models of sexual selection. *Trends Ecol. Evol.*, **19**, 264–271.

Ohno, S. (1970). *Evolution by Gene Duplication*. Springer-Verlag, New York, NY.

Partridge, L., and Hurst, L. D. (1998). Sex and conflict. *Science*, **281**, 2003–2008.

Peichel, C. L., Nereng, K. S., Ohgi, K. A., Cole, B. L., Colosimo, P. F., Buerkle, C. A., Schluter, D., and Kingsley, D. M. (2001). The genetic architecture of divergence between threespine stickleback species. *Nature*, **414**, 901–905.

Pollen, A. A., Dobberfuhl, A. P., Scace, J., Igulu, M. M., Renn, S. C., Shumway, C. A., and Hofmann, H. A. (2007). Environmental complexity and social organization sculpt the brain in Lake Tanganyikan cichlid fish. *Brain Behav. Evol.*, **70**, 21–39.

Tajima, F. (1993). Statistical method for testing the neutral mutation hypothesis by DNA polymorphism. *Genetics*, **123**, 585–595.

α-synuclein. Protein which forms fibrous aggregates associated with the pathology of Parkinson's disease.

β-amyloid. A component of **neuritic plaques** found in post-mortem brains of patients with **Alzheimer's disease**.

β-arrestin. Protein that mediates receptor desensitization by binding to activated β-adrenergic receptor after the receptor is phosphorylated by a **G-protein receptor kinase (GRK)**.

β-galactosidase. An enzyme, encoded in *E. coli* by the *lacZ* gene, that catalyzes the hydrolysis of β-galactosides into monosaccharides, and is often used as a marker or reporter in genetics and molecular studies by converting a substrate such as X-gal into a colored reaction product.

γ-aminobutyric acid (GABA). The major inhibitory neurotransmitter in the brain, generated by decarboxylation of glutamic acid.

κ-opioid receptor. A widely-distributed pharmacological type of opioid receptor in the brain and spinal cord which binds the opioid peptide **dynorphin** to promote reduction in the sensation of pain.

μ-opioid receptor. A pharmacological type of opioid receptor which mediates analgesia, and has high affinity for **enkephalins** and **β-endorphin**, as well as the opium alkaloid **morphine**

Abducens nerve. One of the **cranial nerves** that innervates the muscles that mediate eye movements.

Absolute refractory period. A period immediately following repolarization of an **action potential** during which injection of current into a nerve cannot evoke another action potential.

Accessory gland proteins (Acps). Proteins that are produced in the male accessory gland of the fruit fly *Drosophila melanogaster*, and that are transmitted along with sperm into the female during copulation.

Accessory nerve. One of the **cranial nerves** that controls muscles in the neck.

Accessory olfactory bulb. A distinct region of the brain found in most vertebrates (with the exception of primates) located dorsal and posterior to the main **olfactory bulb** that processes pheromonal information received by the **vomeronasal organ**.

Accessory olfactory system. A division of the olfactory system dedicated to processing social odor cues, such as pheromones

Acetylcholine A neurotransmitter that is released at **neuromuscular junctions** and at **synapses** of the **parasympathetic nervous system**.

Acetyl-CoA. A central metabolite, which consists of an acetyl group linked to a carrier, Co-enzyme A, and which can be metabolized via the Krebs cycle, or serve as precursor for lipid biosynthesis.

Acoustic startle response. A **startle response** that can be evoked by subjecting an animal to a sudden sound that previously had been associated with an electric shock.

Action potential. A nerve impulse that occurs as a result of rapid transient changes in membrane potential, triggered by the sequential opening and closing of voltage-dependent sodium and potassium channels. *See also* **Spike**.

Adaptive landscape. The surface plotted in a three-dimensional graph, with all possible combinations of allele frequencies for different loci plotted in the plane and mean **fitness** for each combination plotted in the third dimension.

Adaptive radiation. The divergence of several new types of organisms from a single ancestral type.

Addiction. Psychological and bodily dependence on a substance or practice which is beyond voluntary control.

Additive/additivity. A condition in which the combined effects of **alleles** at two or more loci are equal to the sum of their individual effects during inheritance of a **quantitative trait**.

Adrenaline. A hormone released from the adrenal medulla into the blood to elicit a **fight or flight response** (also known as **epinephrine**).

Adrenergic. Neural communication mediated by the neurotransmitter **noradrenaline**.

Advanced intercross lines. A population of lines derived from a cross between inbred lines followed by random and sequential intercrossing to increase the probability of recombination and, thus, enhance the resolution for subsequent mapping of **quantitative trait loci**.

Advanced sleep phase syndrome (ASPS). A **circadian rhythm** disorder in which sleep onset occurs in early evening and as a consequence, wakefulness occurs in early morning.

Afferent signals. Neural signals that are carried from the peripheral sense organs towards the central nervous system.

Affiliative behavior. Behavior that strengthens bonds between individuals in a group, such as mutual grooming and parental care.

Affinity chromatography. Purification of a compound based on its binding to a specific ligand that has been conjugated to a resin.

Agonist. A compound that binds to and activates a protein.

Alkylating agents. *See* **base analogs**.

Allele. A sequence variant of a gene at a specific locus.

Allopatric. Present in different non-overlapping geographical locations.

All-or-none events. Events that are driven by a feed-forward process and go to completion once initiated, such as the **action potential**.

Alpha waves. A pattern of smooth, regular electrical oscillations in the human brain that occur when a person is awake and relaxed.

Alzheimer's disease. A progressive neurodegenerative disease of advanced age that results in loss of memory, dementia, and behavioral impairments, and is characterized by **neuritic plaques** and tangles in the brain.

Amiloride. A potassium-sparing diuretic which inhibits salt taste perception when applied to the tongue.

AMPA-type glutamate receptors. *See* **Kainic acid-type glutamate receptors**.

Amplicon. A fragment of DNA amplified by the **polymerase chain reaction**.

Amygdala. A part of the **limbic system**, located in the middle of the brain and connected to the **hippocampus**, which consists of an almond-shaped complex of related nuclei, and relates processes and memories to emotions, including rage, fear, and sexual feelings.

Amyloid precursor protein (APP). A precursor protein that can be proteolyzed to give rise to β-**amyloid**.

Analgesic. Relieving pain.

Analysis of variance (ANOVA). A statistical method that partitions the variance of a set of measurements into different sources of variation.

Anesthesia-resistant memory. An early form of consolidated memory in *Drosophila* that is resistant to anesthesia-induced amnesia of recent events.

Anhedonia. Inability to gain pleasure from enjoyable experiences.

Anorexia. An eating disorder characterized by a pathological tendency to resist food intake.

Anosmia. Inability to smell.

Antagonist. A ligand that blocks the activity of the protein to which it binds.

Antagonistic pleiotropy. A condition in which an **allele** affects the **phenotype** in opposite direction under different conditions (for example, an **allele** being beneficial early in life and deleterious at later age).

Antennae. Paired, flexible, segmented sensory appendages on the head of an insect.

Antennal lobes. The first order neuropil of the insect which receives the input from the olfactory sensory neurons on the antenna.

Anthropocentric. Human-centered.

Apolipoprotein E (ApoE). A cholesterol-carrying protein that may be involved in Alzheimer's disease.

Apoptosis. Programmed cell death.

Aqueduct. A fluid-filled canal that connects the third ventricle and the fourth ventricle in the brain.

Arista. A bristle-like appendage that protrudes from the second antennal segment of the *Drosophila* antenna at the boundary with the third antennal segment, and is thought to have an acoustic and thermosensory function.

Artificial selection. The process in which breeders choose the individuals to be used to produce successive generations.

Ascending. Neural pathways that convey information from the periphery to the brain.

Asperger syndrome. A pervasive developmental disorder, usually of childhood, characterized by impairments in social interactions and repetitive behavior patterns.

Association cortex. Cortical areas that are neither motor nor sensory, but are thought to be involved in higher processing of information.

Association mapping. *See* **Association study, association analysis**.

Association study, association analysis. A method of mapping **quantitative trait loci** by **linkage disequilibrium** with molecular markers in a sample of **alleles** from an outbred population.

Associative learning. A form of learning whereby the subject learns about the relationship between two stimuli, or between a stimulus and a behavior. *See also* **Classical conditioning**.

Assortative mating. Mating between individuals who share more traits in common than would be expected from random mating in the population.

Astrocytes. A type of neuroglial cell in the central nervous system that helps support other nerve cells.

Ataxia. Lack of coordination of voluntary movement.

Auditory transduction. The conversion of sound waves into electrical activity in hair cells of the ear.

Autism. A mental illness that typically affects a person's ability to communicate, form relationships with others, and respond appropriately to the environment.

Autism spectrum disorders (ASD). A spectrum of psychological conditions characterized by widespread abnormalities of social interactions and communication, as well as severely restricted interests and highly repetitive behavior. *See also* **Asperger syndrome**, **autism**.

Autonomic nervous system. The part of the nervous system of vertebrates that regulates involuntary action.

Autophosphorylation. The **phosphorylation** of a molecule by itself via a protein kinase domain within the molecule, a binding site for the ATP that donates the phosphate, and amino acids that undergo **phosphorylation**.

Autosomal. Pertaining to any of the chromosomes of an organism except the sex chromosomes.

Average effect of an allele. The mean value of all progeny that received that allele from a parent, the other allele having come at random from the population, expressed as a deviation from the overall **population mean**.

Avoidance index. A measurement that quantifies the number of individuals within an experimental group that show avoidance behavior, or the percentage of avoidance responses demonstrated by a given individual to a repellent stimulus.

Axo-axonic synapse. **Synapse** between the **axon** of one neuron and the **axon** of another.

Axon. A long process that is responsible for propagating the action potential in a neuron without decrement in signal.

Axon hillock. The region of the axon closest to the cell body where the **action potential** often originates. *See also* **Spike initiation zone**.

Backcross. The cross of an individual with one of its parents.

Balancer chromosome. A chromosome that carries large inversions and serves as a genetic tool to prevent crossing-over between homologous chromosomes during meiosis in *Drosophila*.

Balancing selection. Natural selection in which heterozygotes have increased **fitness** relative to both homozygotes.

Basal forebrain. A collection of structures located ventrally to the **striatum**, which provides the major cholinergic output of the central nervous system.

Basal ganglia. A group of nerve cells located at the base of the brain, composed of the **putamen**, **caudate**, **globus pallidus**, and **substantia nigra**, which participate in the regulation of motor performance.

Base analogs. Chemicals whose molecular structures mimic that of a DNA base and that may, therefore, act as mutagens.

Basiconic sensilla. A type of chemosensory sensilla in the *Drosophila* **antenna** and **maxillary palps**.

Behavioral plasticity. The ability of a behavior to change.

Benthic. Anything associated with or occurring on the bottom of a body of water.

Bicistronic. Two polypeptides encoded by a single messenger RNA and translated together.

Bioamines. Small biologically-active compounds that contain an amine group, such as **dopamine**, **serotonin**, **noradrenaline**, histamine, **octopamine**.

Biological variation. Phenotypic variation within the normal range.

Bioluminescence. A chemical reaction that causes an organism to glow.

Bioluminescent. Having bioluminescence.

Biometrics. The science and technology of measuring and statistically analyzing biological data.

Biotransformation enzymes. Enzymes that detoxify foreign compounds through oxidation, followed by the attachment of water-soluble chemical groups.

Bipolar neuron. A neuron that has two processes arising from opposite poles of the cell body.

Blastocyst. Cleavage stage of an early mammalian embryo, which consists of a hollow ball of cells made of outer trophoblast cells and an inner cell mass.

Blood–brain barrier. A protective layer of **astrocytes** around close-knit blood vessels in the brain, which limit substances that can enter the brain.

Bonferroni correction. A conservative multiple-test correction method, in which one divides the desired false positive rate by the number of tests, and uses that modified number to declare any single change to be significant.

Bootstrapping. A statistical procedure to calculate confidence intervals for fitted values by perturbing the original data set and resolving the model many times.

Bottleneck. A brief reduction in size of a population which usually leads to random **genetic drift**.

Bowman's glands. Mucus-secreting glands in the olfactory neuroepithelium.

Bradykinin. A biologically-active polypeptide, consisting of nine amino acids, that forms from a blood plasma globulin and mediates the inflammatory response, increases vasodilation, and causes contraction of smooth muscle.

Breeding value. The value of a trait transmitted from parents to offspring.

Broad sense heritability. The extent to which variation in **phenotypes** in a population is due to variation in genotypic values, including non-additive genetic variation. *See also* **Narrow sense heritability**.

CA1 neurons. Pyramidal neurons that form part of the neuronal circuit in the **hippocampus**.

Cable properties. Properties of electrical conductivity in neurons that do not depend on **action potentials**, and can be modeled as a capacitor discharging a rapidly spreading and quickly decaying current.

Calcium sensing receptor. A **G-protein-coupled receptor** in cells of the parathyroid glands that monitors the concentration of extracellular calcium to regulate secretion of parathyroid hormone.

Calmodulin. A calcium-binding protein that regulates a wide range of cellular processes, including protein **phosphorylation** at **synapses**.

Calyces. *See* **Calyx**.

Calyx. A component of the **mushroom bodies**, integrative neuropil structures of the insect brain.

Cameleon. A genetically-engineered calcium sensor that can be expressed in transgenic animals to monitor neuronal activation by eliciting **fluorescent energy transfer** as a result of a calcium-induced conformational change within the construct.

Cataplexy cataplectic. A sudden loss of muscle tone and strength, usually caused by an extreme emotional stimulus.

Catecholamines. Amines derived from tyrosine that contain a catechol ring, and that function as neurotransmitters and hormones, including **epinephrine**, **norepinephrine**, and **dopamine**.

Caudate nucleus. A brain structure within the **basal ganglia** responsible for regulating and organizing information being sent to the frontal lobes from other areas of the brain.

CD8-GFP. **Green fluorescent protein** bound to CD8, which directs the fluorescent tag to the cell membrane.

Cell body. The enlarged portion of a neuron containing the nucleus. *See also* **Perikaryon, soma**.

Central canal. A canal filled with cerebrospinal fluid in the center of the spinal cord.

Central complex. A region in the insect brain associated with locomotion.

Central dogma. The notion that genetic information flows from DNA to RNA to protein.

Cerebellar Purkinje cells. Large GABAergic neurons in the **cerebellum**.

Cerebellum. A brain structure in the back of the head between the **cerebrum** and the brain stem, which controls **proprioception** and complex motor functions via inhibitory GABAergic output.

Cerebrum. The largest structure in the brain, which contains **cortex** and forms the cerebral hemispheres.

Cervical. Related to the neck region.

Chaperone. A protein that aids in the folding of a second protein and prevents proteins from taking conformations that would be inactive.

Chemical mutagenesis. Mutations induced by chemicals that alter nucleotides in the DNA.

Chemosensation. The perception of chemicals via the **olfactory**, **gustatory**, **vomeronasal**, and **trigeminal** systems, and combinations thereof.

Chemosensory signals. Molecules that can be perceived and elicit behavioral responses via the **olfactory**, **gustatory**, **vomeronasal**, and **trigeminal** systems, and combinations thereof.

Chemotopic map. A systematic neural representation of chemical structures in the brain.

Chimera. A tissue containing two or more genetically-distinct cell types, or an individual made up of two or more genetically-distinct cell lines.

Cholecystokinin (CCK). A polypeptide hormone produced by the small intestine in response to the presence of fat, causing contraction of the gall bladder, and by the brain as a **neuromodulator** of satiety.

Cholinergic. Neurotransmission mediated by **acetylcholine**.

Chorda tympani. A branch of the **facial nerve** that innervates taste cells on the anterior two-thirds of the tongue.

Chorea. A movement disorder characterized by involuntary jerky movements.

Chromosome substitution lines. Lines in which an entire chromosome has been replaced via genetic crosses by a different homologous chromosome.

Chronobiology. The study of biological rhythms.

Chylomicra (chylomicrons). Large lipoprotein particles that are created by the absorptive cells of the small intestine.

Cichlid fishes. A diverse class of fishes that have undergone rapid **adaptive radiation** during evolution.

Cingulate gyrus. A long curved structure on the medial surface of the cerebral hemispheres that forms part of the **limbic system**.

Circadian, circadian activity, circadian rhythm. A daily activity cycle pattern that repeats at around 24-hour intervals.

***Cis*-regulatory**. Regulatory promoter sequences located adjacent to the gene they control.

Classical conditioning. A form of **associative learning** in which a subject learns the relationship between two stimuli. *See also* **Associative learning**.

Clustergrams. Diagrams in which items with similar features are organized together, giving rise to a pattern of clusters in which items within clusters resemble each other more than items between different clusters.

Coccygeal. Pertaining to the coccyx, i.e. the tailbone, which is the final segment of the vertebral column.

Cocktail party effect. Difficulty in understanding conversations that occur simultaneously.

Co-dominant. The condition in which both **alleles** as a **heterozygote** contribute equally to the **phenotype**.

Coefficient of linkage disequilibrium (*D*). A measure of how much observed **haplotype** frequencies differ from their expected frequencies under random association.

Coeloconic sensilla. A type of chemosensory sensilla in the *Drosophila* antenna.

Co-isogenic. Nearly genetically-identical strains of an organism that vary at only a single locus.

Colony collapse. The sudden disintegration and disappearance of honey bee colonies.

Combinatorial recognition. Identification of the structure of a molecule through the simultaneous recognition of distinct features of the molecule by multiple receptors.

Commissural fibers. Nerve fibers that connect the two hemispheres of the brain.

Common environmental variance. Variance attributable to environmental conditions that cause phenotypes of individuals reared in that environment to be more similar to each other than to individuals reared in different common environments.

Comorbid. Pertaining to a pathological condition that coexists with, but can be independent from, another pathological disorder.

Complementary sex determiner (csd). A polymorphic sex locus in haplo-diploid Hymenopteran insects, where heterozygosity initiates female development.

Complex traits. Traits that are usually normally distributed in a population, and for which the relationship between **genotype** and **phenotype** is complicated, as the **phenotype** is determined through the coordinated action of multiple genes and their interactions with the environment.

Compulsive obsessive behaviors. Behaviors that arise from anxiety, in which a person worries excessively about the circumstances of his or her life over a long period of time, and deals with these thoughts and feelings through ritualized actions.

Conditional neutrality. A situation in which the difference between **quantitative trait phenotypes** is expressed only in some but not in other conditions.

Conditional reflexes. Reflexes that are developed by association with a frequently-repeated stimulus.

Conditioned fear. Anticipatory fear displayed in response to a stimulus previously delivered and associated with a harmful experience.

Conditioned place preference. Preference for a particular location which the subject associates with a positive experience.

Congenic strain. An inbred strain in which a gene or chromosomal region from another specific genetic background has been introduced through repeated **backcrossing**. *See also* **Co-isogenic**.

Contact pheromones. **Chemical signals** that are transmitted between individuals through physical contact to affect the physiology and/or behavior of the recipient.

Contralateral. On the opposite side of the body.

Convergent evolution. Evolution of similar features independently from different unrelated ancestors.

Copy number variations (CVNs). Large deletions and duplications in genomic DNA of which the occurrence varies among individuals in a population.

Co-relation (correlation). The observed systematic relationship between two parameters.

Corpus callosum. A prominent nerve fiber bundle which connects the cerebral hemispheres.

Corpus striatum. A bilateral brain structure in front of the **thalamus** with a striped appearance that forms part of the **basal ganglia**, and is associated with control of locomotion.

Correlation coefficient. A number between $+1$ and -1, which measures the degree to which two variables are linearly related.

Cortex. The outer region of the cerebrum that consists of gray matter, composed of cell layers that contain cell bodies of neurons that integrate information, mediate sensory motor control, and enable cognitive abilities.

Courtship. A behavior aimed at attracting a mating partner.

Courtship rituals. A stereotypic sequence of behaviors aimed at attracting a mating partner.

Covariance. A statistical value measuring the simultaneous deviations of x and y variables from their means.

Cranial nerves. The twelve pairs of peripheral nerves that arise from nuclei in the brain.

CRE binding protein (CREB). A transcription factor that binds to DNA sequences called cAMP response elements and increases or decreases the transcription of certain genes.

Cre-loxP system. Method for the introduction of conditional genetic modifications into specific genes by **homologous recombination** using Cre, a site-specific, bacteriophage P1-derived recombinase, which cuts at loxP sites flanking the target gene.

Cross-fiber pattern theory. The hypothesis that taste qualities are encoded by a neural pattern of activity that arises from different strengths of activation of taste cells by different tastants in a population of nerve fibers.

Crossover interference. Reduction of the probability of coincident occurrence of recombination events at nearby sites.

Cryptochrome. A light-sensitive photoreceptor protein associated with circadian periodicity.

Cyclic AMP response elements (CRE). A sequence in DNA that controls gene expression in response to an increase in cyclic AMP (cAMP) through the **CRE binding protein, CREB.**

Cyclic GMP-dependent protein kinase. An enzyme that phosphorylates protein substrates under the control of cyclic GMP, involved with synaptic modification and implicated in foraging behavior in *Drosophila*, and several other species.

Cycloheximide. An inhibitor of protein biosynthesis in eukaryotic organisms, produced by the bacterium *Streptomyces griseus*.

Cytochrome P450. A diverse family of enzymes that are part of a system that metabolizes foreign substances, and plays a role in steroid hormone synthesis.

Declarative memory. The aspect of memory that stores facts.

Decussate/decussation. The crossing of nerve fibers from one side of the central nervous system to the other.

Deficiency complementation mapping. Using chromosomes that carry deletions (deficiencies) to map loci for a trait by virtue of failure to complement, due to a missing counterpart of that locus in the deficiency region of the sister chromosome.

Deficiency stock. A line containing a chromosome which misses a region that is normally present.

Delayed sleep phase syndrome (DSPS). A **circadian rhythm** disorder in which the onset of sleep and wake times are later than considered normal.

Delirium tremens. A serious, potentially fatal, alcohol-withdrawal syndrome observed in persons who stop drinking alcohol following continuous and heavy consumption, characterized by involuntary muscle spasms, sweating, anxiety, and hallucinations.

Delta waves. High-amplitude brain waves with a frequency of 1-4 Hertz, characteristic of deep sleep. *See also* **Slow wave sleep.**

Dendrites. Neuronal projections that receive signals from other neurons.

Dendritic arborizations. Splitting and ramification of **dendrites.**

Dendritic knobs. Dilatations on **dendrites.**

Dentate gyrus. A region of the **hippocampus.**

Dependence. Physiological and/or psychological addiction to a substance, characterized by adverse mental and/or physical consequences that result from abrupt cessation of usage of the substance.

Depolarization. An electrical state in an **excitable cell**, whereby the inside of the cell is made less negative relative to the outside than at the **resting membrane potential.**

Descending. Neural pathways that convey signals from the brain to the periphery.

Diagonal band. A region in the brain that contains **cholinergic** nerve fibers with projections, among others, to the **hippocampus.**

Diencephalon. The posterior division of the forebrain that includes the **thalamus** and **hypothalamus.**

Dihydroxyphenylalanine (DOPA). The precursor for **dopamine**, which is converted to dopamine by DOPA decarboxylase.

Directional selection. Natural selection for one allele at a locus, and against the other allele. *See also* **Positive selection, negative selection.**

Dizygotic twins. Twins derived from two different eggs that arose from separate and independent fertilization events and, thus, can be either the same or different sex.

Dominant. A condition in which an **allele** determines the value of the **phenotype** irrespective of the second **allele** at that locus.

Dominant-negative mutations. Dominant mutations in which the product of the mutant **allele** interferes with the function of the normal **allele** in the heterozygous state.

Dopamine. An important catecholamine neurotransmitter in the nervous system, associated among others with locomotion, **addiction**, and **schizophrenia.**

Dorsal horn. The dorsal gray matter of the spinal cord, which receives sensory information.

Dorsal paired medial (DPM) neurons. Large **neuromodulatory** neurons in *Drosophila* that project into the **mushroom bodies**, and are associated with memory formation.

Dorsal raphe nuclei. Moderate-size clusters of nuclei in the brain stem that send extensive serotonergic projections to other regions of the brain.

Dorsal root ganglia. Collections of neurons located dorsally, immediately adjacent to the spinal cord, that collect sensory information from the periphery and project **axons** into the spinal cord to convey that information to the central nervous system.

Down syndrome. An abnormal human phenotype, including mental retardation, due to a trisomy of chromosome

21, which is more common in babies born to older mothers.

Drones. Male honeybees whose only function is to mate with the queen.

Dye switching. Switching the fluorescent labels on alternate RNA probes in a hybridization competition assay to EST microarrays to control for the effect of the dye on the efficiency of RNA-DNA hybridization.

Dynorphins. Opioid peptides produced in many regions of the brain and spinal cord that modulate pain responses.

Dyslexia. A learning disability characterized by impaired comprehension of written language, despite normal intelligence.

Dystrobrevins. Components of the dystrophin–glycoprotein complex which recruit signaling proteins.

Ectopic. The occurrence of gene expression in a tissue in which it is normally not expressed.

Effective population size. The number of breeding individuals in a population that give rise to the same variation in allele frequency or average inbreeding as an ideal population of size N.

Efferent signals. Signals that are sent from the central nervous system to the periphery, or away from the **cell body** of a neuron along the **axon**.

Electroencephalogram (EEG). A recording of electrical field activity of the brain.

Electrogenic. Generating a voltage difference.

Electrophoresis. A technique for separating the components of a mixture of charged molecules in an electric field within a gel or other support.

Electroporation. A technique for transfecting cells by the application of a high-voltage electric pulse.

Elevated plus maze. A device consisting of four horizontal, perpendicularly-arranged, alternately open and enclosed runways that is suspended above the ground, and used to measure anxiety as reflected by the willingness of an animal to enter the open arms of the maze.

Ellipsoid body. Part of the **central complex** of the insect brain, as in *Drosophila*, that contributes to control of locomotion.

Embryonic stem cells. Embryonic cells that can replicate indefinitely, transform into other types of cells, and serve as a continuous source of new cells.

Endonuclease. An enzyme that cleaves or hydrolyzes specific phosphodiester bonds within a polynucleotide chain.

Endorphins. Endogenous **opioid** polypeptides that occur naturally in the brain, and have **analgesic** properties.

Endplate. Also known as **neuromuscular junction**; the region of the muscle cell membrane where a **synapse** is formed with the innervating nerve.

Enhancer effect. An effect that is greater than expected from the addition of the sum of two predicted effects.

Enhancer trap. A transgenic construct in which a promoter sensitive to enhancer regulation is fused to a **reporter gene**, such that expression patterns of the reporter gene inserted near a gene on a chromosome identify the spatial regulation of the target gene.

Enkephalins. Endogenous pentapeptides involved in regulating pain and **nociception**.

Entorhinal cortex. A memory center in the brain, which provides the main input to the **hippocampus**.

Entrained. Synchronized with the light–day cycle.

Ephrin A. **Extracellular matrix** component that directs cell adhesion and the formation of axonal projections.

Ephrin B. **Extracellular matrix** component that directs cell adhesion and the formation of axonal projections.

Epidermal growth factor (EGF) receptor. A receptor to which epidermal growth factor binds to promote cell division, and which is present in abnormally high concentrations in many types of cancer cells.

Epinephrine. A hormone released from the adrenal medulla into the blood to elicit a **fight or flight response** (also known as **adrenaline**).

Episodic memory. Memory of events, times, places, and associated emotions in relation to an experience.

Epistasis. The masking or modulation of the phenotypic effect of **alleles** at one gene by **alleles** of another gene through non-additive **enhancer** or **suppressor effects**.

Equilibrium potential. The voltage across a membrane which is established when opposing diffusion and electrochemical forces acting on permeant ions are equal.

ERK (extracellular signal regulated kinase). A member of a family of widely-expressed protein kinase intracellular signaling molecules which are involved in regulation of meiosis, mitosis, and a wide range of functions in differentiated cells.

Essential fatty acid. A fatty acid that cannot be produced in the body, and therefore must be obtained from the diet.

EST expression microarrays, cDNA expression microarrays, spotted arrays. Assemblies of transcribed DNA sequences that are immobilized on a solid matrix, and can be used to probe expression levels of their corresponding mRNAs under different conditions.

Ethogram. A diagrammatic depiction of a behavioral repertoire.

Ethology. The study of behavior

Eugenics. Controlled human breeding based on notions of desirable and undesirable genotypes.

Euphoria. A state of often exaggerated pleasurable feeling.

Evolution. A change in allele frequencies in a population.

Evolutionary arms race. An evolutionary struggle between competing sets of co-evolving genes that develop adaptations and counter-adaptations against each other.

Excitable cells. Cells that can generate electrical currents through **voltage-activated ion channels**.

Excitatory postsynaptic potential (EPSP). An electrical change (depolarization) in the membrane of a postsynaptic neuron caused by the binding of an excitatory neurotransmitter from a **presynaptic cell** to a postsynaptic receptor, which makes it more likely that the postsynaptic neuron will generate an **action potential**.

Exon. A sequence of DNA that codes for the synthesis of a protein, or segment thereof.

Experimental variation. Variation that occurs due to fluctuations in experimental conditions.

Expression microarrays. Assemblies of DNA or oligonucleotides immobilized to a solid matrix that are used to probe expression levels of corresponding mRNAs.

Expressivity. Quantitative variation in expression of a **phenotype** among individuals of the same **genotype**.

Extensor. A muscle that extends a limb by increasing the angle at a joint.

External plexiform layer. A cell layer in the **olfactory bulb** between the superficial glomerular layer and the deeper **mitral cell body layer** that contains inhibitory GABAergic **interneurons**, and modulates incoming neural activity with olfactory information.

Extracellular matrix. A complex network of polysaccharides, proteins, and proteoglycans which are secreted by cells into the extracellular environment to provide structural support, and to contribute to regulation of cell growth, differentiation, and cell–cell communication.

Facial nerve. The seventh **cranial nerve** that controls most of the muscles in the face and receives sensory input from taste cells in the anterior two thirds of the tongue.

False discovery rate (FDR). The expected proportion of false positives among all significant results.

Fan-shaped body. The most prominent structure of the **central complex** of the insect brain.

Fatty liver syndrome. Accumulation of fat in the liver causing enlargement of the liver without obvious pathology, often, but not always, as a consequence of excessive alcohol intake.

Feed-forward process. A process of escalation in which the consequence of an action stimulates further action until an endpoint is reached.

Fight or flight response. The surge in **adrenaline** and cortisol that prepares the body to confront a perceived threat by promoting an increase in the heart rate, slowing digestion, and directing blood to major muscle groups.

Fitness. The genetic contribution of an individual to the next generation.

Fixed action patterns. A behavior that is essentially unchangeable, and usually carried to completion once initiated.

Flehmen response. A mammalian courtship behavior in which the upper lip is curled and the neck is extended, facilitating the reception of pheromonal olfactory cues by the **vomeronasal organ**.

Flexor. A muscle that, when it contracts, causes a joint to bend.

FLP–FRT system. A gene recombination system whereby DNA sequences flanked by FRT sites can be excised through the action of the FLP recombinase.

Fluorescence resonance energy transfer (FRET). A technique for measuring interactions between two proteins *in vivo* in which excitation of a fluorescent group on one protein generates light emission at a wave length that can elicit fluorescence from a nearby fluorescent group on another protein, provided the two fluorescent tags are in close proximity.

FMRFamide. A neuropeptide found in invertebrates, first identified in molluscs, that contains the amino acids phenylalanine, methionine, arginine, and phenylalanine.

Forward genetic screens. A genetic screen aimed at discovering genes for a **phenotype** of interest.

Founder effect. Genetic drift observed in a population founded by a small non-representative sample of a larger population.

Founder population. A group of individuals that establishes a new population, often with a small effective population size likely to experience **inbreeding**.

Fragile X syndrome. The most common form of inherited mental retardation, named for its association with an X-chromosome with a tip that breaks or appears uncondensed.

Frameshift mutation. The insertion or deletion of a nucleotide pair or pairs, causing a disruption of the translational reading frame.

Frontal lobe. One of the four lobes of the brain, divided into motor, pre-motor, and pre-frontal areas, the latter of which is responsible for many cognitive functions.

Functional genomics. The field of research that aims to determine patterns of gene expression, gene functions, and interactions among genes in the **genome**, based on genomic sequences of an organism.

Funiculi, funiculus. A bundle of **axons** that runs along the spinal cord; *also*, the **third antennal segment** of the *Drosophila* antenna.

Gain-of-function mutations (hypermorphic mutations). A mutation that changes the gene product, such that it gains a new and abnormal function, usually with a dominant **phenotype**.

GAL4-UAS. A transgenic expression system in which a tissue-specific promoter is used to express the yeast GAL4 transcription factor, which in turn drives expression of a transgene behind an upstream promoter (UAS) sequence.

GAL80. A repressor of the GAL4 transcription factor, which in certain experimental scenarios can be used for temporal control of tissue-specific expression of transgenes with the **GAL4-UAS** system.

Gamete. A germ cell with a haploid chromosome complement.

Ganglia. Groups of nerve cell bodies in the central or peripheral nervous system.

GCoMP. *See* **Cameleon.**

Gene action. The way in which a gene exerts its effects on a phenotype, for example by being dominant or recessive.

Gene. An ordered sequence of nucleotide bases that encodes a protein or RNA product, with regulatory regions preceding and following the coding region, as well as intervening sequences (**introns**) between individual coding segments (**exons**).

Gene duplications. Duplication of a region of DNA that contains a gene, resulting in an extra copy of that gene.

Gene flow. The movement of genes from one population to another, by way of interbreeding of individuals in the two populations.

Gene–brain–behavior axis. The concept that the **genome** enables the nervous system to express behaviors in a relationship in which gene expression influences nervous system function, and *vice versa*.

Genetic covariance. The extent to which two related individuals are more likely to share **alleles** than are two unrelated individuals.

Genetic drift. Random variation in allele frequency caused by sampling in finite populations.

Genetic load. The relative decrease in the mean **fitness** of a population due to the presence of **genotypes** that have less than the highest fitness.

Genome. The entire complement of genetic material in a chromosome set.

Genome-by-genome interactions. Phenotypic effects that are the result of non-additive interactions between **alleles** at different loci. *See also* **Epistasis.**

Genome-wide association studies (GWAS). Studies which seek to correlate variation in a trait, such as a disease, with **polymorphisms** across the entire **genome** in an outbred population.

Genomic action potential. The transient rapid increase in expression of an immediate early gene in the brain in response to sensory stimulation.

Genomic era. The current stage in the history of science, in which it is possible to sequence whole **genomes** and conduct integrative studies on ensembles of genes across the genome.

Genotype. The specific allelic composition of a certain gene, a set of genes, a cell, or an organism.

Genotype-by-environment interaction (GEI). Variation among **genotypes** in environmental **plasticity**.

Genotype-environment correlation. An association between genotype and environment that occurs if the environment an individual experiences depends on its **genotype**.

Genotypic effect. The effect of a **genotype** on the **phenotype** of a **quantitative trait**; includes breeding values and dominance deviations.

Geotaxis. Movement in response to the earth's gravitational field.

Ghrelin. A polypeptide hormone produced in the gastrointestinal tract (stomach) that stimulates release of growth hormone from the anterior pituitary, and has a role in regulating appetite and energy balance.

Gill withdrawal reflex. An involuntary, defensive reflex of the sea slug *Aplysia* that causes its gill to be retracted when the animal is subjected to a mechanical disturbance.

Glia. The major support cells of the brain, involved in the nutrition and maintenance of the nerve cells.

Globus pallidus. One of the components of the **basal ganglia**, which relays information from the **caudate** and **putamen** to the **thalamus**.

Glomeruli. Spherical structures of neuropil in the insect **antennal lobe**, or vertebrate **olfactory bulb**, which are the sites where the axons of olfactory sensory neurons converge to form the first synaptic relay in the olfactory pathway.

Glossopharyngeal nerve. The ninth **cranial nerve** that innervates the pharynx and mediates sensory input from **taste buds** in the posterior third of the tongue (bitter taste).

Glucagon. A polypeptide hormone secreted by the alpha cells of the islets of Langerhans in the pancreas in response to hypoglycemia.

Glutamatergic neurotransmission. Neurotransmission mediated by the excitatory neurotransmitter glutamate.

Goldman–Hodgkin–Katz equation. An equation that quantifies the voltage potential of resting membranes, based on asymmetric ion distributions across the membrane.

G-protein. A protein that links activation of a membrane receptor to activation of an effector enzyme, while binding and hydrolyzing GTP.

G-protein-coupled receptors (GPCRs). A membrane receptor with seven transmembrane α-helical domains that form three alternating intracellular and extracellular loops with an extracellular and intracellular domain, that signals its activation by activating a **G-protein**.

Granule cell layer. A layer of **interneurons** in the **olfactory bulb**.

Gray matter. An area of the brain that is rich in nerve cell bodies.

Green fluorescent protein (GFP). A small protein, naturally occurring in jelly fish, that spontaneously fluoresces and can be used as a noninvasive fluorescent marker in living cells to trace neural circuits, or monitor expression of certain gene products.

Gustatory receptor. A receptor that is activated by tastants, and which mediates taste sensations.

Gustducin. A **G-protein** that is expressed in taste cells, and mediates stimulation of certain G-protein-coupled gustatory receptors.

Gyri. The crests of convolutions in the cerebral cortex.

Habituation. A behavior in which an animal exhibits a lessened response with repetition of a stimulus.

Hair bundle. The collection of **stereocilia** that project from the apical membrane of sensory cells in the inner ear, that mediate transduction of mechanosensory stimulation into an electrical signal.

Hair cells. Sensory cells that mediate mechanosensory transduction for auditory and vestibular perception.

Haldane's rule. The observation that if hybrids of only one sex are sterile or inviable in a species cross, that sex is likely the one having heterogametic sex chromosomes.

Half-diallel cross. Construction of all possible $n(n - 1)/2$ heterozygous genotypes between n homozygous lines, excluding reciprocal crosses.

Haplotype. A combination of specific **alleles** at two or more loci.

Hardy–Weinberg equilibrium. The stable frequency distribution of genotypes, A_1A_1, A_1A_2, and A_2A_2, in the proportions p^2, $2pq$, and q^2, respectively, (where p and q are the frequencies of the **alleles**, A_1 and A_2) that is a consequence of random mating in the absence of mutation, migration, natural selection, or random drift.

Hebbian synapses. **Synapses** in which persistent presynaptic activation, often as a result of coincident activation of a postsynaptic cell by more than one presynaptic cell, can lead to synaptic modification and strengthening.

Hemizygosity. hemizygous. The condition of loci on the X-chromosome of the heterogametic sex of a diploid species, or a condition in which one part of the **genome**, in a normally diploid species, is present in only one copy.

Heritability. The proportion of phenotypic variation that is explained by genetic variation. *See also* **Broad sense heritability** and **narrow sense heritability**.

Heterogametic. Possessing two heteromorphic (differently-shaped) sex chromosomes (for example, X and Y).

Heterozygote advantage. *See* **Overdominance**.

Heterozygote/heterozygous. Having two different **alleles** at a locus.

Heterozygous effect. The mean value of the **heterozygous genotype**, expressed as a deviation from the mean of the two **homozygotes** at a particular locus.

High density oligonucleotide microarrays. Expression microarrays in which oligonucleotides corresponding to genomic DNA sequences are immobilized to a solid matrix, to serve as probes for nucleic acid hybridization.

Hippocampus. A part of the brain that plays a role in the acquisition of new memories.

Hitchhiking. Change in the frequency of an **allele** due to its linkage with an **allele** at another locus that is undergoing selection.

HKA test. A statistical test for deviation from neutrality, in which the ratios of non-synonymous to synonymous substitutions in a sequence under study and a sequence in the same organism thought to evolve neutrally are compared.

Homologous recombination. Breakage and reunion between homologous lengths of DNA, used in knockout mice to generate defective **alleles** of the target gene.

Homozygote/homozygous. Having two identical **alleles** at a locus.

Homozygous effect. The difference in **phenotype** between individuals with alternative **homozygous genotypes** at a particular locus.

Housekeeping genes. Genes that are expressed in virtually every cell, as they are required for energy metabolism and biomolecular processes necessary for the viability of the cell.

Huntingtin. A protein containing a polyglutamine tract, which when expanded beyond a critical length gives rise to the neurodegenerative disorder **Huntington's disease**.

Huntington's disease. A late, but variable, age-onset lethal human disease of nerve degeneration, inherited as an autosomal dominant **phenotype**.

Hydrophobic. Apolar and non-soluble in water.

Hyperpolarization. An electrical state in an **excitable cell**, whereby the inside of the cell is made more negative relative to the outside than at the **resting membrane potential**.

Hypertrophy. Excessive enlargement of a cell, organ, or tissue.

Hypocretins (orexins). Hypothalamic peptides that affect feeding behavior and regulation of sleep.

Hypoglossal nerve. The twelfth **cranial nerve** that innervates the muscles of the tongue.

Hypomorphic mutations. Mutations that reduce the expression of a **phenotype** without abolishing it.

Hypothalamus. An area of the brain that regulates endocrine activity, as well as somatic functions, such as body temperature, sleep, and appetite.

Identity by descent (IBD). The condition in which two **alleles** at a locus are identical copies of the same ancestral **allele**.

Immediate early gene. A gene which is activated rapidly and transiently in response to a stimulus.

Immunohistochemistry. The use of antibodies as histological tools for identifying patterns of antigen distribution within a tissue.

Imprinting. A learning process in early life, whereby specific patterns of behavior are established; *also,* a genetic phenomenon by which certain genes are expressed in a manner that depends on their parent of origin.

Impulse control disorder. A set of psychiatric disorders that includes intermittent explosive outbursts, stealing, pathological stealing and gambling, pyromania, hair pulling and skin-picking. *See also* **Obsessive compulsive disorder**.

In situ **hybridization**. Use of a DNA or RNA probe to detect the presence of a complementary sequence in tissue sections.

Inbreeding coefficient. The probability that two alleles are identical by descent, i.e. the probability that a zygote obtains copies of the same ancestral gene from both its parents, because they have the same common ancestor.

Inbreeding depression. Change of mean value of a quantitative trait due to inbreeding.

Indels. Insertions or deletions of nucleotide sequences in DNA, or amino acids in proteins, that vary among individuals in a population.

Indirect genetic effects. Effects that arise when genes expressed in one individual affect the expression of traits in other individuals.

Inebriometer. A device used to measure alcohol sensitivity in *Drosophila*.

Infinitesimal model. The theory that **complex traits** arise through interactions among an infinite (or at least very large) number of genes, each of which contributes only a small effect to the **phenotype**.

Inheritance of acquired traits. The theory, proposed by Lamarck, that adaptive changes in a phenotype acquired over the life of an organism could be transmitted to its offspring.

Inhibitory postsynaptic potential (IPSP). An electrical change (**hyperpolarization**) in the membrane of a postsynaptic neuron, caused by the binding of an inhibitory neurotransmitter from a **presynaptic cell** to a postsynaptic receptor, which makes it less likely for the postsynaptic neuron to generate an **action potential**.

Innate releasing mechanism. A neural pathway that sets in motion an instinctive behavioral sequence in response to an environmental stimulus. *See also* **Fixed action pattern**.

Insomnia. Inability to sleep.

Insula. A lobe in the **telencephalon** between the **temporal lobe** and the **parietal** cortex, folded inwards in the **cerebrum** so that it is not visible from the outside of the brain, which provides emotional context to sensory experiences.

Intangible environmental variation. Environmental variation stemming from unknown causes.

Internal ribosome entry site (IRES). A nucleotide sequence in the mRNA that allows initiation of translation during protein synthesis.

Interneurons. A neuron that communicates only with other neurons, but not directly with sensory organs or effector organs.

Interval mapping. A statistical method for mapping **quantitative trait loci** that evaluates linkage to adjacent pairs of polymorphic markers, to simultaneously estimate the effects and map position of the quantitative trait locus.

Intragenomic conflict. A situation that arises when transmission of a particular gene is detrimental to the rest of the **genome**.

Intraventricular pH. The pH of the cerebrospinal fluid that fills the ventricles and central canal of the spinal cord.

Introgression. The introduction of a particular gene, genomic region, or chromosome in the **genome** of a species through repeated back-crossing of an F_I hybrid with one of its parents.

Intron. An intervening DNA segment between **exons** that does not encode a protein sequence.

Inverse PCR. A method by which circularized genomic DNA sequences can be amplified from flanking sequences, using primers that point outward from these sequences.

Inversion. The replacement of a section of a chromosome in the reverse orientation.

Ionotropic receptors. Receptors that undergo a conformational change to open an ion channel when activated by a ligand (e.g. glutamate receptors, odorant receptors of the *Drosophila* IR gene family).

Ipsilateral. On the same side of the body.

Isogenic. Genetically identical (except for sex).

Jacobson's organ. *See* **Vomeronasal organ.**

Johnson's organ. An acoustic, mechanosensory organ in the second antennal segment of *Drosophila.*

Jumping genes. Mobile elements of DNA (**transposons**) that can insert into new locations in the chromosomal DNA, and that can affect the function of genes at or near their insertion sites.

Kainic acid-type glutamate receptors. Classes of glutamate receptors that open monovalent cation channels in response to glutamate.

Karyotype. The entire chromosome complement of an individual.

Kenyon cells. Principal neurons in the **mushroom bodies** of the insect brain.

Kin selection. A form of cooperative behavior, in which individuals help relatives with whom they share genes, since their reproductive success will help perpetuate copies of their genes.

Klüver–Bucy syndrome. A behavioral disorder, characterized among others by altered fear and aggression, that occurs when the function of the **amygdala** in both the right and left medial temporal lobes of the brain is compromised.

Knee-jerk reflex. A reflex extension of the leg resulting from a tap on the patellar tendon.

Knock-out. The process of purposely removing a particular gene from an organism.

Krebs cycle. A series of chemical reactions involved in aerobic respiration that is central to intermediary metabolism.

Labeled line theory. The theory that different qualities of sensory information are transmitted to the brain along separate non-interacting neural circuits.

Lamina. Visual neuropil in the fly brain which receives input from photoreceptor cells R1–R6.

Lamina propria. A vascular layer of connective tissue beneath the epithelium of an organ.

Lateral geniculate nucleus (LGN). A nucleus in the **thalamus** that receives separate projections from retinal ganglion cells from both eyes, and provides a synaptic relay for processing visual information.

Lateral horn. A region of the **lateral protocerebrum** of the insect brain, implicated in higher processing of olfactory information.

Lateral protocerebrum. A region of the insect brain implicated in higher processing of olfactory information.

Lateral septum. A part of the **limbic system** associated with emotions and social behavior.

LD blocks. Regions in the **genome** that have undergone little historical recombination and, hence, contain multiple genes in **linkage disequilibrium** (LD).

Lentiform nucleus. A lens-shaped nucleus associated with control of locomotion that forms part of the **basal ganglia**, and contains the **putamen** and **globus pallidus**.

Leptin. A hormone that has a central role in fat metabolism and in regulating food intake by sensing the amount of stored fat.

Lethal mutation. A mutation that results in death of the individual carrying it.

Lewy bodies. Cytoplasmic inclusions in vacuoles of injured neurons, which serve as histological markers for neurodegeneration, as occurs in **Parkinsons's disease**.

Lickometer. A device that records drinking activity of an experimental animal by registering the number of licks.

Ligand-gated channel proteins. A superfamily of transmembrane proteins that undergo a conformational change on binding a ligand, such as a neurotransmitter, to open an ion channel across the membrane.

Limbic system. Interconnected brain structures that form a loop (limbus) around the brainstem, involved in instinctive behaviors, including feeding and reproduction, as well as olfaction, emotion, motivation, and the control of various autonomic functions.

Limnetic. Living in the open waters of lakes away from the shore.

Linkage. The proximity of two or more markers on a chromosome; markers that are transmitted together with high probability.

Linkage disequilibrium (LD). The condition in which observed frequencies of haplotypes in a population do not agree with haplotype frequencies predicted by multiplying together the frequency of individual genetic markers in each haplotype.

Linkage group. A group of loci known to be linked, such as a chromosome.

Linkage map. A map of chromosomal loci constructed based on recombination frequencies.

Linkage mapping. Identification of **quantitative trait loci** in families from an outbred population, or in segregating generations derived from crosses of inbred lines in model organisms based on co-segregation of phenotypic variation in a quantitative trait and genetic markers, dependent on recombination frequency. *See also* **Linkage study.**

Linkage study. A family-based method to search for a chromosomal location of a gene by demonstrating co-segregation of the disease with genetic markers of known chromosomal location.

Lobula. A region of the insect brain associated with processing of visual information.

Lobula plate. A region of the insect brain adjacent to the **lobula**, associated with processing of visual information.

Local field potentials. Electrical signals that arise from the summated activity of many neurons.

Locomotor reactivity. Movement in response to a disturbance. *See also* **Startle response**.

Locus coeruleus. A nucleus in the brainstem with a bluish appearance from which a widespread noradrenergic projection emanates, implicated in arousal and wakefulness.

Long-term potentiation (LTP). A long-lasting increase in the strength of transmission at a **synapse**, as a result of repetitive or coincident stimulation.

Loop design. An experimental design, commonly applied to the analysis of expression microarray data, in which conditions are compared via a chain of other conditions in an attempt to limit the number of controls normally required for pairwise comparisons.

Loss-of-function mutations. *See* **Null mutations**.

Luciferase. Enzyme used by fireflies to produce light.

Lumbar. The region of the spinal cord below the **thoracic** region and above the **sacral** region.

Major histocompatibility complex (MHC). A group of highly-polymorphic genes whose products appear on the surface of cells, imparting the ability to distinguish self from non-self.

MALDI-TOF (matrix-assisted laser desorption ionization-time of flight). A mass spectrometry method for the identification and structural analysis of small molecules and peptides.

Massed learning. Learning that arises from continuous uninterrupted training sessions.

Maternal environment. The environment experienced *in utero* and postnatally during maternal care.

Maxillary palps. Paired chemosensory appendages that extend from the ventral region of an insect's head.

Maximum likelihood estimation method. A statistical model which fits parameters to a normally-distributed set of data, so that the values of the model parameters make the data "more likely" than any other values of the parameters would make them.

Mechanoelectrical transduction. The conversion of a mechanical stimulus into an electrical signal.

Medial lemniscus. A band of somatosensory fibers that forms an ascending projection from the **nucleus cuneatus** and the **nucleus gracilis** to the contralateral **thalamus**.

Medulla. Visual neuropil in the fly brain that receives input from photoreceptor cells R7 and R8.

Medulla oblongata. A region of the brain continuous with the spinal cord, associated with vital functions like breathing, blood circulation, and swallowing.

Meiosis. A process of division of gametes during which daughter cells are generated so that each contains half the usual number of chromosomes.

Meiotic drive. Preferential production of certain gametes during meiosis, which alters the segregation of genes from Mendelian expectations.

Melanopsin. A photosensitive pigment found in specific retinal ganglion cells that form the retino-hypothalamic tract, which entrains **circadian rhythm** to the light–dark cycle.

Melatonin. A hormone produced by the **pineal gland**, involved in regulating the sleeping and waking cycles.

Membrane potential. The charge difference between the cytoplasm and extracellular fluid in all cells, due to the differential distribution of ions.

Mendelian trait. A categorical trait that segregates in a predictable manner according to expectations from a single locus, as originally formulated by Gregor Mendel.

Mesencephalon. The middle division of the embryonic brain that will develop into the midbrain.

Mesocortical pathway. The dopaminergic pathway that originates from the midbrain **ventral tegmental area**, and innervates areas of the frontal cortex.

Mesolimbic pathway. A dopaminergic pathway associated with reward that comprises the **ventral tegmental area**, the **nucleus accumbens**, and the **prefrontal cortex**.

Metabotropic glutamate receptors. A family of G-protein-coupled receptors that are activated by glutamate.

Metencephalon. The anterior part of the embryonic hindbrain which gives rise to the **cerebellum** and **pons**.

MicroRNAs. Small, highly-conserved 20–23-mer non-coding RNA molecules that regulate protein production by binding to the 3'-untranslated regions of specific mRNAs.

Microsatellite. A specific sequence of nucleotides which contains mono, di, tri, or tetra tandem repeats, also known as simple sequence repeats (SSR), short tandem repeats (STR), or variable number tandem repeats (VNTR), which can be polymorphic at different locations in the **genome**.

Microvilli. Folds of the cell membrane that increase the external surface area.

Middle-term memory. A component of memory that persists longer than short-term memory, but has not yet been consolidated into long-term memory.

Mimicry. The ability of an organism to imitate others, or to merge with its environment.

Miniature endplate potentials (MEPPS). Small postsynaptic **depolarizations** of unitary conductance at the **neuromuscular junction** that occur as a result of the occasional random exocytosis of presynaptic vesicles.

Missense (non-synonymous) mutations. Mutations that change a codon for one amino acid into a codon for a different amino acid.

Mitogen-activated protein kinase (MAPK). A subfamily of protein kinases that regulate cell growth and differentiation.

Mitral cell layer. A layer in the **olfactory bulb** that contains the cell bodies of **mitral cells** that receive olfactory information from olfactory sensory neurons and project axons via the lateral olfactory tract to **piriform cortex**.

Model organisms. Organisms, such as mice, rats, and *Drosophila*, which are suitable and advantageous for laboratory studies to acquire large amounts of data that are likely to be biologically widely-applicable.

Modern evolutionary synthesis. The coalescence of ideas of the early-twentieth century neo-Darwinists that integrated Darwin's theory of natural selection with Mendel's notions of inheritance, and laid the foundation for **population genetics** and **quantitative genetics**.

Molecular receptive fields. The range of chemical structures that can activate a given chemoreceptor.

Monoamine oxidase A. An enzyme that catalyzes the oxidation of monoamines, such as **norepinephrine** and **dopamine**.

Monomorphic. Uniform; containing only one combination of **alleles** at a locus (e.g. presence of only **homozygotes** of one type in a population).

Monozygotic twins. Twins that result from the split of a single fertilized ovum and are, hence, genetically identical.

Morgan. A unit of recombination based on cross-over frequency.

Morpholino antisense oligonucleotides. Synthetic oligonucleotide derivatives that cannot be degraded, but that can be used to bind to and specifically inhibit the translation of certain mRNAs after injecting them into developing embryos.

Morris water maze. A water-filled arena in which mice are trained to use external cues to swim to a hidden platform which is used to study spatial memory.

Mossy fibers. A type of cerebellar and hippocampal neurons with complex ramifications.

Muscarinic acetylcholine receptors. G-protein-coupled **receptors** that are activated by **acetylcholine**.

Muscle spindle. A sensory stretch organ embedded in a muscle that provides proprioceptive information about the state of that muscle to the brain.

Mushroom bodies. Higher-order integrative structures in the insect brain.

Mutation. A change in the DNA sequence.

Mutualism. A relationship between two species from which both derive benefit.

Myelencephalon. The posterior part of the embryonic hindbrain that will develop into the **medulla oblongata**.

Myelin. Material produced by glia, composed of lipids and lipoproteins, that forms protective sheaths around axons which electrically insulate them.

Narcolepsy. A neurological disorder marked by the sudden recurrent uncontrollable induction of sleep.

Narrow sense heritability. The portion of **additive** genetic variance that contributes to the total phenotypic variation and represents the extent to which variation in **phenotypes** is determined by variation in effects of **alleles** transmitted by parents to offspring. *See also* **Broad sense heritability**.

Nature versus nurture. The debate as to whether the manifestation of traits, such as behaviors, are the result of genetic or environmental factors.

Negative geotaxis behavior. Upward movement against the force of gravitation.

Negative reinforcement. Promoting the occurrence of a behavior by training the subject to avoid an unpleasant stimulus which would be delivered in the case of non-compliance.

Negative selection (purifying selection). The selective removal of deleterious **alleles**.

Neofunctionalization. Acquisition of a novel function.

Neomorphic. Acquisition of a new appearance.

Nernst equation. The equation, developed by Nernst, that allows calculation of the membrane potential that results from the asymmetric distribution of a single permeant ion.

Neural crest. Embryonic ectoderm that lies on either side of the neural tube and develops into the cranial, spinal, and autonomic ganglia.

Neural groove. A longitudinal groove that exists between the neural folds before closure of the embryonic neural tube is complete.

Neuritic plaques. Structural abnormalities in the brain characteristic of **Alzheimer's disease**.

Neurofibrillary tangles. Twisted protein filaments within neurons of the cerebral cortex, characteristic of Alzheimer's disease.

Neurogenomics. The study of how the **genome** as a whole contributes to the function of the nervous system.

Neuromodulator, neuromodulatory. A chemical agent that alters the activity of neurons in response to a neurotransmitter.

Neuromuscular junction. The **synapse** between a neuron and the muscle it innervates.

Neuronal nitric oxide synthase (nNOS). A constitutively-expressed enzyme in the brain that catalyzes the conversion of L-arginine to L-citrulline and nitric oxide, a retrograde neurotransmitter that mediates cell communication during **long-term potentiation**.

Neuropeptide Y. A 36-amino acid neuropeptide which, among many other effects, stimulates feeding behavior.

Neuropharmacology. The study of the action of drugs on the nervous system.

Neurotransmitter. A chemical messenger released from the synaptic terminal of a neuron at a chemical **synapse** that diffuses across the synaptic cleft, and binds to and stimulates the postsynaptic membrane.

Nicotinic acetylcholine receptors. Multi-subunit transmembrane proteins that undergo a conformational change to open a large cation-selective ion channel on binding **acetylcholine**.

Nitric oxide (NO). A short-lived free radical gas that activates guanylyl cyclases, and serves as vasodilator in the vasculature and as retrograde neurotransmitter to promote synaptic strengthening in the brain.

NMDA-type glutamate receptor, N-methyl-D-aspartate (NMDA) receptor. A subtype of glutamate receptor that opens an ion channel through which calcium can flow, that can trigger processes of synaptic modification.

Nociception, nociceptive. Perception of noxious, potentially-harmful stimuli.

Nodes of Ranvier. Regularly-spaced gaps of exposed axonal membrane between **myelin** sheaths along the axon of a nerve, where voltage-gated sodium and potassium channels are concentrated to enable **saltatory conduction** of action potentials.

Noduli. Components of the **central complex** of the insect brain, of undefined function.

Nonsense mutation. A mutation that alters a gene so as to produce a truncated product.

Non-syndromic mental retardation. Mental retardation without a known cause.

Non-synonymous mutation. A **mutation** in the DNA that results in a change in the encoded amino acid.

Noradrenaline. One of the principal bioamine neurotransmitters in the central and peripheral aurtonomic nervous system, also known as **norepinephrine**.

Norepinephrine. *See* **Noradrenaline**.

Normal distribution. Any of a family of bell-shaped frequency curves whose relative position and shape are defined on the basis of the mean and standard deviation.

Northern hybridization, Northern blot. Transfer of **electrophoretically**-separated RNA molecules from a gel onto an absorbent sheet, which is then immersed in a labeled probe that will **hybridize** to an RNA of interest to reveal its presence.

NPR-1 pathway. A pathway that mediates feeding behavior in *Caenorhabditis elegans*, which is controlled via a receptor for NPR, a neuropeptide that resembles **neuropeptide Y** in vertebrates.

Nucleus. A cluster of nerve cells in the brain. Plural term is nuclei.

Nucleus accumbens. The largest nucleus in the septal region of the **diencephalon**, which mediates reward sensations and has been implicated in **addiction**.

Nucleus cuneatus. A nucleus in the spinal cord that receives sensory information from the upper part of the body.

Nucleus gracilis. A nucleus in the spinal cord that receives sensory information from the lower part of the body.

Nucleus of the solitary tract . A nucleus in the brainstem that processes gustatory information.

Null mutations. **Mutations** whose effects are either an absence of normal gene product at the molecular level, or an absence of normal function at the phenotypic level.

Occipital lobe. The most posterior lobe of the **cerebrum** that processes visual information.

Octopamine. A bioamine neurotransmitter of invertebrates, functionally similar to **norepinephrine** in vertebrates.

Oculomotor nerve. The third **cranial nerve** that innervates muscles of the eye.

Odorant-binding proteins. Secreted proteins that bind odorants in the aqueous medium that bathes the chemosensory dendrites of olfactory neurons.

Odorant receptors. A superfamily of **G-protein-coupled receptors** that bind chemicals that convey an olfactory sensation.

Odorant response profile. The spectrum of odorants that elicit electrophysiological or behavioral responses in molecular, cellular, or organismal assays.

Oenocytes. Cells in the body cavity of, for example, nematodes, thought to fulfill an immune-protective function.

Olfactory bulb. The most anterior protrusion of the brain, which occurs bilaterally and represents the first synaptic relay and processing stations for olfactory information.

Olfactory cilia. Cilia that protrude from the apical dendritic knobs of olfactory sensory neurons in the nasal lumen, and carry odorant receptors for odor detection.

Olfactory marker protein (OMP). A cytoplasmic protein of poorly-defined function that is expressed almost exclusively in mature functional olfactory sensory neurons.

Olfactory nerve. The first **cranial nerve** that transfers chemosensory information from olfactory neurons to the **olfactory bulb**.

Ommatidia. Repeated units of photoreceptor cells that together form the compound eye of insects.

One gene, one enzyme hypothesis. The notion that the **genome** regulates the expression of proteins, such that a single gene encodes a single protein or enzyme.

Open field behavior. Spontaneous locomotion observed in an experimental enclosure.

Operant conditioning. A type of learning in which behavior is strengthened if it is reinforced, and weakened if it is punished.

Opioid peptides. Endogenous peptides that modulate the perception of pain, which include **enkephalins**. **endorphins**, and **dynorphins**.

Optic chiasma. The point of **decussation** of the optic nerve fibers.

Optic lobes. Structures of neuropil in the insect brain that receive visual information.

Optic nerve. A collection of nerve cells that project visual information from the eyes to the **lateral geniculate nucleus**.

Orbitofrontal cortex. A region of association cortex located at the base of the frontal lobes above the orbits of the eyes, involved in cognitive processes such as decision making, and sometimes considered part of the **limbic system**.

Orexins (hypocretins). *See* **Hypocretins**.

Orphan receptors. Receptors of which the ligands are unknown.

Overdominance. The **fitness** of the **heterozygote** is greater than that of either **homozygote**.

Oviposition. Egg-laying.

Ovulin. A protein transferred in seminal fluid from a male *Drosophila* to a female during mating, which subsequently stimulates egg-laying in the female.

Oxytocin. A 9-amino acid peptide hormone released from the posterior pituitary, implicated in promoting uterine contractions during parturition, milk ejection during lactation, and social bonding.

Pacinian corpuscles. Pressure-sensitive nerve endings in the skin encased in concentric multilayered onion-shaped structures.

Pan-neuronal. Expressed in all neurons.

Papillae (filiform, fungiform, foliate, circumvallate papillae). Evaginations of the lingual surface, which (except for filiform papillae) house **taste buds**.

Parallel evolution. Independent evolution of similar traits in unrelated organisms.

Parallel sequencing. Rapid, large-scale DNA sequencing methods, in which DNA fragments are amplified and amplicons serve as templates for the simultaneous elongation of many *de novo* DNA strands, while enabling identification of each sequentially-added nucleotide.

Parasympathetic, parasympathetic nervous system. Part of the autonomic nervous system that regulates the involuntary activity of glands, smooth muscle, and cardiac muscle, which contains the vagus nerve and nerves arising from the sacral spinal cord.

Paraventricular nucleus. A nucleus of neurosecretory cells in the hypothalamus that produces hormones that are released from the posterior pituitary into the circulation.

Parietal lobe. One of the lobes of the cerebrum involved with cognition, information processing, spatial orientation, and the perception of stimuli related to touch, pressure, temperature, and pain.

Parkin. A ubiquitin protein ligase, likely associated with the removal of misfolded proteins, and implicated in some cases of inherited **Parkinson's disease**.

Parkinson's disease. A neurodegenerative disease in which dopaminergic neurons of the nigrostriatal pathway are compromised, characterized by difficulty in initiating motion, stooped posture, and resting tremor.

Pars intercerebralis. The division between the left and right **protocerebrum** of the insect brain, which in *Drosophila* has been implicated in the regulation of sleep.

Patch-clamp. An electrophysiological technique that enables recording of the activity of single ion channels, through the formation of high-resistance seals between the recording electrode and a membrane patch.

Patellar ligament. A ligament in the knee joint which, when stretched, mediates the **knee-jerk reflex**.

Pearson product–moment correlation coefficient. A measure of the strength of linear dependence between two variables.

Pedicel. The middle segment of the *Drosophila* antenna that houses **Johnson's organ**, which mediates mechanosensory/acoustic perception.

P-element insertional mutagenesis. The introduction of mutations via insertion of a transposable element in or near a gene, a method extensively applied in studies on *Drosophila melanogaster*.

Penetrance. The proportion of the individuals in a population with a given **genotype** that display the **phenotype** associated with that **genotype**.

Periaqueductal gray. Gray matter located around the cerebral aqueduct within the midbrain, which plays a role in modulation of pain.

Perikaryon. The cell body of a **neuron**. *See also* **Soma**.

Perilymph. The aqueous medium inside the insect olfactory sensilla that surrounds the chemosensory neurons and contains **odorant binding proteins** secreted by support cells.

Peripheral endocrine system. Endocrine glands that are not directly part of or anatomically connected to the nervous system, such as those endocrine glands that are regulated by the pituitary.

Permutation analysis. A procedure to correct for multiple-testing, in which an appropriate criterion for statistical significance is established by generating an empirical distribution of *P*-values through repeated disruption and random reassembly of associated parameters in the data set under study.

Permutation tests. *See* **Permutation analysis**.

Phenotype. The detectable morphological, physiological, or behavioral manifestation of a particular **genotype**.

Phenotypic plasticity. The ability of the manifestation of a trait to change under different environmental conditions.

Phenylthiocarbamide (PTC). A substance that tastes intensely bitter to a segment of the human population and is tasteless to others, representing a gustatory experience that segregates essentially as a Mendelian trait.

Pheromone blend. A mixture of pheromones of which multiple components are often required to elicit an effect on the recipient.

Pheromone receptors. Receptors that detect **pheromones**.

Pheromones. Chemical signals that pass information between individuals.

Phonological deficiencies. Deficiencies in the ability to form sounds as part of normal speech.

Phosphatases. Enzymes that hydrolyze esters of phosphoric acid, removing a phosphate group.

Phospholipase A2. An enzyme that specifically hydrolyzes the 2-acyl bond of phospholipids, releasing arachidonic acid as a second messenger.

Phosphorylation. The process of adding phosphate groups to a compound.

Photolithography. A photochemical process used to manufacture **high-density oligonucleotide expression microarrays**.

Phototaxis. Movement towards light.

Phototransduction. The transformation of light energy into electrical activity.

Pineal gland. A small endocrine gland located near the center of the vertebrate brain, between the two hemispheres, that receives innervation from the suprachiasmatic nucleus and produces melatonin, a hormone that modulates wake/sleep patterns.

Pituitary adenylyl cyclase activating peptide (PACAP). A family of neuropeptides that activates adenylyl cyclase, but also acts as neurotransmitter and neuromodulator with regulatory effects on the vascular, endocrine, and immune systems.

Plasticity. Ability to undergo change.

Pleiotropy, pleiotropic effects. Affecting multiple **phenotypes**.

Polymerase chain reaction (PCR). A method for amplifying specific DNA segments by repeated cycles of template denaturation, primer addition, primer annealing, and replication using thermostable DNA polymerase.

Polymodal. Referring to multiple sources of sensory input; *also*, referring to a non-normal distribution that suggests multiple distinct categories.

Polymorphic marker. DNA sequence variants among individuals in a population.

Polymorphisms. Naturally-occurring DNA sequence variants among individuals in a population as a result of spontaneous mutations.

Pons. Region of the brain stem that acts as a relay station between the **cerebellum** and the **cerebrum**, and aids the medulla in the control of breathing.

Pontine reticular formation. A diffuse neural network in the **pons** connected with the reticular formation of the **medulla oblongata**, which controls autonomic functions.

Population admixture. Mating between populations with different **allele** frequencies.

Population genetics. The study of genetic variation in populations.

Population level mutation rate. A **population genetics** parameter that reflects the rate at which mutations occur in a population through estimation of the parameter $4N_e u$, where N_e is the effective population size, and u is the mutation rate per generation.

Population mean. The sum of the products of the effects and frequencies of alleles that contribute to the trait.

Population recombination rate. A **population genetics** parameter that reflects the extent of recombination across a population through estimation of the parameter $4N_e c$, where N_e is the effective population size, and c is the recombination rate per generation.

Positive selection. Increase in frequency in the population of a beneficial **allele**.

Postcentral gyrus. A region of the **parietal lobe** posterior to the central sulcus that contains the **primary somatosensory area**.

Postganglionic. Referring to fibers that project from autonomic ganglia to their target organs.

Postsynaptic cell. The cell on the opposite side of the **synapse** from the synaptic terminal of the stimulating neuron that contains receptor proteins and degradative enzymes for the neurotransmitter.

Postsynaptic potential. A change in the membrane potential of a **postsynaptic cell** that results from activation by a **presynaptic cell**.

Postzygotic isolation. The condition in which genomic incompatibility has developed that prevents two species from giving rise to viable or fertile offspring.

pp90 ribosomal protein S6 kinase (Rsk). A family of kinases associated with the **MAP kinase** pathway, implicated in learning and memory, as well as cell growth and differentiation.

Precentral gyrus. A region of the **frontal lobe** anterior to the central sulcus that contains the primary somatomotor area.

Preference tests. *See* **Conditioned place preference** and **two-bottle preference test**.

Preganglionic. Referring to fibers that provide input to autonomic ganglia.

Preoptic area. A region in the anterior **hypothalamus** which, among others, is involved with thermoregulation.

Prepulse inhibition. The phenomenon in which a low intensity stimulus inhibits the startle-response to a subsequent strong stimulus.

Presynaptic cell. The neuron that releases neurotransmitter onto its target cell.

Prezygotic isolation. Differences between species that prevent them from mating with each other.

Primary somatosensory area. The area of cortex in the postcentral gyrus of the **parietal lobe**, where sensations regarding touch, temperature, propioception, and nociception are mapped along a systematic neural representation of the body surface.

Proboscis extension reflex. An insect's reflex to elaborate its proboscis when gustatory receptors on its front legs are touched with a drop of sugar solution.

Proboscis. The long tubular mouthpart of an insect, used for feeding.

Procedural memory. Memory of skills, and of how to carry out actions.

Programmed cell death (apoptosis). A genetically-determined sequence of events that leads to organized dismantling of the cell.

Proprioception. The perception of the body's movement and posture, arising from sensory stimuli within the body itself.

Prosencephalon. Designation of the embryonic forebrain that will give rise to the **cerebrum** and olfactory lobes.

Proteasome. An assembly of proteolytic enzymes that form the intracellular machinery for protein degradation.

Protein kinase C. A calcium-dependent, diacylglycerol-activated enzyme that phosphorylates protein substrates, involved in signal transduction and cell regulation.

Protein microarrays. Microscopic arrays of immobilized proteins on a solid support, used to identify protein–protein interactions, to identify the substrates of protein kinases, or to identify the targets of biologically-active small molecules.

Protein receptor kinases (GRKs). Enzymes that phosphorylate liganded **G-protein-coupled receptors** during the process of **desensitization**.

Proteome. The complete set of all proteins in a cell.

Proteomics. The large-scale study of all proteins within a cell.

Protocerebral bridge. Part of the **central complex** of the insect brain, important for locomotion, which connects the **fan-shaped body** and the **ellipsoid body**.

Pseudogene. A dysfunctional gene that has accumulated a mutation that prevents the production of a functional gene product.

Pseudostratified neuroepithelium. An epithelium of neuronal cells that appears layered, but in which the cell bodies are distributed at different depths, as in, for example, the olfactory neuroepithelium of vertebrates.

Purifying selection. *See* **negative selection**.

Purinergic receptor (P2X) channels. Transmembrane receptor proteins that undergo a conformational change to open an ion channel on binding of a purine agonist, such as ATP.

Putamen. A portion of the basal ganglia in the middle of the brain which, together with the caudate nucleus, forms the dorsal striatum.

Pyramidal neurons. Corticospinal motor neurons that course through and decussate in the **pyramids**.

Pyramids. Regions near the junction of the **medulla oblongata** and the spinal cord, where fibers of the corticospinal tracts crossover from one side of the central nervous system to the other.

Quanta. Fixed amounts of neurotransmitter contained in, and released from, different synaptic vesicles.

Quantitative complementation test. A statistical test to evaluate whether the difference between alternative **QTL** alleles from two parental strains is greater as **hemizygotes** over a **deficiency** (or **mutant** allele) than as **heterozygotes** with a wild type allele.

Quantitative genetics. The study of the genetics of **complex traits**. *See* **Quantitative traits**.

Quantitative trait locus (QTL). A chromosomal region that contains one or more genes or genetic elements that contribute to phenotypic variation in a **quantitative trait**.

Quantitative trait nucleotide (QTN). A polymorphic marker in a gene shown to be causal to variation in a **quantitative trait**.

Quantitative traits. Traits for which phenotypic variation is caused by multiple segregating genes and environmental variation.

Queen mandibular pheromone (QMP). A **pheromone** blend produced by the honeybee queen that organizes workers in the hive, elicits a **retinue response**, and suppresses their reproductive ability.

Radial glia. Glia that arise during development of the nervous system from the subventricular zone near the ventricle as neural progenitors, and that also migrate toward the surface of the brain forming a scaffold for the subsequent migration of neurons destined to form the cortex.

Rapid eye movement (REM) sleep. A phase of sleep, characterized by rapid eye movements, electroencephalogram waves that resemble the waking state, and the occurrence of dreams.

Reaction norms. Patterns of **phenotypes** produced by a **genotype** under different conditions.

Realized heritability. Heritability estimated from response to phenotypic selection.

Real-time quantitative PCR, quantitative PCR. A method by which the amplification of a DNA sequence can be monitored through the use of fluorescence, which allows quantitative measurement of the amplification product as it is generated.

Receptive field. A segment of the environment or range of a stimulus spectrum that elicits activity in a particular neuron of a sensory pathway.

Receptor tyrosine kinase. Membrane receptors that phosphorylate tyrosine residues when activated, and often play important roles in regulating cell growth, division, and differentiation.

Recessive. The condition in which an **allele** has no effect on the phenotype in the presence of a **dominant allele** at the same locus.

Recombinant inbred (RI) lines. Strains developed by inbreeding the F_2 population, derived from crossing two inbred lines to homozygosity.

Recombination breakpoints. The precise locations in the **genome** where crossing-over has occurred during recombination.

Reconsolidation. Modification of a memory upon recall.

Refractory period. A brief period after stimulation of a nerve, during which the nerve will not respond to a second stimulus.

Regression toward the mean. The tendency of the **quantitative traits** of offspring to be closer to the population mean than are their parents' traits.

Relative refractory period. A brief period after stimulation of a nerve during which the nerve will respond to a second stimulus only if that stimulus is stronger than the first one.

Reporter constructs. A construct that encodes a product which can be readily-assayed when expressed under the control of a particular promoter.

Resident-intruder test. A test for aggression by introducing a strange mouse into the cage of a resident animal, followed by measuring the frequency and/or intensity of attacks on the intruder by the resident.

Resting membrane potential. The voltage difference between the inside and outside of a cell and, in the case of a nerve cell, the potential difference between the inside and outside when the cell is not stimulated.

Restless legs syndrome. An uncomfortable sensation in the legs while sitting or lying down that gives rise to an uncontrollable urge to move them.

Restriction enzyme, restriction endonuclease. An enzyme that cuts DNA at a specific short sequence of nucleotides.

Reticular activating system. A neural network in the brainstem that maintains an alert state in the cerebral cortex, and is concerned with arousal.

Retinal ganglion cells. Cells of the innermost layer of the retina, which project axons through the optic nerve to the brain.

Retinohypothalamic tract. The projection of **retinal ganglion cells** to the **suprachiasmatic nucleus** in the **hypothalamus**, which mediates synchronization of circadian activity with the day–night cycle.

Retinue response. Aggregation of worker bees around the queen in response to **queen mandibular pheromone**.

Retrograde. Movement from the **axon** terminal toward the cell body in the reverse direction of **action potentials**.

Retrotransposons. Transposable elements that can replicate themselves and accumulate in the host **genome** by reverse transcription through an mRNA intermediate.

Rett syndrome. An inherited developmental mental disorder with characteristic hand movements, observed only in females.

Reverse genetic screens. Approaches aimed at discovering the function of a gene by investigating the **phenotype** that results when this gene is mutated.

Rhinencephalon. Those regions of the cerebrum that receive and process olfactory information.

Rhombencephalon. The most caudal of the three primary vesicles formed during embryonic development of the brain, which divides into the **metencephalon** and **myelencephalon**, and will give rise to the **cerebellum**, **pons**, and **medulla oblongata**.

RNA interference (RNAi). The process by which foreign, double-stranded RNA is recognized and degraded, and which may be an evolutionarily- conserved defense mechanism against RNA viruses and transposable elements.

RNA-induced silencing complex (RISC). A ribonucleoprotein complex that cleaves specific mRNAs which are targeted for degradation by homologous double-stranded RNAs as a result of **RNA interference**.

Rotating rod. A behavioral assay used to test balance and **proprioception**.

Running wheel. A treadmill for measuring spontaneous locomotor activity in rodents.

Sacral. Related to a lower pelvic segment of the spinal cord, below the **lumbar** region and above the **coccygeal** segment.

Saltatory conduction. Propagation of **action potentials** at the **nodes of Ranvier** in a jumping fashion by allowing current to spread intracellularly between the nodes.

Saturation screens. Extensive genetic screens, to the point where multiple introduced mutations affect the same genes more than once.

Scape. The most proximal segment of the *Drosophila* **antenna**.

Schizophrenia. A severe mental disorder characterized by delusions, hallucinations, incoherence, and physical agitation.

Scolopedia. Mechanosensory/acoustic cells in the insect antenna.

Seasonal affective disorder (SAD) (winter depression). A common form of depression in northern countries, induced by long dark winters.

Second messenger. A chemical signal, such as calcium ion or cAMP, that relays a hormonal message from a cell's surface to its interior.

Selective sweep. The rapid spread of an allele and linked loci through the population, as a result of positive selection.

Semantic memory. Storage of general information, such as names or facts.

Senile dementia. Decline in cognitive ability as a result of age-dependent neurodegeneration.

Sensitization. Enhancing sensitivity through prior stimulation.

Sensitized background. A genetic background that carries a mutation.

Sensory–motor integration. The process of determining appropriate motor output in response to sensory perception.

Serotonin (5-hydroxytryptamine, 5-HT). A monoamine neurotransmitter in the brain that, among many functions, plays a role in regulating mood, including depression and anxiety, sleep and learning.

Sex determination. The genetic or environmental process by which the sex of an individual is established.

Sex-linked. The inheritance pattern of loci located on the sex chromosomes (usually the X-chromosome in XY species), or referring to the loci themselves.

Sexual antagonism. The situation in which an **allele** is favored in one sex, and selected against in the other.

Sexual dimorphism. A **phenotype** that differs between males and females.

Sexual selection. The forces determined by mate choice acting to cause one **genotype** to mate more frequently than another genotype.

Shibire. A gene that encodes a dynamin protein necessary for synaptic vesicle recycling, a temperature-sensitive dominant negative **allele** which can be utilized for reversible synaptic silencing.

Silent (synonymous) mutation. A mutation in which the function of the protein product of the gene is unaltered.

Single cohort. A colony of honey bees that are synchronized in their developmental stage.

Single marker analysis. A statistical method that assesses whether variation at polymorphic sites in the **genome** is associated with phenotypic variation in a quantitative trait, by considering each polymorphic marker independently.

Single nucleotide polymorphisms (SNPs). Natural mutations that segregate in a population, and arise from alterations of single nucleotides in the **genome**.

Slave oscillators. Systems that display rhythmicity, dictated and coordinated by a central pacemaker.

Sleep apnea. A cessation of airflow through the oral and nasal cavities during sleep.

Sleep consolidation. Fusion of brief sleep bouts into longer periods of uninterrupted sleep.

Sleep homeostasis. A compensatory increase in sleep following a period of sleep deprivation.

Sliding window analysis. The analysis of a DNA sequence that focuses on overlapping segments along the sequence.

Slow-wave sleep, deep sleep. A state of usually dreamless sleep that is characterized by delta waves and a low level of autonomic physiological activity.

Small interfering RNAs (siRNAs). Small double-stranded RNAs that are used as a research tool to inhibit protein production.

SNP genotyping. Identification of sequence variants (single nucleotide polymorphisms) in DNA.

Social engineering. The notion of creating an optimal society by determining which people should interbreed. *See also* **Eugenics**.

Sociogenomics. The study of the genetic underpinnings of social organization and social interactions.

Soma. The cell body of a **neuron**. *See also* **Perikaryon**.

Somatosensory. Sensory information from the body surface, including touch, pressure, vibration, proprioceptive, nociceptive and temperature sensations.

Song crystallization. Development of the final mature song of a song bird, following a period of song learning.

Southern blot. Transfer of electrophoretically-separated fragments of DNA, after denaturation, from a gel to an absorbent sheet of material, such as nitrocellulose, to which the DNA binds and to which labeled probes can subsequently be hybridized to identify fragments of interest.

Spaced learning. Learning induced by temporally discontinuous training sessions.

Speciation. The process by which new species that cannot interbreed arise from a common ancestor.

Sperm competition. Competition between the sperm of two or more males from multiple matings for the fertilization of an ovum.

Spermatheca. Storage organs for sperm in female insects.

Spermatogonial stem cells. Cells in the male gonads that serve as a reservoir for the generation of sperm cells through ongoing cell divisions.

Spike. A nerve impulse that occurs as a result of rapid transient changes in membrane potential, triggered by the sequential opening and closing of voltage-dependent sodium and potassium channels. *See also* **Action potential**.

Spike initiation zone. A region of a nerve cell at the boundary of the cell body and the **axon**, where **action potentials** can be initiated in response to **depolarization**.

Spinal nerves. Peripheral nerves that arise from the spinal cord.

Standard deviation. Square root of the variance of individual sample data.

Standard error. Square root of the variance.

Startle response. Arousal, locomotion, or escape behavior in response to a sudden unexpected event.

Stereocilia. Actin containing filaments that protrude from the apical membranes of hair cells to form pivoting bundles of filaments that when displaced transduce mechanosensory information.

Sticklebacks. Species of fish named for their characteristic spiny body armor that have undergone extensive **adaptive radiation**.

Striate cortex. The primary visual cortex in the **occipital lobe** of the cerebrum.

Student's *t*-test. A statistical test to determine whether the difference in mean between two groups is significant.

Subesophageal ganglion. A region of the insect brain to which gustatory neurons project.

Subfunctionalization. The acquisition of complementary functions of two genes after gene duplication.

Subiculum. The most inferior component of the **hippocampus**.

Substance P. An 11-amino acid neurotransmitter peptide that mediates pain sensations in dorsal root ganglia spinal cord nociceptive neurons.

Substantia nigra. A group of large-pigmented dopaminergic neurons in the midbrain that project to the striatum to regulate locomotion, the destruction of which gives rise to **Parkinson's disease**.

Subventricular zone. A neurogenic region that lines the lateral ventricles and gives rise to **radial glia** and cortical neurons during embryonic development, and postnatally to neurons that continue to migrate to the **olfactory bulb**.

Sulci. The grooves of the wrinkles of the cerebral hemispheres.

Superior colliculus. A region in the midbrain that controls eye movements.

Suppressor effect. A non-linear interaction whereby one locus diminishes the effect of another on a **phenotype**; a form of **epistasis**.

Suppressor/enhancer screens. Screens designed to identify loci that increase or diminish the effect of a mutation on a phenotype. *See also* **Sensitized background, epistasis**.

Suprachiasmatic nucleus. A **hypothalamic** nucleus located above the **optic chiasma** that innervates the **pineal gland** and plays a role in managing **circadian** rhythmicity.

Supraoptic nucleus. A neurosecretory nucleus located in the **hypothalamus** above the lateral border of the optic tract, which produces hormones released from the posterior pituitary into the circulation.

Survival of the fittest. The driving force of Darwinian evolution, based on the theory that only those organisms best able (fittest) to obtain and utilize resources will survive and reproduce.

Sustentacular cells. Glia-like supporting cells in olfactory neuroepithelium.

Sympathetic nervous system. The division of the peripheral autonomic nervous system that mediates the **"fight or flight" response** by accelerating the heart rate, constricting blood vessels, and raising blood pressure.

Sympatric. Present at the same geographical location.

Synapse. The specialized region at the **axon** terminal where the **presynaptic cell** signals to a **postsynaptic cell** through the release of neurotransmitter.

Synaptic cleft. The space between the **presynaptic cell** and the **postsynaptic cell**.

Synaptic remodeling. Activity-dependent changes in the structure of a **synapse** or the number of **synapse**s in a pathway.

Synaptic transmission. The release of a neurotransmitter from the **presynaptic cell**, its diffusion across the **synaptic cleft**, and its binding to and activation of receptors on the **postsynaptic cell**.

Synaptobrevin-GFP. **Green fluorescent protein** conjugated to synaptobrevin, a component of synaptic vesicles that targets this fluorescent marker to synaptic terminals.

Synapto-pHluorin. A pH-sensitive fluorescent marker attached to synaptic vesicle membranes that increases fluorescence during synaptic exocytosis.

Synonymous mutation. A **mutation** in the DNA that does not result in a change in the encoded amino acid.

Synteny. The condition of two or more genes being located on the same chromosome even without demonstrable linkage; *also*, chromosomal regions that contain corresponding orthologous genes between different species.

T1R receptors. Receptors that mediate sweet and umami taste perception.

Tagging SNP. A single nucleotide polymorphism that is representative of a DNA region (commonly human DNA), by virtue of being in **linkage disequilibrium** with, and hence, predictive of, other polymorphisms in that region.

Tamoxifen-sensitive promoter. A DNA element that becomes activated after binding the anti-estrogen drug tamoxifen that can be used to control temporal expression of **transgenes**.

Tap-withdrawal response. The defensive withdrawal reflex of the gill in *Aplysia* in response to a mechanical stimulus.

Tarsi. An insect's front legs.

Tastants. Molecules that elicit a gustatory sensation.

Taste buds. Clusters of taste cells.

Taste pore. An opening at the top of the **taste bud**, through which taste cell sensory membranes can be exposed to substances on the tongue.

Tau-GFP. **Green fluorescent protein** bound to microtubule-associated tau protein as a neuronal marker.

Tau-lacZ. β-galactosidase bound to microtubule-associated tau protein as a neuronal marker.

Telencephalon. The most anterior part of the forebrain, including the cerebral hemispheres and olfactory lobes.

Temperature compensation. Maintenance of the same circadian rhythm at different temperatures.

Temporal lobe. The lobe of the cerebral hemisphere located on the side in front of the **occipital lobe**, which among others contains the auditory cortex.

Tequila. A gene that encodes a proteolytic enzyme that is transiently up-regulated in *Drosophila* during learning, possibly reflecting synaptic modification.

Testosterone. A male steroid sex hormone produced by Leydig cells in the testes.

Thalamus. A central structure in the brain that relays information from the periphery to the cortex.

Theory of neutral molecular evolution. The hypothesis that most mutations that occur have no effect on reproductive fitness, and their population genetics is dominated by random drift.

Theta waves. A low-frequency, low-amplitude brainwave of a person who is awake but relaxed and drowsy.

Thigmotaxis. The tendency to move close along the wall of an enclosure.

Third antennal segments. The major olfactory organs of *Drosophila*.

Thoracic. Pertaining to the region of the chest.

Threshold potential. The voltage across the membrane at which a depolarizing current triggers **an action potential**.

Tight junctions. Intercellular junctions between epithelial cells in which the outer layers of the cell membranes fuse, generating an impermeable barrier between the cells.

Tiling arrays. **High density oligonucleotide microarrays** composed of oligonucleotides that cover the entire **genome** of both coding and non-coding sequences at closely spaced intervals.

Tip-links. Filaments that connect the distal ends of **stereocilia** on hair cells, and are instrumental in mediating mechanosensory transduction.

Tolerance. Relative resistance of an individual to the adverse effects of a substance compared to the rest of the population, or an increase in drug resistance as a consequence of previous exposure that requires higher dosages to achieve the same effect.

Transcriptome. The set of all mRNAs present at a particular moment in one cell or a population of cells.

Transgenic. Containing genetic material from another species.

Transient receptor potential (TRP) channels. A superfamily of ion channels that are involved in various types of sensory reception.

Transition. A point mutation involving substitution of one base pair for another by replacement of one purine by another purine, or one pyrimidine by another pyrimidine.

Translocation. Movement of a chromosomal segment to a different location, often on a different chromosome.

Transmission disequilibrium. Family-based linkage; genetic linkage between a genetic marker and a trait.

Transposable elements/transposons. DNA segments that can move to different insertion sites in the **genome** when acted on by a **transposase**.

Transposase. An enzyme that mediates transposition of DNA segments.

Transposon-mediated mutagenesis (*P*-element insertional mutagenesis). Mutations made by insertion of a transposon in the DNA that disrupts the function of the gene near its insertion site.

Transversion. A point mutation in which a purine is replaced by a pyrimidine, or *vice versa*.

Trichoid sensilla. A type of sensilla on the **third antennal segment** of *Drosophila*, containing olfactory sensory neurons.

Trigeminal nerve. The fifth **cranial nerve** that detects nociceptive sensations, and innervates, among others, the face, teeth, jaws, lingual, and nasal cavities.

Trigeminal nucleus. A nucleus in the brainstem that receives nociceptive information via the **trigeminal nerve**.

Trisomy. Presence of three copies of a chromosome, as in the case of **Down syndrome**.

Trochlear nerve. The fourth cranial nerve, which controls the superior oblique muscles of the eyes.

Tuning curves. Graphs that depict neuronal activity as a function of sound frequency in auditory neurons.

Turbinates. Cartilage protrusions in the nose that are covered with olfactory and respiratory epithelium.

Two-bottle preference test. A test in which an experimental animal is given a choice to drink from two bottles containing different solutions.

Tyrosine. An amino acid with a phenol group, which serves as precursor for the synthesis of **catecholamines**.

Tyrosine hydroxylase. The rate-limiting enzyme in the biosynthesis of **dopamine**.

Umami. The taste of monosodium glutamate, readily perceptible as an independent taste modality by individuals of Asian origin.

Underdominance. The fitness of the heterozyote is less than both homozygotes.

Unrooted phylogenetic tree. A representation of evolutionary relatedness without assumptions about common ancestry.

Upstream Activator Sequence (UAS). A DNA promoter sequence for the yeast transcription factor GAL4.

Usher syndrome. A genetic disorder characterized by hearing impairment and an eye disorder called retinitis pigmentosa, in which vision worsens over time.

V1R receptors. *See* **V1R-type vomeronasal receptors**.

V1a receptors. Receptors for **vasopressin**.

V1R-type vomeronasal receptors. Pheromone receptors expressed in neurons in the apical layer of the **vomeronasal** neuroepithelium.

V2R receptors. *See* **V2R pheromone receptors**.

V2R pheromone receptors. Pheromone receptors expressed in neurons in the basal layer of the **vomeronasal** neuroepithelium.

Vagus nerve. A prolific cranial nerve that mediates many responses of the **parasympathetic nervous system**, such as slowing of the heart rate.

Vanilloid receptors. A family of receptors related to Trp channels that are involved in nociception, including hot temperature, capsaicin (the hot component of chili peppers), and low pH.

Variable number tandem repeat (VNTR). A locus that is hypervariable, because of tandemly repeated DNA sequences, presumably arising from unequal crossing-over or slippage during replication.

Vasopressin. A polypeptide hormone secreted by the posterior lobe of the pituitary gland, and also by some neurons in the **hypothalamus**.

Ventral horn. The front or ventral gray column of the spine, through which the **axons** of motor neurons leave the spinal cord.

Ventral nerve cord. A major bundle of **axons** that runs along the ventral side of invertebrate animals, contains mainly motor neuron axons and some interneuron **axons**, and is probably homologous to the vertebrate spinal cord.

Ventral pallidum. A component of the **basal ganglia** of the brain.

Ventral tegmentum, ventral tegmental area. Component of the reward pathway in the brain, located near the top of the brainstem.

Ventricles. Cerebrospinal fluid-filled spaces in the brain, interconnected with one another and continuous with the central canal of the spinal cord.

Vestibulocochlear nerve. The **cranial nerve** that transfers acoustic information and proprioceptive information from the inner ear to higher brain areas.

Voltage-clamp method. A technique for holding the membrane potential constant while measuring currents across the cell membrane.

Voltage-gated ion channels. Membrane proteins that change their conformation to form ion channels in response to changes in the membrane potential.

Vomeronasal organ. An elongated chemosensory organ located above the vomer bone in the roof of the palate that mediates the recognition of **pheromones**.

Western blotting. A technique for identifying a particular protein using antibodies after **electrophoretic** separation in a gel and transfer to a membrane.

White matter. Neural tissue in the brain and spinal cord composed primarily of **myelin**- covered **axons**.

Winter depression, seasonal affective disorder (SAD). A form of depression most often associated with the lack of daylight in extreme southern and northern latitudes during the winter months.

Withdrawal. Any of a group of physical and psychological symptoms occurring in an individual deprived of an accustomed dose of an addicting agent.

Zeitgeber time. An experimental measure of time standardized to the onset of an entraining stimulus.

A